Land Use and Land Cover Mapping in Europe

Remote Sensing and Digital Image Processing

VOLUME 18

For further volumes:
http://www.springer.com/series/6477

Ioannis Manakos • Matthias Braun
Editors

Land Use and Land Cover Mapping in Europe

Practices & Trends

Springer

Editors
Ioannis Manakos
Information Technologies Institute
Centre for Research
and Technology Hellas
Thessaloniki, Greece

Matthias Braun
Department of Geography
University of Erlangen-Nürnberg
Erlangen, Germany

Responsible Series Editor: A. Marçal

ISSN 1567-3200
ISBN 978-94-007-7968-6 ISBN 978-94-007-7969-3 (eBook)
DOI 10.1007/978-94-007-7969-3
Springer Dordrecht Heidelberg New York London

Library of Congress Control Number: 2014931649

Cover Illustration: CORINE Land Cover: http://www.eea.europa.eu/legal/copyright
Print file generation: FÖMI, Budapest, Hungary

Printed on acid-free paper

Springer is part of Springer Science+Business Media (www.springer.com)

Preface

Land use and land cover (LULC) is a core information layer for a variety of scientific activities and administrative tasks (e.g. hydrological modeling, climate models, land use planning). In the last two decades, land use cover change (LUCC) became an additional irreplaceable observation feature not only within Europe but on a global context. LULC mapping products constitute mandatory baseline datasets, which are required over large areas in different levels of detail and shall be provided in a homogeneous and reliable way. To this end, space- and air-borne remote sensing techniques coupled with field information are gaining ground against large-scale statistical surveys based on in situ observations.

Europe has a long heritage on land use cover mapping activities. CORINE land cover currently experiences its fourth update, as part of the GIO land (GMES/ Copernicus Initial Operations Land) project, with an intended update every 5 years. Under the umbrella of the Copernicus Program of the European Space Agency and the European Commission, a Fast-Track-Service on Land with regular European-wide coverage and updates is anticipated. It forms the base for subsequent so-called nationally funded downstream services.

The aim of the proposed book is to synthesize recent and current activities on land cover mapping in Europe and from Europe. It shall provide an overview on activities and projects covering large-scale mapping from an operational point of view (state-of-the-practice) and state-of-the-art analysis techniques from the scientific point of view. It is complemented by additional review papers and best-practice examples covering various specific aspects of LULC as e.g. degradation, deforestation or nature conservation, but also gives perspectives of data use and integration such as the integration into LULC modeling.

The editors are aware that due to the multitude of LULC and LUCC studies on local, national and European level – performed and initiated from science, industry, and public administration – this book can only cover a subset of contemporary observations and activities, as an indication of the pulse of science, applications, and perspectives in its era. An equivalent multifold thematic compilation on Remote Sensing advancements in LULC and LUCC mapping is not yet available for Europe. The editors wish to raise awareness, discussion points, and set challenges, indicating the pace of progress along with dead-ends and bottlenecks. Please, enjoy reading.

Thessaloniki, Greece Ioannis Manakos
Erlangen, Germany Matthias Braun

Acknowledgements

The editors wish to thank all authors for their intensive work, patience, and involvement both in the writing and the reviewing processes, and their Organizations for agreeing to support our initiative.

The editors are also grateful to both the European Association of Remote Sensing Laboratories (EARSeL) and Springer Verlag for the opportunity to act as a platform for dissemination and discussion of such spearhead topics. Ideas for the structure of this book were gathered by the Workshops of the EARSeL Special Interest Group "Land Use & Land Cover", to whose members and contributors the editors feel in debt.

Their special thanks go to the Book Series Editor of EARSeL, André Marçal, and to the former Chairman of EARSeL, Rainer Reuter, for their continuous support and promotion of our initiative during the preparation of this book.

The Editors received support in their authors' coordination and editing role by Dimitrios Biliouris and Zisis Petrou, without whose active engagement this book could not have been realized. We share therefore with you our high appreciation for their involvement.

As none of these could have become reality without the consensus of their employing Institutes, the editors would like to express their gratitude to their host Institutes, the Centre for Research and Technology Hellas, and the University of Erlangen-Nürnberg, for their support.

Sincerely,
Ioannis Manakos & Matthias Braun

Contents

Part I
Framework Conditions

Chapter 1
Remote Sensing in Support of the Geo-information in Europe

Ioannis Manakos and Samantha Lavender

1.1 How Policy Feeds into the Development of Information Services

The primary goal of European States' policies is the preservation and, wherever possible, improvement of the citizens' quality of life. However, challenges remain in relation to the conservation of natural resources, reduction of risks and threats, sustainability of urban and rural development and resource security including food and water. The human/natural environment interaction needs to be managed in 4 dimensions (4D), 3D spatial and temporal, which requires an underpinning Information Service including a system to assimilate data and model scenarios. The Millennium Ecosystem Assessment (Hassan et al. 2005) has paved the way by assessing the consequences of ecosystem change for human well-being (security, resources needed for a good life, health, and good social relations leading to freedom of choice and action) leading to the definition of ecosystem services (supporting, provisioning, regulating, and cultural ones). Direct drivers include: changes in local Land Use & Land Cover (LULC); species introduction or removal; technological adaptation and use; consumption of resources; climate change; various natural, physical, and biological drivers besides climate change. Indirect drivers would include: demographics; economics; socio-political; Science & Technology; cultural & religious. The Condition and Trends Working Group found that over the past 50 years, humans have changed ecosystems more rapidly and extensively than in any comparable period of time in human history; largely to meet rapidly growing demands for food, fresh water, timber, fibre and fuel.

I. Manakos (✉)
Centre for Research and Technology Hellas, Institute of Information Technologies, Thessaloniki 57001, Greece
e-mail: imanakos@iti.gr

S. Lavender
Pixalytics Ltd, 1 Davy Road, Plymouth Science Park, Plymouth, Devon PL6 8BX, UK
e-mail: slavender@pixalytics.com

I. Manakos and M. Braun (eds.), *Land Use and Land Cover Mapping in Europe: Practices & Trends*, Remote Sensing and Digital Image Processing 18,
DOI 10.1007/978-94-007-7969-3_1, © Springer Science+Business Media Dordrecht 2014

Strategic areas of intervention are recognized by the European Environmental Agency (EEA) (Dufourmont 2011), with topics being categorised as:

- **Environmental:** such as air quality, air pollutant emissions, biodiversity, greenhouse gas emissions and freshwater availability and condition.
- **Cross-Cutting:** such as climate change impacts, vulnerability and adaptation of ecosystems, environment and health, maritime issues, sustainable consumption including production of waste, land use, agriculture, forestry, energy and transport.
- **Integrated Environmental Assessment:** such as integrated environmental assessment, regional and global assessment, decision support systems, economics and strategic design.
- **Information Services and Communication:** such as shared environmental information systems and communications.

European Union (EU) policies are driven by the aforementioned considerations; being supported and iteratively improved through the results of directed research undertaken according to EU's funding frameworks. Ultimately, the aim is to operationalize the research results, through information services, in addition to dissemination and awareness raising campaigns. Feedback from beneficiaries and end users across the Member States is sought, which feeds back into policy and there-by closes the innovation circle new funding calls are opened.

Supporters of the development and implementation of Europe's policy in terms of geospatial data acquisition, processing and distribution include the EEA, European Space Agency (ESA), Copernicus Programme (formerly called GMES, an acronym for Global Monitoring for Environment and Security), and networks of excellence including commercial and scientific associations acting at both National and European levels. European networks include the European Association of Remote Sensing Companies (EARSC), European Association of Remote Sensing Laboratories (EARSeL), European environment and information network (EIONET) and EURISY. These networks have complimentary and overlapping missions with activities related to:

(a) Space infrastructure development which supports core services that address strategic areas of intervention.
(b) Methodological advancement that includes the ability to standardized information and provides it in near real time (NRT).
(c) Definition and expansion of the downstream services to create value added (VA) products for the end users, including sophisticated products with a simple user interface for everyday life.
(d) Dissemination and promotion of data, information and techniques.
(e) Training and capacity building.

1.2 Current Status and Challenges

Internationally, initiatives are increasingly being taken by agencies acting at a national, regional, and continental or global level in an effort to establish a benchmark for assessing LULC changes. However, it's important to quantify the reliability of the information received, and to enhance the potential of space applications by improving hardware and software technology in support of new scientific discoveries and ultimately end user requirements; the aim being the provision of the most relevant information in a form that is of use to decision and policy makers.

During the last decade, the EEA and ESA through Copernicus and wider activities have provided many hundreds of millions of Euros of funding to support the development of science leading to operational applications. This has included the geoland, geoland2, and BOSS4GMES series of projects that provided a prototype land core service (http://land.copernicus.eu/) that lead into the GMES Initial Operation (GIO) contract for Europe (EEA 2011). The prototype marine core service is currently the MyOcean2 7th Framework Programme (FP7) project, which provides a wide range of temporal datasets including: monitoring (encompassing NRT), multi-year, time invariant and forecast. The parameters are both physical (e.g. salinity, sea level, temperature and sea ice thickness) and biological (e.g. optical characteristics and phytoplankton biomass) in nature. MACC-II – Monitoring Atmospheric Composition and Climate – is FP7 project currently delivering regional and global pre-operational atmosphere services with recent (historical) data, present conditions and forecasts including air quality, climate forcing, stratospheric ozone, UV radiation and solar-energy resources. In February 2013 the EU Council agreed to mobilize around €3.8Bn for Copernicus through the Multiannual Financial Framework agreement (2014–2020), which the European Parliament approved in July 2013, with the ocean and atmospheric services due to become operational in 2014.

Products from the Copernicus services and projects rely on the provision of satellite imagery from contributing missions, with the first mission called Sentinel-1 (a polar-orbiting satellite carrying a C-band Synthetic Aperture Radar, SAR) due for launch in 2014; agricultural applications are expected to benefit greatly from Sentinel-1's all-weather images. In July 2013 it was agreed that data and information produced in the framework of the Copernicus programme should be made available to the users on a full, open and free-of-charge basis, in order to promote their use and sharing, and to strengthen Earth Observation (EO) markets in Europe; on the assumption that any harm to private-sector satellite operators will be outweighed by the expected growth in value-added services derived from the data. A challenge for users will be 'big data' as it's expected that the Sentinel missions will provide at least tenfold increases in data volume compared to Envisat comparable instruments e.g. the average Sentinel-1 scenario will produce over 500 Gb per day of NRT data.

Asian and American Organizations, acting together under common initiatives such as the Group on Earth Observation (GEO) and the International Society of Digital Earth (ISDE) or UNESCO Natural and Cultural Heritage Programmes, have been seeking partnerships and solutions in an effort to generate LULC products with the highest possible precision (i.e. GEO 2011). Questions arising in LULC Special Interest Groups (SIGs), such as those of EARSeL and the International Society of Photogrammetry and Remote Sensing (ISPRS), are always focused around the same keywords: methodology improvement; efficient homogenization of the production; precision of information retrieved; regional model adaptation and adjustment whilst having standardized data assessment procedures. In addition, the International User Community is setting (with scientific support) its input requirements so that a series of satellites can be launched (Copernicus Sentinel missions plus recently launched European PROBA-V and Chinese ZY-3 among others) to guarantee a continuation of data provision.

Within the aforementioned framework, one may notice from the literature and ones' own experience that systematically acquired ground truth data is missing, whilst EO datasets still suffer from mistrust in terms of both reliability and applicability. The EEA supports the in-situ activity within Copernicus, but focused on infrastructure metadata generation rather than directly supporting the data collection itself that remains the responsibility of National agencies. There are also important debates in the scientific community about the quantification of accuracy assessment rules and (for land products in particular) the influencing factors of topography, projection systems, and rule set definition for LULC and change detection mapping. The final outcome of all these discussions is that enough data and methodologies exist, but more coordination and homogenization is needed for the ultimate goal of end user acceptance to be reached. The INSPIRE Directive (INSPIRE 2007) supports this endeavour and Europe's networks are closely following.

Therefore, remote sensing of land surfaces faces the following challenges driven by recent research and technological developments (Manakos 2013):

(a) **Image Classification**: very high resolution (VHR) and hyperspectral sensors require the development of a new generation of classification techniques. Two different operational scenarios are suggested: (i) definition of training sets by interactive labelling of unlabelled samples carried out by photointerpretation, and (ii) definition of training set by using active learning techniques to drive in-situ data collection campaigns. Therefore, new strategies that integrate semi-supervised learning with active learning need to be investigated (Bruzzone and Marconcini 2009), as well as techniques that leverage on previous existing knowledge and datasets.
(b) **Change Detection (CD) Analysis:** from the simple post classification comparisons undertaken in the 1970s up to the complex algebra transformations and classifications of the 2000s (e.g. Texture-based Algebra, Robust Change Vector Analysis (CVA), Transformation Kernel Principal Component Analysis (PCA), Fast Fourier Transform, Object-based Post-Classification Comparison (PCC), Multisource PCC Support Vector Machines (SVMs)) challenges remain, such

as the: pre-processing issues (geometry & radiometry); influence of the CD algorithm; segmentation approach and threshold selection; accuracy of the change mask; influence of number and type of sensors; influence of surface features (also in 3D). Further investigation is needed if optimal approaches are going to be defined for operational users.

(c) **Data Fusion from optical, radar, and thermal infrared sensors, operating at different spatial and temporal scales, and from multifaceted products:** information retrieval potential relies on the synergy and complementarity of combining remotely sensed data from multiple sources; especially when the radiation interacts with the surface in very different ways. However, the fusion of such data remains a challenge; since the launch of Envisat the community has discussed the synergistic usage of the sensors on-board, but the number of scientific publications addressing this remains low as considerable effort has been required to understand and improve the data coming from individual sensors. In addition, the increased availability of processed image analysis products as continuous data sets for various land use/land cover bio-geophysical parameters (e.g. biomass, vegetation composition, tree heights, percentage of tree/shrub cover) rather than fixed classes requires the further development of legends and classification schemes that allow for their interpretation and fusion in the data-, and knowledge-base for an area. Overall, the primary objective remains the same i.e. a performance improvement in capturing spatio-temporal variation of surface elements.

(d) **Accuracy Assessment:** Ground data quality is of major importance when estimating the accuracy of LULC extent and change detection. Quality impacts vary with the nature of errors and often with prevalence. Challenges may be identified in terms of the: genuine difficulty in discriminating classes (definition) i.e. biological variability; technical problems such as misregistration and pre-processing; use of inappropriate reference targets i.e. leading to spatial autocorrelation; use of misleading measures of accuracy; use of a biased approach to accuracy assessment. In addition, one has to recognize that sources of error and uncertainty originate from error in the ground data (Foody 2010), which is often not assessed. Recently there is an effort to find ways to utilize the plethora of available increasing amount of in-situ images from citizen sensors acquired for arbitrary reasons to increase the training capacity of the classifier, and accuracy of the derived products (e.g. 'Mapping and the Citizen Sensor' COST Action TD1202).

In addition, the challenges facing Copernicus and other multi-faceted programmes such as Galileo is conveying their importance (and ultimately value for money) to the citizens so that continued underpinning financial support is available; in July 2013 Copernicus received the European Parliament's approval for its inclusion in the Multiannual Financial Framework (MFF) budget for 2014–2020 with provision of €3 786 million (at 2011 economic conditions). This approval follows several months of difficult negotiations, and so is a significant political milestone. The Committee of the Regions (CoR) is the voice of Europe's local and region authorities that is ready to assume the role of intermediary and coordinator between themselves and the relevant bodies involved in Copernicus (Stahl 2012).

1.3 Future Trends and Conclusions

Based on more than four decades of innovation, developments and achievements in EO technologies, methodologies and applications, Europe is proceeding from the islands of pure research towards multi-modal and multi–source data assessment including processes automation, data harmonization, web downstream service development and tailor made solutions (Manakos 2013). Discussions within the community are focusing on the importance of Quality Assurance e.g. the Quality Assurance framework for Earth Observation (QA4EO, http://qa4eo.org/) that was established and endorsed by the Committee on Earth Observation Satellites (CEOS) as a direct response to a call from GEO and recent discussions co-ordinated by EARSC/ESA (April 2013) on a certification scheme for the EO Industry.

Environmental and agricultural applications include LULC change, disaster response, detailed mapping for monitoring purposes and 3D mapping with expectations directed towards the combination of observations from diverse instruments (Radars, Lidars, radiometers, optical sensors, etc.) in intelligent ways (Freeman 2012). There are high expectations for data acquisition and abundance from the upcoming fleet of Sentinel missions (the first three are expected to follow within 12 months of the MFF budget approval), which together with complimentary mission (including TerraSAR-X, Pleiades, RapidEye, DMCii, the US JPSS missions and Japanese GCOM series) aim to supply the demand from and for most EO applications.

Completed and on-going projects have paved the way towards GMES Initial Operations and Pan-European coverage plus been promoting capacity building and enhancing member states' engagement. With the last call for Space related proposals under FP7 having closed (November 2012), the EU agency expects new projects to establish a basis for the development of innovative products, applications with improved performance and services that combine existing and upcoming sensor data with in-situ sources in a novel manner. In return, results need to feed into end user decision support system and EO methodological/technological developments should take full advantage of the next generation of satellite missions.

Copyright issues are also a "hot" debate topic, addressed within the new EU Framework Programme for Research and Innovation (Horizon 2020). Sawyer and de Vries (2012) suggest that data from the upcoming Sentinel missions should be regarded as Public Sector Information, increasing their value for money. The report notes "GMES may well be Europe's goose capable of laying golden eggs. But how can we ensure a steady sustainable business model: do we take one egg (direct returns from sales of data) or do we allow the egg to hatch, hoping more golden-egg-laying geese will follow?" The free and open data policy for Sentinel data is expected to foster data reuse. From other missions (e.g. change in the U.S. Landsat data policy) there is already evidence that economic benefits are magnified when the data are made available at low or marginal cost so that barriers to entry are minimised; the entire Landsat archive became freely available in December 2008 and since then downloads have been increasing exponentially with one million downloads achieved in August 2009 and 12 million in July 2013.

Within the content of scientific research, one must look at the latest developments and advances of human activities to understand what will be the future requests from the environmental and agricultural remote sensing communities. Today, the land is covered (in general) by artificially sealed and urban areas, arable and permanent crops, forests and wetlands, semi-natural and altered landscapes, open and bare soils, and pastures. In the future we expect to see increasing urban sprawl, bio-fuel crops, food crops, soil degradation, rehabilitation and reforestation activities. The availability of water resources is increasingly worrying to both the scientific community and society, and in addition, climate change impacts shall be identified, confronted and mitigated. Biodiversity, food security, natural resource depletion, deforestation, soil degradation, disaster management, and urban sprawl are among the most important keywords for future EO applications.

Still, whatever the developments will be, the main issues remain as:

- **Engagement of Member States:** Local and Regional governments, e.g. through CoR, need to remain interested and aware of the potential of remote sensing (and Copernicus specifically) for supporting civil security and enhanced quality of life to their citizens. In addition, they need to promote the usage of the new advancements into everyday life once value is proven.
- **Research Direction**, its documentation, and promotion to the wider public of actors and policy implementers. It's expected that funding will be increased for the Operational Program of the EU and reduced for the research and development sector.
- **Standardization, Harmonization and Usability:** There is an urge and strategy to produce thematic layer products in a standardized and homogenized way, for which quality and credibility remain stable across wider geographical areas so that administrative and projects' implementation borders do not hinder joined-up utilization.

References

Bruzzone L, Marconcini M (2009) Toward an automatic updating of land-cover maps by a domain adaptation SVM classifier and a circular validation strategy. IEEE Trans Geosci Remote Sens 47(4):1108–1122

Dufourmont H (2011) Requirements in LULC monitoring in Europe. EEA keynote presentation at the 4th EARSeL SIG WS on LU/LC in Prague

EEA (2011) DK-Copenhagen: GMES initial operations 2011–2013—land monitoring services: high resolution land cover characteristics of 5 main land cover types—6 lots, 2011/S 149–247068.

Contract Notice, Services, http://ted.europa.eu/udl?uri=TED:NOTICE:247068–2011:TEXT:en:HTML&src=0. Accessed 15 Sept 2012

Foody GM (2010) Assessing the accuracy of land cover change with imperfect ground reference data. Remote Sens Environ 114(10):2271–2285

Freeman A (2012) How we will view the Earth from Space in 2025. Delegate handbook of RSPSoc 2012 conference in Greenwich, UK, pp 29–31

GEO (2011) GEO 2012–2015 WORK PLAN, Revision 1, p 82

Hassan R, Scholes R, Ash N (eds) (2005) Ecosystems and human well-being: a framework for assessment – current state and trends, vol 1. Island Press, Washington, DC, pp 1–25

INSPIRE (2007) Directive 2007/2/EC of the European Parliament and of the Council of 14 March 2007 establishing an Infrastructure for Spatial Information in the European Community (INSPIRE), p 14

Manakos I (2013) Remote sensing in Europe: status analysis and trends focusing on environment and agriculture. J Aeronaut Space Technol 6(1):1–5

Sawyer G, de Vries M (2012) GMES and data: geese and golden eggs. Available from the EARSC website. p 80

Stahl G (2012) Regions and GMES – a promising alliance. Windows on GMES publication "Discover what GMES can do for the European regions and cities", p 168

Chapter 2
Global Land Cover Mapping: Current Status and Future Trends

Brice Mora, Nandin-Erdene Tsendbazar, Martin Herold, and Olivier Arino

2.1 Introduction

The observation of global-scale land cover (LC) is of importance to international initiatives such as the United Nations Framework Convention on Climate Change (UNFCCC) and Kyoto protocol, governments, and scientific communities in their understanding and monitoring of the changes affecting the environment, and the coordination of actions to mitigate and adapt to global change. As such, reliable and consistent global LC (GLC) datasets are being sought. For instance, GLC datasets are used as an input for many Global Circulation Models, Earth Systems Models and Integrated Assessment Models used for global and regional climate simulations, dynamic vegetation modelling, carbon (stock) modelling, ecosystem modelling, land surface modelling, and impact assessments (Hibbard et al. 2010; Herold et al. 2011).

The selection of GLC datasets and their quality have a significant influence on the outcomes of these models (Hibbard et al. 2010; Nakaegawa 2011). However, the existing GLC datasets are often selected without considering their quality and suitability for a specific application (Verburg et al. 2011). This is due, notably, to the lack of interoperability and inter-comparability between the datasets (Jung et al. 2006; Herold et al. 2008). Uncertainties of LC datasets also result in considerable differences in modelling outcomes (Hibbard et al. 2010; Nakaegawa 2011; Verburg et al. 2011). For instance, Benitez et al. (2004) have noted that the choice of GLC dataset influenced the model results by as much as 45 %. Moreover, lower

B. Mora (✉) • N.-E. Tsendbazar • M. Herold
Laboratory of Geo-Information Science and Remote Sensing, Wageningen University,
P.O. Box 47, 6700 AA Wageningen, The Netherlands
e-mail: brice.mora@wur.nl; nandin.tsendbazar@wur.nl; martin.herold@wur.nl

O. Arino
Earth Observation Science and Applications Department, ESA/ESRIN,
Via Galileo Galilei, Casella Postale 64, 00044 Frascati, Italy
e-mail: olivier.arino@esa.int

I. Manakos and M. Braun (eds.), *Land Use and Land Cover Mapping in Europe: Practices & Trends*, Remote Sensing and Digital Image Processing 18,
DOI 10.1007/978-94-007-7969-3_2,

quality LC datasets (e.g., <80 % overall accuracy) have strong effects on atmospheric simulations (Ge et al. 2007; Sertel et al. 2010). The need for GLC datasets with better quality and increased interoperability and inter-comparability has also been highlighted by GLC dataset user surveys for GlobCover maps and the LC Climate Change Initiative (LC-CCI) (Herold et al. 2011; Verburg et al. 2011).

In response to this need, international bodies such as Group on Earth Observation (GEO) and Global Climate Observation System (GCOS) were initiated to coordinate global cooperation to advocate and foster the establishment of an operational and continuous global-scale LC observing system (GCOS 2012; GEO 2012). Earth observation (EO) communities in Europe have been involved in the developments in GLC observation. For example, the European Commission Joint Research Centre, the Université Catholique de Louvain (UCL), Wageningen University and other partners are actively working on the production of GLC maps such as GLC 2000 (Bartholomé and Belward 2005), GlobCover (Arino et al. 2007), and LC-CCI (Defourny et al. 2011a, b, see Sect. 2.4 in this book) and on the integration, harmonization and validation of GLC datasets via their participation to other international initiatives such as the Global Observation of Forest Cover and Land Dynamics (GOFC-GOLD) initiative, and GEO (GEO 2012).

This chapter reviews the current status in GLC mapping and foresees upcoming developments within the field. The existing GLC maps and their characteristics are briefly summarized in Sect. 2.2.1. Section 2.2.2 highlights current issues that need to be overcome in GLC mapping initiatives. Sections 2.3 and 2.4 discuss upcoming solutions and recommendations, respectively.

2.2 Status and Improvements for Land Cover Maps

2.2.1 Existing Land Cover Maps

Advancements in remote sensing technologies during the last two decades have enabled the production of several GLC datasets supporting their extensive use in scientific research on modelling notably. The first attempts to map GLC using remote sensing produced 8 km and 1° of latitude coarse spatial resolution maps for years 1984 and 1987 respectively (DeFries and Townshend 1994; DeFries et al. 1998). Following these efforts, International Geosphere-Biosphere Programme Data and Information System's GLC map (IGBP – DISCover) and University of Maryland (UMD) datasets, the first 1 km resolution GLC datasets, were produced for the 1992–1993 period (Hansen et al. 2000; Loveland et al. 2000). Moderate-resolution Imaging Spectroradiometer (MODIS), GLC2000, and GLC by National Mapping Organizations (GLCNMO) products were also developed afterwards with data acquired around 2000, with the same spatial resolution (1 km) (Friedl et al. 2002; Bartholomé and Belward 2005; Tateishi et al. 2011). Moreover, 300 m and 500 m spatial resolution GlobCover and MODIS GLC maps were produced with the recent development of higher resolution time series satellite data for different periods (Table 2.1) (Arino et al. 2007; Friedl et al. 2010).

Table 2.1 Description of previous Global Land Cover (GLC) maps

Spatial resolution/ pixel size	IGBP-DISCover 1 km	UMD	MODIS	MODIS 5 500 m	GLC 2000 1 km	GLCNMO	Glob Cover 300 m	Glob Cover v2
Input data	AVHRR: Monthly NDVI from 10 day composites	AVHRR: Monthly NDVI and 5 bands from 10 day composites	MODIS: 16 day composites of 7 bands and EVI	MODIS: Monthly EVI, LST and 7 bands from 8 day composites	SPOT-Vegetation: Monthly to 3 monthly NDVI composites	MODIS: 16 day composites of NDVI and 7 bands	MERIS: Bi-monthly from 10 day composites	
Time of data collection	1992–1993		2001	2001–2008	Nov 1999- Dec 2000	2003	2005–2006	2009
Classification method	Unsupervised clustering	Supervised classification tree	Supervised decision tree	Supervised decision tree boosting	Optimal classification methods	Supervised classification	(Un)supervised spatio-temporal clustering	
Classification scheme	IGBP 17 class	Simplified IGBP 14 class	IGBP, UMD and other	5 different LC classification system including IGBP, UMD	LCCS 22 class	Modified LCCS 20 class	LCCS 22 class	
Validation data	Independent validation datasets from HR satellite data	Evaluated using other dataset	Cross validated using HR satellite data		Independent validation datasets from HR satellite data	Independent validation datasets from HR satellite data and other datasets	Independent validation datasets from VHR satellite data and other datasets	
Absolute positional accuracy (RMSE)	~1 km		1–1.5 km	50–100 m	300 m ~1/3 pixel	141–277 m	77 m	

(continued)

Table 2.1 (continued)

Spatial resolution/ pixel size	IGBP-DISCover 1 km	UMD	MODIS	MODIS 5 500 m	GLC 2000 1 km	GLCNMO	Glob Cover 300 m	Glob Cover v2
Area weighted thematic overall accuracy (%)	67		71.60 ± 2.5	74.8 ± 1.3	68.6 ± 5	81.20	73.10	67.50
Reference	Scepan et al. (1999), and Loveland et al. (2000)	Hansen et al. (2000)	Friedl et al. (2002, 2010)		Bartholomé and Belward (2005) and Mayaux et al. (2006)	Tateishi et al. (2011)	Bontemps et al. (2011), and Defourny et al. (2011b)	

Table 2.2 User distribution for the GLOBCOVER map by thematic field and organization type

	Carto graphy (%)	Climate/ meteorology/ hydrology (%)	Information technology/ GIS (%)	Natural resources (Agriculture, forestry, biodiversity) (%)	Remote sensing (%)	Total (%)
Commercial sector	2.69	2.42	9.41	3.48	2.96	20.97
Government organization	1.88	1.88	2.96	3.50	3.76	13.98
Non-government organization	2.69	2.96	4.30	6.45	0.81	17.20
University/ Research	3.23	8.87	10.22	13.98	11.56	47.85
	10.48	16.13	26.88	27.42	19.09	100.00

Source: GLOBCOVER user survey, N = 372, Herold et al. (2011)

Mid to coarse spatial resolution sensors such as AVHRR, SPOT-VEG, MODIS and MERIS are the main source for the existing GLC datasets. As shown by Chander et al. (2010) calibration of top of atmosphere reflectance EO data has improved over the recent years. GLC mapping initiatives benefit from these advances notably for LC change analysis. Different categories of classification algorithms (unsupervised/supervised, parametric/non-parametric) were applied to characterize GLC using IGBP and LCCS classification schemes (Loveland et al. 2000; Di Gregorio and Jansen 2005). GLC maps have been validated using varying approaches that comprised different reference datasets, sample selection scheme, sample unit size, minimum mapping unit, and reference data classification procedure *etc*. (Scepan et al. 1999; Hansen and Reed 2000; Mayaux et al. 2006; Friedl et al. 2010; Bontemps et al. 2011; Tateishi et al. 2011).

2.2.2 *What Needs to Be Improved*

User requirements surveys for GlobCover and the upcoming LC-CCI GLC datasets were conducted to address the needs of general and key users (e.g. the climate modelling community) (Herold et al. 2011). As highlighted in Table 2.2, the users of existing GLC maps are diverse, coming from different thematic fields and different organization types. While almost half of the users are coming from a university/research background, there is also significant use in governmental, non-governmental and commercial sectors across several disciplines.

The user survey for observing LC as Essential Climate Variable (ECV) has highlighted that LC remains a key dataset that serves as a base for many land surface parameters and associated temporal variability (Bontemps et al. 2011). The users stressed some requirements in terms of accuracy, stability, spatial resolution, and thematic content that are not met by the GLC datasets currently available (Bontemps et al. 2012; Herold et al. 2011). In addition, further investigation and advancements on consistency issues across GLC datasets and validation efforts for GLC monitoring are also emphasized by the mapping communities (Herold et al. 2008; Olofsson et al. 2012).

Table 2.1 shows the existing GLC maps have around 70 % (varying from 67 to 81 %) overall area-weighted correspondence with reference datasets. However, GLC map-like users have stressed that such datasets should have a maximum error of 5–15 % as a target, or at least higher than current quality, to be further used in modelling applications (Herold et al. 2011). Thus, there is a clear need to improve the current quality of GLC maps. Moreover, the relative importance of different class accuracies varies significantly depending on the users. Commonly, evergreen broadleaf trees, snow/ice, barren land classes show high accuracy (Giri et al. 2005; McCallum et al. 2006). On the other hand, general inability of GLC mapping approaches to clearly discriminate mixed trees, shrubs, and herbaceous vegetation due to low spectral separability has been noted. More attention is needed to improve the accuracy of these classes and the overall quality of the maps (Herold et al. 2008; Fritz et al. 2011).

Consistency and comparability of different GLC maps needs to be further analysed for a better understanding of their suitability and limitations for specific applications. Currently, the use of differing methodological approaches (e.g., classification scheme, data sources and algorithms) for GLC map production raises consistency issues and makes comparisons difficult. Consistency and comparability studies are commonly implemented using per pixel spatial (dis)agreement analysis (Hansen and Reed 2000; Göhmann et al. 2009; Fritz et al. 2011). These analyses show good overall agreement on spatial pattern, but limited agreement for some classes in specific areas (Giri et al. 2005; Herold et al. 2008). Disagreement is mostly observed in transition zones where a mixture of main vegetation components like shrub, tree grass (Hansen and Reed 2000; Herold et al. 2008). Unfortunately, LC change primarily occurs in transition zones, which makes it difficult to observe from differences between GLC datasets (Herold et al. 2008). Temporal instability of multi-year GLC products is also regarded as a major challenge in GLC change observations (Herold et al 2012; Bontemps et al 2012). This situation calls for strengthened international cooperation between GLC mapping communities to agree on a common set of harmonized GLC mapping procedures.

As indicated, landscape heterogeneity is one main driver of inconsistencies between the LC datasets, and it is identified as a major challenge for GLC mapping (McCallum et al. 2006; Herold et al. 2008; Wu et al. 2008). In addition, the use of coarse spatial resolution datasets ($\geq$300 m) induces the presence of several LC types in one pixel especially in transition zones. Current spatial resolution of GLC maps can be sufficient for some users such as climate modelling community. However, Landsat-type fine resolution datasets are also required for some model parameters and for description of change (Herold et al. 2011). Thus, the use of fine resolution satellite dataset will not only increase the usability of GLC datasets, but also help to ensure higher quality of LC characterization in heterogeneous and transition zones. Nevertheless, data availability of such fine resolution satellite data with high temporal frequency, particularly in consistent cloud covered areas is the biggest constrain for this.

Several statistically rigorous assessments of GLC maps were done using independent validation datasets (Scepan et al. 1999; Herold et al. 2008; Bontemps et al. 2011). As GLC maps are used for a large number of applications, user-oriented accuracy reporting can help understanding the uncertainty and limitations of LC datasets for specific applications (DeFries and Los 1999). Such accuracy reporting from GLC map user perspectives are limited (DeFries and Los 1999; Mayaux et al. 2006). More work is needed to improve flexibility of user oriented accuracy assessment methods as current overall accuracy and class-specific methods cannot provide comprehensive information addressing varying specific end-user needs. Validation datasets used for the GLC map quality assessment also calls for an international cooperation and requires significant effort to reach high-quality reference datasets. Thus, a comprehensive approach making best use of existing resources to develop an operational integrated and flexible reference dataset is sought (Herold et al. 2011). However, varying methodical approaches (e.g. sampling design, sample unit, legends, and classification approaches) applied for current reference datasets makes it a challenge (Olofsson et al. 2012).

An operational GLC observing system must provide LC change estimates for a comprehensive delivery of societal benefits. Coarse-resolution LC change observation provides useful information on long-term trends, inter-annual versus intra-annual dynamics, and the indication of large and cumulative land change, and hot spots; however, the reliability of this information is often questioned particularly in transitional and heterogeneous areas. On the other hand, fine-scale (i.e. Landsat-type) satellite data are currently the most suitable data sources for observing a large array of LC/land use change processes with confidence, but only a few examples have demonstrated operational feasibility (Kennedy et al. 2010; Goodwin et al. 2013). Thus, a combined approach using coarse and fine scale satellite observations, and in-situ observations seems the most suitable avenue for global and regional scale LC change studies (Bontemps et al. 2012). The need for such operational approaches is currently emphasized in starting or strengthening national forest monitoring activities in many developing countries to build capacity for a global participation in the Post-2012 Agreement on Climate Change (GLCA 2009). Progress in monitoring forest loss using the combination of coarse and fine scale satellite images at global level can be observed now (Hansen et al. 2010). Successful implementation and technical credibility of a GLC change assessment require agreement, dedication, collaboration and coordination among countries and this, from the supply of consistent observation data to the delivery of harmonized LC products.

2.3 Moving Forward

The development of new sensors is aimed to ensure continuity and increased frequency for consistent and continuous LC observations. Furthermore the necessity to provide supplementary and new sources of information has been urged since the failure of the Landsat-5 platform (Fall 2011) and the failure of ENVISAT MERIS mission (April 2012). The concomitant development of improved data processing methods, as well as the establishment of standardized or harmonized data processing procedures, demonstrates an accelerating trend towards the production of sound, and consistent global products. We present the main national and multi-national initiatives currently being led to overcome the aforementioned issues and meet the needs expressed by the users of LC information. We present also the emerging trends in terms of services, tools, applications, and the new users associated to GLC products.

2.3.1 Satellite Missions Allow Moving to Inclusion of Multiple Sensors, Finer Scale and Longer Time-Series Products

Looking forward from the progress of the last four decades in satellite observation the European Global Monitoring for Environment and Security (GMES) (now Copernicus) programme is aimed at providing information on Earth and its climate to better understand the role of human activities on the changes being observed at the global scale. The GMES programme provides a range of services among which satellite, airborne, and *in-situ* data for EO (Aschbacher and Milagro-Pérez 2012). As part of this programme, the launch of a series of EO *Sentinel* satellites is scheduled for the coming years. The first series will include a Synthetic Aperture RADAR (SAR) sensor (Sentinel-1), a high resolution optical sensor (Sentinel-2) (Drusch et al. 2012), and a moderate spatial resolution (300 m) optical sensor, and microwave sensors (Sentinel-3). Each of these satellite missions will encompass a pair of satellites to improve revisit time period, geographical coverage and rapid data dissemination (Berger et al. 2012). The launch of the first Sentinel-2 satellite is currently scheduled for mid-2014. In addition to Copernicus programme, the Pléiades constellation is another satellite constellation that is designed by France and Italy under the Optical & Radar Federated EO (ORFEO) programme (Lamard et al. 2008; CNES 2012). The satellites are designed to provide multi-spectral optical images with a two meter spatial resolution. Commercial distribution of images from Pléiades-1A is effective while images from the second satellite (1B was launched in December 2012) will start during 2013. Furthermore, a constellation of two new high-resolution (8 m for multi-spectral bands), optical imaging satellites from the Système pour l'Observation de la Terre (SPOT) series is also expected. First satellite (SPOT-6) was launched in September 2012 and launch of SPOT-7 is scheduled for 2014 (Astrium 2012). In the United States of America (USA), the National Aeronautics and Space Administration (NASA) and the United States Geological Survey (USGS) lead the Landsat Data Continuity Mission (LDCM). As part of this initiative the Landsat-8 satellite was launched in February 2013. The new satellite provides images of similar characteristics compared to its latest predecessor. First data is now available for download.

Existing EO systems combined with the scheduled arrival of new space-born sensors, especially embedded in platform constellation will facilitate the mitigation of atmospheric constraints inherent to the acquisition of optical images in tropical and boreal areas. For instance a 5-day revisit time period is expected for a given location when Landsat-8 and Sentinel-2 constellation satellites will be operational and combined. Positive outcomes are also expected regarding global-scale change detection monitoring with the generation of more complete time-series data. Building upon the existing archives of Landsat, MODIS, MERIS, AVHRR, and ERS/ASAR data are instrumental for long-term consistency and continuity of tracking land surface dynamics.

2.3.2 *Novel Global Land Cover Products Are Being Developed*

A clear trend towards the use of satellite data of higher spatial resolution for GLC analysis can be observed (Table 2.1). This dynamic is further reinforced by the GLC mapping projects from scientists in China and the USA. A GLC mapping project from Tsinghua University (Beijing) based on Landsat, Hun Jin (HJ), and Beijing (BJ) satellite data aims to provide GLC map products with an emphasis on water bodies, wetlands, and human settlements (Liao 2013; Chen 2012). Map products should be finalised and made available by the end of 2013. A Landsat-based GLC map product has been released (early 2013) by another team from Tsinghua University (Gong et al. 2013). The product depicts Earth's LC circa year 2010. While the first Chinese project relies on an automatic classification procedure and significant manual checking and editing, the second project is based on automatic classification procedures solely. In the USA, the NASA and USGS support a 30-m spatial resolution GLC mapping project based on Landsat data ($n \simeq 10{,}000$) acquired around 2010 (Stone 2010; Lee-Ashley and Moody 2010). These two GLC maps are expected to be released within the next 2 years and will meet the recommended requirements for GLC products expressed in terms of spatial resolution (Herold et al. 2009). For instance, Landsat-type data has been proven to be efficient at providing sufficient information for LC and LC change mapping at national scale with Minimum Mapping Units (MMU) comprised between 1 and 5 ha (Herold et al. 2009). Global characterization of tree cover using Landsat data is also recently released (Sexton et al. 2013; Townshend et al. 2012).

The European Space Agency (ESA) has initiated the Climate Change Initiative (CCI), a programme for the monitoring of Essential Climate Variables (ECV) (Sect. 2.4). Besides providing satellite data, data processing algorithms and methods, the ESA CCI will also produce a suite of spatially explicit ECVs. LC is one of the 14 terrestrial ECVs. ECV monitoring is to be conducted in 3 phases. After consulting the scientific community and dressing the detailed list of requirements and specifications in phase 1, the systems were developed and first maps were produced during Fall 2012 (phase 2). Phase 3 will consist in assessing the trends of the generated products, optimizing model calibration and validation, and quality assessment procedures, in close collaboration with the climate research community (Food and Agriculture Organization 2007). Maps for three epochs (2000, 2005, and 2010) will be released during fall 2013. Thus, the trend towards deriving more accurate GLC products targeted at the need for specific user community is obvious and a logical development given that there is a large variety of users whose needs cannot be all met by the current products.

To achieve the goal of producing sound and consistent GLC products, the data acquisition, the processing chain, and the implementation of the mapping procedures to make the products available to the user community need to comply with a series of standardized or harmonized practices that facilitate global-scale coordination between the stakeholders. The establishment and acceptance of such guidelines is an on-going process involving a range of institutions and persons coming from universities, public research centres non-governmental organizations (NGOs), private sector, and governments. These efforts need to include the option to ingest data on land change in near-real time (Verbesselt et al. 2012).

2.3.3 International Coordination and Harmonization Remain Vital

On-going international initiatives offer opportunities to improve relevance, acceptance, and approaches to operationalize and coordinate global and regional LC mapping surveys. Efforts are currently made via four major thematic areas: (a) standards for LC characterization, (b) standards methods for LC accuracy assessment, (c) GLC observations and applications and (d) LC change monitoring (Herold et al. 2012). Several initiatives that take lead on such efforts are summarized below.

The Group on Earth Observation (GEO) is one of the most prominent scientific and technical processes specifically concerned with EO sponsored by a partnership of 88 governments and 64 international organizations (as of March 2012). The GEO has recently recognized the importance of LC information to contribute to the nine GEO societal benefits (see Sect. 2.3.7 in this chapter: citation *to be fused later with:*) (GEOSS 2005). A specific GEO Task for GLC and LC change is aimed at providing recommendations for the production of consistent GLC datasets and services. The current trend is to move forward the development of products of higher spatial resolution (<50 m) and to emphasize the use of time-series products to characterize LC change and its dynamics. Such LC products are meant to be available to the user community through the Global Earth Observation System of Systems (GEOSS) infrastructure. Following the GEO 2009–2011 Work Plan (GEO 2010), the 2012–2015 Work Plan is being developed by a range of international bodies among which the GOFC-GOLD initiative (GEO 2011).

The GOFC-GOLD initiative is a panel of the Global Terrestrial Observing System (GTOS) sponsored by the Food and Agriculture Organization (FAO), the United Nations Educational, Scientific and Cultural Organization (UNESCO), the World Meteorological Service (WMO), the International Council for Science (ICSU), and the United Nations Environment Programme (UNEP). Specifically, the GOFC-GOLD LC Project Office is a major international body funded by ESA that contributes to the advancements in the four aforementioned thematic areas (gofcgold.wur.nl). The GOFC-GOLD LC Office is currently engaged in (1) ensuring continuity and consistency of observations, (2) promoting harmonization, interoperability and synergy of LC products, (3) developing validation standards and supporting their implementation, (4) improving adequacy and advocacy of land information products, and (5) supporting capacity development. The GOFC-GOLD LC Implementation Team (IT) has contributed to large series of international LC programmes, such as working on the development of standard reports for the LC and Biomass ECVs, the validation framework and implementation of GlobCover products and doing comparative validation studies between GLC products (Herold et al. 2008; Bontemps et al. 2011). The GOFC-GOLD LC-IT has taken lead roles in the implementation of several GLC-related GEO tasks (GEO 2011). The GOFC-GOLD LC Office and REDD + Working Group have played also a leading role in the development and update of the Reducing Emissions from Deforestation and forest Degradation (REDD+) Sourcebook (version 18 released in Fall 2012) in which LC information remains crucial (gofcgold.wur.nl/redd). The Sourcebook provides methods and procedures for monitoring and reporting anthropogenic greenhouse gas emissions and removals associated with deforestation, gains and losses of carbon stocks in remaining forests, and reforestation. Future updates of the Sourcebook are expected regarding good practices on LC map accuracy assessment. The subsequent presentation of international harmonization and standardization initiatives for GLC mapping further highlights the active role of the GOFC-GOLD LC Office in this field.

The UNFCCC and the International Panel on Climate Change (IPCC) among other United Nations (UN) bodies support initiatives to implement systematic observations of ECVs (IPCC 2006). There are currently 50 ECVs comprising LC as one of them (GCOS 2010). Under the supervision of the GTOS, the report on the LC ECV dresses the list of current data, products and capabilities for operational GLC mapping (Sect. 2.4). A series of recommendations are also provided among which to strengthen continuity and availability of data at different observation scales, the production of a flexible and continuous reference data in support of the calibration and validation of the models, the need for further international development and adoption of LC and LC change mapping standards, and a better coordination of the efforts between the stakeholders.

The outcomes of the GlobCover user survey show a good match between user requirements and the broader requirements from relevant international panels, e.g., those presented in the report on the LC ECV (Herold et al. 2009). The user groups express a need for stable LC data, increased capacity for time-series analyses, consistency among the model parameters, capacity to discriminate anthropogenic vegetation from natural vegetation, and establish the history of disturbance. LC products should also allow flexible use to serve at different scales and purposes. A general need for transparent information on the processing steps and the quality of LC products is expressed as well. Specifically, the availability of a multi-date accuracy assessment system and the use of the LC Classification System (LCCS, Di Gregorio and Jansen 2005) is advocated. The LCCS has been developed by FAO and UNEP as a comprehensive and standardized classification system designed for mapping purposes. The system is independent from the mapping scale and allows a dynamic creation of classes without obliging the user to relate to a pre-defined list of names by a dynamic combination of LC diagnostic attributes called *classifiers*. The last version of the LCCS, i.e., the LC Metadata Language (LCML – LCCS v.3), is proposed as a standard by the International Organization for Standardization (ISO) under the reference ISO 19144–1. Complementary specifications are under development under the reference WI 19144–2. Some GLC map products already use the LCCS (see Table 2.1). The outcomes of the GlobCover user requirements analysis were used as input for the product specification of the ESA LC-CCI in addition to LC-CCI user survey (Herold et al. 2011).

2.3.4 GLC Validation Is Becoming Operational

The Land Product Validation sub-group of the Working Group (WG) on Calibration and Validation (Cal/Val) from the Committee on Earth Observation Satellites (CEOS) aims to address the challenges associated with the validation of GLC products (NASA 2012). Accordingly, the CEOS Cal/Val WG compiled a document on recommended practices for validation of regional and GLC maps (Strahler et al. 2006). Moreover, CEOS Cal/Val WG in collaboration with GOFC-GOLD LC-IT initiated an operational GLC validation effort. This effort aims to develop a "living" dataset of validation sites to be used for statistically rigorous validation of GLC maps. Such dataset should have probability sampling scheme independent of any specific LC map and support statistically rigorous accuracy estimation. Currently, a research group at Boston University is developing an independent LCCS-compliant validation dataset consisting of 500 globally distributed sites for GLC products (Fig. 2.1) (Olofsson et al. 2012). A suite of multi-spectral very high spatial resolution (<1 m) satellite images is being acquired, segmented, classified, and visually checked. The production of validation data at such spatial resolution follows the trend observed for GLC products now being generated at higher spatial

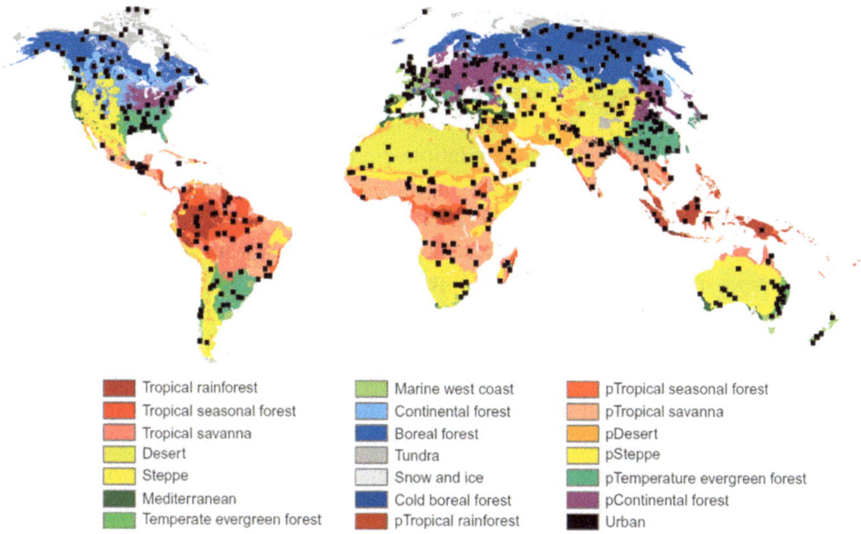

Fig. 2.1 Spatial distribution of validation sites from the Boston University database (Source: Olofsson et al. 2012). Land cover classes are derived from the Köppen climate classes (Peel et al. 2007)

resolution (30 m). The GOFC-GOLD LC-IT is also working on the development of an online information system to make these validation sites available to the community along with a set of recommendations to guide the user to the most appropriate dataset and usage (good practices). A beta version of the web portal has been released (gofcgold.wur.nl/sites/gofcgold_refdataportal.php) and the platform is expected to be operational in late 2013, hosting a dozen of reference datasets.

The ESA-CCI follows the reporting standards that are being developed under the lead of the GTOS, the Global Climate Observing System (GCOS), and its panels. The overall objective of the ESA initiative is to revisit the algorithms required to generate the GLC maps, design and implement a system that can provide GLC products derived from various EO sensors to the climate change community. In the frame of the ESALC-CCI an independent product validation and comparison will be performed to provide a robust assessment of LC product accuracy and precision. Strengthened user confidence in these LC products, acceptance, and legitimacy of the products are also expected within the international user and producer community. As such, a review of the GlobCover product validation sites (Bontemps et al. 2011; Defourny et al. 2011b) is undergoing under the lead of the Université Catholique de Louvain. The UCL is assisted by the GOFC-GOLD LC-IT for this task and the dataset will be made available on the aforementioned reference data portal.

2.3.5 New Services and Tools

The outcomes of the Sentinel-2 Preparatory Symposium (s2symposium.org/) stressed specific demands in terms of services and tools from different LC user communities. A series of recommendations and requests were addressed to ESA showing what current and future needs of the GLC mapping community as a whole are: open and free access to data, higher revisiting capacity, availability of procedures to process large data loads (corrections, cloud masking, mosaicking, classification, time-series analysis). Interoperability between data sources from different sensors (inter-sensor calibration, archive linkage) has been stressed as well, to enhance temporal revisit. In addition capacity development, in Non-Annex 1 countries in particular, still needs to be further strengthened. Note the GEO Global Forest Observation Initiative (GFOI) and the GOFC-GOLD Regional Networks among other initiatives, play an active role to reduce capacity gaps through different training programs. While stakeholders recognized the feasibility of GLC mapping with existing data and tools, on-going research notably focuses on techniques allowing the integration of different and complementary sources of information such as optical, Radar and Lidar data (Lucas et al. 2006; Bork and Su 2007; Lu et al. 2011; Li et al. 2012) and time-series analysis for change detection (Gutman and Masek 2012; Verbesselt et al. 2012).

2.3.6 The First Global Assessments of Land Cover Change

Characterization of change and dynamics of LC is a developing research area in the EO community especially since the advent of new processing techniques and a facilitated access of EO data (Sect. 2.3.5). Intra-annual LC dynamics can be characterized through the observations of vegetation phenology, seasonal snow coverage, flooding, fire occurrence, *etc.*, (Defourny et al. 2012; Bontemps et al. 2012). Many EO initiatives such as ESA-CCI, NASA-MODIS Land Program are actively working on the monitoring of such variables. Daily to yearly products characterizing the GLC condition are being produced using time-series analysis with MODIS, MERIS, SPOT and AVHRR data. Similarly, large-area LC change and longer-term trends in vegetation and fire characteristics can also be estimated using time-series analysis (Huang et al. 2002; Verbesselt et al. 2012). As an example, the University of Maryland produces an annual Vegetative Cover Conversion product, which consists in a global-scale LC change detection system at a 250 m spatial resolution (Carroll et al. 2006) that have now been used as input to the estimation of deforestation carbon emission patterns globally (Harris et al. 2012).

However, LC change mostly occurs at a smaller scale than this coarse resolution data can observe. It is now increasingly possible to overcome such a limitation as finer spatial resolution earth observation data (e.g. Landsat imagery) becomes more openly available at global scales. First serious attempts are being made using now freely available Landsat data provided by the Global Land Survey collection from the USGS and NASA (Gutman et al. 2012). The USA and China are currently using this Landsat archive to produce GLC maps (see Sect. 2.3.2). It is envisioned that the LC information of the American products will be updated every year or every 5 years, depending on the product (Stone 2010; Lee-Ashley and Moody 2010). Focused on forest cover and land use change, a sample-based global remote sensing survey is being conducted as part of the Global Forest Resources Assessments (FRA) led by the FAO (Gerrand et al. 2009) in cooperation with EU Joint Research Centre's TREES III project (ies.jrc.ec.europa.eu/index.php?page=70). Forest area and change rates have been calculated for years 1990, 2000, and 2005 using samples ($\simeq$13,500) from classified Landsat scenes validated by national experts. The same assessment exercise is planned for year 2010. The first results have been presented with more comprehensive analysis (Anonymous 2011).

2.3.7 New Users and Applications

The existing and upcoming LC products presented in this chapter can provide useful information for a wider variety of users and for a wider range of applications. The advent of new sources of EO data, improved processing techniques, standards, and services give a glimpse on the added value of such products. For instance, the proposed products and services associated to the nine societal benefits identified by GEO (GEO 2011) show how the scientific community, NGOs, private sector, governments and society as a whole can benefit from LC products. For **Disasters** (fire, earthquakes, flooding), LC information can help short term action planning; for **Health**, LC characteristics can help the identification of favourable conditions for disease vectors; for the **Energy** sector LC information can be useful to characterize the location of energy consumption spots and suitable areas for renewable energies such as wind turbines and solar panels; for **Climate** modellers, LC information can help modelling greenhouse gas emissions cause by LC change and phenology; for **Water** resources, LC information can help optimizing consumption and protect water bodies and wetlands; for **Weather**-related activities, information on LC change can help modelling radiation balance and sensible heat exchange, and provide information on land surface roughness; for **Ecosystems,** LC information can help the characterization of human alterations, monitoring ecosystem conservation, vegetation characteristics and change, as well as driving processes; for **Agriculture**, LC information can help monitoring crop production and cultivation practices and potentially associated land degradation. LC information can help monitoring **Desertification** and plan actions to mitigate and adapt to the phenomenon. Finally **Biodiversity** understanding, monitoring and conservation can benefit from LC information with the characterization of ecosystems, habitats, land fragmentation and connectivity. Thus, the potential use of LC and change data is large and the trend to deriving more targeted products for specific users is already obvious with a current focus on monitoring LC as ECV and for climate-change related, carbon emission assessment, purposes. But it is also expected that many other users will directly benefit from progress made for a specific use application; perhaps not in the full possible scale but to have a starting point to derive more specific products to meet their requirements.

2.4 Conclusion

GLC datasets remain a key input for scientific communities, NGOs, private initiatives, and governments. The need for an operational and continuous GLC observation is emphasized by different user communities. Therefore, the quality and consistency assessments of existing and up-coming GLC datasets should be highlighted for a better understanding of their suitability and limitations for specific applications. Reliable observation of LC is sought by GCOS. For this purpose, GLC dataset producers are working closely with climate modelling user groups (e.g., Climate Modelling User Group (CMUG)) to reflect their requirements. However, long term sustained interactions is not guaranteed. Current approaches on GLC dataset generation, thematic contents and validation still needs to be harmonized. A good documentation of GLC datasets generation and inter-comparison of different LC-ECV products are required for understanding incompatibilities with other datasets. The CEOS Cal/Val working group is actively working on GLC validation and good practice guidelines LC and LCC validations are introduced. New robust validation datasets from Boston University, the LC-CCI, and the Tsinghua GLC validation dataset are coming up while the importance of crowdsourcing validation datasets is also emphasized by the producer community (Fritz et al. 2009).

A number of initiatives from Europe (GMES, SPOT, Pléiades programmes) and the USA (Landsat continuity programme) will secure EO data supply continuity in the years to come. The recent end of life cycle of some satellite platforms (Landsat-5, Envisat, Advanced Land Observing Satellite (ALOS)) act as supplemental incentive to make these programmes operational. A clear trend towards higher spatial resolution map products is observed with the on-going Landsat-scale Chinese and USA GLC mapping projects, and the European GMES projects (Sentinel constellations). In parallel a series of international coordinated efforts to ease data access, to standardize (LCCS) or harmonize (GEO, GOFC-GOLD LC-IT, CEOS Cal/Val WG) mapping procedures, are underway. The GOFC-GOLD initiative as one major international body fosters free access to data and products. This coordination process that involves both GLC information producers and users is now crucial as the emerging new services and tools associated to the availability of new EO data sets broaden the scope of applications and concern a growing number of user communities (GOFC-GOLD 2013).

The Rio+20 – the UN Conference on Sustainable Development – organized in June 2012 in Rio de Janeiro, Brazil, tackled a range of topics embracing the green economy in the context of sustainable development and poverty eradication and the institutional framework for sustainable development. The need for monitoring carbon emission notably due to deforestation and forest degradation was highlighted. As a result this UN initiative represents an important internationally coordinated political incentive to ensure the development of new sensors, the continuity and increased frequency of Earth LC observations, and concomitant development of improved data processing methods as well as the establishment of globally standardized/harmonized data processing procedures.

References

Anonymous (2011) Global forest land-use change from 1990 to 2005 – initial results from a global remote sensing survey. Rome, Italy

Arino O, Gross D, Ranera F, Bourg L, Leroy M, Bicheron P, Latham J et al (2007) GlobCover: ESA service for global land cover from MERIS. In: Geoscience and remote sensing symposium – IGARSS 2007, 23–28 July 2007, Barcelona, Spain. IEEE International, Piscataway, pp 2412–2415

Aschbacher J, Milagro-Pérez MP (2012) The European Earth monitoring (GMES) programme: status and perspectives. Remote Sens Environ 120(2012):3–8

Astrium (2012) SPOT-6, -7 programme. http://www.astrium-geo.com/en/147-spot-6-7

Bartholomé E, Belward AS (2005) GLC2000: a new approach to global land cover mapping from Earth observation data. Int J Remote Sens 26(9):1959–1977

Benítez P, McCallum I, Obersteiner M, Yamagata Y (2004) Global supply for carbon sequestration: identifying least-cost afforestation sites under country risk consideration. International Institute for Applied System Analysis, Laxenburg, Austria

Berger M, Moreno J, Johannessen JA, Levelt PF, Hanssen RF (2012) ESA's sentinel missions in support of Earth system science. Remote Sens Environ 120:84–90

Bontemps S, Defourny P, van Bogaert E, Arino O, Kalogirou V, Perez JR (2011) GlobCover 2009, products description and validation report. European Space Agency, Frascati, Italy, and Université Catholique de Louvain, Louvain-la-Neuve, Belgium.

Bontemps S, Herold M, Kooistra L, van Groenestijn A, Hartley A, Arino O, Moreau I, Defourny P (2012) Revisiting land cover observation to address the needs of the climate modeling community. Biogeosciences 9(6):2145–2157

Bork EW, Su JG (2007) Integrating LIDAR data and multispectral imagery for enhanced classification of rangeland vegetation: a meta analysis. Remote Sens Environ 111(1):11–24

Carroll ML, DiMiceli CM, Townshend JRG, Sohlberg RA, Hansen MC, DeFries RS (2006) Vegetative cover conversion MOD44A, deforestation. In: Burned vegetation – collection 4, ed. University of Maryland, College Park, Maryland. http://glcf.umiacs.umd.edu/data/vcc/

Chander G, Xiong X, Choi T, Angal A (2010) Monitoring on-orbit calibration stability of the Terra MODIS and Landsat 7 ETM + sensors using pseudo-invariant test sites. Remote Sens Environ 114:925–939

Chen J (2012) China 30 m-resolution Global Land Cover Map in 2012. GIM International. http://www.gim-international.com/issues/articles/id1838-China_mresolution_Global_Land_Cover_Map_in.html

CNES (2012) ORFEO Pleiades. http://smsc.cnes.fr/PLEIADES/index.htm

Defourny P, Bontemps S, Martin B, Brockman C, Fomferra N, Grit K, Kruger O (2011a) CCI land cover project – product specification document, version 1.2. www.esa-landcover-cci.org

Defourny P, Bontemps S, Schouten L, Bartalev S, Cacetta P, De Wit A, Di Bella CM et al (2011b) GLOBCOVER 2005 and GLOBCOVER 2009 validation: learnt lessons. In: GOFC-GOLD global land cover & change validation workshop, Laxenburg, Austria

Defourny P, Mayaux P, Herold M, Bontemps S (2012) Global land-cover map validation experiences: toward the characterization of quantitative uncertainty. In: Giri C (ed) Remote sensing of land use and land cover – principles and applications. CRC Press – Taylor and Francis, Boca Raton, pp 207–224

DeFries RS, Los SO (1999) Implications of land-cover misclassification for parameter estimates in global land-surface models: an example from the simple biosphere model (SiB2). Photogramm Eng Remote Sens 65(9):1083–1088

DeFries RS, Townshend JRG (1994) NDVI-derived land cover classifications at a global scale. Int J Remote Sens 15(17):3567–3586

DeFries RS, Hansen MC, Townshend JRG, Sohlberg R (1998) Global land cover classifications at 8 km spatial resolution: the use of training data derived from Landsat imagery in decision tree classifiers. Int J Remote Sens 19:3141–3168

Di Gregorio A, Jansen LJM (2005) Land cover classification system: classification concepts and user manual: software Version 2. Food and Agriculture Organization of the United Nations, Rome

Drusch M, Del Bello U, Carlier S, Colin O, Fernandez V, Gascon F, Hoersch B et al (2012) Sentinel-2: ESA's optical high-resolution mission for GMES operational services. Remote Sens Environ 120(May):25–36

Food and Agriculture Organization (2007) FAO website. www.fao.org/gtos/topcECV.html

Friedl MA, McIver DK, Hodges JCF, Zhang XY, Muchoney D, Strahler AH, Woodcock CE, Gopal S, Schneider A, Cooper A (2002) Global land cover mapping from MODIS: algorithms and early results. Remote Sens Environ 83(1–2):287–302

Friedl MA, Sulla-Menashe D, Tan B, Schneider A, Ramankutty N, Sibley A, Huang XM (2010) MODIS collection 5 global land cover: algorithm refinements and characterization of new datasets. Remote Sens Environ 114(1):168–182

Fritz S, McCallum I, Schill C, Perger C, Grillmayer R, Achard F, Kraxner F et al (2009) Geo-wiki. Org: the use of crowdsourcing to improve global land cover. Remote Sens 1(3):345–354

Fritz S, See L, McCallum I, Schill C, Obersteiner M, van der Velde M, Boettcher H, Havlík P, Achard F (2011) Highlighting continued uncertainty in global land cover maps for the user community. Environ Res Lett 6:44005

GCOS (2010) Implementation plan for the global observing system for climate in support of the UNFCCC, GCOS-138, vol 138

GCOS (2012) Global climate observing system. http://www.wmo.int/pages/prog/gcos/index.php?name=AboutGCOS

Ge J, Qi J, Lofgren BM, Moore N, Torbick N, Olson JM (2007) Impacts of land use/cover classification accuracy on regional climate simulations. J Geophys Res 112(D5), D05107

GEO (2010) GEO 2009–2011 work plan – revision 3

GEO (2011) GEO 2012–2015 work plan – revision 1

GEO (2012) Group on earth observations. http://www.earthobservations.org/about_geo.shtml

GEOSS (2005) The Global Earth Observation System of Systems GEOSS 10-year implementation plan. Available at: www.earthobservations.org

Gerrand AM, Lindquist EJ, Wilkie M, Shimabukuro Y, Cumani R, Hansen MC, Potapov P, Achard F (2009) The 2010 global forest resource assessment remote sensing survey. In: Proceedings of the 33rd international symposium on remote sensing of environment ISRSE, 2–5, Stresa, Italy

Giri C, Zhiliang Z, Reed B (2005) A comparative analysis of the Global Land Cover 2000 and MODIS land cover data sets. Remote Sens Environ 94(1):123–132

GLCA (2009) Toward a post-2012 agreement on climate change: recommendations of global leadership for climate action. Global leadership for climate action. http://www.globalclimateaction.com/images/pdf/glca_recomm_post2012_agreement_climatechange.pdf

GOFC-GOLD (2013) Third GOFC-GOLD symposium. Wageningen University, Wageningen, The Netherlands, 15–19 April 2013. http://www.gofcgold.wur.nl/sites/Gofcgold_Symposium2013.php

Göhmann H, Herold M, Jung M, Schultz M, Schmullius CC (2009) Prototyping a probability-based Best Map Approach for global land cover datasets at 1 km resolution using MODIS, GLC2000, UMD and IGBP. In: 33rd ISRSE, Stresa, Italy

Gong P, Wang J, Yu L, Zhao Y, Zhao Y, Liang L, Niu Z et al (2013) Finer resolution observation and monitoring of global land cover: first mapping results with Landsat TM and ETM + data. Int J Remote Sens 34(7):2607–2654

Goodwin NR, Collett LJ, Denham RJ, Flood N, Tindall D (2013) Cloud and cloud shadow screening across Queensland, Australia: an automated method for Landsat TM/ETM + time series. Remote Sens Environ 134:50–65

Gutman G, Masek JG (2012) Long-term time series of the Earth's land-surface observations from space. Int J Remote Sens 33(15):4700–4719

Gutman G, Justice C, King LA (2012) The NASA land-cover and land-use change program – research agenda and progress (2005–2011). In: Giri C (ed) Remote sensing of land use and land cover – principles and applications. CRC Press – Taylor and Francis, Boca Raton, pp 379–396

Hansen MC, Reed BC (2000) A comparison of the IGBP DISCover and University of Maryland 1 km global land cover products. Int J Remote Sens 21(6–7):1365–1373

Hansen MC, DeFries RS, Townshend JRG, Sohlberg R (2000) Global land cover classification at 1 km spatial resolution using a classification tree approach. Int J Remote Sens 21(6/7):1331–1364

Hansen MC, Stehman SV, Potapov PV (2010) Quantification of global gross forest cover loss. Proc Natl Acad Sci 107(19):8650–8655

Harris NL, Brown S, Hagen SC, Saatchi SS, Petrova S, Salas W, Hansen MC, Potapov PV, Lotsch A (2012) Baseline Map of carbon emissions from deforestation in tropical regions. Science 336(6088):1573–1576. doi:10.1126/science.1217962

Herold M, Mayaux P, Woodcock CE, Baccini A, Schmullius CC (2008) Some challenges in global land cover mapping: an assessment of agreement and accuracy in existing 1 km datasets. Remote Sens Environ 112(5):2538–2556

Herold M, Woodcock CE, Cihlar J, Wulder MA, Arino O, Achard F, Hansen MC et al (2009) Assessment of the status of the development of the standards for the terrestrial essential climate variables – land cover. Rome

Herold M, van Groenestijn A, Kooistra L, Kalogirou V, Arino O (2011) Land Cover CCI user requirements document. Louvain-la-Neuve, Belgium

Herold M, Kooistra L, van Groenestijn A, Defourny P, Schmullius CC, Kalogirou V, Arino O (2012) Building saliency, legitimacy, and credibility towards operational global and regional land cover observations and assessments in the context of international processes and observing Essential Climate Variables (ECV'S). In: USGS/Earth Resources Observation and Science (EROS) Center, Giri CP (eds) Remote sensing of land use and land cover: principles and applications. CRC Press, Sioux Falls, pp 397–414

Hibbard K, Janetos A, van Vuuren DP, Pongratz J, Rose SK, Betts R, Herold M, Feddema JJ (2010) Research priorities in land use and land-cover change for the Earth system and integrated assessment modelling. Int J Climatol 30(13):2118–2128

Huang C, Wylie B, Yang L, Homer CG, Zylstra G (2002) Derivation of a tasselled cap transformation based on Landsat 7 at-satellite reflectance. Int J Remote Sens 23(8):1741–1748

IPCC (2006) Guidelines for national greenhouse gas inventories, vol 4 AFOLU (Agriculture, Forestry and Other Land Use). Kanagawa, Japan

Jung M, Henkel K, Herold M, Churkina G (2006) Exploiting synergies of global land cover products for carbon cycle modeling. Remote Sens Environ 101(4):534–553

Kennedy RE, Yang Z, Cohen WB (2010) Detecting trends in forest disturbance and recovery using yearly Landsat time series: 1. LandTrendr — temporal segmentation algorithms. Remote Sens Environ 114(12):2897–2910

Lamard JL, Frecon L, Bailly B, Gaudin-Delreeu C, Kubk P, Laherrere JM (2008) The high resolution optical instruments for the pleiades HR Earth observation satellites. In: International Astronautical Federation (ed) 59th International Astronautical Congress. Glasgow, pp 2650–2662

Lee-Ashley M, Moody J (2010) United States launches new global initiative to track changes in land cover and use. Lee-Ashley M, Moody J (ed). US. Department of the Interior

Li G, Lu D, Moran E, Dutra L, Batistella M (2012) A comparative analysis of ALOS PALSAR L-band and RADARSAT-2 C-band data for land-cover classification in a tropical moist region. ISPRS J Photogramm Remote Sens 70(06):26–38. doi:10.1016/j.isprsjprs.2012.03.010

Liao A (2013) Global land surface water product at 30 m resolution. ISPRS/GEO workshop on high resolution global land cover mapping, 24 April 2013, Beijing, China

Loveland TR, Reed BC, Brown JF, Ohlen DO, Zhu Z, Yang L, Merchant JW (2000) Development of a global land cover characteristics database and IGBP DISCover from 1 km AVHRR data. Int J Remote Sens 21(6/7):1303–1330

Lu D, Li G, Moran E, Dutra L, Batistella M (2011) A comparison of multisensor integration methods for land cover classification in the Brazilian Amazon. GISci Remote Sens 48 (3):345–370. doi:10.2747/1548-1603.48.3.345

Lucas RM, Cronin N, Moghaddam M, Lee A, Armston J, Bunting P, Witte C (2006) Integration of radar and Landsat-derived foliage projected cover for woody regrowth mapping, Queensland, Australia. Remote Sens Environ 100(3):388–406

Mayaux P, Eva H, Gallego J, Strahler AH, Herold M, Agrawal S, Naumov S et al (2006) Validation of the Global Land Cover 2000 map. IEEE Trans Geosci Remote Sens 44 (7):1728–1739

McCallum I, Obersteiner M, Nilsson S, Shvidenko A (2006) A spatial comparison of four satellite derived 1 km global land cover datasets. Int J Appl Earth Obs Geoinf 8(4):246–255

Nakaegawa T (2011) Uncertainty in land cover datasets for global land-surface models derived from 1-km global land cover datasets. Hydrol Process 25(17):2703–2714

NASA (2012) CEOS LPV website. http://lpvs.gsfc.nasa.gov/

Olofsson P, Stehman SV, Woodcock CE, Sulla-Menashe D, Sibley AM, Newell JD, Friedl MA, Herold M (2012) A global land-cover validation data set, part I: fundamental design principles. Int J Remote Sens 33(18):5768–5788

Peel MC, Finlayson BL, McMahon TA (2007) Updated world map of the Köppen–Geiger climate classification. Hydrol Earth Syst 11:1633–1644

Scepan J, Menz G, Hansen MC (1999) The DISCover validation image interpretation process. Photogramm Eng Remote Sens 65(9):1075–1081

Sertel E, Robock A, Ormeci C (2010) Impacts of land cover data quality on regional climate simulations. Int J Climatol 30(13):1942–1953

Sexton JO, Song X-P, Feng M, Noojipady P, Anand A, Huang C et al (2013) Global, 30-m resolution continuous fields of tree cover: Landsat-based rescaling of MODIS vegetation continuous fields with lidar-based estimates of error. Int J Digit Earth 6(5):427–448

Stone R (2010) Earth-observation summit endorses global data sharing. Science 330(6006):902. http://www.sciencemag.org/content/330/6006/902.short

Strahler AH, Boschetti L, Foody GM, Friedl MA, Hansen MC, Herold M, Mayaux P, Morisette JT, Stehman SV, Woodcock CE (2006) Global land cover validation: recommendations for evaluation and accuracy assessment of global land cover maps. Luxembourg

Tateishi R, Uriyangqai B, Al-Bilbisi H, Ghar MA, Tsend-Ayush J, Kobayashi T, Kasimu A et al (2011) Production of global land cover data – GLCNMO. Int J Digit Earth 4(1):22–49

Townshend JR, Masek JG, Huang C, Vermote EF, Gao F, Channan S et al (2012) Global characterization and monitoring of forest cover using Landsat data: opportunities and challenges. Int J Digit Earth 5(5):373–397

Verbesselt J, Zeileis A, Herold M (2012) Near real-time disturbance detection using satellite image time series. Remote Sens Environ 123:98–108

Verburg PH, Neumann K, Nol L (2011) Challenges in using land use and land cover data for global change studies. Glob Chang Biol 17(2):974–989

Wu W, Shibasaki R, Yang P, Ongaro L, Zhou Q, Tang H (2008) Validation and comparison of 1 km global land cover products in China. Int J Remote Sens 29(13):3769–3785

Chapter 3
The Users' Role in the European Land Monitoring Context

Núria Blanes Guàrdia, Tim Green, and Alejandro Simón

3.1 A Bit of History About the European Land Monitoring Process

In 1985 the European Commission started a programme to collect information relating to the environment named CORINE (COoRdination of INformation on the Environment). In the case of land cover data, national databases and maps were found to be too meager and diverse to be able to collect, harmonize and present the information in a comparable way from all the member states at a European level. Provided this situation, a programme to collect land cover and land use data at EU level (the CORINE Land Cover (CLC) programme) was started, centralizing the remote sensing based land cover mapping efforts in a single European land cover dataset, resulting in the CLC-1990. Since 1994, the European Environment Agency (EEA) has integrated CLC into its work program, and two more campaigns, the CLC-2000 and CLC-2006, have been produced. With these three time series of land cover datasets, maintaining its centralized specifications and standardization (although the production is decentralized at the member state level), CLC has become the *de facto* standard for a pan-European land monitoring system (HELM 2011).

In 2001, the GMES (Global Monitoring for Environment and Security) Action Plan (COM (2001) 609) started an overall discussion on how to deal with Land Cover Change exercises in Europe among other topics. Several expert groups

N. Blanes Guàrdia (✉)
European Topic Centre on Spatial Information and Analysis,
Universitat Autònoma de Barcelona, Barcelona, Spain
e-mail: nuria.blanes@uab.cat

T. Green
European Forest Institute Joensuu, Finland

A. Simón
European Topic Centre on Spatial Information and Analysis,
Universidad de Malaga, Malaga, Spain

I. Manakos and M. Braun (eds.), *Land Use and Land Cover Mapping in Europe: Practices & Trends*, Remote Sensing and Digital Image Processing 18,
DOI 10.1007/978-94-007-7969-3_3,

(the GMES Steering Committee, the GMES Advisory Council, the Implementation Group on the Land Monitoring Core Service, and since 2011, the GMES User Forum) became involved in discussions with the GMES Bureau, EEA, and with the Directorate Generals of the European Commission concerning how the GMES Land program could tackle the land monitoring program in Europe by using semi-automatic classification of remote sensing images to produce land cover and land use information at pan-European, European and local levels.

Based on these discussions and on the rich heritage of outcomes of the different GMES related projects funded by the European Space Agency and the European Commission (such as GSE AquaSoil/SAGE, GSE GMES Urban Services, GSE Forest Monitoring, GSE Land, geoland, GNU, geoland2 and HELM), the GMES Land monitoring program at European level has evolved from previous specifications formulated to develop mapping products (e.g. full automatic land cover classification for Europe with 21 classes) to the current specifications creating services to support current mapping activities like the production of five High Resolution Layers (imperviousness, forest, grasslands, water bodies and wetlands). These High Resolution Layers (HRL) will be operationally produced during the GIO (GMES Initial Operations) program, providing information such as degree of imperviousness, tree cover density or a mask of wetland areas. The data can be used for instance to better specify the content of the CORINE Land Cover (also included in GIO funding) classes at European level and also, to provide more content base to national land cover classifications.

It should be highlighted that, in parallel to the GMES developments at EU level, Member States are investing in the development of their own land monitoring programs to fulfill their specific national and regional needs. In order to explore the compatibility of national needs and systems with each other and with European products, several countries have experience concerning the potential of a bottom-up approach to develop a European land monitoring program based on the harmonization of data from existing national systems. Such an approach would be in line with the Infrastructure for Spatial Information in the European Community (INSPIRE) and Shared Environmental Information System (SEIS) principles of producing information once and at the level where it is most appropriate. However, its straightforward implementation is still to be proved.

Such collaborative initiatives have already started, and the diverse stakeholders that should contribute to the whole process are establishing the roles and the mechanisms to work together in order to ensure compatibility between different components.

Figure 3.1 provides an overview of the EU land monitoring in the last 20 years. It displays the relationships between the different GMES activities until 2012, focusing on the research and pre-operational phases of the GMES program, at European and at Member State level, which is also the focus of this article.

In December 2012 Copernicus became the new name of the European Commission's Earth Observation Programme, previously known as GMES (Global Monitoring for Environment and Security). Copernicus aims to set up informative operational services in six domains: land, marine, atmosphere, security, emergency response, and climate change.

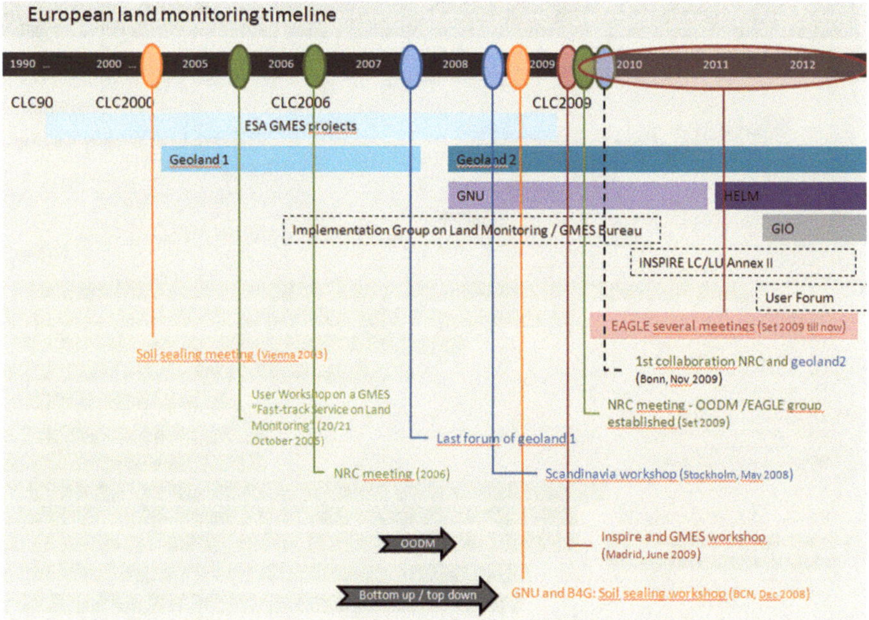

Fig. 3.1 Timeline of the European land monitoring programme until the end of the pre-operational phase of the GMES programme

3.2 Policies Influencing the Establishment of a European Land Monitoring Program

Within the various European Directives there is no legal obligation to derive land cover information. However, information about land cover and land use is a key requirement for many directives (e.g. Water Framework Directive, Habitats Directive, . . .) and other reporting obligations including international ones that should be accomplished by the Member States (e.g. Kyoto Protocol under the United Nations Framework Convention on Climate Change, Convention on Biological Diversity, Forest Europe – Ministerial Convention on Protection of Forests in Europe). Figure 3.2 shows the most relevant policies and reporting obligations for Europe. The implementation of the INSPIRE directive is a key step to ensure that the spatial data infrastructures of the Member States are compatible and usable in a European and international context, addressing 34 spatial data themes needed for environmental applications (including land cover and land use).

In 2008 the European Commission published a Communication to the European Parliament, the Council, the European Economic and Social Committee and the Committee of the Regions on *Global Monitoring for Environment and Security (GMES): we care for a safer planet* (COM(2008) 748 final) (EC 2008a) with the objectives to:

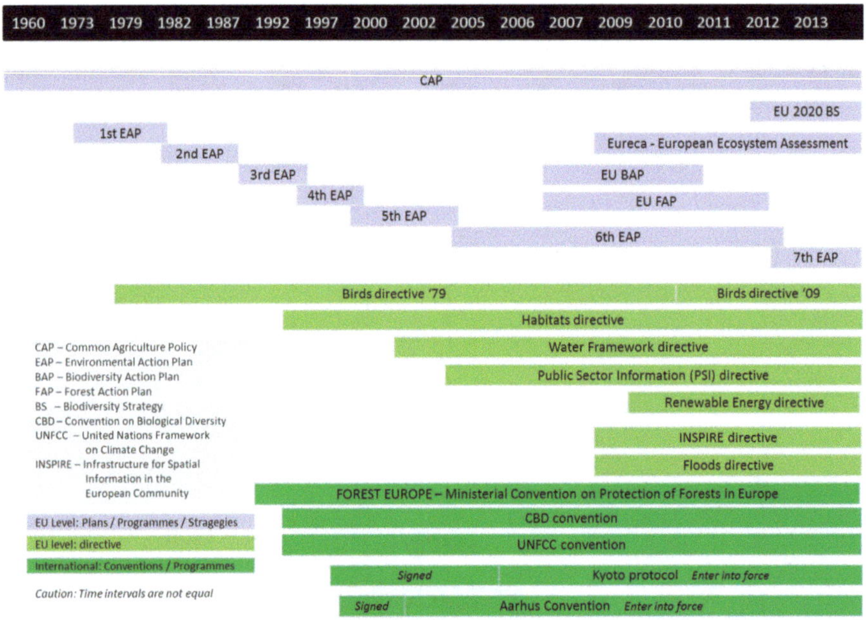

Fig. 3.2 Policies and EU actions related to land monitoring

1. Define a transparent and sustainable governance framework that contains a clear division of the roles of the *partners* in the GMES partnership, based on the principle that GMES should use to the largest extent possible existing capacities;
2. Guarantee user uptake, in particular through constant involvement of users so that GMES remains *user driven*;
3. *Reassure stakeholders* about the EU commitment to GMES in the sensitive phase of demonstration which precedes the move to operation;
4. Outline how the governance and financing framework can be *implemented in a reasonable timeframe* (EC 2008b).

3.3 Which Users Have Been Involved in the European Land Monitoring Program and What Have Been Their Roles?

From the different communications published by the EU, policy makers and public authorities have been defined as the major users of the GMES land initiative. In parallel, national mandated bodies and research institutions have been in charge of the development of the national land monitoring program and of the development of the CORINE Land Cover program at the EU level, which has been included as one of the GMES activities.

Defined by their geographical influence, the users in the current European land monitoring context are divided into:

- High-level European and International user organizations, which influence and support the production of uniform European wall-to-wall products;
- National and regional/local user organizations, which are more focused on determining the added value of the services being supplied, establishing what would be their general needs and provide access to in-situ data.

In the GMES program, there has been variable involvement of user organizations linked to the different projects resulting in variable levels of satisfaction with the project outcomes. Involvement ranged from partnerships between service providers and users where users were actively engaged in establishing requirements and service specifications, service reviews and final assessment of the utility of the product; to a more theoretically and political involvement, where users would have little opportunities to participate in the development of the services.

An evaluation concerning the user involvement in different GMES domains (Land, Forestry, Atmosphere and Geosciences) and related projects running until June 2009 was carried out by the GNU project (GNU 2010); where it was stated that the GMES projects are generally initiated and led by service providers -as described within the framework for these projects laid out by the ESA and the European Commission (Table 3.1). Only a minor proportion of the budget was available to involve the users and follow-up user uptake of the products and services being developed.

The provision of direct funds to user organizations could increase their capacity and motivation to participate more actively in the projects and in further (and perhaps also foreseen) developments. In contrast, if most of the project budget is dedicated to production, users have to provide their inputs often using their own time and resources. Thus, the outcomes will mainly depend on the motivation of the person/organization being involved. Hence, efforts have increased to establish a more representative user involvement in the research and implementation projects being undertaken towards a new and harmonized land monitoring program at EU level. Attempts have been made to involve users in different GMES projects and GMES related forums to discuss future approaches. Nevertheless, there is still much room for improvements.

Regarding the European institutions, it is important to highlight the two main roles they play in these processes: (1) they are the institutions providing the funding to develop European research programs or projects (e.g. FP7 – GMES Land – geoland2) and operational services (e.g. GIO 5 Pan-European High Resolution Layers); and (2) they determine the requirements or specifications of the products to be developed and also the specifications to validate those products.

Only the European Commission through its Directorates General as well as the GMES Office can be considered as mandated bodies to take decisions in respect to new developments of the European land monitoring program. The formal mandated and therefore, decision-making body for GMES is the GMES/Copernicus Committee, which in fact approves the regulations related to GMES.

Table 3.1 Key user organisations and their role within the GMES/Copernicus programme

User organizations	Roles and activities: Legislate	Endorse policy	Develop projects	Provide funds	Development of services/applications	Provision of recommendations	Data provision	Data compilation	Control	Monitor	Implement	Use data/service
European institutions												
European Commission – Directorates General (DGs)	■			■					■			■
European Environmental Agency			■		■	■		■		■		■
European Spatial Agency			■	■			■					■
GMES Office/GMES Committee			■	■					■	■		■
National (or sub-national) organizations												
GMES User Forum						■						
NRCs for Land Cover and Spatial Analysis (EIONET – Member States)			■			■	■	■		■	■	■
NFP Working Group on GMES		■	■			■			■	■		■
National ministries (Environment, Planning, Rural development, Agriculture,...)	■	■		■		■	■	■		■	■	■
National agencies/offices (e.g. national geographic institutes, national forest agencies, national forest inventories)			■		■					■	■	■
Others												
Research Institutes			■		■	■				■		■
SMEs			■		■		■				■	■
Spatial planners			■		■	■						■
Expert groups for EU directives						■				■		■
The public												■

On 25 May 2011, a Delegation Agreement was signed between the European Union and the European Environment Agency (EEA). The agreement tasks EEA with the technical coordination of the pan-European and local components of the GIO land monitoring service. EEA was the mandated body for the initial operations of the GMES/Copernicus land component, due to its aim of supporting the development and implementation of sound environmental policies in the EU and other EEA member countries by delivering timely, targeted, relevant and reliable information to policy-makers and the public. To accomplish this aim, through different information channels EEA collects environmental data from:

- EIONET (the European Environmental Information and Observation Network) – Partnership network of the EEA and its member and cooperating countries to provide timely and quality-assured data, information and expertise for assessing the state of the environment in Europe;
- GMES/Copernicus – The European Commission's Earth Observation Program provides data to help deal with a range of disparate issues including climate change and border surveillance on land, sea and atmosphere;
- GEOSS (The Global Earth Observation System of Systems) – It promotes common technical standards so that data from the thousands of different instruments can be combined into coherent data sets.

Therefore, EEA can be seen as one of the main European users. It informs about the EU needs and provides the technical user requirements to the services being currently developed within the European land monitoring program (e.g. 5 High Resolution Layers), as well as to deliver key information to the policy makers and to the public. It could also propose the next steps to be achieved by the program, highlighting the challenges that need to be overcome.

Future stages of the implementation of the European land monitoring program should be guided by the mandated bodies and authorized users. Currently the mandated body is the GMES/Copernicus Committee. However, it would also be important to broaden the list of authorized users setting up a comprehensive list to be endorsed by the GMES User Forum. By doing so, the responsibility to propose next steps and decisions would be shared among a large number of stakeholders. This would ensure that a broader range of interests would be taken into account to achieve the final aims of (1) having a land monitoring program harmonized at European level, and (2) GMES using existing capacities to the largest extent possible (EC 2008a).

3.4 Shaping the Role of the Users in the Mid-to Long-Term

There have already been attempts to change the involvement of the user organizations and public institutions into the "projects environment" and into the GMES program itself. The attempts address issues such as allowing users a more active role based on

- a direct participation in the products' development,
- in the provision of data, and
- last but not least, to undertake the quantitative validation and verification of the services being developed.

In fact, many users highlighted the need to link the update of national products with European products, and vice versa, in terms of thematic content and acquisition dates. This shall avoid duplication of work (principle of subsidiarity (EC 2009) and aims to provide comparable and interlinked data.

All those activities require funding, either as in-kind contribution from the user side in anticipation of the development of services that they better presume will be useful for them, or through planned participation of the users. Insufficient consideration of user involvement and user funding leads to a situation within some projects, where users have been asked to provide user requirements when the project has already started. Instead it would be better for users to be involved in the project planning, a prerequisite for the design of truly useful products in later stages of the project.

In order to achieve a more representative involvement and interaction on GMES governance with users from member states, the GMES User Forum was established by the Commission in 2011. It is the latest GMES user initiative aiming to ensure systematic consultations with authorized users, including users from Member States. The consultations will tackle the issue of a more active user involvement and should end in the establishment of agreed user needs (EC 2008a). However, the Forum has an advisory role only providing mainly recommendations for the decisions taken by the EU mandated bodies.

At the first official meeting of the GMES User Forum, May 2011 in Brussels (GMES User Forum 2011b), the role of obtaining harmonized land cover data was discussed and the importance of the INSPIRE Thematic Working Groups on Land Cover and Land Use was also highlighted. With regard to data policy and data distribution, the need for free, open and interoperable land products, in line with GEOSS data sharing principles, was stressed. Several delegations were concerned about the need for clear service specifications for the land products allowing repeatability, compatibility with national datasets and transparent communication among the stakeholders involved in the land domain.

Moreover, besides workshops being held at national level to promote the GMES program, other relevant meetings related to the land domain were organized under the GMES User Forum umbrella, to discuss for instance issues related to access to the reference/in situ data (September 2011, Brussels) (GMES User Forum 2011a), global land (December 2011, Lisbon), on data and information policy (January 2012, Brussels), and the 5th GMES User Forum (March 2013).

As an outcome of these meetings a serious user uptake effort is required. Within this user uptake, users would be actively involved

- in the current development and testing phases of the products,
- in understanding the administrative logistics behind, and
- in enabling the integration of the GMES services in national workflows and reporting activities.

Therefore, it is important to build up real and active Copernicus user groups, supporting effective federating and including all different administrative levels stakeholders' identified, which should actively interact as equal partners with the service providers and become involved in steering the process.

By establishing common working agendas, common initiatives and efforts between all relevant stakeholders involved in the process, it would be possible to change perspectives and expectations that have been established long time ago.

Several working groups are currently steering and contributing to the evolution of the European land monitoring program, such as:

- the EIONET NRC Land Cover (EEA network),
- the EAGLE working group (EIONET Action Group on Land Monitoring in Europe) or
- the INSPIRE Thematic Working Groups on Land Cover and Land Use.

Nevertheless, the GMES Committee (now Copernicus Committee) as the formal mandated body should be enabled to coordinate and structure the contributions from those users' organizations and working groups.

3.5 Current Situation and Aspects Still to Be Addressed in the Future

Provided that this article has been mainly focused on the research and pre-operational phase of the GMES program, the authors have considered that it may be interesting to include some hints on the current situation considering the GIO regulation, in place from September 2010, and the Copernicus framework, being adopted at the end of 2012 (EC 2013).

In 2010 Regulation EU No 911/2010 was issued on the European Earth monitoring programme (GMES) and its initial operations (2011–2013). GMES initial operations (GIO) are considered as a transition phase between the research phase (based mainly on FP6, FP7 and ESA funds) and the full exploitation of GMES capabilities in an operational phase. The GIO call established the work program for the European Earth monitoring program (GMES) in order to allow an operational GMES system by 2014 (EC 2011).

> Its primary objective is to provide, under Union control, information services which give access to accurate data and information in the field of the environment and security and are tailored to the needs of users. In doing so, GMES should foster better exploitation of the industrial potential of policies of innovation, research and technological development in the field of Earth observation. (EC 2010).

The GIO program was in fact, the culmination of many years of consultations and research with various experts and stakeholders.

While the European remote sensing services industry leads the production and design phase of the five HRL, EEA ensures the continuity of CLC 2012 in a cooperation framework with its member states. The HRL development was split thematically by layer and also geographically, and awarded to various project consortia in 2011. The initial work started in 2011 and the production will be finished in 2014. Likewise, GIO was also aimed to support HRL workshops to promote cooperation between countries and service providers boosting verification and enhancement activities and finding synergies between EU and national land cover products.

Previously in GMES and afterwards in GIO regulation (2010), user uptake has already being foreseen and implemented, and it is expected that it will be also included in the operational phase 2014–2020 of the Copernicus program. It is important to entrust a transparent mechanism for users' consultation at different administrative levels. The GMES User Forum has already addressed some of the issues (and it is expected that it continues as such):

- The Commission is responsible for user and service requirements (endorses)
- The Commission consults the User Forum before endorsing user requirements
- The Commission consults the GMES committee before endorsing the service requirements

Nevertheless, taking into account all the initiatives already being done, the balance in the involvement of service providers and different types of users is a key issue to continue the development of the Earth Observation programme at EU level, especially for the land component.

A sustained effort and funding is still needed to provide services, to build the community of users, and to develop and improve the services according to the user needs. Efforts should also be focused on establishing a more representative interaction with users, as it was recognized that the interactions earlier in the GMES program were variable. Therefore, there is still the need to establish and develop long-term partnerships among users, the EU institutions and service providers. In addition it is necessary to boost mandated bodies and organizations, as well as addressing the role that each type of user organization should perform, in order to guarantee the success of the Copernicus GMES programme.

And last but not least, an intensification of communication is required among stakeholders involved to define and consolidate the data policy. This simple but important step allows, on the one hand, the stakeholders (users to service providers) to know the conditions of supply (e.g. cost, frequency of production), and on the other hand, to support the provision of the services in well-known conditions in the European and in the global market. The intentions to support full, open and free access to information services have also been stressed and encouraged during the last years.

References

EC (2008a) Global Monitoring for Environment and Security (GMES): we care for a safer planet. Communication from the Commission to the European Parliament, the Council, the European economic and Social Committee and the Committee of the regions. Retrieved 15 July, 2012. Web site: http://eur-lex.europa.eu/LexUriServ/LexUriServ.do?uri=COM:2008:0748:FIN:en:PDF

EC (2008b) "Global Monitoring for Environment and Security (GMES): we care for a safer planet" – Impact Assessment. Retrieved 15 July, 2012. Web site: http://eur-lex.europa.eu/LexUriServ/LexUriServ.do?uri=SEC:2008:2808:FIN:EN:PDF

EC (2009) The principle of subsidiarity. European Parliament, the Council, the European economic and Social Committee and the Committee of the regions. Retrieved 30 May, 2012. Web site: http://www.europarl.europa.eu/ftu/pdf/en/FTU_1.2.2.pdf

EC (2010) Regulation (EU) No 911/2010 of the European Parliament and of the Council of 22 September 2010 on the European Earth monitoring programme (GMES) and its initial operations (2011–2013). Official Journal L 276 , 20/10/2010 P. 0001–0010. Web site: http://eur-lex.europa.eu/LexUriServ/LexUriServ.do?uri=CELEX:32010R0911:EN:PDF

EC (2011) On the European Earth monitoring programme (GMES and its operations (from 2014 onwards). Communication from the commission to the European Parliament, The Council, The European Economic and Social Committee and the Committee of the Regions. Retrieved 15 July, 2012. Web site: http://ec.europa.eu/enterprise/policies/space/files/gmes/com(2011)831_en.pdf

EC (2013) Copernicus in brief. European Commission. 20 April 2013. http://copernicus.eu/pages-principales/overview/copernicus-in-brief/

GNU (2010) GNU Experiences Report. Retrieved 5 May, 2012. FP6 – GMES Network of Users. Web site: http://www.gmes-network-of-users.eu/fileadmin/inhalte/gnu/pdf_files/Deliverables/GNU_Experiences_Report_rev.pdf

GMES User Forum (2011a) GMES User Forum – Workshop access to reference data/in situ. Brussels, 26 September 2011. Web site: http://ec.europa.eu/enterprise/newsroom/cf/itemlongdetail.cfm?item_id=5362&lang=en&tpa_id=141&displayType=calendar

GMES User Forum (2011b) Outcomes and recommendations from the GMES User Forum Preparatory workshop on Access to Geospatial Reference Data for GMES Land monitoring and other services. Brussels, 17 May 2011. Web site: http://ec.europa.eu/enterprise/newsroom/cf/_getdocument.cfm?doc_id=7046

HELM (2011) Panorama of the European Land Monitoring. Retrieved 6 August, 2012. FP7 – Harmonized European Land Monitoring (HELM). Web site: http://www.fp7helm.eu/fileadmin/site/fp7helm/HELM_Panorama.pdf

Chapter 4
Towards an European Land Cover Monitoring Service and High-Resolution Layers

Steffen Kuntz, Elisabeth Schmeer, Markus Jochum, and Geoffrey Smith

4.1 GMES/Copernicus – The European Contribution to Global Environmental Monitoring

The current global economic crisis forces nations to pinch and scrape on all their expenses. At the same time the demand for more and better information on the state of environment is increasing, together with more obligations to report on development and to estimate the impacts of environmental and spatial planning policies. Both needs may motivate a much stronger use of remote sensing, as so-called "free data" covering large regions or even continents become more and more important.

The advantages of remote sensing have been acknowledged by many stakeholders world-wide. In Europe this has led to a dedicated effort called "Global Monitoring for Environment and Security – GMES" (nowadays "Copernicus"). It is a joint initiative of the European Commission and the European Space Agency. GMES can be considered as a major European activity to make available timely and high quality information, services and knowledge, and to provide autonomous and independent access to information in relation to environment and security. Copernicus is also the European Union contribution to the Global Earth Observation System of Systems – GEOSS.

Starting in 1998, today Copernicus is fast moving towards an operational phase. It builds upon four pillars:

1. The space component (observation satellites and associated ground segment with missions observing land, atmospheric and oceanographic parameters). This comprises two types of satellite missions: ESA's five families of dedicated

S. Kuntz (✉) • E. Schmeer • M. Jochum
Airbus Defense and Space/Infoterra GmbH, Immenstaad, Germany
e-mail: Steffen.Kuntz@astrium.eads.net

G. Smith
Specto Natura Ltd., College Road, Impington, Cambridge CB24 9PL, UK
e-mail: geoffsmith@specto-natura.co.uk

I. Manakos and M. Braun (eds.), *Land Use and Land Cover Mapping in Europe: Practices & Trends*, Remote Sensing and Digital Image Processing 18,
DOI 10.1007/978-94-007-7969-3_4, © Springer Science+Business Media Dordrecht 2014

Sentinel space missions and those from other space agencies, called Contributing Missions.
2. In-situ measurements (ground-based and airborne data gathering networks providing information on oceans, continental surface and atmosphere).
3. Data harmonisation and standardisation.
4. Services to users.

> Users will be provided with information through services dedicated to a systematic monitoring and forecasting of the state of the Earth's subsystems. Six thematic areas are developed: marine, land, atmosphere, emergency, security and climate change. A land monitoring service, a marine monitoring service and an atmosphere monitoring service contribute directly to the monitoring of climate change and to the assessment of mitigation and adaptation policies. Two additional GMES services address respectively emergency response (e.g. floods, fires, technological accidents, humanitarian aid) and security-related aspects (e.g. maritime surveillance, border control). GMES services are all designed to meet common data and information requirements and have global dimension

http://www.gmes.info/pages-principales/overview/gmes-in-brief/

In the following chapter the geoland2 project, which addresses the "Land Monitoring Core Service (LMCS)" will be briefly described emphasising the high spatial resolution layers providing basic land surface information with wall-to-wall coverage of Europe.

4.2 geoland2

4.2.1 Background

The European CORINE Land Cover (CLC) approach[1] – the only existing harmonised European land data base available today – comprises 44 thematic classes with a minimum mapping unit (MMU) of 25 ha for stock, and 5 ha for changes, respectively. Since 1990 the CLC products have demonstrated the overall value of a harmonised data base at the European level. CLC is an excellent tool for strategic analysis and planning, as well as for general evaluations and overviews of the state-of-the-art and for facilitating policy making at European level. However, CLC's thematic content comprises a mixture of land cover and land use classes. Also, its MMU serves well the needs of the European Commission and the European Environmental Agency (EEA) but is not suited for national or regional planning activities. In addition, the previous update rate of 10 years had not been

[1] On a proposal from the Commission in 27 June 1985 the European Council adopted a decision on the CORINE programme (Coordination of information on the environment) to compile information on the state of the environment; to coordinate the compilation of data and the organization of information within the Member States or at international level; and to ensure that information is consistent and that data are compatible.

http://www.eea.europa.eu/publications/COR0-landcover

sufficiently rapid to track fast changes, which may appear, for instance, when new member states join the European Union.

To improve or support CLC former research projects in the framework of GMES funded by the European Commission and the European Space Agency were carried out. As a result, the "GMES Land Monitoring Core Service (LMCS)" shall provide accurate and cross-border harmonised geo-information at global to local scales. The service produces geographical information on land cover including its seasonal and annual changes and monitors variables such as the vegetation state or aspects of the water cycle. It has a wide range of applications in land use/land cover change, soil sealing, water quality and availability, spatial planning, forest monitoring and global food security. The pre-operational land monitoring service of GMES was developed by the European Union (EU) funded project geoland2 (http://www.gmes-geoland.info).

4.2.2 The Project

geoland2 was an EU funded 7th Framework Programme (FP7) Research Project, which was responsible for the development and pre-operational validation of the GMES LMCS. The project started in September 2008 and finished in 2012. It comprised 51 project partners and more than 80 major international user organisations.

Building on the results achieved by its predecessors[2] geoland2 is the last brick towards the implementation of a fully mature GMES LMCS. Based on land cover, land use or bio-physical information derived from Earth Observation satellite data, the service provides decision-makers with relevant information on the changing conditions of natural resources (e.g. water quality information across catchments basins). geoland2 aimed to:

- Organise a qualified production network;
- Build, validate and demonstrate operational processing lines in representative European and global test sites and on European level;
- Set-up a user driven product quality assurance process.

geoland2 addresses two different service levels: Core Mapping Services (CMS) and Core Information Services (CIS). Core Mapping Services provide:

- Land cover/land use data;
- A range of bio-physical parameters describing the continental vegetation state, the radiation budget at the surface and the water cycle;
- Seasonal and annual change monitoring at a range of scales and extents (Jochum 2010).

[2] i.e. mainly ESA GMES Service Element projects (2003–2007) and the FP 6 projects geoland (2004–2006) and BOSS4GMES (2006–2008) together with national initiatives.

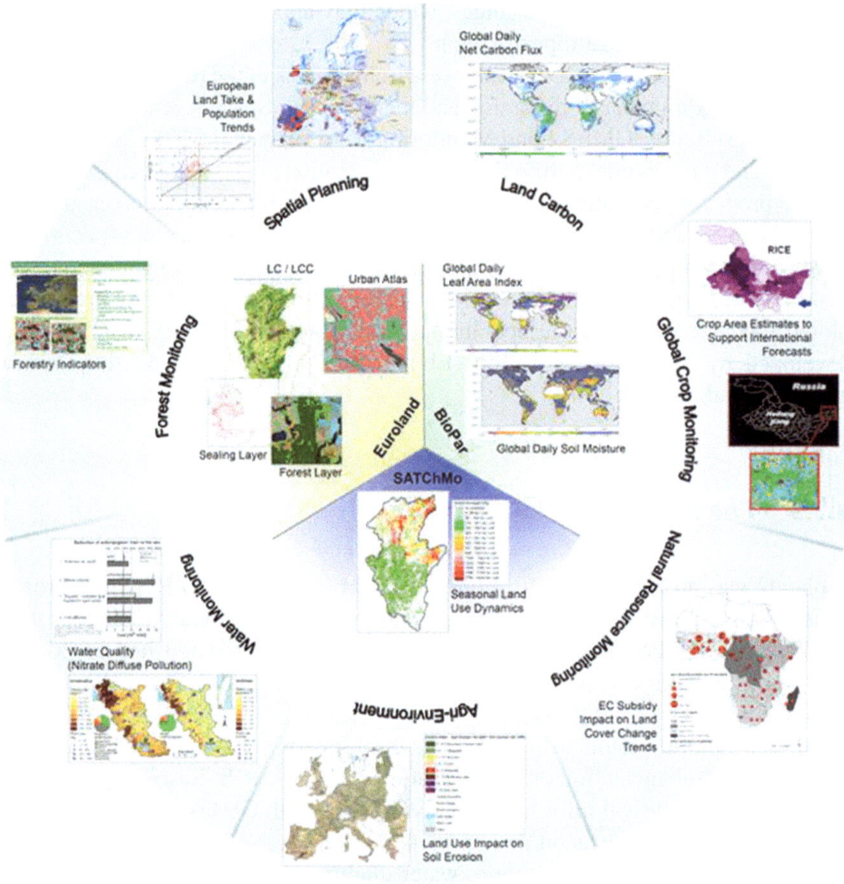

Fig. 4.1 Overview of the geoland2 service approach. In the center three "Core Mapping Services" are addressing local to regional land cover and continental to global bio-geophysical parameters. They serve the seven Core Information Services with basic input data for environmental modeling

The CMS products cover a wide variety of thematic content, spatial scales from local to global, and update frequency, from 1 day to several years.

The Core Information Services (CIS) aim at demonstrating the added-value that can be built on the Core Mapping Service in various fields. They propose a set of more specific, focused and context sensitive thematic products related to forest monitoring, spatial planning, land carbon monitoring, global crop monitoring, natural resource monitoring, agri-environmental monitoring and water monitoring applications. The CISs support reporting to European Environmental Policies and international treaties on climate change, food security and the sustainable development of Africa. Figure 4.1.[3]

[3] All data can be accessed free of charge under the following link: http://www.geoland2.eu/portal/

The CMS within the GMES LMCS are divided into three groups related to the scale at which they will be implemented: local, continental and global.

4.3 The Continental LMCS

The continental component of the LMCS comprises at present high spatial resolution biophysical parameters and the five high resolution layers. The following chapters describe the process of the service definition and the consolidated service specification.

A full description of the service specification can be found in geoland2- CMS Euroland (2011).

4.3.1 Service Definition Development

In order to overcome the shortcomings of CORINE for regional management and reporting, the FP6 project geoland set up a Core Service Land Cover (CSL) in 2004. Its goal was to achieve a consensus on a new European land cover data base and demonstrate its benefits offering improved spatial detail and thematic content compared to CORINE. The data base was designed to serve common land monitoring needs providing the status quo and the possible changes of Europe's landscape for European users and Member States enabling a wide range of downstream sectorial applications and user uptake.

Based on this approach, in 2007 the geoland2 proposal was made towards a service evolution offering more thematic content in certain areas (i.e. more forest classes and agriculture features) and a 1 ha MMU. The GMES Implementation Group Land supported this approach in its Strategy Document[4] and it was accepted in the geoland2 negotiations with DG Enterprise.

However, as a result of the intensified discussion with Member States since October 2008, the GAC paper 13–02, and based on recent feedback from EIONET members at several meetings between 2009 and 2011 this approach is not valid anymore. For political reasons – and here mainly the principle of subsidiarity – Member States do not want GMES services to interfere with their national mapping and monitoring obligations. Hence, the current approach for the local and continental LMCS emphasises the provision of intermediate products which are suitable for a large variety of applications and can actually support national obligations. The current way forward is:

[4] Strategic Implementation Plan of GMES Fast Track Land Monitoring Core Service, final Version, 24/04/2007; DG ENTR/IG Land.

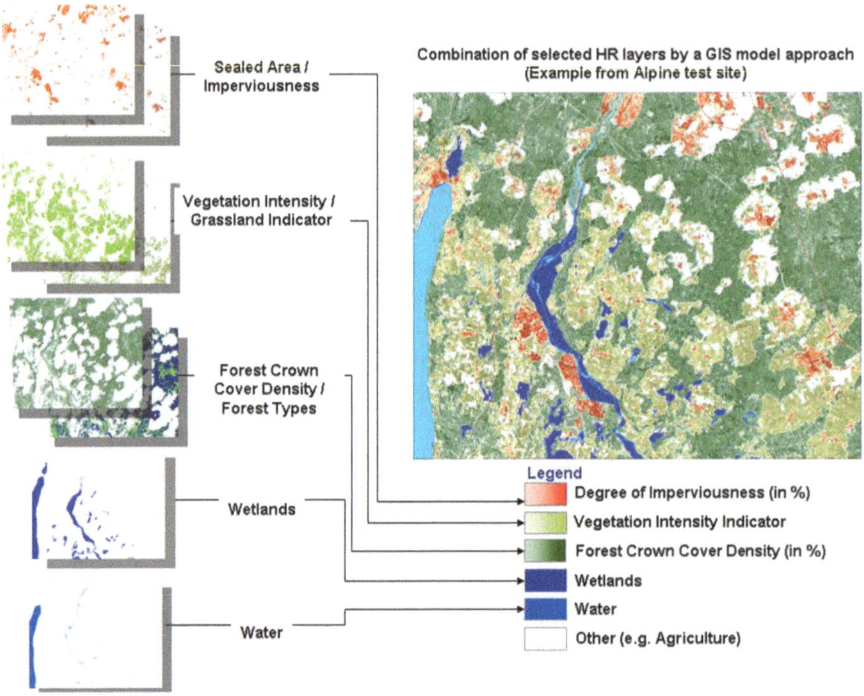

Fig. 4.2 Example for the five HR layers

- Urban Atlas monitoring is continued on behalf of DG Regional Policies.
- CLC change mapping shall be continued to assure European-wide harmonised time series on LC/LU changes.
- High resolution biophysical parameters shall be made available on a regular basis for the whole of Europe.
- A set of five high resolution thematic land cover layers (HR layers) shall be produced. These layers shall comprise information on imperviousness (sealed areas update), forests, grasslands (arable land/pasture), wetlands and small water bodies (Fig. 4.2).

4.3.1.1 Service Specification

The set of biophysical parameters offered from geoland2 includes an estimate of green vegetation, brown vegetation and soil cover fractions, LAI.[5] FAPAR,[6] chlorophyll content, a shadow factor, water and snow cover fractions, and water

[5] Leaf Area Index.

[6] Fraction of Absorbed Photosynthetically Active Radiation.

height for each pixel. These parameters are produced every 10 days from MERIS data (details in: geoland2- CMS Biopar 2011). In addition, production of these parameters at high spatial resolution (20–30 m pixels), as well HR data sets from Landsat, SPOT or IRS have been found to be very useful; in the future even higher quality from Sentinel-2 imagery will be available.

These high spatial resolution primary or "support layers" are provided usually as:

1. 20 m by 20 m pixels carrying information that can be perceived as "probabilities" or "densities" (for imperviousness, crown cover, etc.). These full resolution HR layers are not an end product (in the sense of traditional land cover maps), but a support tool to populate/characterise land cover objects, e.g. CORINE Land Cover or statistical grids. Hence, they are more geared towards a statistical use of the information.
2. High resolution layer "secondary products", usually aggregated to 1 ha MMU, derived from the "support layers" and ancillary information (e.g. build-up layer derived from the degree of sealed areas or changes in tree cover density).

The usage of the HR Layers in the context of CLC is still debated and discussions are not finalised. While some users see more the use of the HR Layers in populating CLC polygons, others see their applicability also during the CLC update process.

The "support" or "primary" layers shall provide comparable measures across Europe to:

1. Support characterisation of any (meaningful) user specified units (e.g. CLC polygons, reference grids, etc.);
2. Allow users (or value added providers) to derive additional layers (or products) based on user defined thresholds which can be driven by local context and shall provide already interpreted information.

The following Table 4.1 provides a short overview on the current service specification developed by geoland2 and presented to and discussed with European experts from the EAGLE[7] group.

In the framework of the GMES Initial Operations (GIO) phase the wall-to-wall mapping of HR biophysical parameters and the five HR layers has been initialised. However, according to the open public tender issued by the EEA in summer 2011,[8] for cost reasons, the HR layers and biophysical parameters are produced with a reduced service specification.

[7] Eionet Action Group on Land monitoring in Europe (EAGLE).

[8] Open call for tender No. EEA/SES/11/004.

Table 4.1 Specification overview of the Urban Atlas and the five high resolution (HR) layers

Urban Atlas	Information content	Specific benefits
	Land use and land cover data with 19 classes for Large Urban Zones with more than 100.000 inhabitants	Development of a cost efficient update methodology of the existing Urban Atlas data contributing for instance for the European Spatial Development Perspective and the Urban Environment Thematic Strategy
Data type:	Vector	
MMU	0.25 ha (artificial); 1 ha (non-artificial)	
Update frequency	3 years	
HR layer	Information content	Specific benefits
Imperviousness	Built-up areas including continuous degree of imperviousness ranging from 0 to 100 %	Input to State of Environment Report: Land-take trend in Europe (vs. Fast Track Service Sealing 2006) Input to various reporting & management obligations (e.g. Water Framework Directive, Soil Thematic Strategy, Convention on Long-range Transboundary Air Pollution, Kyoto Protocol, Urban Environment Thematic Strategy; European Spatial Development Perspective; Streamlining European Biodiversity Indicators; Common Agricultural Policy; Common Database on Designated Areas and national sustainability strategies)
Forest	Continuous Forest Crown Cover Density and Forest Type compositions	International: Environment for Europe Ministerial Conference (EfE); Ministerial Conference on the Protection of Forests in Europe (MCPFE). EU level: EEA State of Environment Report, SEBI2010* indicators, support to the EU Forest Action Plan implementation. National level: support to national forest inventory and monitoring
Grassland	Grassland areas with a continuous degree of intensity	Input to Habitats Directive Common Agriculture Policy (cross-compliance aspects: agri/forest conversion, environmentally friendly farming, maintenance of grasslands) Global Warming impact monitoring (desertification in the South, spread of humid grassland in the North)

(continued)

Table 4.1 (continued)

HR layer	Information content	Specific benefits
Wetlands	Wetland areas according to RAMSAR definition, wetness indicator	Provision of first pan-European data set on wetlands, Improve National digital data to RAMSAR, Birds directive, Common Database on Designated Areas data, Habitats and NATURA 2000 sites
Water	Small inland water bodies such as lakes, water reservoirs, river, streams	Input to various reporting & management obligations (Water Framework Directive, Flood Directive, Climate Change, Aarhus Convention, CAP)
Data type	Raster	
MMU[a]	Pixel level (20 m), validated to 1 ha	
Update frequency	3–5 years	

[a]*MMU* minimum mapping unit
[*]SEBI - Streamlining European Biodiversity Indicators. http://ec.europa.eu/environment/nature/knowledge/eu2010_indicators/index_en.htm

4.3.2 *Expected Benefits*

Like all GMES Core Services once implemented in a sustainable wall-to-wall monitoring scheme over the European mainland the benefits resulting from the local and continental LMCS will be on different levels and will differ according to the needs of the various customers interested in land cover information and changes in land cover over time.

As frequently stated, not only from international organisations and user DGs, but from scientists as well, coherent transboundary information on land cover and land use is important. It not only provides measures to assess the impact of European policies and directives across the member states (e.g. in international river basins such as the Rhine or Danube), but it can also support new reporting obligations (e.g. coming from UNFCC or UNEP).

For Member States the HR and VHR layers can support their reporting obligations by reducing the overall costs, as shown in Germany where the sealing layer was successfully used for CLC production. In addition, they allow the upgrade of national data bases; e.g. again in Germany where the sealing layer for the national topographic landscape model (DLM-DE) was deployed. The availability of the other HR layers could also improve such initiatives

It is expected that the soon available HR and VHR layers for the whole of the European area will motivate member states to invoke national programmes to take best benefit from the new data sources.

On a regional level the HR Layers may allow the upgrading and/or updating of environmental indicators and cross-border planning.

The possibility to connect in-situ monitoring with spatially explicit information may allow the improvement of the remote sensing based information by calibration of models, leading to better environmental indicators and improved statistics, as has been demonstrated for instance by the Core Information Service Spatial Planning.

Of critical importance for all governmental users is the long-term sustainability of the local and continental LMCS services with appropriate levels of quality. Only by providing reliable and repetitive time series of data for the monitoring of urban and regional developments will GMES/Copernicus create the highest benefits and lead-in investments required by national and regional entities for the whole service chain to become economically feasible.

References

geoland2- CMS Biopar (2011) User interface dossier, Issue 1.2

geoland2- CMS Euroland (2011) User interface dossier, Issue 1.00

Jochum M (2010) GMES Land monitoring service as developed by geoland2 – product and service portfolio; http://www.gmes-geoland.info/service-portfolio/land-cover-and-land-use-monitoring-products.html

Kuntz S, Jochum M, Leo O, Brink A, and Combal B (2011) Remote sensing and food nutrition security. In: Carlos Antonio Alvares Soares et al (eds) 1. international symposium on food security and poverty reduction, Vicosa, 16–18 Nov 2011, ISBN 978-85-60249-02-2, pp 01–16

Related Web pages: http://www.gmes.info/fileadmin/files/4.%20GMES%20Services/GMES_Land_Service_Portfolio_19Nov10.pdf

Part II
Operational European Mapping and Monitoring Services

Chapter 5
CORINE Land Cover and Land Cover Change Products

György Büttner

5.1 Introduction

From 1985 to 1990, the European Commission implemented the CORINE Programme (Co-ordination of Information on the Environment): an experimental information system on the state of the European environment was established; and nomenclatures and methodologies were developed and agreed at European level. CORINE Land Cover (CLC) was specified to standardize data collection on land in Europe to support environmental policy development. CLC data provide information on the bio-physical characteristics of the earth surface. Images acquired by Earth Observation (EO) satellites are used as the main source data to derive land cover and land use information (EEA Task Force 1992). The implementation of CLC follows a bottom-up approach,[1] meaning that national teams are producing the database for their own country, and these data are integrated at the European level. Project management is provided by the European Environment Agency (EEA). Despite its limitations in spatial resolution, CLC has become the primary spatial data source on land for EEA. CLC is widely used for indicator development, environmental modelling and land cover/land use (LC/LU) change analysis in the European context. Other Commission Services (e.g. DG ENV, DG AGRI) also rely on CLC. Page-view statistics (8,600 page-views per month average in 2011) show that CLC is one of the most popular databases of EEA.

[1] As opposed to the top-down approach, where data processing is centralised.

G. Büttner (✉)
European Environment Agency (EEA), Copenhagen, Denmark
e-mail: Gyorgy.Buttner@eea.europa.eu

I. Manakos and M. Braun (eds.), *Land Use and Land Cover Mapping in Europe: Practices & Trends*, Remote Sensing and Digital Image Processing 18,
DOI 10.1007/978-94-007-7969-3_5, © Springer Science+Business Media Dordrecht 2014

5.2 Technical Specification

The basic parameters of CLC have not changed during its lifetime (Table 5.1), thus maintaining the comparability between consecutive inventories.

Table 5.1 Evolution of CORINE Land Cover

	CLC1990	CLC2000	CLC2006	CLC2012
Satellite data	Landsat-5 MSS/TM single date	Landsat-7 ETM single date	SPOT-4/5 and IRS LISS III dual date	IRS LISS III and RapidEye dual date
Time consistency	1986–1998	2000 +/− 1 year	2006+/− 1 year	2011–2012
Geometric accuracy, satellite data	≤50 m	≤25 m	≤25 m	≤25 m
Min. mapping unit/width	25 ha/100 m	25 ha/100 m	25 ha/100 m	25 ha/100 m
Geometric accuracy, CLC	100 m	Better than 100 m	Better than 100 m	Better than 100 m
Thematic accuracy, CLC	≥85 % (probably not achieved)	≥85 % (achieved (Büttner and Maucha 2006))	≥85 % (not checked)	≥85 %
Change mapping (CLCC)	Not implemented	Boundary displacement min. 100 m; change area for existing polygons ≥5 ha; for isolated changes ≥25 ha	Boundary displacement min.100 m; all changes ≥5 ha are to be mapped	Boundary displacement min.100 m; all changes ≥5 ha are to be mapped
Thematic accuracy, CLCC	–	Not checked	≥85 % (achieved (Büttner et al. 2011))	≥85 %
Production time	10 years	4 years	3 years	2 years
Documentation	Incomplete metadata	Standard metadata	Standard metadata	Standard metadata
Access to the data (CLC, CLCC)	Unclear dissemination policy	Dissemination policy agreed from the start	Free access for all users	Free access for all users
Number of countries involved	26 (27 with late implementation)	30 (35 with late implementation)	38	39

5.2.1 *Minimum Mapping Unit and Minimum Mapping Width*

The Minimum Mapping Unit (MMU) is 25 ha; this means that objects having less than 25 ha area cannot be present in the database. The Minimum Mapping Width (MMW) of linear elements is 100 m; this means that objects (most typically highways and rivers) having less than 100 m width cannot be present in the database (EEA Task Force 1992). The explanations for these two values are: (1) The satellite images at that time had coarse resolution and poor geometric accuracy (namely 57 m $\times$ 79 m for Landsat MSS); (2) Producing CLC with photo-interpretation is a labour intensive process, so a compromise had to be found between mapping detail and production costs; and (3) In CLC1990, when interpreters drew and coded CLC polygons on plastic overlay at scale 1:100.000, the 25 ha MMU was considered the smallest object that could be mapped. Similarly, the 100 m MMW was considered as the narrowest linear element that could be drawn (0.1 cm width at a scale of 1:100.000).

Land cover objects having a size smaller than the MMU are generalized. The generalization is based on the 'similarity' between the small object (size <MMU) and the valid objects in the neighbourhood (Bossard et al. 2000) (e.g. a small wetland is joined to a neighbouring water body rather than to a forest). Generalization is usually straightforward for the experienced photo-interpreter, but it is not so evident for automation.

In CORINE Land Cover Change (CLCC) mapping, it was necessary to reduce the MMU for changes down to 5 ha to produce policy relevant information at the European scale. This resulted in a much more detailed CLCC layer than is possible in the CLC status layers (MMU ratio is 25/5 = 5) (Büttner et al. 2002a).

The use of a 25 ha (CLC) and 5 ha (CLCC) MMU is obligatory in the European CLC datasets. However, there are a few examples of more detailed national CLC and CLCC databases (e.g. Finland and Sweden apply a semi-automatic methodology, producing national land cover data with << 25 ha MMU, which is then generalised to yield the European CLC dataset (CLC2000 Finland Final Report 2005), (Engberg 2005)).

5.2.2 *Nomenclature*

The standard CLC nomenclature (Table 5.2) is hierarchical, including three levels of thematic detail in five major groups (Heymann et al. 1994): (1) artificial surfaces; (2) agricultural areas; (3) forests and semi-natural areas; (4) wetlands; and (5) water bodies. In addition to pure land cover classes, the nomenclature includes land use classes (especially within the artificial surfaces group) and some classes have mixed LC/LU character as well. Altogether there are 44 classes on level-3. The description of the CLC classes has become more detailed during the last 20 years. These enhancements have the main aim to further improve some of the class definitions or to discuss certain special cases (Feranec et al. 2007), so the results of inventories are kept comparable over time. The use of level-3 classes is obligatory in the European CLC

Table 5.2 The standard CORINE Land Cover nomenclature

Level 1	Level 2	Level 3
1. Artificial surfaces	1.1. Urban fabric	1.1.1. Continuous urban fabric
		1.1.2. Discontinuous urban fabric
	1.2. Industrial, commercial and transport units	1.2.1. Industrial or commercial units
		1.2.2. Road and rail networks and associated land
		1.2.3. Port areas
		1.2.4. Airports
	1.3. Mine, dump and construction sites	1.3.1. Mineral extraction sites
		1.3.2. Dump sites
		1.3.3. Construction sites
	1.4. Artificial, non-agricultural vegetated areas	1.4.1. Green urban areas
		1.4.2. Sport and leisure facilities
2. Agricultural areas	2.1. Arable land	2.1.1. Non-irrigated arable land
		2.1.2. Permanently irrigated land
		2.1.3. Rice fields
	2.2. Permanent crops	2.2.1. Vineyards
		2.2.2. Fruit trees and berry plantations
		2.2.3. Olive groves
	2.3. Pastures	2.3.1. Pastures
	2.4. Heterogeneous agricultural areas	2.4.1. Annual crops associated with permanent crops
		2.4.2. Complex cultivation patterns
		2.4.3. Land principally occupied by agriculture, with significant areas of natural vegetation
		2.4.4. Agro-forestry areas
3. Forest and semi-natural areas	3.1. Forests	3.1.1. Broad-leaved forest
		3.1.2. Coniferous forest
		3.1.3. Mixed forest
	3.2. Scrub and/or herbaceous associations	3.2.1. Natural grassland
		3.2.2. Moors and heathland
		3.2.3. Sclerophyllous vegetation
		3.2.4. Transitional woodland-scrub
	3.3. Open spaces with little or no vegetation	3.3.1. Beaches, dunes, sands
		3.3.2. Bare rocks
		3.3.3. Sparsely vegetated areas
		3.3.4. Burnt areas
		3.3.5. Glaciers and perpetual snow
4. Wetlands	4.1. Inland wetlands	4.1.1. Inland marshes
		4.1.2. Peat bogs
	4.2. Marine wetlands	4.2.1. Salt marshes
		4.2.2. Salines
		4.2.3. Intertidal flats
5. Water bodies	5.1. Inland waters	5.1.1. Water courses
		5.1.2. Water bodies
	5.2. Marine waters	5.2.1. Coastal lagoons
		5.2.2. Estuaries
		5.2.3. Sea and ocean

datasets. However, there are many examples of more detailed (e.g. level-4, level-5) national nomenclatures, where one or more of the standard level-3 classes are hierarchically subdivided (e.g. Estonia, Hungary, Portugal, Spain, Sweden, Turkey).

5.3 History of CORINE Land Cover

In its 25 year history, CLC has maintained its basic technical specifications (e.g. nomenclature, geometric resolution (Heymann et al. 1994)), but the process of technical implementation has significantly changed since its beginning.

The first CORINE Land Cover project ('CLC1990') was implemented in most of the EU12 member countries in the 1990s, as well as in the 13 partner countries in Central and Eastern Europe (EEA Task Force 1992). The nomenclature was finalized during the project.

In CLC1990, ortho-correction was not routinely applied in producing the base image map document for photo-interpretation. Today, with the availability of a DEM with appropriate spatial and vertical resolution, ortho-correction of satellite imagery is a standard process, providing higher geometric precision of the imagery. In CLC1990, mapping technology entailed photo-interpretation by drawing manually on a plastic overlay covering a 1:000.000 scale printout of a satellite image map (i.e. Landsat TM, sometimes Landsat MSS). Drawings on the plastic overlay had to be digitized manually to create the final database. The lack of ortho correction and the deformation of the plastic often caused geometric distortion of the resulting land cover data.

In-situ (ancillary) data in CLC1990 were mainly topographic maps and black-and-white photographs as hardcopy. Quality control and assurance were difficult tasks in CLC1990 as the checking of photo-interpretation had to be carried out on the plastic overlay.

Since the setting up of EEA in 1990 (EEC regulation 1210/1990), and the establishment of the European Environment Information and Observation Network (Eionet), EEA has had the responsibility of managing the CORINE databases.

The second CLC inventory was implemented within the 'IMAGE&CLC2000' project managed by the EEA and the Joint Research Centre (JRC). There were two main aims with similar priorities: (1) produce an updated CLC database based on lessons learnt during the 1st inventory with improved geometry (CLC2000) and (2) derive a database of land cover changes (CLCC) between 1990 and 2000 (Büttner et al. 2004). In CLC2000 the technology of drawing on transparencies was discarded, and fully replaced by computer-assisted photointerpretation (CAPI).

CLC2000 also included the removal of problems caused by the plastic overlay applied in the first inventory. Bulk geometric mistakes were removed by polynomial transformation or rubber-sheeting and residual geometric errors exceeding 100 m and coding mistakes were removed by editing. An additional step was to generalise all polygons smaller than 25 ha. This was a heavy task in some countries, where MMU <25 ha was applied systematically in CLC1990 (e.g. MMU = 10 ha in Belgium, resulting hundreds of <25 ha polygons). In some countries correcting CLC1990 needed significantly more efforts than the subsequent updating.

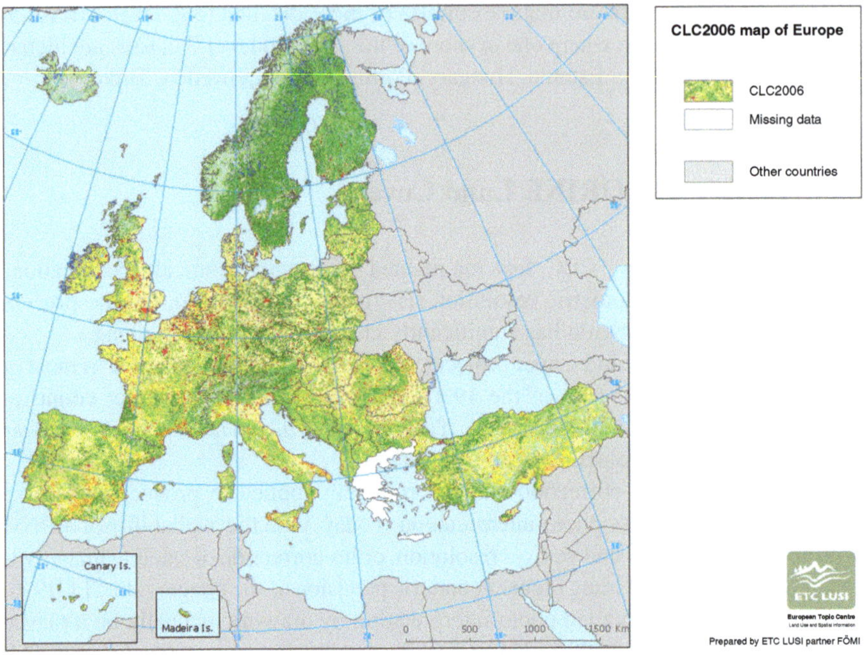

Fig. 5.1 Coverage of the CLC2006 inventory (Büttner et al. 2010)

CLC2000 has provided the opportunity for mapping CLC changes. Majority of countries first updated CLC1990 to CLC2000, then the two status layers were intersected to derive CLCC. However, due to the different MMUs in CLC and CLCC the result included lots of noise and false changes. Changes of objects having size under the 25 ha MMU were naturally omitted by this method. A few countries used the other way of update, where first changes were mapped then the new status layer was produced in GIS (Maucha et al. 2004). This method was later selected as the preferred method by the EEA. All in all CLCC database production for CLC2000 was not homogeneous across Europe.

The third CLC inventory, 'CLC2006' (Steenmans and Büttner 2006), was the result of EEA's collaboration with the European Commission (EC) and the European Space Agency (ESA) on the implementation of the Fast Track Service on Land Monitoring (FTS LM) in line with the communication: 'Global Monitoring for Environment and Security (GMES):[2] From Concept to Reality' (European Commission 2005). The CLCC database was considered as the primary product, and a uniform change mapping methodology was applied. Dual date satellite imagery (SPOT-4/5 and IRS P6 LISS III) provided enhanced change mapping capabilities. A significant increase in the number of participating countries took place (Fig. 5.1). Some of the countries newly entering the project also produced CLC2000 based on CLC2006 and CLCC(2000, 2006).

[2] GMES = Global Monitoring for Environment and Security (named Copernicus since late 2012).

At the time of CLC2006 scanned topographic maps and national coverages of digital colour aerial photographs (ortho-photos) are commonly available. Thematic maps, such as LPIS are also frequently accessible. Computer-assisted quality control provides written, geo-located explanations regarding the problems and supports harmonized production of the database all over Europe.

Data dissemination has also been improved. Since the second inventory (CLC2000) data have had dual ownership (EEA and the country). Today CLC data are freely accessible from EEA to any person or legal entity.

5.4 Components of CORINE Land Cover Inventories

In this chapter the commonalities regarding the major components of CLC inventories are described (Table 5.3).

5.4.1 Satellite Image Acquisition and Processing

Satellite images to support CORINE Land Cover mapping are provided by ESA. These image data sets are often referred to as IMAGE[year], where 'year' refers to the characteristic year of the image acquisition period (e.g. IMAGE2006 was taken in 2005–2007). These satellite images are also usually used for purposes other than the CLC inventory. At the time of the first four CLC inventories, ESA did not have its dedicated satellite(s) suitable for continental land monitoring, therefore ESA established agreements with appropriate satellite image providers for image acquisition. Since 2009, ESA is leading a federation of EO missions, named GMES Space Component, which has the objective to ensure a comprehensive and sustainable supply of data from space-based observations in response to GMES Service needs. The GMES Space Component provides harmonised access to data from different EO sources, ensuring seamless access to different EO data coming from

Table 5.3 Typical work packages of a CORINE Land Cover inventory

WP	Tasks	NRC	EEA	ESA	JRC	Data & service providers
1.1	Satellite data acquisition	o		X		o
1.2	Ortho-correction			X		o
1.3	Satellite image mosaic				X	o
2	In-situ data collection	o	X			o
3.1	CORINE land cover change mapping	X	X			
3.2	CORINE land cover production	X	X			
4	Validation		X			
5	Data dissemination	o	X	o		
6	Project management	X	X	o	o	o

X leading organisation, *o* organisation involved, *NRC* National Reference Centre, responsible for CLC multiple missions including ESA, national and other third party missions (GMES Space Component Data Access Portfolio, Data Warehouse 2011–2014). ESA's Sentinel-2 satellites, planned for launch in 2014 (http://en.wikipedia.org/wiki/Sentinel_2) are expected to provide the primary data support for the fifth and future CLC inventories.

Ortho-correction of satellite imagery is provided by service providers. A large amount of imagery has to be processed (e.g. in IMAGE2006, a total of 2,416 SPOT 4&5 and 1283 IRS P6 images were ortho-rectified (Lima 2009)). The satellite image mosaic is a general purpose product manufactured by using the pan-European coverage of high spatial resolution satellite imagery. For instance, using IMAGE2000 data the JRC has produced three types of mosaics: combination of Landsat TM bands 321; 453; and 752 (RGB), which can be accessed via WMS (http://image2000.jrc.ec.europa.eu/index.cfm/page/product_characteristics/p/p5).

5.4.2 In-Situ Data

The ortho-rectification of satellite imagery needs topographic information (Ground Control Points, Digital Terrain Model). As this information can be reused in subsequent CLC projects, new data collection is needed only for new participating countries.

In CLC and CLCC mapping, which are implemented by national teams, highly relevant in-situ data are provided by the countries. These valuable data are considered as in-kind contributions to the project. Digitized topographic maps, up-to-date digital ortho-photos and ground survey are the most commonly used type of in-situ information. For the semi-automatic production of CLC, data stored in national land use databases (national Spatial Data Infrastructure) are of utmost importance. On the other hand, Eurostat's LUCAS projects (The Lucas survey 2003) providing field based land use and land cover data and photographs in a regular grid are examples of centralized in-situ data support.

5.4.3 CLCC Mapping and CLC Production

CORINE Land Cover Change (CLCC) mapping has been carried out since the second inventory. CLCC is considered as the primary product since the third inventory. The new CLC status layer is derived by adding together the CLCC and the old CLC status layer. Photo-interpretation was applied by a majority of countries to derive the first CLC status layer, as well as the CLC Change layer, because semi-automatic methods were not considered mature enough to handle the large number of CLC classes in the diverse geographic environment of Europe. However, in the Scandinavian countries and recently in Germany (Federal Environment Agency 2012) and in Ireland (Irish National Project Plan 2011–2013) a combination of using land use data, regression estimation for forestry, image processing and/or generalisation have been applied to yield semi-automatically produced CLC and to some extent CLCC. Change mapping Guidelines (Büttner and Kosztra 2007) were designed to primarily support the photo-interpretation approach because: semi-automated methods are not yet widespread and their standardisation is difficult given the dependency on available national input data.

5.4.3.1 Methodology of Change Mapping

The 'change mapping first' approach has been recommended to follow as a method of deriving CLCC (Büttner and Kosztra 2007). The aim is to produce European coverage of real land cover changes that:

- are larger than 5 ha and wider than 100 m;
- occurred between $year_1$ and $year_2$;
- reflect real evolution processes (e.g. urban sprawl, new forest plantation, forest fire damage, new water reservoir).

Experts interpret CLC changes directly on screen, by comparing IMAGE_$year_1$ and IMAGE–$year_2$ data in a dual-window environment. The delineation of changes must be based on CLC_$year_1$ polygons in order to avoid creating sliver polygons and false changes when producing a CLC_$year_2$ database. An interpreter must give two CLC codes to each change polygon: code_$year_1$ and code_$year_2$. These codes must represent the land cover status of the given polygon in the 2 years, respectively. A change code pair thus shows the process that occurred in reality (Fig. 5.2) and may be different from the codes occurring on the CLC_$year_1$ map and/or in the final CLC_$year_2$ map (due to generalization applied in producing CLC).

With the three variables: CLC_$year_1$, CLC_$year_2$ and CLCC($year_1$, $year_2$), each polygon can have either of these two statuses: valid (if >MMU) and not valid (if <MMU), and in theory, $2^3 = 8$ different mapping cases can occur (Maucha et al. 2004). A 'technical change' attribute has been introduced to be able to delineate non-real 'changes' in cases when a land cover patch existed in $year_1$, it could not be mapped as a separate polygon due to the 25 ha limit in $year_1$, and the interpreter wished to include in the database in $year_2$ (Büttner and Kosztra 2007). In the course of change mapping, recognized mistakes of the old status layer have to be corrected. This process yields a CLC_$year_1$(revised) layer.

The main benefits of the 'change mapping first' approach are: (1) changes are interpreted directly (the interpreter has to think about what the real process was); and (2) all changes larger than 5 ha can be easily delineated regardless of their geometric position (whether attached to an existing polygon or not). The weakness is that some small (<5 ha) deficiencies in CLC_$year_2$ cannot be avoided (Maucha et al. 2004). Table 5.4 gives insight to the European CLCC (2000, 2006) database (Büttner et al. 2010).

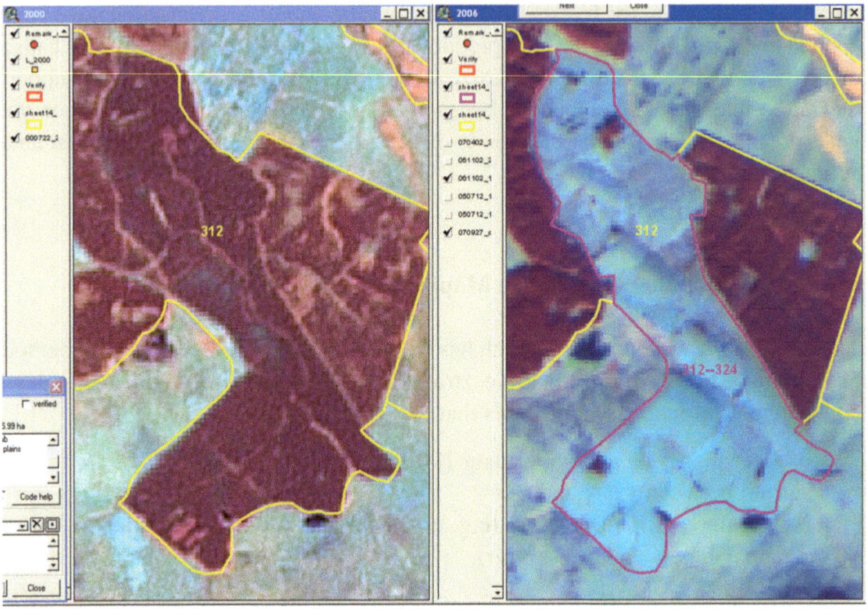

Fig. 5.2 Change mapping example (Ireland). Coniferous forest (code = 312) in 2000 (*left*) is clearcut by 2006 (*left*). The loss of forest cover is characterized by the light color of the ground due to the disappearance of crown cover. Forestry regulation protects forest cover by requiring replanting after harvesting. The loss of forest cover is therefore temporary. The transitional woodland area (code = 324) will sooner or later be replaced by forest. This is the most widespread change type in CLCC (2000, 2006)

Table 5.4 Some features of the CLCC(2000, 2006) Europe database (V15) (Büttner et al. 2010)

Total changed area:	70,824 km^2	
Part of Europe (without sea and ocean) that changed between years 2000 and 2006	1,24 %	
Number of change polygons	358,969	
Number of change types occurring	935	
Number of change types altogether providing 90 % of total change area	73	
Number of sporadic change types (each giving less than 0.1 % of total change area)	853	
Change types providing 50 % of total change area	312–324	24,547 km^2
	324–312	6,311 km^2
	311–324	5,729 km^2
Largest change type in artificial surfaces classes	133–112	2,492 polygons
Largest change type in agriculture classes	231–211	3,210 polygons
Largest change type in forests and semi-natural classes	312–324	146,596 polygons
Largest change type in wetlands and water classes	412–324	1,017 polygons
Country with the largest amount of changes in CLCC(2000, 2006)	Portugal (1,4 %/year)	

5.4.3.2 Change Mapping by Means of CAPI

Software developed by ESRI (ArcGIS and ArcView) were the most widely used tools to support CLC change mapping by means of computer-aided photo-interpretation (CAPI) technology. About half of the participating countries have used InterChange software (Büttner et al. 2002b) running under ArcView 3.x to implement CLC2000 and CLC2006. For CLC2012 InterChange has been rewritten as a stand-alone application (http://clc2012.taracsak.hu/). Its improved functionality significantly facilitates updating, change mapping, quality control and the correction of CLC databases by means of CAPI.

5.4.3.3 Production of the New Status Layer (CLC_year2)

With the CLCC($year_1$, $year_2$) database completed, CLC_$year_2$ is generated through an automated process:

$$\text{CLC\_year}_2 = \text{CLC\_year}_1(\text{revised})(+)\text{CLCC}(\text{year}_1, \text{year}_2)$$

(+) means the following operation: revised CLC_$year_1$ and CLCC($year_1$, $year_2$) databases are intersected, then CLCC($year_1$, $year_2$) polygons' $code_{year1}$ is replaced by $code_{year2}$, and finally neighbours with similar code are unified. Small (<25 ha MMU) polygons are generalized according to a priority table (Bossard et al. 2000). Table 5.5 includes CLC2006 statistic for Europe.

The consequence of the above methodology (and eventually that of the different MMUs) is that the difference (intersect) between two consecutive status layers (e.g. CLC2000 and CLC2006) will differ from the corresponding CLCC layer (e.g. CLCC(2000, 2006)). The magnitude of difference depends on the size distribution of change polygons. If there are many changes in the size range of 5–25 ha, the difference can be significant. If all changes were larger than 25 ha, then there would be no difference (Fig. 5.3, (Büttner and Kosztra 2007)).

5.4.3.4 Semi-automatic Approaches Applied in CLC

Some countries have applied semi-automatic solutions to map CLC and CLCC. The main aims of these solutions are to replace labour-intensive photo-interpretation, increase accuracy and reproducibility and provide better compatibility with national databases. The implementation of a semi-automatic approach heavily depends on the type of LU/LC data available in the country. Brief overviews of the methodologies are presented below.

In Finland, physics-based pre-processing of satellite images, classification of stratified imagery, regression estimation for forestry parameters, visual interpretation, integration of national land use data, and generalization of high-resolution national data to European CLC are applied. Level-4 CLC nomenclature is used for national

Table 5.5 CORINE Land Cover 2006 statistics for Europe (V15) (Büttner et al. 2010)

CLC code	Short class name	No. of polygons	Area (km^2) level-3	Area (km^2) level-1	% of total
111	Continuous urban fabric	6,041	6,727	214,938	3.75
112	Discontinuous urban fabric	140,338	153,544		
121	Industrial or commercial units	31,193	23,710		
122	Road and rail networks	3,224	2,568		
123	Port areas	1,130	1,147		
124	Airports	1,586	3,379		
131	Mineral extraction sites	10,306	7,213		
132	Dump sites	1,518	1,120		
133	Construction sites	2,832	1,899		
141	Green urban areas	4,650	3,099		
142	Sport and leisure facilities	15,131	10,533		
211	Non-irrigated arable land	180,133	1,216,467	2,441,791	42.65
212	Permanently irrigated land	9,666	81,841		
213	Rice fields	969	8,074		
221	Vineyards	20,314	40,441		
222	Fruit trees and berry plantations	17,185	28,822		
223	Olive groves	11,589	37,870		
231	Pastures	184,484	395,863		
241	Annual crops with permanent crops	5,492	9,563		
242	Complex cultivation patterns	184,242	302,529		
243	Agriculture land with significant natural vegetation	242,572	287,390		
244	Agro-forestry areas	6,286	32,930		
311	Broad-leaved forest	195,289	549,314	2,784,970	48.64
312	Coniferous forest	175,634	741,147		
313	Mixed forest	182,630	342,001		
321	Natural grassland	67,968	208,283		
322	Moors and heathland	42,309	163,149		
323	Sclerophyllous vegetation	27,319	86,846		
324	Transitional woodland-scrub	255,631	338,577		
331	Beaches, dunes, sands	4,204	8,081		
332	Bare rocks	17,040	90,667		
333	Sparsely vegetated areas	56,825	239,253		
334	Burnt areas	584	1,169		
335	Glaciers and perpetual snow	1,711	16,484		
411	Inland marshes	9,444	14,203	135,021	2.36
412	Peat bogs	58,575	104,624		
421	Salt marshes	1,675	3,318		
422	Salines	200	632		
423	Intertidal flats	2,092	12,244		
511	Water courses	2,109	13,584	148,520	2.59
512	Water bodies	49,931	126,495		
521	Coastal lagoons	487	5,818		
522	Estuaries	261	2,625		
Total		2,232,799	5,725,240	5,725,240	100.00

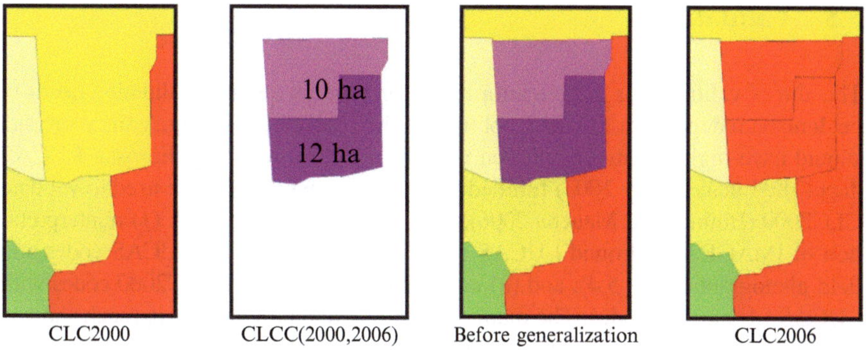

Fig. 5.3 The effect of generalization on producing the new status layer. A new construction site (*pink*) and industry (*lilac*) on former arable land (*yellow*) are appearing next to a settlement (*red*). In CLC2006, the construction site and industry polygons being below MMU (25 ha) are generalized into the existing settlement. The difference between CLC2000 and CLC2006 will not be the same as CLCC, which shows real changes (Büttner and Kosztra 2007)

purposes with pixel-resolution (25 m). Forest and semi-natural areas and wetlands in particularly have more thematic detail (CLC2000 Finland Final Report 2005). In Iceland, data are provided by relevant national authorities and institutions. Several classes are derived from remotely sensed data. Applied tools are: GIS harmonization and generalization (CLC2006, CLC2000 and CLC-Changes in Iceland 2009). Norway applies generalization and the merging of data from land resource maps and national registers (Sjolbørg Flo Heggem and Strand 2010), (Aune-Lundberg and Strand 2010). In Sweden, theme-wise classification of satellite imagery (interactive thresholding or automatic classification), visual interpretation, forest classification calibrated by National Forest Inventory data, and generalization of high-resolution national data to European CLC are applied. Sweden has produced 58 level-6 classes for national purposes (called SMD) with pixel-resolution (25 m) and MMU = 1, 2, 5 or 25 ha (Engberg 2005). In the United Kingdom, the semi-automatic generalization of national Land Cover Map (itself produced by a semi-automatic process) followed by the interactive editing of results are the applied tools (Balzter et al. 2011).

5.4.3.5 Quality Control

An important element of CLC project management on behalf of EEA is quality control of the work done by national teams. Usually two verifications are organised in each country during a CLC inventory. The main aims of these missions are to discuss progress with the team, check intermediate results, and give advice on the application of nomenclature and mapping of changes. Verification procedures have a corrective purpose with feedback from the team responsible for production. For checking the CLC and the CLCC databases, dedicated software (InterCheck) is used (http://clc2012.taracsak.hu/).

5.5 Validation

The aim of validation is to give information on product quality. Validation needs to be done on independent EO data, not used in the course of mapping. Validation data should have higher spatial resolution than that used for mapping. Eurostat LUCAS data (The Lucas survey 2003) fulfilled these requirements and were used to validate CLC2000 (Büttner and Maucha 2006). Two approaches were used: (1) reinterpretation of IMAGE2000 around LUCAS sampling points based on LUCAS codes and field photographs (Fig. 5.4); and (2) automatic comparison of CLC2000 codes with LUCAS LU and LC codes. The main result of the reinterpretation approach showed that the overall reliability of CLC2000 was 87.0 ± 0.8 %, i.e. the 85 % specified accuracy requirement was fulfilled. The result of the automatic comparison showed that agreement between CLC2000 and LUCAS LU/LC was 74.8 ± 0.6 %. This lower accuracy was attributed to the resolution difference between the two datasets. The highest class-level reliability (> 95 %) was obtained for rivers (511), lakes (512), industrial and commercial units (121) and discontinuous urban fabric (112). The analysis revealed that the subjectivity of photo-interpretation could be noticed in 18.2 % of the samples. The most subjective CLC classes were as follows: agriculture with significant amount of natural vegetation (243), transitional woodland, shrub

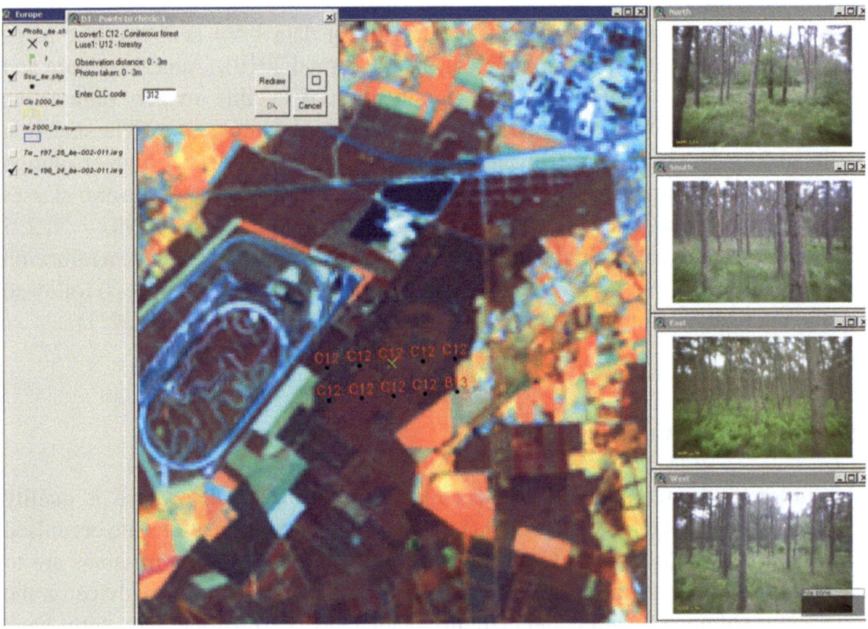

Fig. 5.4 Validation of CLC data by means of LUCAS (Büttner and Maucha 2006). The CLC code had to be estimated around the yellow dot over the satellite imagery, by using LUCAS field photographs (*right*) and LU/LC codes obtained on the field (*red codes*) (example is from Belgium.)

Table 5.6 The European CORINE Land Cover products, distributed by EEA (Version 15)

Products	Type	Characteristics
CLC1990	Raster	100 and 250 m grid
CLC2000	Raster	100 and 250 m grid
CLC2006	Raster	100 and 250 m grid
CLC2000	Vector	by CLC codes, 44 classes/files
CLC2006	Vector	by CLC codes, 44 classes/files
CLCC(1990, 2000)	Vector	
CLCC(1990, 2000)	Raster	100 m grid
CLCC(2000, 2006)	Vector	
CLCC(2000, 2006)	Raster	100 m grid

Projection: ETRS89 LAEA (EPSG: 3035)

(324), complex cultivation patterns (242) and mixed forest (313). The lowest class-level reliability (below 70 %) was obtained for the sparse vegetation class (333).

Stratified random sampling was used for validating the CLCC(2000, 2006) database. This was the first change validation effort in the history of CLC (Büttner et al. 2011). The exercise proved to be difficult because the amount of CLC-Changes was small (Table 5.4). Therefore a special sampling strategy was applied: 100 sample points were selected from inside each of the 25 level-1 change types, thus representing the whole change polygon population. The obtained 87.8 ± 3.3 % overall accuracy (calculated using commission errors only) based on 2,405 samples is satisfactory. The omission error was not possible to measure due to the very large sample size that is required. The large number of participating countries made it unrealistic to collect very high spatial resolution orthophotos or satellite imagery and even topographic maps for the purposes of validation. Therefore validation was executed through the re-interpretation of IMAGE2000 and IMAGE2006, supported through the use of Google Earth imagery. 17 of the 25 change type groups showed accuracy higher than 85 %, 13 types of which had accuracy higher than 90 %, including the largest level-1 change class (i.e. internal changes in forest and semi-natural vegetation). Significant change types with accuracies below 85 % were: (1) forest/semi-natural area changed to agriculture; and (2) artificial area changed to forest/semi-natural cover (e.g. reclamation of mineral extraction sites).

5.5.1 Data Dissemination

Most of the countries participating in CLC have an internal data dissemination service, which distributes data in national projection. Table 5.6 presents the dissemination of European CLC data by the EEA Data Service (http://www.eea.europa.eu/data-and-maps/find/global#c12=Corine+Land+Cover).

Rights of use are explained through EEA standard re-use policy (http://www.eea.europa.eu/data-and-maps/find/global#c12=Corine+Land+Cover), which states that, unless otherwise indicated, re-use of content on the EEA website for commercial or non-commercial purposes is permitted free of charge, provided that the source is acknowledged (http://www.eea.europa.eu/legal/copyright). The copyright holder is the European Environment Agency.

5.6 CORINE Land Cover 2012

The fourth CLC inventory (CLC2012) is implemented as part of the GMES[3] Initial Operations (GIO) initiated by DG ENTR of the European Commission. The coordination of the GIO land monitoring is delegated to EEA for implementation (REGULATION (EU) No 911/2010). At the time of writing of this article the project was in the start-up phase. The ESA Data Warehouse (GMES Space Component Data Access Portfolio 2011–2014) provides a satellite image catalogue and download system for all GMES-related activities, including CLC2012. Two satellite image coverages have been acquired (primarily IRS/ResourceSat LISS III and RapidEye) in 2011–2012. Gap filling is targeting those areas which were not covered by imagery during the planned 2 year image acquisition period. The technical implementation of CLC2012 will be similar to the CLC2006 inventory. Majority of countries will apply CAPI technology mapping the CLCC layer first. Germany (Federal Environment Agency 2012) and Ireland (Irish National Project Plan 2011) joined the Scandinavian countries by applying a semi-automatic methodology based on the integration of existing land use data, satellite image processing and generalization.

39 European countries, covering an area of approximately about 5.8 Mkm^2 plan to participate in CLC2012:

- All EU27 countries (members of EEA as well): Austria, Belgium, Bulgaria, Cyprus, Czech Republic, Denmark, Estonia, Finland, France, Germany, Greece, Hungary, Ireland, Italy, Latvia, Lithuania, Luxembourg, Malta, the Netherlands, Poland, Portugal, Romania, Slovakia, Slovenia, Spain, Sweden, United Kingdom;
- All other EEA member countries: Liechtenstein, Iceland, Norway, Switzerland, Turkey;
- EEA cooperating countries: Albania, Bosnia-Herzegovina, Croatia, the Former Yugoslavian Republic of Macedonia, Kosovo under UNSCR 1244/99, Montenegro.

Results are expected by the end of 2014.

[3] GMES = Global Monitoring for Environment and Security (named Copernicus since late 2012).

5.7 Future of CORINE Land Cover

The main criticism against CLC is that: (1) the MMU is large, and databases are not spatially detailed enough; (2) the nomenclature consists of land cover and land use classes; (3) some of the classes (e.g. mixed agriculture) are difficult to translate to other systems, e.g. the LCCS (Land Cover Classification System) (Herold et al. 2009); and (4) CLC data cannot be used for statistical comparison with other surveys (Sjolbørg Flo Heggem and Strand 2010).

There are two on-going activities that pave the way towards future of European land monitoring: EAGLE (Eionet Action Group on Land monitoring in Europe (http://sia.eionet.europa.eu/EAGLE/#Activities)); and HELM (Harmonised European Land Monitoring) FP7 project (http://www.umweltbundesamt.at/ms/fp7helm/fp7helm_home/?zg=). EEA encourages building the European Land Monitoring System on national and sub-national land monitoring systems, i.e. an improved bottom-up solution. Countries would like to have a long-term European plan to harmonise national programmes with the European one. The GIO land project (REGULATION (EU) No 911/2010), having the high-resolution layers (HRL) component (Imperviousness density, Tree Cover density, Forest Type, Permanent grassland, Wetland and Water) and the CLC component, is the first step towards integration of theme-wise high spatial resolution information and landscape level mapping provided by CLC. This gives the possibility of populating CLC polygons with more precise land cover information (Fig. 5.5).

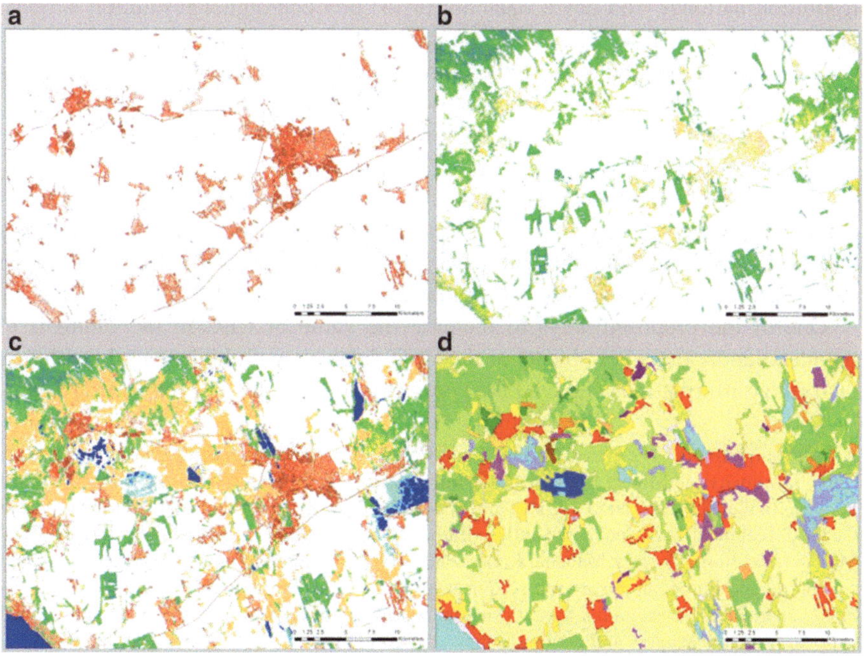

Fig. 5.5 Comparison of HRLs and CLC. (**a**) Imperviousness density; (**b**) Tree Cover density; (**c**) Integrated High Resolution Layers (Imperviousness, Forest Type, Grassland, Wetland and Water; (**d**) CORINE Land Cover (example is from Hungary)

5.8 Conclusions

CORINE Land Cover (CLC) was specified to standardize data collection on land in Europe to support environmental policy development. Since the late 1980s, three European CLC inventories have been realised (timed around 1990, 2000 and 2006). The 4th inventory recently commenced, as part of the GIO land (GMES/Copernicus Initial Operations Land) project. The number of participating countries has been increasing – currently 39 with a total area coverage of 5,8 Mkm^2. The project is mainly co-financed by the European Commission and participating countries and implemented by national teams under the management of EEA.

Ortho-corrected high spatial resolution satellite images (Landsat TM/ETM in the past, IRS, SPOT, RapidEye recently, and Sentinel-2 in the future) provide the geometrical basis for mapping. Multitemporal imagery improves the accuracy of classification. In-situ data (topographic maps, ortho-photos, and ground survey data, etc.) are essential ancillary data.

The basic technical parameters of CLC (i.e. nomenclature, minimum mapping unit and minimum mapping width) have not changed since the beginning of the project; therefore the results of the different inventories are comparable. The method of mapping has, however, changed significantly. Working on plastic overlay in the 1990s' was fully replaced by computer assisted photo-interpretation. Nowadays semi-automatic methodologies are introduced in order to (1) replace at least in large part the labour-intensive photo-interpretation (2) increase reproducibility and objectivity, (3) build upon national inventories including a higher resolution national land cover database or (4) produce higher resolution national land cover data parallel with European data.

Two European validation studies have shown that the achieved accuracy is above the specified minimum accuracy of 85 % for CLC, as well as for CLCC. Results of the CLC inventories can be downloaded from the EEA Data Service free of charge for all users.

An increased use of semi-automatic solutions in replacement of photointerpretation is expected in future CLC inventories. National land monitoring programmes will be harmonised with the European CLC by translating LC/LU information between different nomenclatures (http://sia.eionet.europa.eu/EAGLE/#Activities).

One of the proposed ways to increase the value of CLC in the future is to populate landscape level objects of the CLC database with high spatial resolution land cover information. The GIO land project is making the first steps towards this.

Acknowledgements The first CORINE Land Cover inventory was technically coordinated by Chris Steenmans and Michel Bossard. The nomenclature was finalised by Michel Bossard and Jan Feranec. In the CLC2000 and CLC2006 projects, the following experts contributed to the work of the CLC Technical Team led by the author: Barbara Kosztra, Jan Feranec, Gabriel Jaffrain, László Mari (verification), Gergely Maucha, Robert Pataki (validation) and Tomas Soukup (final technical control and European integration). Advices given by Geoff Smith to restructure the article are highly appreciated. Last but not least, the author would like to acknowledge the contribution of all national project managers and national teams.

References

Aune-Lundberg L, G-H Strand (2010) CORINE Land Cover 2006. The Norwegian CLC2006 project. Report from the Norwegian Forest and Landscape Institute, 11/2010

Balzter H et al (2011) GMES Initial Operations (GIO) Land Monitoring 201 –2013 Pan-EU Component. National Project Plan for the UK, University of Leicester, December 2011

Bossard M, Feranec J, Otahel J (2000) CORINE Land Cover technical guide – addendum 2000. EEA technical report no. 40

Büttner G, Kosztra B (2007) CLC2006 Technical Guidelines, EEA, technical report 17/2007

Büttner G, Maucha G (2006) The thematic accuracy of CORINE land cover 2000. assessment using LUCAS. EEA technical report no 7/2006. ISSN 1725-2237

Büttner G, Feranec J, Jaffrain G (2002a) CORINE Land Cover update 2000, technical guidelines, EEA technical report no. 89

Büttner G, Maucha G, Taracsák G: Inter-Change (2002b) A software support for interpreting land cover changes. In: Proceedings of the 22nd EARSeL symposium, Prague, 4–6 June 2002; pp 93–98, Millpress, 2003

Büttner G, Feranec J, Jaffrain G, Mari L, Maucha G, Soukup T (2004) The CORINE Land Cover 2000 Project. http://www.eproceedings.org/static/vol03_3/03_3_buttner2.pdf

Büttner G, Kosztra B, Maucha G, Pataki R (2010) Implementation and achievements of CLC2006, final report, EEA, 2010

Büttner G, Maucha G, Kosztra B (2011) European validation of Land Cover changes in CLC2006 project, EARSeL symposium, Prague 2011, Remote Sensing and Geoinformation, not only for Scientific Cooperation, Lena Halounová (Editor)

CLC2000 Finland final report (2005) Finnish Environment Institute (SYKE)

CLC2006, CLC2000, CLC-Changes in Iceland (2009) Final report, National Land Survey of Iceland

EEA Task Force (1992) CORINE Land Cover. A European community project

EEC regulation (1210/1990)

Engberg A (2005) Swedish CLC2000 final report, Lantmäteriet

European Commission (2005) Global Monitoring for Environment and Security (GMES: From Concept to Reality), COM 2005 565 final

Feranec J, Büttner G, Jaffrain G (2007) Illustrated guide, "CORINE Land Cover nomenclature" (Corrections of the Addendum 2000)

GMES Space Component Data Access Portfolio, Data Warehouse (2011–2014). http://gmesdata.esa.int/web/gsc/dap_document

Herold M, Hubald R, Di Gregorio A (2009) Translating and evaluating land cover legends using the UN Land Cover Classification System. http://nofc.cfs.nrcan.gc.ca/gofc-gold/Report%20Series/GOLD_43.pdf

Heymann Y, Steenmans C, Croissille G, Bossard M (1994) CORINE Land Cover. Technical guide. EUR12585 Luxembourg, office for official publications of the EC

Implementation of Pan-EU Component of GIO Land in Germany. Federal Environment Agency Germany, June 2012

Irish National Project Plan for implementation of GIO Land Monitoring Activities (2011–2013) Environmental Protection Agency, January 2012

Lima V (2009) Report on progress WP1.3, 3rd Steering Committee meeting, EEA, 22 April 2009

Maucha G, G Taracsák, G Büttner (2004) Methodological questions of CORINE Land Cover change mapping. In: Proceedings of the 2nd international workshop on the analysis of multi-temporal remote sensing images, MultiTemp-2003 workshop. World Scientific Publishing Co., Singapore, pp 302–313

REGULATION (EU) No 911/2010 of the European Parliament and of the Council of 22 September 2010 on the European Earth monitoring programme (GMES) and its initial operations (2011 to 2013) http://www.google.com/search?q=EU+Regulation±%28EU%29±n%C2%B0911%2F2010

Sjolbørg Flo Heggem E, Strand G-H (2010) CORINE Land Cover 2000. The Norwegian CLC20000 project. Report from the Norwegian Forest and Landscape Institute, 10/2010

Steenmans C, Büttner G (2006) Mapping land cover of Europe for 2006 under GMES. In: Proceedings of the 2nd workshop of the EARSeL SIG on land use and land cover. Bonn, Germany, 28–30 September, 2006, pp 202–207

The Lucas survey. European statisticians monitor territory, Updated edition (2003) EC, http://www.uni-mannheim.de/edz/pdf/eurostat/03/KS-AZ-03-001-EN-N-EN.pdf

Chapter 6
European Area Frame Sampling Based on Very High Resolution Images

Marek Banaszkiewicz, Geoffrey Smith, Javier Gallego, Sebastian Aleksandrowicz, Stanislaw Lewinski, Andrzej Kotarba, Zbigniew Bochenek, Katarzyna Dabrowska-Zielinska, Konrad Turlej, Andrew Groom, Alistair Lamb, Thomas Esch, Annekatrin Metz, Markus Törmä, Vassil Vassilev, and Gedas Vaitkus

6.1 Introduction

A sampling frame is a representation of a population to be sampled. If it consists of a set of geographic units, it can be called an Area Sampling Frame (Faulkenberry and Garoui 1991). Spatial sampling, also called Area Frame Sampling (AFS) is used in many fields as an alternative to list frame sampling. For environmental estimates it becomes an essential tool (Stein and Ettema 2003; de Gruijter et al. 2006). In particular it is important to infer information about land characteristics in the whole

M. Banaszkiewicz (✉) • S. Aleksandrowicz • S. Lewinski • A. Kotarba
Space Research Centre of the Polish Academy of Sciences, Earth Observation Group, Bartycka 18A, 00-716 Warsaw, Poland
e-mail: marekb@cbk.waw.pl; stlewinski@cbk.waw.pl

G. Smith
Specto Natura Ltd., College Road, Impington, Cambridge CB24 9PL, UK
e-mail: geoffsmith@specto-natura.co.uk

J. Gallego
Joint Research Centre, Institute for Environment and Sustainability, Via Fermi 2749, 21027, Ispra, VA, Italy
e-mail: javier.gallego@jrc.ec.europa.eu

Z. Bochenek • K. Dabrowska-Zielinska • K. Turlej
Institute of Geodesy and Cartography, Modzelewskiego 27, 02-679, Warszawa, Poland
e-mail: bochenek@igik.edu.pl

A. Groom
Astrium Services, Europa House, The Crescent, Farnborough, Hampshire GU14 0NL, UK
e-mail: andrew.groom@infoterra-global.com

A. Lamb
Astrium GEO-Information Services – Infoterra Ltd., Atlas House 41 – Wembley Road, Leicester LE3 1UT, UK
e-mail: alistair.lamb@infoterra-global.com

I. Manakos and M. Braun (eds.), *Land Use and Land Cover Mapping in Europe: Practices & Trends*, Remote Sensing and Digital Image Processing 18,
DOI 10.1007/978-94-007-7969-3_6,

investigated area (e.g. land cover or land use). Two fundamental techniques of data acquisition for land use or land cover estimation are field inspection and remote sensing. The advantage of the first one is that a well trained surveyor can unambiguously determine a (large) number of parameters, perform in-situ measurements and collect samples for further analysis in a specialized laboratory. Remote sensing, on the other hand, can provide uniform data from large areas, but the collected information will be less thematically detailed. The decisive factor in choosing one approach or another is the cost to benefit assessment.

An AFS scheme can be either random or systematic. There are several examples of AFS that have been employed on national, continental and global scale. The UK Countryside Survey is a prime example where a clearly thought out strategy to AFS design has consistently produced useful results and has been able to adapt to changing political requirements over 30 years. Apart from these random schemes, two systematic approaches should be mentioned: Firstly, the Land use/cover area frame survey (LUCAS) was designed to collect agricultural and environmental data and photographs by field observation of a specific geographically referenced point to obtained harmonized information across the European Union (EU) about land cover (LC) and land use (LU) (Gallego and Delincé 2010), and secondly, FRA2010 a FAO survey of forests, that considers a non-stratified systematic sample of 20 × 20 km in geographical coordinates. Each site is located on the intersection of integer-valued meridian and parallel (Eva et al. 2010).

The Seasonal and Annual Change Monitoring Service (SATChMo) is one of three Core Mapping Services (CMS) developed in geoland2 project. It is focused on LC/LU change detection on annual and seasonal time scales and comprises four

T. Esch
Department Land Surface, German Aerospace Centre (DLR), Earth Observation Centre (EOC), German Remote Sensing Data Centre (DFD), Oberpfaffenhofen, Germany
e-mail: thomas.esch@dlr.de

A. Metz
Institute for Geoinformatics and Remote Sensing, University of Osnabrueck, Barbarastrasse 22b 49076, Osnabrueck, Germany
e-mail: ametz@uni-osnabrueck.de

M. Törmä
Finnish Environment Institute SYKE, PO. Box 140 00251, Helsinki, Finland
e-mail: markus.torma@ymparisto.fi

V. Vassilev
Remote Sensing Application Centre (ReSAC), 61 TzarAssen Street, 1463, Sofia, Bulgaria
e-mail: vassil.vassilev@resac-bg.org

G. Vaitkus
Institute of Aerial Geodesy, Applied Research Centre, Pramones pr.13 LT-51327, Kaunas, Lithuania
e-mail: gedas.vaitkus@gmail.com

different groups of products: (i) AFS Europe based on very high spatial resolution (VHR, <5 m) images acquired annually, the subject of this chapter, (ii) AFS Africa, dedicated to change detection in LC in sub-Saharan Africa in 5 year intervals; the high spatial resolution (HR, 20–30 m) Landsat-type data are collected on a regular grid, (iii) seasonal vegetation changes and annual crop area estimates based on medium spatial resolution (MR ~300 m)/HR images, and (iv) global land cover change indicators derived from MR multi-temporal data series. The product groups are linked through change detection methodology.

This chapter deals with the AFS Europe products only. It starts with the description of AFS objectives and presents the design of the sampling scheme tailored to the amount of available VHR data. Next, the classification scheme and algorithm are shown, the production chain including product validation is described and the results of a statistical analysis are given. The following section is dedicated to the change detection method deployed by SATChMo and shows first results of the chosen approach. The chapter is concluded with a preliminary assessment of the developed products and with presentation of possible downstream products that can be derived from the AFS Europe output data.

6.2 AFS Objectives and Design

The SATChMo AFS scheme was designed after a careful analysis of possible solutions, taking into account the following issues: (i) shall a random sampling of test sites across Europe be preferred over a systematic sampling on a pre-defined grid, (ii) will it be beneficial to introduce strata for instance urban or coastal areas, (iii) how to define the size of the sample unit (single site), and (iv) how to assess the statistical accuracy of the area or the area change estimators.

For the first issue, there is a large amount of literature showing that in cases when the spatial correlation is a decreasing function of the distance, the systematic sampling with a random starting point is superior to random sampling (Cochran 1977; Bellhouse 1988). This is true in particular for land cover data obtained from remote sensing (Dunn and Harrison 1993). The systematic sampling performs well, because it ensures a good spatial distribution of the samples. Its main drawback is that there is no unbiased estimation of the variance. The usual formulas, valid for random sampling, overestimate the variance if applied to systematic sampling, but alternative formulas based on local variances can be introduced to substantially reduce the bias (Wolter 1984). A more significant drawback of the straightforward systematic sampling is the difficulty of accommodating the sample size to the available budget without rerunning the whole design process (Stehman 2009). For SATChMo a version of systematic sampling based on multiple replicates was used that keeps the good spatial distribution and subsequent standard error reduction and is quite flexible to accommodate sample size changes. The scheme is very similar to the one used for the Eurostat LUCAS 2006 survey (Gallego and Delincé 2010).

When considering stratification, the original design of SATCHMO supported requirements of other geoland2 services as well as the need to complement the

LUCAS survey and address European Environment Agency (EEA) issues. The strata were defined as:

- High lands: Defined as the set of sites with more than 50 % above 1,200 m above sea level. This stratum was chosen to enhance the collaboration with LUCAS, since LUCAS excludes the field visits in points above 1,200 m.
- Euroland: Sites with more than 50 % in Euroland sites (Kuntz et al. 2014 & http://www.gmes-geoland.info/fileadmin/geoland2/redakteur/pdf/Project_Documentation/User_Requirements/g2_EL-RP-D_EL010-1_UserRequirements_I1%2000.pdf)
- Urban Atlas: Sites that intersect some of the Large Urban Zones covered by the GMES Urban Atlas (http://www.eea.europa.eu/data-and-maps/data/urban-atlas)
- Coastal areas: Sites that touch a 10 km buffer of the coast. This stratum and the Urban Atlas stratum were considered because more important land cover changes are expected in these areas.
- Cyprus and Malta: Countries not covered by LUCAS survey.
- Other areas: The remaining area not covered by the above strata.

The choice of the size of sampling unit (area) should be done based on two main criteria: obtained coefficient of variation and effective cost of processing including data acquisition. The first one favors smaller units, the other larger ones.

The main constraint to be considered is that sampling units should be compatible with VHR images. Most VHR images have a size between 10 km × 10 km and 20 km × 20 km. However the range is larger if we include in the VHR category SPOT super-mode images with 2.5 m resolution. The assessment of these criteria can be done with the help of CLC (CORINE Land Cover, JRC-EEA 2005) as pseudo-truth. This allows simulating and comparing any type of sampling scheme, although the behavior of CLC is not quite the same as the behavior of land cover observed on the field.

A way of assessing the size that optimizes the cost-efficiency is considering a cost function of the type $C = \alpha + \beta n + n\gamma s = \alpha + n(\beta + \gamma s)$, where C is the total cost, α is the fixed cost (management, etc) that has no influence at all on optimization, n is the number of sites, β is the part of the cost per site that approximately does not depend on the site size (image ordering and reception, correction, etc.), s is the area of each site (we express it in number of 10 × 10 km units) and γ is the marginal cost when s increases by one unit. γ has a relatively high value in the case of visual photo-interpretation. In this case smaller sites (e.g. 10 × 10 km) are more efficient. If image processing is mainly automatic, γ has a lower value and larger sites (e.g. 30 × 30 km) are more efficient.

The efficiency of a sampling site can be measured in terms of "equivalent number of points". A value Q for the equivalent number of points means that n sites are equivalent to nQ unclustered points. It can be proved that $Q \approx 1/\rho_M$, where ρ_M is a weighted average of the spatial correlation between points inside the same site (Gallego 2011; Gallego and Stiebig 2012).

Both approaches lead to the same conclusion. Table 6.1 illustrates some results of the simulations with sites of 10 and 30 km compared with the sampling accuracy

Table 6.1 Sampling accuracy for area estimation of some land cover classes and changes

		Coefficient of variation (CV)		
		100 images 10 × 10 km (%)	100 sites 30 × 30 km (%)	LUCAS2006 (11 countries) (%)
Based on CLC2000	Artificial	17.4	11.2	1.1
	Arable rainfed	9.5	7.1%	0.4
	Arable irrigated	63.6	45.1	1.9
	Forest	9.7	7.5%	0.45
	Shrubland	18.0	13.0	1.35
Based on changes CLC1990–2000	New artificial	30.8	20.5	
	New agriculture	49.1	28.0	
	Agricultural abandonment	35.6	20.1	
	Other changes	22.1	16.4	

of LUCAS 2006 for some land cover classes (first rows) and changes (last rows). For major land cover classes, the sampling error of a VHR-image based sample is much higher than the sampling error of LUCAS. Even if this table does not include a cost analysis, it shows and it will be difficult for an image-based approach to be competitive with field visits in the EU, except in the areas with difficult access, such as mountains. The situation is radically different for land cover change: even if the Coefficients of Variation (CV) shown by the table look modest, they would be a major contribution in a field in which field surveys still have major problems, mainly because it is very difficult to distinguish between real land cover changes and fake changes due to location errors or different interpretations of the nomenclature by different surveyors. These simulation results lead to some suggestions:

- SATCHMO should focus on land cover changes as the first priority, although the land cover status remains an important target.
- For land cover status, the sampling scheme should give priority to areas with difficult access.
- If the image analysis is mainly automatic (small value of γ) and the acquisition of large images is feasible, they are more cost-efficient than small sites.

Based on above mentioned analysis the original design of AFS Europe comprised 350 sites of 30 × 30 km to be imaged by SPOT with 2.5 m spatial resolution. This original design had to be changed when the number of scenes to be acquired was reduced to 198 due to logistic reasons and the SPOT sensor was replaced by the Kompsat-2 system. The smaller number of scenes was split in only 3 strata: urban, high altitude and other. However, even this scheme did not survive the confrontation with the acquisition offer and finally a set of 114 sites were selected to be imaged by Kompsat-2 15 × 15 km scenes taken in 2009. Most of the sites were located in the spots, where 198 sites were chosen, but some were taken in neighboring locations (Fig. 6.1).

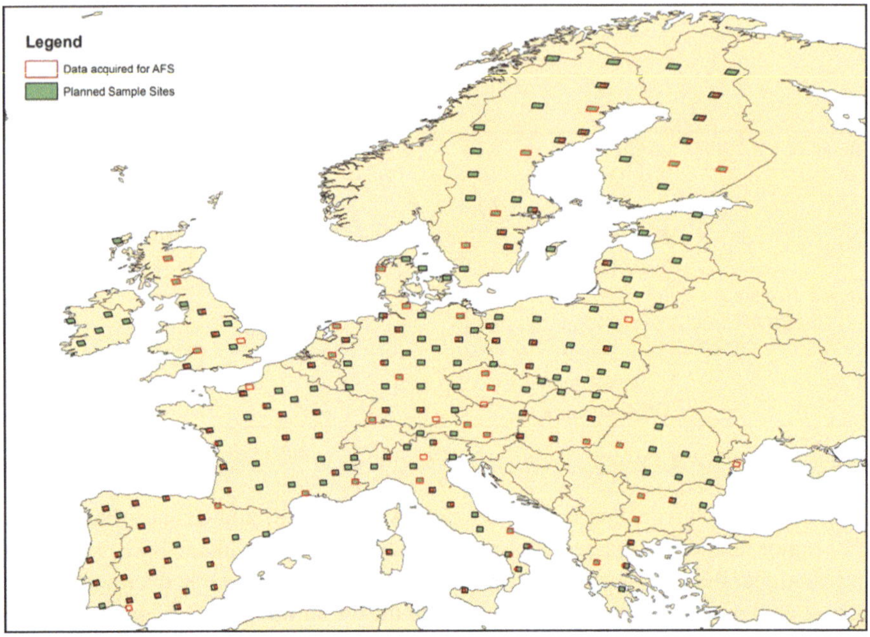

Fig. 6.1 Distribution of 198 planned sites (*green rectangles*) and 114 acquired sites (*rectangles* with *red lining*)

6.3 Classification

The land cover classification algorithm was developed at the Institute of Geodesy and Cartography (IGiK) with certain elements delivered by Infoterra UK. The classification approach was prepared for the discrimination of ten generic land cover classes that correspond to main land cover types in Europe and could be recognized by relatively automated processing independently of the date (season) when the images are acquired. Incidentally, the proposed classes are very similar to several Level 1 classes used in LUCAS (Table 6.1).

The algorithm was developed as a generic tool for processing VHR images. Since the main type of SATChMo acquired images were from KOMPSAT-2, the algorithm was tailored to cope with this kind of data and later extended to include FORMOSAT-2 images. Both sensors provide a panchromatic channel and four less spatially detailed multispectral channels (blue, green, red, infrared). The spatial resolution of the two image types were not the same; Kompsat-2(1 m for PAN, 4 m for MS data) and FORMOSAT-2 (2 and 8 m, respectively).

The main difficulty was to develop a solution that was effective for images throughout the whole of Europe. As a result of the research study a semi-automatic object-oriented method, based on eCognition software was prepared for the operational mapping. The algorithm has been named SATChMo-K2 (Lewinski et al. 2010).

The basic feature of the algorithm is the assumption that agricultural areas and water class are characterized by low values of the texture while the remaining classes are associated with high texture values (de Kok and Wezyk 2008). Division of the content of images on high and low texture is made on the basis of panchromatic channel processed by Sigma filters. All classification steps are executed sequentially and the whole process was divided into four main stages: (1) division of study area into two object groups, first group characterized by high texture and the second by low texture measures; (2) classification of high texture group, including urban/artificial class, forests/woodland/trees, sparse woody vegetation and bare non-cultivated ground; (3) classification of low-texture group, comprising agricultural areas, grasslands, snow and ice (if existing) and water; (4) re-classification of existing classes to refine classification output. The individual classes are identified on the basis of distinguishing features of panchromatic and multi-spectral data. The only class that is not classified directly is "agriculture areas". These areas are classified at the end as non-classified to the other classes. This procedure helped to avoid the complicated process of identification of various forms of agriculture, which occur in Europe.

Out of the ten land cover classes three are not classified automatically. To this group belong grasslands, other vegetation and clouds. Grasslands are not classified because it is impossible to distinguish them using just one satellite image and additionally when the time of an image acquisition could be unrelated to the whole vegetation period.

The classification process has been developed within object-oriented eCognition software environment using the "architect" interface with the support of Infoterra UK. The classification is performed interactively by an operator, who has access to selected functions, however most of discrimination rules are hidden and all parameters of segmentations are predefined and fixed.

The last steps of classification process concerns generalization and manual editing. According to assumptions applied to land cover maps in SATCHMo, MMU was set to 0.25 ha. The operator has access to handy tools which allow to correct the automated results. The classification interface including tools for manual editing was prepared in collaboration with Infoterra UK. The final results are exported to raster and vector files. Implemented functions allow the saving of classification parameters to allow them to be used on subsequent scene of a similar landscape type. Thanks to this feature, classification of images representing similar geographical regions can be perform automatically or only some minor modifications are required.

The total number of scenes acquired from June 2009 to June 2010, the baseline interval for classification and later change detection, amounted to 114. The images were distributed among five production teams that processed them using the common SATChMo-K2 tool. The images selected for processing by each team belonged to a geographically uniform part of Europe.

The classical method of product validation was employed, i.e. 500 points were randomly chosen on each images and interpreted by an operator. In that way the error matrix was generated for each image. The designed accuracy of the product

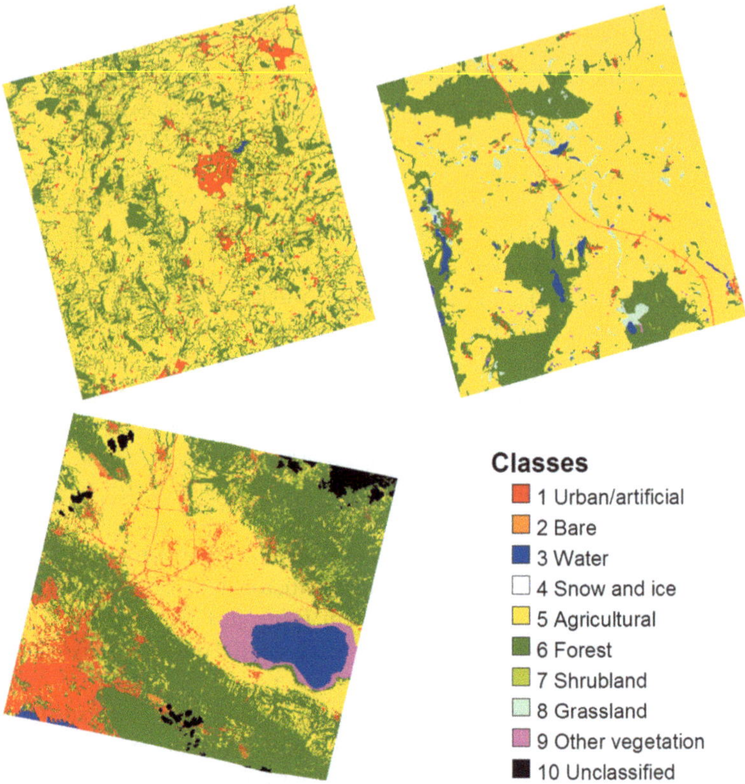

Fig. 6.2 Examples of land cover classification for three different sites: UK (mixed agriculture, *top left*), Poland (agriculture, *top right*), Greece (mixed, near Thessaloniki, *bottom left*)

was set to 85 % and the products (maps) that did not pass this threshold were considered as wrongly classified and were corrected. Three examples of products from UK, Poland and Greece are shown in Fig. 6.2.

Overall accuracy of products is 89 % (Fig. 6.3), with 21 sites not reaching the threshold (85 %). The preliminary statistical analysis of the classification products, provided on the assumption that the sites are chosen randomly, gives the results presented in Table 6.2. However, due to the problems with data acquisition that resulted in uncontrolled choice of scenes with respect to the designed pattern, the statistical analysis will be repeated using classes from CLC data base as co-variables (Table 6.3).

The estimated class LC proportion was calculated by simple average over the total number of sites. For calculating the coefficient of variance of the Class Estimator a bootstrapping method was employed. Specifically, 500 replicas were randomly generated from the original 114-site sample. The obtained CV can be improved by applying regression method with CLC proportions as a covariable.

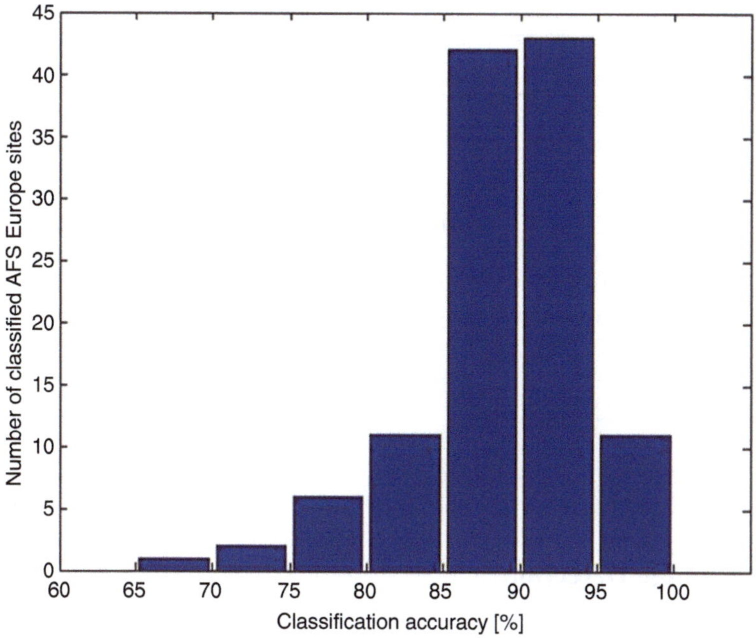

Fig. 6.3 Accuracy of validated land cover classification for 116 images. The overall accuracy amounts to 89 %

Table 6.2 Comparison of SATChMo and LUCAS classes

Class no	SATChMo nomenclature	LUCAS nomenclature
1.	Urban and artificial areas	ARTIFICIAL LAND
2.	Bare non-cultivated ground; soils, rock, sand dunes, dry lake beds, inter-tidal mud, rock, soil, sand dunes and, Inter-tidal mud	BARE LAND
3.	Water	WATER AREAS
4.	Snow and ice	
5.	Agricultural areas; irrigated and non-irrigated Arable cropped areas, permanent crops (orchards, vineyards, olive trees), pastures and set-aside fallow land	CROPLAND
6.	Forest/woodland/trees; broadleaf and coniferous trees	WOODLAND
7.	Sparse woody vegetation; shrubs and bushes	SHRUBLAND
8.	Grassland	GRASSLAND
9.	Other vegetation; moorland, reed beds, saltmarsh and other not specified vegetation	
10.	Unclassified; clouds and clouds shadows, voids	

Table 6.3 Statistics of LC classes obtained from 107 images

Class no	Class description	Estimated class LC proportion	Relative standard error (CV) (in %)	No of images with LC class
1	Urban impervious	0.05466	15.0	107
2	Bare non-cultivated ground	0.01547	27.5	107
3	Water	0.02772	21.8	103
4	Snow and ice	0.00194	51.3	12
5	Agricultural areas	0.41052	7.0	107
6	Forest/woodland/trees	0.28194	7.8	107
7	Sparse woody vegetation	0.14292	13.2	107
8	Grasslands	0.02582	18.9	81
9	Other vegetation	0.02763	28.4	69
10	Cloud, voids etc.	0.01139	30.8	63

6.4 Change Detection

The approach of the AFS Europe team was to make a survey of existing change detection methods, choose the most appropriate and efficient one concerning the product requirements, modify it to the point when automatic processing is guaranteed and the manual corrections are reduced to minimum, and, finally, cast it into the form of a ready-to-use eCognition-based tool.

After a careful analysis of methods available in the literature the Multivariate Alternate Direction (MAD) transform was chosen as a baseline (Nielsen et al. 1998). If multispectral images of a scene acquired at times *t0* and *t1* are represented by random vectors $\boldsymbol{X}$ and $\boldsymbol{Y}$, which are assumed to be multivariate normally distributed, the difference D between the two images is calculated by $D = \boldsymbol{a}T\boldsymbol{X} - \boldsymbol{b}T\boldsymbol{Y}$ (Nielsen et al. 1998). Analogously to a principal component transformation, the vectors $\boldsymbol{a}$ and $\boldsymbol{b}$ are sought subject to the condition that the variance of D is maximized and subject to the constraints that $\text{var}(\boldsymbol{a}^T\boldsymbol{X}) = \text{var}(\boldsymbol{b}^T\boldsymbol{Y}) = 1$. Determining the vectors $\boldsymbol{a}$ and $\boldsymbol{b}$ in this way is a standard statistical procedure which considers a generalized eigenvalue problem. For a given number of bands N, the procedure returns N eigenvalues, N pairs of eigenvectors and N orthogonal (uncorrelated) difference images, referred to as to the MAD components.

The Iteratively Reweighted Multivariate Alteration Detection (IR-MAD) method (Nielsen 2007) for detecting the change makes a step forward with respect to the standard MAD method and first calculates the ordinary canonical and original MAD variates to assign different weights to the observations in the following iteration steps. Small changes get a higher weight than bigger changes. This allows delimiting regions in which changes occur from regions without changes. Iterations are performed until a defined termination criterion is reached.

The main advantages of this approach are: (i) radiometric normalization of images is not necessary, (ii) the method is very sensitive to changes, hence it rarely happens that changes are missed. The main drawback is that MAD generates alarms that do not correspond to land cover changes but rather to seasonal variations. To improve the algorithm in these points two additional characteristics were introduced in processing of a pair of images from different dates: NDVI and texture. Both of them improve the detection of real thematic (i.e. not seasonal) changes.

SATChMo change detection algorithm is suitable tool for the analysis of VHR satellite images. Two sets of data from periods: T1 and T2 constitute the input to the algorithm. Each set consist of four multispectral channels (blue, green, red and near infrared) and a panchromatic channel. In addition, the algorithm uses the classification made for the time T1 using Satchmo-K2 algorithm, and the results of the IR-MAD transformation (Nielsen and Canty 2005).

The main part of the algorithm, based on the MAD transformation, was developed by the Space Research Centre of Polish Academy of Sciences (SRC PAS) and next complemented with functions for identifications the direction of changes proposed by IGiK.

The algorithm for change detection is a three-stage approach. In the first step, the high spatial resolution data: from the time T1 and T2, the T1 classification result and the result of the IR-MAD transformation consisting of four components and $\chi2$ layers should be entered into the algorithm. Then additional layers, such as texture layers created using the Lee-sigma filter, as well as the Normalized Difference Vegetation Index (NDVI) layers, are generated. They allow for filtering of changes occurring in vegetation, which are recognized by the MAD algorithm. According to the design goal of the algorithm (which is the maximum automation of the process), the thresholds distinguishing the occurrence of changes in the individual layers are automatically generated using statistical values, or, if it is not possible, with the use of T1 classification. The process results in the creation of the mask of changes, which is then used to identify the type of change.

In parallel information on change directions (types of change) is created. The process of detection of change directions starts from scene segmentation, based on original image data (multispectral + panchromatic) and on thematic classification of image from T1. After obtaining homogeneous objects first classification is performed, which divides image into two process groups: re-vegetation and de-vegetation. The criteria of determination of these two groups are based on NDVI thresholding. Next, in order to delineate anthropogenization group, a feature for extracting built-up areas is applied. As a result of this two-stage procedure final map containing three classes of change directions is produced. This information is inserted into the mask of general changes, produced at first stage of algorithm. The areas not detected as changes but still marked as change directions, subject to further discrimination of changes of land cover types. Next both thematic layers are combined using a union procedure to create resultant map of change types and directions. This map is verified and corrected with the use of the tool for manual editing, prepared specially for land cover change method. As a result of the whole

Fig. 6.4 An example of change detection processing performed on eCognition platform. *Top*: images acquired at dates T0 and T1. *Bottom left*: change detected automatically, *bottom right*: change accepted manually. Automatically detected changes (*red* patches in the *bottom left* panel) are assessed by the operator as real (new building in the *middle*) or seasonal (agricultural parcels in the *left*). Only the first one is accepted in the final change/no change map (*bottom right*)

workflow three outputs are produced: map of change directions, map of change types (all possible combination of ten classes) and classification map for T2.

The map of change directions, in conjunction with land cover classification map produced for T1, determined some changes in land cover types, especially those manifesting in urbanization process. The validation method, elaborated by the Remote Sensing Application Center in Bulgaria, is based on the LPIS Quality Assurance developed at the EU Joint Research Centre (JRC) and it consists of: (i) generation of a sample of randomly selected objects (items of inspection or inspection units) from the change product, (ii) evaluation of the selected objects (inspection units) in respect of the correctness of the "change/no change" status they represent, (iii) analysis of observations and assessment of change direction in reference to producer's values. Three different metrics were introduced to quantify the validation process: (i) Difference between the total area changed for the whole scene, as reported by the inspection and the total area changed for the whole scene, as reported by the change product (acceptance threshold is 15 %), (ii) Number of erroneously classified inspection units – separately for the two types (change and no change), (iii) Histogram of the erroneously classified inspection units in relation to the land cover found. An example of the change detection processing is shown in Fig. 6.4.

6.5 Conclusion

The Area Frame Sampling based on VHR images on continental scale (Europe) was developed by the SATChMo team in the framework of geoland2 project. The three main elements of the whole processing chain comprise:

1. Development of the cost-optimal sampling scheme that takes into account available resources (cost of acquisition + processing time) as well as the expected outcome (statistics of LC and land cover changes (LCC) classification),
2. Semi-automatic classification algorithm for VHR images that generates ten classes for any image acquired during a snow-free season and gives good accuracy reaching on an average 88 %,
3. Change detection algorithm that is based on the MAD method employed for multispectral VHR images but uses also NDVI and texture to eliminate seasonal changes.

The results of the processing of 114 images are very promising and show how much could be gained concerning local scale LC classification and assessment of changes in Europe when 5–10 % of the surface area of the continent is covered annually with VHR images. Since the algorithms and programs were tested on images acquired by several sensors (Ikonos, Kompsat-2, Formosat) and proved to be robust and reliable, they can be easily modified to accept any VHR data source. The potential users of SATChMO AFS products include Eurostat (as service complementing LUCAS), DG Env and EEA (concerning change detection in hot-spot areas).

The maps of AFS sites are used to derive new products, in particular environmental indicators, such as patch density, patch richness or different diversity indices. The local component of GMES, whatever will be its main objective, should benefit from the results and products delivered by AFS Europe team within SATChMo.

References

Bellhouse DR (1988) Systematic sampling. In: Krisnaiah PR, Rao CR (eds) Handbook of statistics, 6th edn. Elsevier Science Publishers, Amsterdam, pp 125–146

Cochran W (1977) Sampling techniques. Wiley, New York

de Gruijter JJ, Brus DJ, Bierkens MFP, Knotters M (2006) Sampling for natural resource monitoring. Springer, Berlin

de Kok R, Wezyk P (2008) Principles of full autonomy in image interpretation. The basic architectural design for a sequential process with image objects. In: Blaschke TH, Lang S, Hay GJ (eds) Object-based image analysis. Series: Lecture notes in geoinformation and cartography. Springer, Berlin, pp 697–710. ISSN: 1863–2246

Dunn R, Harrison AR (1993) Two-dimensional systematic sampling of land use. J R Stat Soc: Ser C: Appl Stat 42(4):585–601

Eva H, Carboni S, Achard F, Stach N, Durieux L, Faure JF, Mollicone D (2010) Monitoring forest areas from continental to territorial levels using a sample of medium spatial resolution satellite imagery. ISPRS J Photogramm Remote Sens 65(2):191–197

Faulkenberry GD, Garoui A (1991) Estimating a population total using an area frame. J Am Stat Assoc 86(414):445–449

Gallego FJ (2011) The efficiency of sampling very high resolution images for area estimation in the European Union. Int J Remote Sens 33(6):1868–1880. doi:10.1080/01431161.2011.602993

Gallego FJ, Delincé J (2010) The European Land Use and Cover Area-frame statistical Survey (LUCAS). In: Benedetti R, Bee M, Espa G, Piersimon F (eds) Agricultural survey methods. Wiley, Chichester, pp 151–168

Gallego FJ, Stiebig HJ (2012) Area estimation from a sample of satellite images: the impact of stratification on the clustering efficiency. J Appl Earth Obs Geoinf 22:139–146. doi:10.1016/j.jag.2012.03.003

JRC-EEA (2005) CORINE land cover updating for the year 2000: image2000 and CLC2000. In: Lima V (ed) Products and methods. Report EUR 21757 EN, Ispra, Italy

Kuntz S, Schmeer E, Jochum M, Smith G (2014) Towards an European land cover monitoring service and high-resolution layers

Lewinski S, Bochenek Z, Turlej K (2010) Application of object-oriented method for classification of VHR satellite images using rule-based approach. Geoinf Issue 2(1):19–26

Nielsen AA (2007) The regularized iteratively reweighted MAD method for change detection in multi- and hyperspectral data. IEEE Trans Image Process 16:463–478

Nielsen AA, Canty MJ (2005) Multi- and hyperspectral remote sensing change detection with generalized difference images by the IR-MAD method. IEEE, MultiTemp2005, Biloxi, Mississippi, USA

Nielsen AA, Conradsen K, Simpson JJ (1998) Multivariate alteration detection (MAD) and MAF processing in multispectral, bitemporal image data: new approaches to change detection studies. Remote Sens Environ 64:1–19

Stehman SV (2009) Sampling designs for accuracy assessment of land cover. Int J Remote Sens 30 (20):5243–5272

Stein A, Ettema C (2003) An overview of spatial sampling procedures and experimental design of spatial studies for ecosystem comparisons. Agric Ecosyst Environ 94(1):31–47

Wolter KM (1984) An investigation of some estimators of variance for systematic sampling. J Am Stat Assoc 79(388):781–790

Chapter 7
European Forest Monitoring Approaches

Markus Probeck, Gernot Ramminger, David Herrmann, Sharon Gomez, and Thomas Häusler

7.1 Introduction

Consistent, accurate, reliable and up-to-date information on the state of forests in Europe is required by European countries for reporting and policy making in the frame of several European and international forest- and environment-related policies, action plans and international agreements in the fields of environmental protection, protection of biodiversity and ecosystems, conservation planning, sustainable use of natural resources, climate change mitigation actions and environmental modelling at both national and international levels. In the European context, relevant policies and action plans comprise for example the EU Biodiversity Strategy to 2020, the Europe 2020 Strategy and related Greenhouse Gas reduction targets, the European Environment Agency's 5-yearly State of the Environment Reporting (SOER), the SEBI2010 (Streamlining European Biodiversity Indicators on progress towards the target of halting the loss of biodiversity by 2010), Forest Europe (the Ministerial Conference on the Protection of Forests in Europe – MCPFE) or the Alpine Convention. The most relevant international policies comprise the United Nations Framework Convention on Climate Change (UNFCCC) and the Kyoto Protocol, the United Nations Convention on Biological Diversity (UNCBD), the United Nations Forum on Forests (UNFF), the FAO's Forest Resources Assessment (FRA) and the associated Temperate and Boreal Forest Resources Assessment (TBFRA).

Highly accurate forest information is typically available on national level, mainly from National Forest Inventories (NFI's) and partially from national land use/land cover classification systems. Therefore, data harmonisation and improvement of access is one of the major challenges (JRC 2008). On European level, such national forest data were in the past frequently suffering from a lack of harmonised

M. Probeck (✉) • G. Ramminger • D. Herrmann • S. Gomez • T. Häusler
GAF AG, Arnulfstr. 199, D-80634 Munich, Germany
e-mail: forestry@gaf.de

I. Manakos and M. Braun (eds.), *Land Use and Land Cover Mapping in Europe: Practices & Trends*, Remote Sensing and Digital Image Processing 18,
DOI 10.1007/978-94-007-7969-3_7,

definitions, spatial detail and timeliness, however recent harmonisation efforts initiated by FAO and followed up by the NFI's in the framework of the European Network of National Forest Inventories (ENFIN) with support of EC's Joint Research Centre (JRC) are leading the way towards better comparability and large-area usability of NFI data.

Since several years, such national forest information has been increasingly complemented with remote sensing-based observations, which are considered an adequate and accepted means for providing reproducible, reliable and spatially explicit information on forest cover and characteristics over a large spatial extent in a cost-efficient and objective manner (Seebach et al. 2011; UNECE/FAO 2001). Up to date there have been several efforts for mapping European forests from regional to continental scales, with the overarching aim to develop and establish a systematic and sustainable operational monitoring capacity.

This article provides a comprehensive overview of earth observation-based forest monitoring approaches in Europe from early approaches using low- and medium-resolution satellite data to operational monitoring systems developed in the frame of the European earth observation programme Copernicus (formerly called GMES). The latter comprise mainly the pan-European land use/land cover information provided by CORINE Land Cover (CLC), the regional, national and pan-European-scale Forest services and products implemented by the ESA-funded GMES Service Element Forest Monitoring (GSE FM), the pan-European forest products provided by the Joint Research Centre's (JRC) European Forest Data Centre (EFDAC), pre-operational forest monitoring services developed by the European Commission-funded FP7 Land core project geoland2 and the first-time operational mapping of a pan-European High-Resolution Forest Layer in the frame of the GMES Initial Operations (GIO) Land, being implemented by the European Environment Agency (EEA) since 2012. Additionally, customised regional forest Downstream services as designed by the FP7 project EUFODOS are presented. Conclusions and prospects for future forest monitoring approaches in the upcoming Sentinel-2 era are provided.

7.2 Legacy of European Forest Monitoring

Many regional- and national-scale land use/land cover and forest mapping approaches have applied remote sensing techniques to map forest resources with high thematic accuracy and high spatial resolution. Examples are the Land Cover and Land Use Information System of Spain – SIOSE (e.g. Valcarcel et al. 2008), DLM-DE in Germany (Arnold 2009) or the UK Land Cover map (Smith et al. 2007; Fuller et al. 2005). High-resolution national forest area maps have been produced e.g. for Austria (Gallaun et al. 2007) or Switzerland (Romero et al. 2007). However, since such approaches have been explicitly designed to fulfil in first instance specific regional/national user-defined requirements, they have been making use of a variety of different mapping approaches, different national forest definitions and diverse sources of input data (Seebach et al. 2011). Therefore, such data are typically lacking thematic homogeneity beyond the national borders, making any use for international policy, reporting or scientific purposes difficult.

For this reason, a range of initiatives with different specifications has been set up for mapping forests and land use/land cover types from global to continental scales in a large-area harmonised and consistent manner. In order to select those products, which are best suited for specific applications, it is important to understand the characteristics and specific strengths of such maps, since the selection of land/forest cover products greatly influences the results of further value-added thematic applications such as environmental, climate or hydrological modelling (e.g. Ludwig et al. 2003). In order to derive spatially explicit forest maps at continental scales, earth-observation data are considered the most cost-efficient and large-area consistent option (Pekkarinen et al. 2007).

On European level, a methodology for mapping information on land cover/land use based on earth observation satellite data had been developed for the first time by the Commission of the European Communities (CEC) CORINE (Co-ordination of Information on the Environment) Land Cover (CLC) project (EEA Task Force 1992). The main objective of this comprehensive land use/land cover classification system was to derive consistent and compatible land cover (including forest) information for all of Europe by means of visual interpretation of Landsat images. The CLC nomenclature combines land cover and land use elements for a total of 44 land cover/land use classes with a consistent Minimum Mapping Unit (MMU) of 25 ha (Bossard et al. 2000) and 5 ha for change area updates, respectively. CLC discriminates three different forest classes: Coniferous, Broadleaved and Mixed Forest, plus additionally Agro-Forestry Areas and Transitional Woodland/Shrub. It has meanwhile been repeatedly updated and is available as a well-established time series of land-use/land-cover information from EEA's Land Use Data Centre (LUDC) for the years 1990, 2000 and 2006 (Heymann et al. 1994; Büttner et al. 2004; Büttner et al. 2010) in different data formats together with many derived products and user applications. The CLC 2012 update is currently being carried out and a further update is foreseen for the reference year 2018 (Langanke et al. 2013). The overall thematic accuracy of CLC was found to be generally over 85 %, with the reliability of forest classes being slightly higher (85–90 %) (Seebach et al. 2011). Although CLC has become a well established LU/LC standard that is meanwhile widely and successfully used for various environmental applications, its specifications do also set limits to the usability for more sophisticated applications that require e.g. high spatial resolution, frequent updates or unambiguous class definitions without mixed classes (Langanke et al. 2013).

Therefore other remote sensing-based approaches have been conceived and implemented to overcome such shortcomings. In the following, the development of those initiatives focusing on deriving European-scale forest information is presented, which gradually have migrated from using low-resolution NOAA-AVHRR (Advanced Very High Resolution Radiometer) and medium-resolution IRS-WiFS (Wide Field Sensor) imagery (GAF 2001) to the use of high-resolution Landsat, IRS-LISS III and RapidEye data. The presumably first consistent remote sensing-based pan-European forest area map had been the ESA Digital Forest Map, which was prepared as a contribution to the World Forest Watch project of the International Space Year 1992, using NOAA-AVHRR imagery from the years 1989–1992. The map had a reported overall classification accuracy of 82.5 %, assessed with Landsat-MSS data (Häusler et al. 1993). It discriminated the classes Forest, Non-Forest and Water. This pan-European forest database had been used in the following for the creation of another Forest/Non-Forest dataset of Europe, established in the frame of the FIRS (Forest Information from Remote Sensing) project (Kennedy et al. 1994). Roy et al. (1997) investigated also the use of the Normalized Difference Vegetation Index (NDVI) in combination with surface temperature and a regional bio-geographic stratification approach.

In order to overcome the limitations of low-resolution satellite data while taking advantage of their large area coverage, spectral mixture modelling approaches had been developed to derive sub-pixel fractional land and forest cover information in a spatially explicit manner (e.g. Oleson et al. 1995; Foody and Cox 1994; Cross et al. 1991). Such approaches were also applied to derive forest proportion maps from NOAA-AVHRR images with 1 km^2 spatial resolution. Häme et al. (2001) classified pixel-based "forest probabilities" for a pan-European mosaic of 49 NOAA-14 AVHRR images from 1996/1997 stratified into three bio-geographic regions. Päivinen et al. (2001) as well as Schuck et al. (2003) combined these results with forest inventory statistics of 15 EU countries to derive adjusted forest maps corresponding to the official regional/national level forest area statistics, resulting in the Calibrated European Forest map (CEFM1996). On regional scales, the feasibility of sub-pixel land cover class de-composition approaches has been demonstrated for sets of multiple land cover classes – including distinct Broadleaved and Coniferous forest proportion discrimination – by applying spectral mixture modelling together with GIS information and fuzzy rule sets (e.g. Probeck et al. 2004).

Satellite image acquisitions from the Medium-Resolution Imaging Spectrometer (MERIS) on board the ENVISAT satellite at full 300 m spatial resolution (FR) have been used by the European Space Agency in partnership with EEA, FAO, GOFC-GOLD, IGBP, UNEP and JRC for creating global image composites and producing the global land cover classification GlobCover. It provides more than 20 land cover classes and is made available by ESA for two reference periods, i.e. December 2004 – June 2006 (Defourny et al. 2009) and January – December 2009 (Arino et al. 2010) under http://due.esrin.esa.int/globcover/. Though GlobCover provides a spatial resolution that is significantly better than all previous satellite-based global land cover maps, the class nomenclature is not straightforward to handle for potential forest-related applications, since broadleaved and coniferous forest cover is distributed over 12 partially mixed classes.

The most significant European initiative that makes systematic and operational use of Earth Observation (EO) imagery for land monitoring purposes is the Copernicus programme, formerly called Global Monitoring for Environment and Security (GMES). Copernicus is a European flagship initiative jointly implemented by the European Commission (EC), the European Space Agency (ESA) and the European

Environment Agency (EEA). It is considered one of the most ambitious EO programmes ever undertaken, aiming to establish an autonomous and operational European EO capacity. Its main objective is to support policy makers and public authorities in developing and implementing policies and legislation related to environment and security at the European level through systematic acquisition and evaluation of multi-source EO and in-situ data. Additionally, Copernicus has the broader objective of contributing to international environmental and climate change related monitoring efforts by providing services also on a global scale (Farquhar 2011). It comprises a dedicated space component, an in-situ component and a service component.

The Land Monitoring service as part of the Copernicus service component has been systematically prepared and demonstrated through several phases and projects, and has finally entered its Initial Operations Phase in 2010 with the adoption of Regulation no. 911/2010 on the European Earth Monitoring programme (GMES) and its initial operations (2011–2013) by the European Parliament and the Council on 22 September 2010 (European Commission 2010). The following sections provide information on the GMES/Copernicus Land services with highest relevance for European forest monitoring: the pan-European forest products provided by JRC's European Forest Data Centre (EFDAC), the services provided by the ESA GMES Service Element Forest Monitoring (GSE FM), pre-operational forest services and products developed and demonstrated by the FP7 land core project geoland2 and the pan-European high-resolution forest products being implemented in the frame of the GMES/Copernicus Initial Operations (GIO). Additionally, the current stage of development with respect to customised and regional forest downstream services is presented.

7.3 Pan-European Products from the European Forest Data Centre

The European Commission and EEA agreed in 2005 to establish "European Data Centres" for the provision of data and information in some environmental fields. The Joint Research Centre of the European Commission (JRC) acts as data centre for soil (European Soil Data Centre – ESDAC) and forest (European Forest Data Centre – EFDAC). The EFDAC is organisationally attached to the JRC's Institute for Environment and Sustainability (IES), Land Management and Natural Hazards (LMNH) Unit, FOREST Action. It aims at providing a single focal point for forest data and information relevant for policy reporting, and at hosting and connecting to relevant forest products (including datasets, documents or statistical data) and providing web-based tools. Further, the EFDAC provides an infrastructure for hosting of data collected under EU forest-related regulations and provides access to existing forest databases of the EC and EU member states. It also hosts the European Forest Fire Information System (EFFIS).

EFFIS provides a multitude of up-to-date forest fire related information for over 30 Mediterranean and other European countries in support of services and institutions in charge of forest protection against fires in the EU, both as download and online viewing services. Amongst other services, the EFFIS website (http://forest.jrc.ec.europa.eu/effis/) provides latest information on the actual fire season in Europe and especially in the Mediterranean. It comprises meteorological fire danger maps on a daily basis and forecasts for several days, a satellite image history of the last 7 days, and daily updated maps of most recent hot spots and fire occurrences. It also provides historic fire and burnt area information, which reach back more than 25 years for some countries (San Miguel-Ayanz et al. 2012). All EFFIS map products are currently derived from the daily processing of MODIS satellite imagery with 250 m spatial resolution.

The European Forest Data Centre also hosts high-resolution pan-European forest maps for the years 2000 and 2006 (http://efdac.jrc.ec.europa.eu/), which had been produced by JRC in order to overcome the limitations of previously available European forest maps, which were typically available only with limited spatial detail, either because being based on low-/medium-resolution satellite input data (AVHRR, MERIS, WIFS, see above) or because applying an MMU of 25 ha (as in case of CLC, see above).

The Forest Map 2000 was derived by an automated scene-by-scene nearest neighbour classification of 415 Landsat ETM+ images from the Global Land Cover Facility (GLCF) and Image2000, using adapted CORINE Land Cover 2000 data as training inputs, and applying a subsequent mosaicking to a seamless European product (Kempeneers et al. 2012; Seebach et al. 2011; Pekkarinen et al. 2009). In order to retrieve the Forest Map 2006, the Image2006 dataset consisting of more than 3,800 IRS-LISS III and SPOT scenes (2 coverages) was classified. Additionally, MODIS satellite imagery was incorporated into the mapping process to enable proper Forest Type discrimination (Kempeneers et al. 2012).

Whereas the Forest Map 2000 distinguishes the classes Forest and Non-Forest based on LANDSAT ETM + data (Pekkarinen et al. 2009), the Forest Map 2006 provides an additional discrimination of Broadleaved and Coniferous Forest (Kempeneers et al. 2011, 2012). The Forest Maps 2000 and 2006 have the following technical main characteristics:

- Consistent datasets independent of national boundaries, covering EEA-39 (2000: without Turkey) countries
- Uniform coordinate system ETRS89-LAEA
- Thematic Classes: Forest/Non-Forest (2000, 2006), Broadleaved/Coniferous Forest (2006)
- Spatial resolution: 25 m; no Minimum Mapping Unit
- High geometric and thematic accuracy

The derived Forest Cover Maps for 2000 and 2006 were validated with independent field survey point data from the European Land Use/Cover Area Frame Statistical Survey – LUCAS (Kasanko 2010), randomly selected GoogleEarth reference points and other sources. An overall accuracy of 90.8 % was reported for the Forest Map

2000, while the overall accuracy for the Forest Map 2006 reached 84 %. The notably lower accuracy of the more recent map is mainly caused by a lower achieved Producer's Accuracy of the Forest class (Kempeneers et al. 2012).

The continuity of this time series of pan-European Forest products with 25 m spatial resolution has been ensured through the GMES Initial Operations (GIO) phase where a dedicated so-called "Service Element 2" is included in the forest product specifications (see below). This will provide another update of the Forest-/Non-Forest as well as the Forest Type product for the reference years 2011/2012.

7.4 The GMES Service Element Forest Monitoring

The GMES Service Elements (GSE's) are one component of the ESA Earthwatch Program which was conceived to make use of the next generation of ESA operational EO satellite missions. The main aim of all GSE's was to deliver policy-relevant, operational information services, generally using such data sources that best meet user needs and requirements. In most instances, this meant a combination of EO data, in-situ data, ancillary data and analytical models.

The GSE Forest Monitoring (GSE FM) forms an integral part of the ESA GMES Service Elements (GSE) programme. It had two main stages of development: the first Stage was considered to be a consolidation phase (2003–2005), which focused on consolidating, aggregating, integrating and improving existing European pre-cursor service capacities, systems and precursor services, in order to make them operational within a reasonable time frame. The next phase (2005–2009) was the implementation phase and targeted the roll-out and expansion of the extensive portfolio of regional, national and pan-European-scale Forest services and products to users worldwide (Häusler et al. 2009). GSE FM has been conducted by an international consortium of European forest service providers, various users and expert consultants since 2003, and has been routinely supplying users with a service portfolio of reliable, timely and effective information products and services on the state of global forests to support decision-making and improved implementation of policies that enable sustainable forest management, compliance with specific protocols and binding conventions, and related user- and/or policy-driven activities.

In order to identify sustainable forest services in preparation of the GMES operational implementation phase, GSE FM had undertaken a thorough policy foundations review (van Brusselen and Schuck 2005) among potential forest user organisations. This identified that forest-related information needs are often similar and sometimes identical for several users across policy sectors and processes. It was found that in case of overlapping requirements, duplication of efforts can be avoided by taking into account such requirements in a simultaneous and coordinated way. Three major categories of forest information needs resulted from the policy review. These were:

1. "area": surface area and specification of location and boundaries of an area and changes per land cover/forest class;

2. "biomass": biomass as such, volume, changes of biomass and volume, growth and increment;
3. "disturbances": biotical damage, abiotic damage – particularly forest fires.

Information needs related to these variables are supported across the whole forest and environmental policy spectrum. The potential user organisations, which were addressed in a stakeholder survey, indicated that they could be helped first and foremost with information on 'forest area', 'forest area changes' (in terms of afforestation, reforestation and deforestation as well as land use and land use change between different land use classes), and 'aboveground vegetation biomass and changes therein'. Users pointed out that products and services should target to serve different reporting obligations simultaneously (van Brusselen and Schuck 2005).

As a consequence, the key policies addressed by GSE FM were the United Nations Framework Convention on Climate Change (UNFCCC) and the Kyoto Protocol (KP), the UN Convention on Biological Diversity (UNCBD), the Ministerial Conference on the Protection of Forests in Europe (MCPFE) and related Criteria & Indicator processes, the United Nations Forum on Forests (UNFF) and national Forest Programmes (Gomez et al. 2008). Accordingly, the GSE FM offered the following services addressing specific policy areas on different spatial scales (Häusler et al. 2009):

Pan-European Scale

- PAN European Forest Monitoring Service

National Scale

- Support to National UNFCCC and Kyoto Protocol Reporting on Land Use, Land Use Change and Forestry (LULUCF) Activities

National and Sub-national Scales

- Forest Information Up-date
- Support to Environmental Monitoring
- Detection and Post-monitoring of Natural and Human Induced Forest Disturbances

Local Scale

- Support to Management and Reporting Obligations of LULUCF Clean Development Mechanism (CDM) Projects

Mainly the Forest products generated by the pan-European and the national scale services can be considered as Core service (precursor) inputs to Downstream services, such as IRS-AWiFS-based Forest maps of pan-European service nature generated for several European countries (see Fig. 7.1 – left), or national-scale Forest Area and Change products generated in several regions of Europe (see Fig. 7.1 – right). All forest products produced by GSE FM are accessible on request.

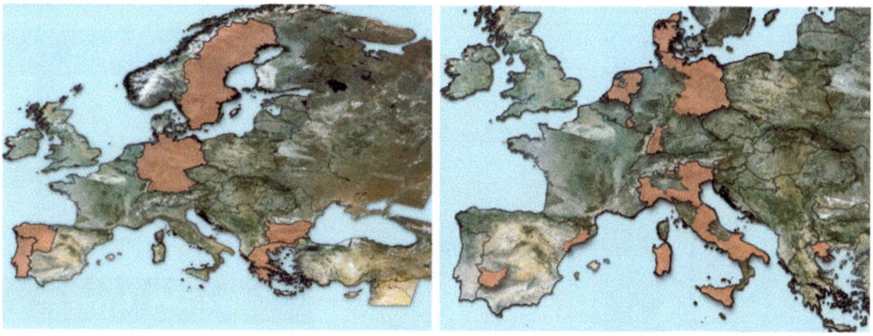

Fig. 7.1 GSE Forest Monitoring service areas mapped for the Pan-European Forest Monitoring service (*left*) and the Support to National UNFCCC and Kyoto Protocol Reporting on LULUCF service (*right*) (Häusler et al. 2009)

Based on these policy foundations, the GSE FM service network offered services including land use/cover and land use/cover change maps, forest cover and forest cover change maps, clear cut/disturbance maps and related databases, stand-type maps that support sub-national forest Geographic Information Systems (GIS), forest fragmentation and structural diversity maps, stem volumes, biomass and carbon statistics and corresponding change data, as well as user-customised versions and combinations thereof (Häusler and Gomez 2007). Figure 7.2 provides a comprehensive overview of the products' spatial scales and information levels addressed by GSE FM. Whereas the outer information layers (EO data, general land cover) represent the more generic product portfolio constituents which served the information needs of national- and international-level users in congruence with the concept of a core service, the innermost layers (Tree Species, Biodiversity) represent more value-added and customised, downstream-type applications serving regional-level users. Most products have been implemented and provided to the users both in a mono-temporal (Fig. 7.2, left) and a multitemporal change detection mode (Fig. 7.2, right).

By the end of the GSE FM Stage 2 (May 2009), about half of the European GSE FM users confirmed to apply the GSE FM forest products in one or more of the international policy processes that are listed below. The list covers all important international and most key national forest-related policy processes.

- European Environment Agency 'SEBI2010' (Streamlining European Biodiversity Indicators on progress toward the target of halting the loss of biodiversity by 2010) in the framework of the UN Convention on Biological Diversity
- European Environment Agency 5-yearly State of the Environment Report
- Forest Europe/the Ministerial Conference on the Protection of Forests in Europe
- Environment for Europe ministerial process
- United Nations Framework Convention on Climate Change and Kyoto Protocol
- Global Forest Resources Assessment (FRA) of the Food and Agriculture Organisation (FAO)

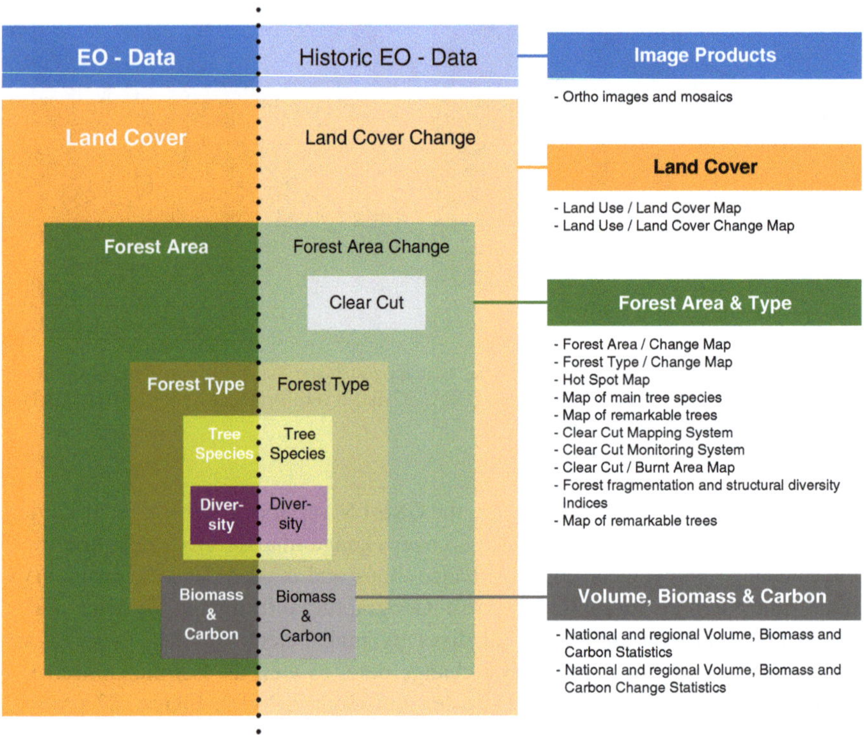

Fig. 7.2 GSE Forest Monitoring service portfolio of mono-temporal (*left*) and multitemporal change detection (*right*) products. The outer layers represent generic overarching, the inner layers further value-added and specialized service products (Häusler et al. 2009)

- Convention on Long-Range Trans-boundary Air Pollution (CLRTAP)
- The EEA EURECA report: Europe's contribution to the update of the Millennium Ecosystem Assessment
- Alpine Convention – State of the Alps
- Indirect use for NFI and other national monitoring and reporting tasks
- National forest mapping and hot spots location monitoring
- Detection of changes in forest area

These policies will continue to determine forest-related information needs both for international as well as national user organisations. With this extensive portfolio of policy-based and user-accepted services from pan-European to regional scale, GSE FM has laid the foundations for many currently pre-operational and operational European Forest Monitoring services, which are being implemented in the frame of current GMES/Copernicus core and downstream services (see next sections).

7.5 Development of Pre-operational Services: geoland2

The project geoland2 (www.gmes-geoland.info) has been the most influential GMES/Copernicus Land research & development project under the EU 7th Framework Programme (FP7), from September 2008 to December 2012. It constituted a major step forward in the implementation of the Copernicus Land Monitoring Core Service (LMCS) by successfully demonstrating the three components (Local, Continental and Global) of the LMCS. It dealt with a wide range of topics including land use, land cover change, forest monitoring, water quality, spatial planning, or global carbon monitoring. The main objectives of geoland2 were to consolidate the LMCS specifications and prepare, validate and demonstrate pre-operational service chains and products, giving convincing proof of the LMCS's feasibility by setting up a pre-operational prototypic production environment ready for operations. More than 50 European Service Providers and over 80 major international user organisations have been assembled by geoland2 (Jochum et al. 2011).

In terms of EO data, geoland2 has been widely making use of the GMES Space Component Data Access (GSC-DA) mechanism which had been put in place specifically for the GMES Pre-operations Phase based on a grant agreement between the European Commission (EC) and the European Space Agency (ESA). The GSCDA's main aim was to serve pre-operational GMES services such as geoland2 with a harmonised and sustainable access to space-based data from various Earth Observation missions, including ESA, national and Third-Party Missions. Specifically, the data sets DAP_MG2-3_01 "European wall-to-wall coverage 2009 (also called 'Image2009') and the access to the previously produced Image2006 (Müller et al. 2009) have been of primary importance for the development and demonstration of pre-operational Forest core services in geoland2. These data sets comprise IRS-P6 LISS-III and SPOT 4/5 HRVIR/HRG images resampled to 20 m pixel resolution.

The basic structure of geoland2 consists of three 'Core Mapping Services' and seven thematically specialized 'Core Information Services' (Jochum et al. 2011):

- The 'Core Mapping Services' (CMS) have been set up based on the critical review of existing user needs carried out in previous GMES Land projects, in order to produce multi-purpose generic geo-information products on land cover and land use and its annual and seasonal changes, as well as a variety of additional biophysical variables. The CMS undertook to design, set up, test and demonstrate pre-operational prototypes which should become ready for operational implementation.
- The CMS acted as the basis for more specialised thematic applications by the 'Core Information Services' (CIS) and potential future Downstream applications. The CIS had set out to define specific information products (e.g. for forest and environmental indicators, impact analysis and forecast models) and assess the utility of products and services coming from the CMS together with their involved end-users.

Further detail on the specific geoland2 tasks is given in Lacaze et al. (2010). In terms of Forest Monitoring, geoland2 represents a direct continuation and evolution of the GMES/Copernicus forest core service constituents that had been previously designed and demonstrated by the ESA GSE Forest Monitoring (see above). According to their partially more core service- and partially more downstream service-like nature, Forest Monitoring services have been included both in geoland2's continental CMS task 'Euroland' (European Land Cover) and in the dedicated CIS task "Forest Monitoring".

The continental LMCS component as conceptualised in the Euroland task was designed as a set of high-resolution (HR) pan-European thematic land cover layers (Imperviousness, Forest, Grassland, Wetland and Water). These 'HR Layers' have been shaped in response to the existing gap in the process of European-level harmonisation and integration of national mapping and land monitoring activities, with the aim to provide important complementary information (e.g. enabling various downstream applications, or updating and attributing new information to existing classification systems). The HR Layers were designed to support EEA and Member States in their policy reporting obligations as well as assist national mapping/monitoring and Downstream service activities. Further detail on the geoland2 HR Layers is provided by Kuntz et al. (2011).

Specifically, the 'HR Forest Layer' has been suggested to bridge the gap between current forest area/cover definitions, which differ not only between European countries but also depend on the specific purpose the Forest information is used for. Methods and algorithms have therefore been designed, tested and demonstrated to produce continuous and spatially explicit pixel-based estimates on Forest Crown Cover Density and Forest Type Composition (in terms of Coniferous vs. Broadleaved Forest fractions) that can subsequently be flexibly converted into any required forest categories defined by different national- or international-level users according to their specific requirements. Such customisation can be done e.g. in terms of defined crown cover thresholds, forest type classes, minimum mapping units etc. Forest Crown Cover Density is defined as the percentage of the forest area covered by vertically projected tree crowns (Sirro et al. 2012) and is typically required in many national and international forest definitions (e.g. FAO 2000).

Depending on the individual user requirements, the HR Forest Layer products are provided with a consistent 0.5 ha or 1 ha Minimum Mapping Unit (MMU), with respect to Forest Area. The full HR Forest Layer contains as 'primary' information layers a Forest Crown Cover Density product (see Fig. 7.3) and a Forest Type product (see Fig. 7.4). As 'secondary' (derived) products a customised Forest Area/Cover product is retrieved (applying a forest-definition based threshold to crown cover density) as well as a Forest Area Change product, using multi-temporal change detection.

Figures 7.3 and 7.4 provide examples of these two products from a randomly selected sub-area in Southwest Poland near the border to the Czech Republic, as had been produced as part of the geoland2 trans-boundary demonstration site EU-16e Poland/Czech Republic in 2011.

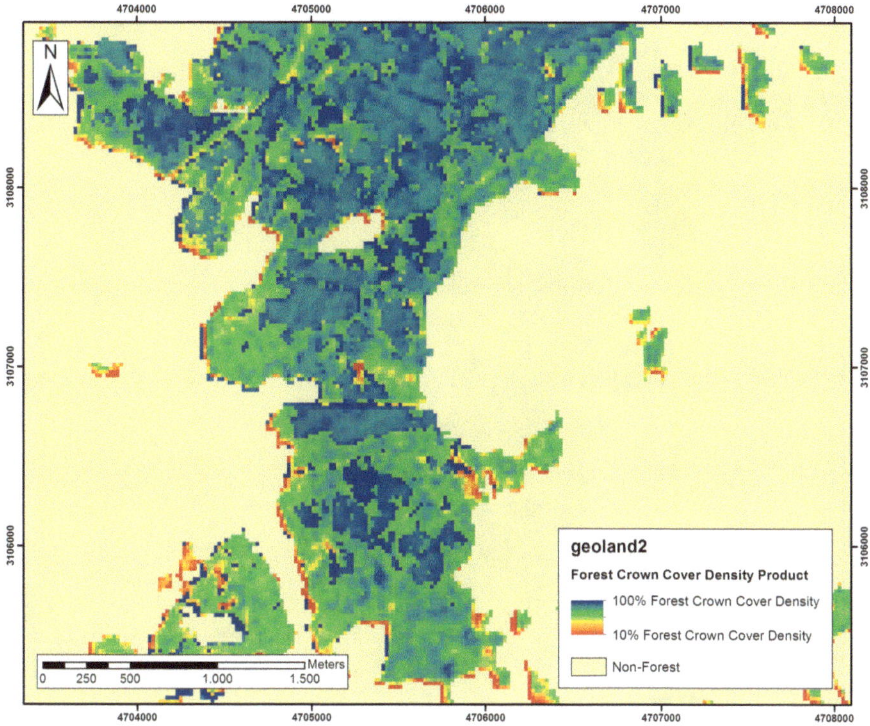

Fig. 7.3 The geoland2 Forest Crown Cover Density product in an area of SW-Poland, derived from an IRS-P6 LISS-III scene of 17 July 2006 (Credits: geoland2/GAF AG. Produced using products © Antrix Corporation Limited 2006; distribution by Euromap GmbH, Germany, all rights reserved; provided under EC/ESA GSC-DA)

For all demonstration sites, ortho-rectified Image2006 data with all spectral bands (Green, Red, NIR, and SWIR) were available with 20 m pixel size. Several classification algorithms were tested and benchmarked for mapping the above quantitative continuous forest characteristics in a cost-efficient, large area operational manner. Two principal methods were found to be suited for meeting these multiple requirements:

- A multiple linear regression model describing the relation between grey values in the HR optical satellite image bands and the continuous forest characteristics, providing for each pixel a percentage of broadleaved/coniferous forest proportion as well as a fractional crown cover. (Schardt et al. 2014);
- The probability estimation method of Häme et al. (2001), also resulting in pixel-based estimates of broadleaved/coniferous forest proportion and fractional crown cover (Sirro et al. 2012).

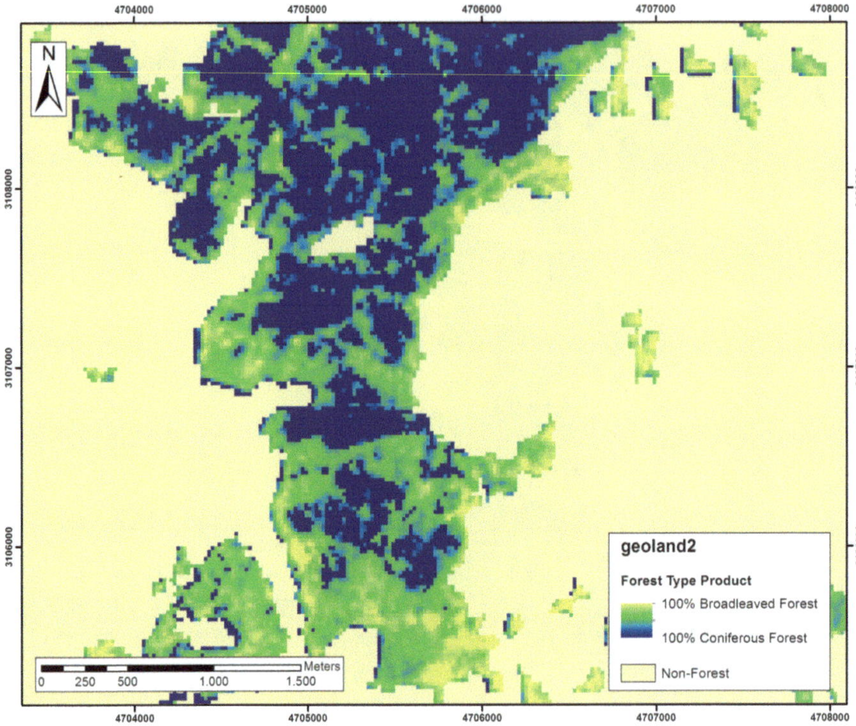

Fig. 7.4 The geoland2 continuous Forest Type composition product in an area of SW-Poland, derived from an IRS-P6 LISS-III scene of 17 July 2006 (Credits: geoland2/GAF AG. Produced using products © Antrix Corporation Limited 2006; distribution by Euromap GmbH, Germany, all rights reserved; provided under EC/ESA GSC-DA)

In geoland2, both methods have been successfully implemented and demonstrated by forest service providers in several demonstration sites across Europe. Independent validation with in-situ and/or very-high resolution (VHR) satellite/aerial imagery confirmed a high to medium accuracy of the classification results (e.g. Sirro et al. 2012) with the best results under Western, Central and Eastern European conditions. It proved that the best suited measure for quantifying the accuracy of such continuous-scale data is the coefficient of determination R^2 derived from scatterplots between VHR-interpreted stratified random samples and the HR-based classification results. Typically an $R^2 > 0.65$ was reached in good conditions.

Specifically, the geoland2 Forest Crown Cover Density and the Forest Type products have migrated to the subsequent Copernicus initial operational phase, though with modified specifications (see below). The above shown geoland2 forest products (Figs. 7.3 and 7.4) are located in the Southwest of Poland where also GIO HR Forest Layer products were available at the time of writing. The production of intermediate GIO HR Forest Layer products has been concluded meanwhile

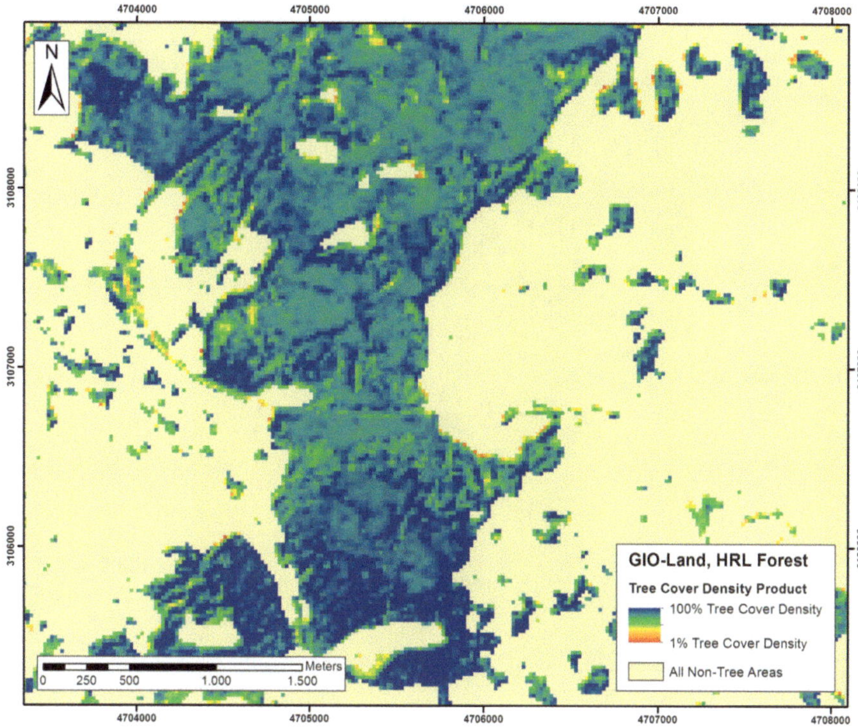

Fig. 7.5 GIO-Land HR Forest Layer: Tree Cover Density product in an area of SW-Poland, derived from an IRS-P6 LISS-III scene of 25 September 2011 (Credits: GIO-Land/GAF AG. Produced using products © Antrix Corporation Limited 2011; distribution by Euromap GmbH, Germany, all rights reserved; provided under EC/ESA GSC-DA)

(see Figs. 7.5 and 7.6). Despite an acquisition time difference of 5 years, the figures allow to investigate the systematic differences between the products, which are in fact primarily caused by the different product specifications between geoland2 and GIO. A detailed discussion of the differences between the geoland2 and the GIO forest products is provided in the next section.

Additional to these CMS products of the HR Forest Layer, geoland2 has been also establishing a range of forest indicator products in the frame of the CIS Forest task, which make use of the CMS forest products as inputs to derive grid-based forest indicators and metrics providing various measures to support characterization of forest fragmentation and connectivity as well as environmental and biodiversity status and changes.

All geoland2 Forest tasks pursue an open data access policy; therefore all mapped demonstration data are made available free-of-charge for non-commercial use and can be explored and accessed via the geoland2 data portal (http://www.geoland2.eu/portal/).

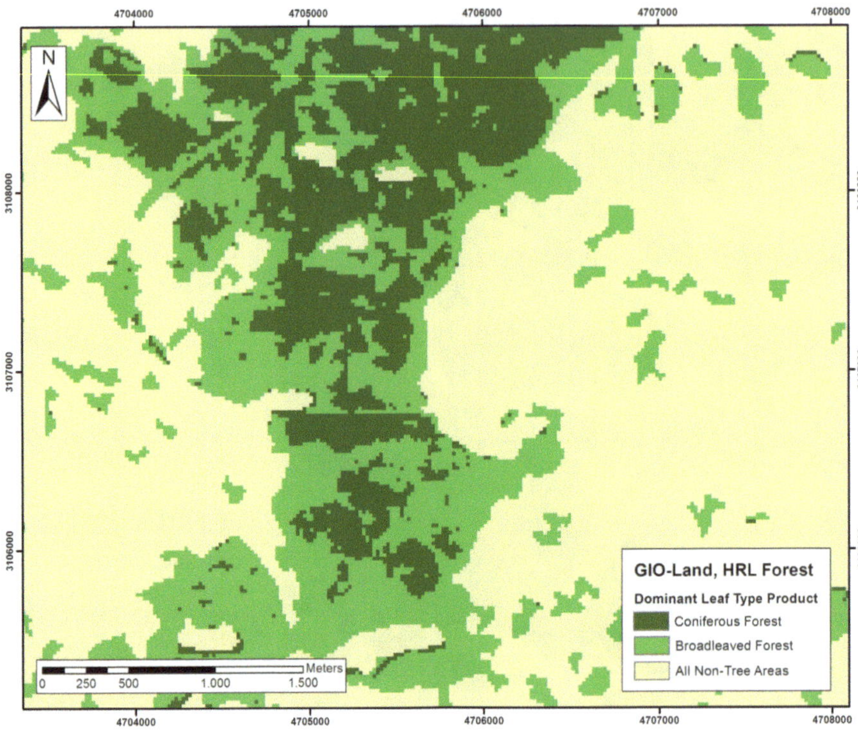

Fig. 7.6 GIO-Land HR Forest Layer: Forest Type – Dominant Leaf Type product in an area of SW-Poland, derived from an IRS-P6 LISS-III scene of 25 September 2011 (Credits: GIO-Land/ GAF AG. Produced using products © Antrix Corporation Limited 2011; distribution by Euromap GmbH, Germany, all rights reserved; provided under EC/ESA GSC-DA)

7.6 Operational Forest Monitoring: The GIO HR Forest Layer

In order to bridge the transition period between the Copernicus/GMES pre-operational phase and the full operations starting in 2014, the EC had put forward a proposal for GMES and its Initial Operations (GIO) in 2009. The necessity to provide such bridging support has been recognised especially for the Copernicus/GMES Land and Emergency Core Services.

The Copernicus/GMES Land Monitoring service led by the EC's DG Enterprise and Industry, entered its Initial Operations (GIO) phase with the final adoption of the EC Regulation (EU) n°911/2010 on the GMES programme and its Initial Operations (2011–2013) by EU Ministers during the General Affairs Council meeting and the entry into application on 22 September 2010, thereby reaching EU law status. The regulation has the overarching objective of contributing to the establishment of GMES as an operational programme by providing sustainable funding to the initial operational phase of GMES, enabling a gradual build-up of fully operational capabilities until 2014, as well as to put into place the necessary structures for GMES governance (European Commission 2010). During this transition phase, the development of the GMES-dedicated Sentinel satellites will continue.

Thematically, the Land Monitoring service focuses on the topics defined by the Land Monitoring Core Service Implementation Group (IG LMCS) and the results of various user consultations. It builds on precursor activities (geoland and geoland2), ESA's GSE Land and Forest Monitoring projects, CLC, the Land Fast Track precursor service, on the GMES Preparatory Action on Reference Data Access and on other activities at European and national levels (Langanke et al. 2013). The particular objective of the Copernicus Land Monitoring service is to make use of EO data in combination with other sources of data (e.g. national in-situ) for providing users in the field of land use/land cover, forestry- and environment-related applications with up-to-date and free-of-charge multi-purpose information that is common to a large community of users. It addresses a wide range of environmental, agricultural, regional development, transport and energy policies at EU level, as well as several European commitments to international Conventions.

In December 2011, EEA has contracted industry consortia of European service providers to map the GIO HR Layers Forest, Imperviousness, Grassland, Wetland and Water by means of (semi-)automated classification of HR satellite images in the frame of the "GMES Initial Operations 2011–2013 Land Monitoring Services: High resolution land cover characteristics of five main land cover types" (EEA/SES/11/004) in five separate regional Lots for all EEA-39 countries. Before start of operational production in mid-2012, an initial Streamlining Phase was carried out to consolidate product specifications for all HR Layers, as well as to ensure product consistency across the regional Lot boundaries.

The HR Forest Layer is primarily being mapped on basis of the ESA Data Warehouse (DWH) wall-to-wall HR image coverage of the CORE_01 'Optical HR pan-EU coverages 2011/2012' dataset, comprising predominantly IRS-P6/Resourcesat-2 LISS-III and RapidEye imagery (ESA 2013). Additionally, Medium Resolution (MR) pan-European IRS-AWiFS coverages (CORE_08) as well as VHR satellite imagery are needed to support image data pre-processing and HR image classification. All HR satellite imagery have to undergo a radiometric pre-processing procedure to account for sensor and radiometric differences as well as atmospheric and topographic effects, in order to allow a large-area consistent retrieval of the observed quantitative forest characteristics.

The specific objective of the HR Forest Layer production is to map forest cover characteristics, i.e. tree cover density and forest type based on the HR satellite images. This shall also support deducing a forest cover map with an MMU of 0.5 ha in line with Member States' requirements for their reporting obligations towards e.g. UN-ECE/FAO. Therefore, the HR Forest Layer processing chain comprises a per-pixel classification of the spatially explicit and continuously-scaled Tree Cover

Table 7.1 HRL forest specifications (EUFODOS 2013 and Langanke et al. 2013, both modified)

Tree cover density (20 m × 20 m)	Tree cover density values from 1 to 100 %
Forest type (20 m × 20 m) consisting of two layers	(1) Dominant leaf type Coniferous and broadleaved tree cover (plus non-forest); MMU of 0.5 ha and 10 % tree cover density threshold applied (2) Support layer: map of trees under agricultural use and in urban context from CLC and HR layer imperviousness 2009
Tree cover density (100 m × 100 m)	Tree cover density values from 1 to 100 %
Forest type (100 m × 100 m)	Coniferous, broadleaved and mixed forest. Trees under agricultural use and urban context from additional support layer are removed

Density and of a Dominant Leaf Type (Broadleaved, Coniferous) product in original 20 m pixel resolution and in an aggregated 100 m × 100 m version of Tree Cover Density and Forest Type. The latter includes also a 'Mixed Forest' class (see Table 7.1). Table 7.1 provides the latest available overview of the GIO Land HR Layer Forest specifications by EEA.

The above product specifications constitute a certain modification as compared to the products suggested and demonstrated by geoland2 (see above), such that:

- the GIO Tree Cover Density product is a pixel-based "tree" product without application of an MMU or a density threshold, in contrast to the corresponding geoland2 "Forest" Crown Cover Density product;
- the GIO Forest Type product consists of two sub-products, in order to take into account on the one hand all pixels with tree cover ≥1 %, and on the other hand comply to a stricter "Forest" definition (FAO), according to which trees with predominantly agricultural use and trees in urban context need to be excluded from forest areas;
- both the GIO Tree Cover Density product and the Forest Type product will be provided with 20 m × 20 m original pixel resolution and additionally in a 100 m × 100 m spatially aggregated version

The production of the HR Layers is organised as a mix of industrial production of 'intermediate products' and decentralized contributions to product verification and enhancement by the EEA member and cooperating countries. These countries are further expected to contribute to the process in the form of in-kind provisions of nationally available in-situ data. However, the highly variable availability and quality of in-situ data across the EEA-39 countries has proven a considerable challenge, as well as the timely availability of HR image data as indispensible basis for all classification steps. Remaining gaps in the 2011/2012 coverage over Northern Europe and the British Isles have necessitated additional image acquisitions in 2013 (Langanke et al. 2013). Another challenge for large-area harmonized mapping is the effect of seasons and vegetation phenology on the satellite images' spectral

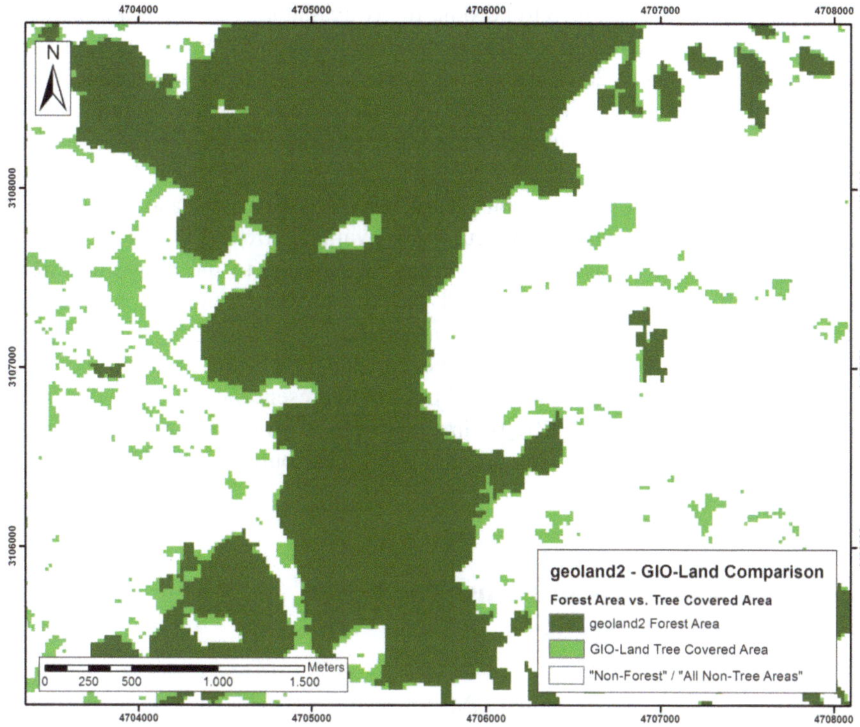

Fig. 7.7 Comparison between the geoland2 Forest Area of 2006 (forest crown cover density >10 %, MMU 1 ha) and the additional GIO-Land HR Forest Layer's Tree-Covered Area of 2011 (tree cover density ≥1 %, no MMU) in an area of SW-Poland (Credits: geoland2 & GIO-Land/ GAF AG. Produced using products © Antrix Corporation Limited 2006 & 2011; distribution by Euromap GmbH, Germany, all rights reserved; provided under EC/ESA GSC-DA)

characteristics as well as the atmospheric disturbances, i.e. clouds and haze (Sirro et al. 2012).

Figures 7.5 and 7.6 provide intermediate product examples for the GIO HR Forest Layer in SW-Poland in the same area as in case of Figs. 7.3 and 7.4, which allows for a direct product comparison. Algorithmic improvements from geoland2 to GIO comprise amongst others a better algorithmic de-coupling between the forest density and type characteristics (see Fig. 7.3/Fig. 7.4 vs. Fig. 7.5/Fig. 7.6), which are considered to be independent variables.

The geoland2 Forest Type product contains considerably more information content and spatial detail within forest areas than the corresponding GIO product due to its continuous-scale representation of coniferous and broadleaved tree species proportions per pixel (see Fig. 7.4 vs. Fig. 7.6). In terms of the geoland2 Forest Crown Cover Density vs. the GIO Tree Cover Density product, the situation is exactly the other way round due to mapping without MMU and algorithmic improvements in GIO (see Fig. 7.3 vs. Fig. 7.5). This causes a significant difference in the maximum tree cover outline, as is separately shown in Fig. 7.7.

All HR Layers are foreseen to become freely available both as harmonized and independently validated 100 m spatial resolution product versions and in their original (non-validated) 20 m spatial resolution versions. A regular update of the HR Forest Layer together with the other HR Layers is currently envisaged for 2015 as integral part of the pan-European component of the Copernicus Land services (Langanke et al. 2013).

The HR Layer products are considered to complement and add value to the pan-European LU/LC information currently available from CLC as well as to national-level LU/LC classification systems. Additional to the above described 'standard' HR Forest Layer, a so-called 'Service Element 2' will be mapped for further monitoring purposes, matching as closely as possible the pan-European forest products that the JRC's European Forest Data Centre provides (see above). It shall provide a per-pixel classification of tree cover presence/absence and dominant leaf type, using the DWH CORE_01 imagery with 25 m resolution and European ETRS89_LAEA projection as the primary input data source.

7.7 Customised Applications: The Forest Downstream Sector

In addition to the above described GMES/Copernicus forest core services, many other value-added, so-called 'Downstream services' are expected to emerge in the near future, being tailored to more specific public or commercial users' needs (e.g., forecasting services with a local/regional scope, further thematic value-adding on top of the Core service products, services integrating natural scientific and socio-economic data, etc.), possibly even for markets that are up to now not making use of any EO imagery (Farquhar 2011).

A variety of Downstream service applications has already been consolidated and demonstrated by previous FP6 and GSE projects (e.g. GSE Forest Monitoring, see above). Such Downstream services typically use local/regional in-situ data and/or modelling approaches in addition to the EO-based core service products, in order to support legally mandated organisations in their regular reporting and decision making on national and regional scales with very detailed information products. Future customised Downstream applications might address commercial and/or export spin-off opportunities, building on available Copernicus/GMES products and capacities. It is expected that especially for small- and medium-sized enterprises (SME's) which are serving regional/local demands or specific Downstream market sectors, a broadly usable multi-purpose Land core service will offer excellent opportunities for business development, and substantially stimulate the Downstream service market (Jochum et al. 2011).

Since Forests play a vital role in the European economy and environment, the protection of the ecological and economic functions of forest resources is one of the key aspects of forest-related Copernicus/GMES Downstream services. A range of such services has been developed towards an operational level and demonstrated in several service cases together with regional forest users by a consortium of European research organisations and commercial service providers in the frame of the EU FP7 research project EUFODOS (European Forest Downstream Services) from 2011 to 2013. EUFODOS has established close links to an extensive forest user community that is well connected to other GMES User Groups, which is expected to facilitate the roll-out of the prototypic services to larger scales, and create a socio-economic benefit (Schardt et al. 2012).

EUFODOS has been prototypically setting up services by means of state-of-the-art satellite and laser scanning (LIDAR) technology to serve forest authorities with cost-effective, timely and comprehensive information on forest structure and damages, since these authorities have both the legal mandate and the obligation to take appropriate countermeasures against forest damages in order to ensure a long-term sustainable forest management (Schardt et al. 2012). Available services comprise the assessing of insect-infested damages, windfall events, storm or snow damage, and similar events.

Specifically, the use of space- and air-borne optical and radar systems allows acquiring up-to-date EO imagery within short time intervals during or after a damage event. The combined utilization of such 'after-event' imagery together with existing GMES/Copernicus forest core service products (such as the GIO HR Forest Layer's Tree Cover Density product) and operational processing chains is expected to allow undertaking a rapid forest damage detection, delineation and impact assessment in a cost-efficient manner, on an operational basis.

7.8 Conclusions & Outlook

Together with the evolution of optical satellite sensors in terms of spatial, temporal and spectral capabilities over the past decades, also pan-European forest monitoring approaches have developed from coarse-scale forest/non-forest maps towards operational high-resolution monitoring systems of spatially explicit forest characteristics in Europe. All above described forest monitoring approaches were following slightly different product specifications and achieved slightly different accuracy levels, typically determined by the applied sensors' physical characteristics and limitations as well as different user requirements.

The GIO Land HR Forest Layer in combination with the upcoming European fleet of EO satellites (ESA Sentinels) designed for long-term sustainable operation from 2014/2015 onwards, is considered to provide a new, sustainable and widely applicable standard in European EO-based forest monitoring that will be serving the EO data needs of the GMES/Copernicus programme and the wider user community. Specifically the Sentinel-2 satellites are eagerly awaited, as they are expected to boost Copernicus Land and forest applications with their systematic global delivery of HR multispectral images in unprecedented quality and frequency. In essence, Sentinel-2 will combine a large swath width with frequent revisit times and a systematic global acquisition scheme at high spatial resolution and at a large number of spectral bands. Thus Sentinel-2 will provide enhanced continuity of SPOT-, IRS- and LANDSAT-type imagery. Sentinel-2 will carry an optical payload with 13 bands in the Visible, NIR and SWIR spectrum at 10, 20 and 60 m spatial resolution with a swath width of 290 km. A revisit time of only 5 days at the equator (under cloud-free conditions) and 2–3 days at mid-latitudes will be achieved with two satellites. This short revisit time together with the large swath width of Sentinel-2 will constitute a major step forward towards much improved data availability compared to the current optical HR satellite acquisition schemes. The 13 spectral bands of Sentinel-2 should additionally enable a more reliable compensation of atmospheric influences and an improved cloud masking also in case of thin clouds (Sirro et al. 2012). This is expected to facilitate future updates of the current GIO HR Layers.

The launch of the first Sentinel-2 satellite unit is expected in late 2014 / early 2015. Data continuity and sustainability is planned to be ensured by a series of further Sentinel-2 satellite units to be launched in regular intervals. Especially the short revisit time will allow the detection and monitoring of rapid changes, e.g. for disaster control or vegetation monitoring during the growing season. In terms of Forest Downstream Services, Sentinel-2 will specifically support operational monitoring of Forest damages caused by fires, storms, etc.

Since the Sentinel satellites are being funded by ESA and EU Member States, a 'free and open' data access policy has been announced (Aschbacher and Milagro-Perez 2012). Together with observations from the Landsat-8 satellite, cloud-free optical satellite data availability will presumably not remain a major limiting factor in operational forest monitoring applications. However, the operational handling of the associated very large volumes of satellite data needs to be properly addressed. Available funding (and a sufficient extent of land cover/land use change) given, even moving to higher monitoring frequencies for forest core service products might become an option in the future.

Acknowledgments Part of the research leading to these results has received funding from the European Union's Seventh Framework Programme (FP7) under the grant agreements n°218795 (geoland2) and n°262786 (EUFODOS). Further operational service development and implementation has been funded by the European Space Agency (ESA) under the ESA/ESRIN contract n°19277/05/I-LG (GSE Forest Monitoring) and the European Environment Agency (EEA) under the Framework Contract n° EEA/SES/11/004-Lot2/Lot3 (GIO-Land HR Layers) with funding by the European Union. The opinions expressed in this publication are those of the authors only and do not represent the above mentioned Agencies' official positions.

References

Arino O, Ramos J, Kalogirou V, Defourny P, Achard F (2010) GlobCover 2009. In: Proceedings of the ESA living planet symposium, SP-686, June 2010

Arnold S (2009) Digital Landscape Model DLM-DE – deriving land cover information by integration of topographic reference data with remote sensing data. In: Heipke C, Jacobsen K, Müller S, Sörgel U (eds) High-resolution earth imaging for geospatial information. International Archives of Photogrammetry, Remote Sensing and Spatial Information Science, vol 38, Part 1_4_7-W5

Aschbacher J, Milagro-Perez MP (2012) The European earth monitoring (GMES) programme: status and perspectives. Remote Sens Environ 120:3–8

Bossard M, Feranec J, Otahel J (2000) CORINE land cover technical guide – Addendum 2000. Technical report no. 40. Copenhagen (EEA)

Büttner G, Feranec J, Jaffrain G, Mari L, Maucha G, Soukup T (2004) The CORINE land cover 2000 project. EARSeL eProc 3(3):331–346

Büttner G, Kosztra B, Maucha G, Pataki R (2010). Implementation and achievements of CLC2006. Final report, Copenhagen (EEA)

Cross AM, Settle JJ, Drake NA, Päivinen RTM (1991) Subpixel measurement of tropical forest cover using AVHRR data. Int J Remote Sens 12(5):1119–1129

Defourny P, Bicheron P, Brockmann C, Bontemps S, Van Bogaert E, Vancutsem C, Huc M, Leroy M, Ranera F, Achard F, Di Gregorio A, Herold M, Arino O (2009) The first 300-m Global Land Cover Map for 2005 using ENVISAT MERIS time series: a product of the GlobCover System. In: Proceedings of the 33rd international symposium on remote sensing of environment (ISRSE), Stresa, Italy, May 2009

EEA Task Force (1992) CORINE Land Cover – a European community project. Brochure prepared as contribution to European conference of the international space year, Munich, 30.03.-04.04.1992, 22 pp

ESA (2013) GMES Space Component data access portfolio: data warehouse 2011–2014. Issue 2.8, 144 pp, URL: http://gmesdata.esa.int/c/document_library/get_file?uuid=9f57e0f4-af57-43ca-aa26-b9418fbf40ea&groupId=10725

EUFODOS project (2013) Newsletter issue no. 9 (Jan. 2013), 2p, URL: http://www.eufodos.info/sites/default/files/downloads/EUFODOS_Newsletter09_0113.pdf

European Commission (2010) Regulation (EU) No. 911/2010 of the European Parliament and of the Council of 22 September 2010 on the European Earth Monitoring Programme (GMES) and its Initial Operations (2011 to 2013). Official Journal of the European Union, vol 53, OJ L 276 of 20 October 2010

FAO (2000) On definitions of forest and forest change. Forest resources assessment programme, working paper 33. FAO, Rome, Italy. 15 pp

Farquhar C (2011) GMES: what's it all about? GEOconnexion Magazine, September 2011, pp 44–46

Foody GM, Cox DP (1994) Sub-pixel land cover composition estimation using a linear mixture model and fuzzy membership functions. Int J Remote Sens 15(3):619–631

Fuller R, Cox R, Clarke R, Rothery P, Hill R, Smith G et al (2005) The UK land cover map 2000: planning, construction and calibration of a remotely sensed, user-oriented map of broad habitats. Int J Appl Earth Obs 7:202–216

GAF (2001): IRS WIFS image mosaic classification of forest classes for the European Union. Final report, Contract Nr. 17240-2000-12 F1ED ISP DE. Munich: GAF, 26 pp

Gallaun H, Schardt M, Linser S (2007) Remote sensing based forest map of Austria and derived environmental indicators. In: Proceedings of ForestSat 2007, Montpellier, France, 5–7 November 2007

Gomez S, Häusler T, Probeck M (2008) GMES helps countries fulfil their reporting obligations in the field of climate change. Window on GMES (2):49–55 (ISSN 2030–5419)

Häme T, Stenberg P, Andersson K, Rauste Y, Kennedy P, Folving S, Sarkeala J (2001) AVHRR-based forest proportion map of the Pan-European area. Remote Sens Environ 77(1):76–91

Häusler T, Gomez S (2007) Achievements of the GMES service element forest monitoring: progress towards sustainable service delivery. In: Proceedings of ForestSat 2007, Montpellier, France, 5–7 November 2007

Häusler T, Saradeth S, Amitai Y (1993) NOAA-AVHRR forest map of Europe. In: Van Genderen JL, Zuidam RA, Pohl C (eds) Proceedings of the international symposium operationalization of remote sensing, 19–23 April 1993, ITC Enschede, vol 8, pp 37–48

Häusler T, Gomez S, Ramminger G, Siwe Ngamabou R (2009) GMES Service Element Forest Monitoring: Final report 2003–2009. Munich, 116 pp; Download URL: http://www.gmes-forest.info/sites/default/files/downloads_with_images/GSEFM_Final_Report.pdf

Heymann Y, Steenmans C, Croisille G, Bossard M (1994) CORINE land cover. Technical guide. Office for Official Publications of the European Communities, Luxembourg, 136 pp

Jochum M, Kuntz S, Lacaze R, Banaszkiewicz M, Olsson B, Bicheron P, Probeck M, Leo O, Calvet JC, Brink A, Goor E (2011) geoland2 – operational monitoring services for our changing environment. In: European Commission (ed) Let's embrace space – space research achievements under the 7th Framework Programme, pp 30–38. ISBN 978-92-79-19704-8

Joint Research Centre (2008) Work programme 2008 – forest data and information systems. FP7 – WP2008 – Action n 22003. Institute for Environment and Sustainability, Joint Research Centre.

Kasanko M (2010) LUCAS – a multi-purpose land use survey. In: SIGMA – the Bulletin of European Statistics, no. 01/2010 (ISSN 1018–5739), pp 18–19

Kempeneers P, Sedano F, Seebach L, Strobl P, San Miguel-Ayanz J (2011) Data fusion of different spatial resolution remote sensing images applied to forest-type mapping. IEEE Trans Geosci Remote Sens 49(12):4977–4986

Kempeneers P, Sedano F, Pekkarinen A, Seebach L, Strobl P, San Miguel-Ayanz, J (2012) Pan-European forest maps derived from optical satellite imagery. IEEE Earthzine 5 (2nd quarter theme)

Kennedy P, Folving S, McCormick N (1994) European forest ecosystems mapping and forest statistics: the FIRS project. In: Kennedy P, Päivinen R, Roihuvuo L (eds) Proceedings of the workshop on designing a system of nomenclature for European forest mapping, Joensuu, Finland, 13–15 June 1994 (EUR 16113 EN). CEC, Luxembourg

Kuntz S, Probeck M, Garzon A, Soukup T, Smith G, Kleeschulte S, Hoffmann C, Gallaun H, Littkopf A, Manunta P, Sedano F, Radutu A, Lönnqvist A, Aifantopoulou D, Gitas I, Vaitkus G, van Wolvelaer J, Green T, Rosengren M (2011) geoland2 – Continental component of the land monitoring service. In: European Commission (ed) Let's embrace space – Space Research achievements under the 7th Framework Programme, pp 45–53. ISBN 978-92-79-19704-8

Lacaze R, Kaptein A, Banaszkiewicz M, Bartholomé E, Bicheron P, Calvet J-C, Goor E, Hoffmann C, Kuntz S, Léo O, Olsson B, Probeck M (2010) geoland2 – Operational monitoring services for our changing environment. In: Lacoste-Francis H (ed) Proceedings of ESA living planet symposium, Bergen, Norway, 28 June–2 July 2010 (ESA-SP-686), ISBN 978-92-9221-250-6

Langanke T, Büttner G, Dufourmont H, Steenmans C (2013) High resolution land cover mapping on a continental scale for 39 European countries. Context – status – applications – future developments. In: Proceedings of the 35th international symposium of remote sensing of environment, Beijing

Ludwig R, Probeck M, Mauser W (2003) Mesoscale water balance modeling in the Upper Danube watershed using sub-scale land cover information derived from NOAA-AVHRR imagery and GIS techniques. Phys Chem Earth 28:1351–1364

Müller R, Krauss T, Lehner M, Reinartz P, Forsgren J, Rönnbäck G, Karlsson Å (2009) IMAGE 2006 European coverage, methodology and results, DLR report, 55p. URL: http://earth.esa.int/pub/ESA_DOC/Image2006-v1_01.pdf

Oleson KW, Sarlin S, Garrison J, Smith S, Privette JL, Emery WJ (1995) Unmixing multiple land-cover type reflectances from coarse spatial resolution satellite data. Remote Sens Environ 54:98–112

Päivinen R, Lehikoinen M, Schuck A, Häme T, Väätäinen S, Kennedy P, Folving S (2001) Combining earth observation data and forest statistics. EFI Research Report 14. European Forest Institute, Joensuu, Finland. Joint Research Centre of the European Commission, EUR 19911 EN, 101 pp

Pekkarinen A, Reithmaier L, Strobl P (2007) Pan-European Forest/Non-Forest Mapping based on Landsat data. In: Proceedings of the 10th AGILE international conference on geographic information science 2007. Aalborg University, Denmark

Pekkarinen A, Reithmaier L, Strobl P (2009) Pan-European forest/non-forest mapping with Landsat ETM+ and CORINE land cover 2000 data. ISPRS J Photogramm Remote Sens 64:171–183

Probeck M, Ludwig R, Mauser W (2004) Spectral unmixing of Sub-Scale land cover from Multitemporal NOAA-AVHRR imagery using a combined GIS- and fuzzy logic approach. In: Smits PC, Bruzzone L (eds) Analysis of multi-temporal remote sensing images: proceedings of the second international workshop on MultiTemp 2003, Ispra (Italy) 16–18 July 2003. World Scientific, Singapore, pp 80–88

Romero J, Häusler T, Volden E, Renda O (2007) Space for UNFCCC and the Kyoto Inventory and GSE forest monitoring projects. In: Proceedings of the ENVISAT symposium 2007, Montreux, Switzerland. ESA SP-636, July 2007

Roy D, Kennedy PJ, Folving S (1997) Combination of the normalized difference vegetation index and surface temperature for regional scale European forest cover mapping using AVHRR data. Int J Remote Sens 18:1189–1195

San Miguel-Ayanz J, Schulte E, Schmuck G, Camia A, Strobl P, Liberta G, Giovando C, Boca R, Sedano F, Kempeneers P, McInerney D, Withmore C, de Oliveira SS, Rodrigues M, Durrant TH, Corti P, Oehler F, Vilar L, Amatulli G (2012) Comprehensive monitoring of wildfires in Europe: the European Forest Fire Information System (EFFIS). In: Tiefenbacher J (ed) Approaches to managing disaster – assessing hazards, emergencies and disaster impacts. In Tech, Rijeka, p 162

Schardt M, Granica K, Astola H (2012) EUFODOS – European forest downstream services improved information on forest structure and damage – Project presentation. In: Möttus M (ed) (2012): Finnish remote sensing days 2012. Book of abstracts. Department of Geosciences and Geography C7. University of Helsinki, 58 pp

Schardt M, Gallaun H, Steinegger M, Probeck, M (2014) Large area estimation of continuous forest characteristics by means of satellite remote sensing data. Int J Appl Earth Obs Geoinf, in preparation

Schuck A, Päivinen R, Häme T, Van Brusselen J, Kennedy P, Folving S (2003) Compilation of a European forest map from Portugal to the Ural mountains based on earth observation data and forest statistics. For Policy Econ 5:187–202

Seebach LM, Strobl P, San Miguel-Ayanz J, Gallego J, Bastrup-Birk A (2011) Comparative analysis of harmonized forest area estimates for European countries. Forestry 84(3):285–299

Sirro L, Häme T, Ahola H, Lönnqvist A (2012) Forest crown cover estimation in Northern Boreal and temperate European forests. In: Proceedings of 'First Sentinel-2 Preparatory Symposium', Frascati, Italy (ESA SP-707, July 2012)

Smith G, Beare M, Boyd M, Downs T, Gregory M, Morton D (2007) UK land cover map production through the generalisation of OS MasterMap. Cartogr J 44(3):276–283

UNECE/FAO (2001) Forest Resources of Europe, CIS, North America, Australia, Japan and New Zealand. (Industrialized temperate/boreal countries). UN-ECE/FAO contribution to the Global Forest Resources Assessment 2000. Main Report. ECE/TIM/SP/17. New York and Geneva. 445 p

Valcarcel N, Villa G, Arozarena A, Garcia-Asensio L, Caballero ME, Porcuna A, Domenech A, Peces JJ (2008) SIOSE, a successful test bench towards harmonization and integration of land

cover/use information as environmental reference data. The International Archives of the Photogrammetry, Remote Sensing and Spatial Information Sciences, vol XXXVII, Part B8, pp 1159–1163, Beijing
van Brusselen J, Schuck A (2005): Policy foundations review for geographically explicit information needs on forests. EFI internal reports No. 18. European Forest Institute, Joensuu, Finland

Chapter 8
The European Urban Atlas

Enrique Montero, Joeri Van Wolvelaer, and Antonio Garzón

8.1 Overview and Main Characteristics

The origin of the Urban Atlas answers to a European Commission need for detailed and comparable information related to cities. The urban information is so far provided by local and national statistics or using data extracted from land cover maps as CORINE Land Cover. CORINE Land Cover provides a resolution of 1:100,000 (25 ha) and is clearly unsatisfactory for the demand of the Commission services. As in Europe, urban areas accommodate more than three-quarters of the population and these areas have grown rapidly in recent decades, there is an urgent need for pan-European, reliable and inter-comparable urban planning data. The Urban Atlas, developed as part of GMES (Global Monitoring for Environment and Security) brings exactly that.

In order to achieve this goal, the Urban Atlas brings together thousands of images from European satellites and provides detailed and cost-effective mapping of larger urban zones, yielding accurate land cover and usage data.

One of the main purposes of the Urban Atlas is to provide more detailed information to Urban Audit: the main requirement is information with more resolution and comparable.

The current specifications of Urban Atlas fulfill these conditions: 0.25 ha in urban areas and the same methodology and product in all the major agglomerations of Europe.

Besides the Urban Audit, the Urban Atlas is a source of accurate information to use in the framework of ESPON and INTERREG program.

E. Montero • A. Garzón (✉)
INDRA Company, Alcobendas, Spain
e-mail: agarzon@indra.es

J. Van Wolvelaer
EUROSENSE Company, Wemmel, Belgium
e-mail: joeri.van.wolvelaer@eurosense.com

I. Manakos and M. Braun (eds.), *Land Use and Land Cover Mapping in Europe: Practices & Trends*, Remote Sensing and Digital Image Processing 18,
DOI 10.1007/978-94-007-7969-3_8,

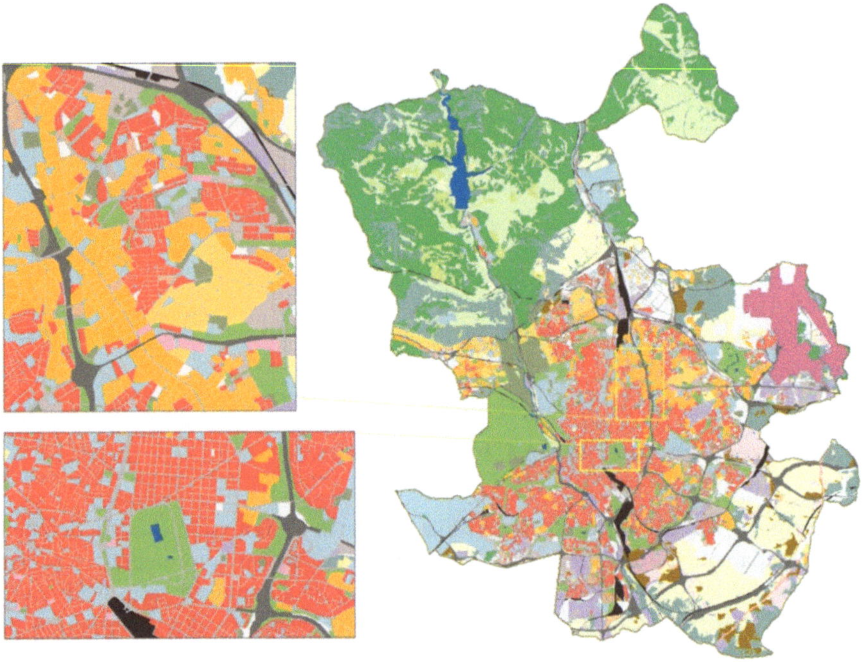

Fig. 8.1 Example of the UA of the city of Madrid (Legend in Fig. 8.4)

The Urban Atlas concept has been present in the GMES initiative from the beginning since one of the first projects started in GMES Service Element (funded by ESA) was the GMES Urban Services (GUS) project where the first services on thematic Land Use and the legend based in CLC and MOLAND were provided to regional and local users.

The Urban Atlas is coordinated with related European projects, as GMES Core Services (Local component of the Land Monitoring Core Service), GSELand (third yearly cycle), Geoland-2 and Boss4GMES (Fig. 8.1).

Urban Atlas provides coverage for a detailed and cost-effective mapping of larger urban zones, giving accurate land cover and usage data. The Urban Atlas mission is to provide high-resolution hotspot mapping of changes in urban spaces and indicators for users such as city governments, the European Environment Agency (EEA) and European Commission departments. More than 300 major cities in the EU were covered by mid 2011. All EU capitals are included in the exercise plus a large sample of large and medium-sized cities participating in the European Urban Audit (Fig. 8.2).

The main source of the Urban Atlas is SPOT 5 satellite imagery (or other available VHR imagery), the first operational production was performed using imagery with a reference date of 2006 (+/−1 year).

The Urban Atlas cities are mapped at 1:10.000, using in total 20 classes, of which the 17 urban classes have a minimum mapping unit (MMU) of 0.25 ha and the 3 non-urban classes a MMU of 1 ha. The minimum accuracy of the data provided is 85 % for artificial surfaces and 80 % for the other classes.

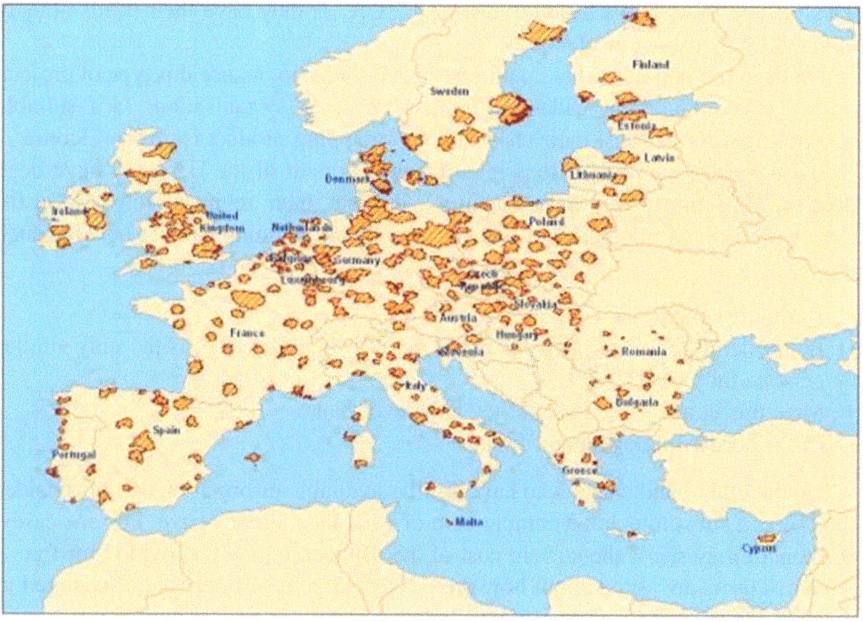

Fig. 8.2 Cities and LUZ (large urban zone) of UA

The urban fabric (CLC classes 1.1.1 and 1.1.2) are differentiated by their degree of imperviousness which is integrated from the LMCS (Land Monitoring Core Service) high resolution soil sealing layer. The production is based on a mix of CAPI (Computer Aided Photo-interpretation) and object oriented classification.

8.2 Analysis of User's Requirements

To date, a very wide range of local users have got to know and interact with the Urban Atlas product. These organisations deal with different topics (e.g. environmental policies, urban planning, etc.) and therefore they can have different needs hardly fulfilled with a single geo-information product. Consequently, users have shown very different levels of satisfaction.

A very important factor when the product is evaluated by the user is the country of origin and the amount of available geodata for their working area. For example, Spain invests high amounts of money in geo-information at national and regional level, producing tailored products, mostly in-house, to suit their particular needs and generating regular updates of the information. Therefore, many organisations have exactly what they need by developing their own geo-information projects.

This type of user shows little interest in the UA, as they have their needs already covered by their own products.

On the other hand, in other countries, such policies to finance this type of project, do not exist or are not widespread. In these cases, organisations lack suitable geo-information to assist their tasks and any mapping product is most welcome.

If we look at user's reactions regarding the benefit of the UA, they have been diverse. Those users who have no similar product, have found it really useful for their tasks in land management and environmental planning, stating benefits such as:

- Harmonised approach.
- Better detail than CORINE for urban areas (when CORINE is the only similar geodata they have).
- Most useful and Use product they have ever had.
- Cross-border homogeneous data.

On the other hand, users who have similar available information do not consider the UA as a substitute, but a complement of what they already have. In some cases, the benefit they see is the current cost of the product, which is provided for free.

When users are asked about how the product should be improved or changed to make it more useful and useable to better cover their needs, answers are quite similar regardless their degree of satisfaction with the current specification. Users who need to work at local level consider the **current scale insufficient**, as the current degree of detail does not separate individual buildings and geometric accuracy should be finer.

The other main shortcoming found at the moment by most users is the **thematic detail**. Some classes (both urban and non-urban) are seen as too aggregated and they should be separated into several individual components. For example, class 12130 (Public military and private services) is generally considered too broad and should be disaggregated in their individual components, otherwise this introduces a lot of uncertainty making this class quite useless. Other users, more involved with environmental issues, would like more detail for non-urban classes, disaggregating, for example, class 31000 (Forest) into deciduous and evergreen forests.

Those users who have had the opportunity to see their cities produced at local and European level (as defined by DGRegio) have shown disappointed with the result of the latter, as the loss of thematic content makes the geo-data much less useful for local purposes, not seeing much compatibility between the product and the tasks it could be useful for. Also they appreciate a very **high unbalance between scale, MMU and number of classes** (Fig. 8.3).

Users' expectations for the future are focused on the **regular updating** of the product (1–2 years). For land management and urban planning change assessment is a very important issue and most users lack products or projects devoted to a regular monitoring of changes.

At the moment, the most involved European user is DGRegio. As soon as DGRegio showed interest in the UA, they got involved in the technical discussions and definition of the European UA, approving a final version of the product,

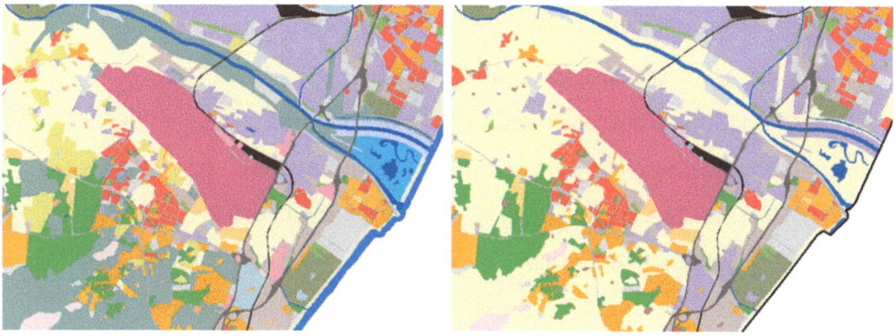

Fig. 8.3 Detail of the Urban Atlas of Málaga (Spain) at local level (*left*) and European level (*right*) (Legend in Fig. 8.4)

which guidelines are being applied in the current production. Therefore, **user's requirements at European level are considered to be fulfilled**, as, at the moment, the "only" European-level user has stated its needs and consequently the product has been tailored to fit these requirements.

8.3 Urban Atlas: Main Features

The Urban Atlas concept has been developed in such a way that the maps serve multiple purposes: the Urban Atlas should help urban planners for better assessing risks and opportunities, ranging from threat of flooding and impact of climate change, to identifying new infrastructure and public transport needs.

In order to do so, the maps are created with following specifications:

- Provides harmonised land cover/land use maps at scale 1:10,000 and according to a common classification
- Designed to measure urban land use at high resolution and at high/low levels of soil sealing
- Covers 305 major European agglomerations, based on Urban Audit's Larger Urban Zones
- Minimum mapping unit: 0.25 ha in urban and 1 ha in rural areas
- Minimum linear width: 10 m
- Positional Accuracy : ± 5 m
- Thematic classes based on CORINE LC nomenclature and GUS Legend (based on MOLAND) – 20 thematic classes
- Thematic accuracy: 85 % urban areas, 80 % outside
- Vector format
- Use of commercial off the shelf (COTS) navigation data for the road network
- Degree of sealing for 11 classes based on GMES FTS (Fast Track Service) Soil Sealing Layer specifications

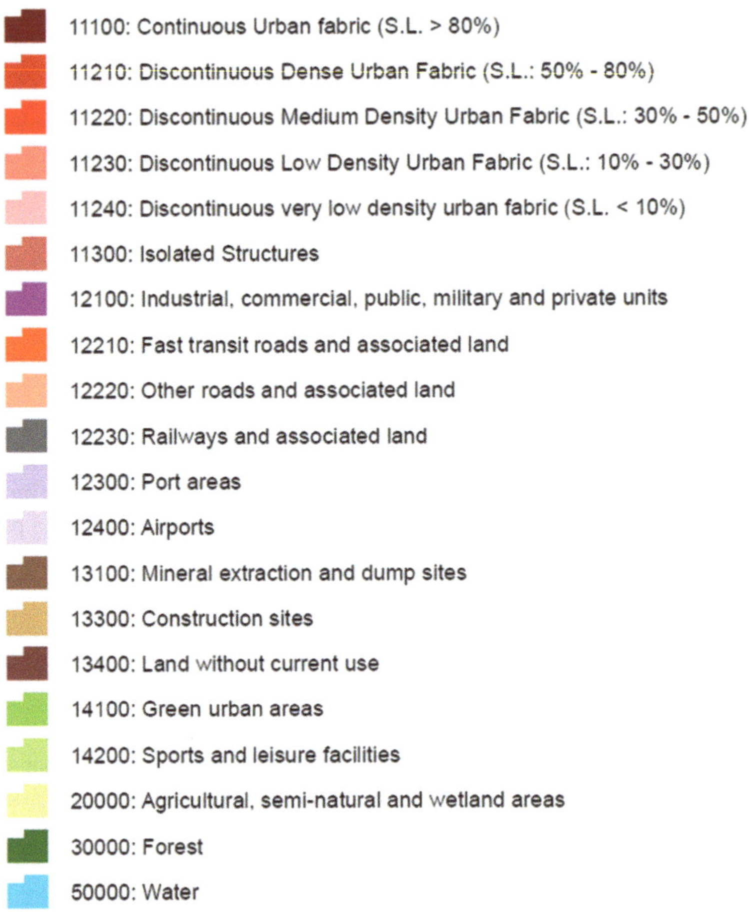

Fig. 8.4 Urban Atlas legend

- All input data described by metadata according to the INSPIRE metadata profile specifications and guidelines (Fig. 8.4)

8.4 Methodology Description for UA Update

In the framework of Geoland2, the European Land Monitoring Service (EUROLAND) addresses the local (i.e. the Urban Atlas) component of the Land Monitoring Core Service (LMCS).

One of the goals within the project was to develop an Urban Atlas update procedure based on a mapping of changes. This actual change delineation and

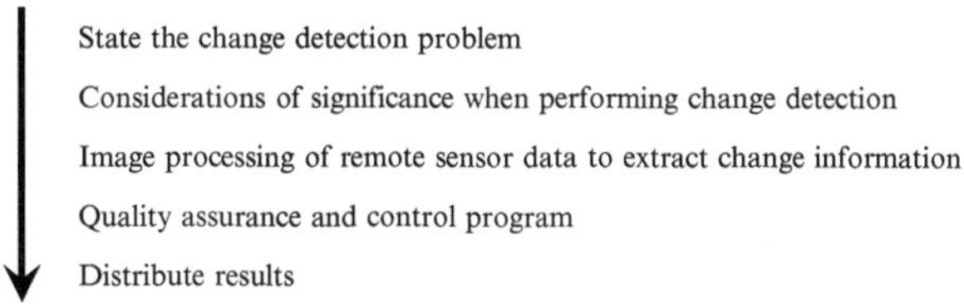

Fig. 8.5 General steps used to conduct digital change detection using remote sensor data (from Jensen 1996)

interpretation is a manual process. However, this effort was supported by the results of automated change detection. In the next paragraphs, the possible approaches for change detection considered are described.

8.4.1 General Change Detection Considerations

The first step in change detection is to determine the nature of the problem in terms of substance, time, and space. Then the desired remote sensing imagery/data can be defined, along with supporting ancillary data. Processing procedures may then be outlined, including classification schemes. A sequence of steps needed to perform change detection using digital remote sensing data is outlined (Fig. 8.5).

8.4.2 Change Detection Algorithms

For the detection of change pixels, several statistical techniques exist, calculating e.g. the spectral or texture pixel values, estimating the change of transformed pixel values or identifying the change of class memberships of the pixels (Lu et al. 2004; Baatz and Schäpe 2000). But when adapted to high-resolution imagery, the results of these pixel-based algorithms are sometimes limited. Especially if small structural changes are to be detected, object-oriented procedures seem to be more precise and meaningful. Object-oriented change detection and analysis techniques can in addition estimate the changes of the mean object features (spectral colour, form, etc.), assess the modified relations among neighbouring, sub- and super-objects and find out changes regarding the object class memberships.

Image differencing is one of the most effective methods in image algebra and can result in a meaningful change/non-change map if appropriate threshold values are chosen.

A common method in image differencing is to consider the pixels' brightness magnitude. The measure of the change across all the bands in an image can be derived and then the relative difference between the two images is. In image-to-image change detection, selecting a threshold is a trade-off between real changes and the noise level.

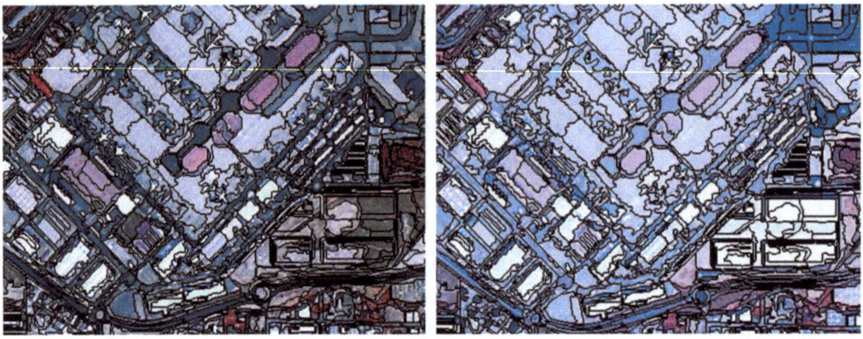

Fig. 8.6 Segmentation of SPOT5 images (2005 and 2008) based on both images gives a good delineation of changed and non-changed features. (Madrid, ES) (from van Wolvelaer et al. 2010)

One motivation for the object-oriented approach is the fact that, in many cases, the expected result of most image analysis tasks is the extraction of real world objects, proper in shape and proper in classification. The proposed approach relies on three steps: first is segmentation of the image to produce image objects, second is the detection of changes by comparing both images and by the use of suitable features, and finally a post-processing using some rules to reduce the vagueness and increase the reliability of the results.

For image data taken at two different dates, the image segmentation could be performed in three different ways, depending on the input data (Niemeyer et al. 2008):

(a) On the basis of the bi-temporal data set, i.e. using a data stack that consists of both scenes.
(b) Based on the image data of one acquisition time; the generated object boundaries are then simply assigned to the image data of the second acquisition time without segmentation.
(c) Separately for the two times, i.e. the two data sets are segmented independently.

The variation of objects shape features is in any case an important indicator for real object changes and needs to be taken into account for object-based change detection.

Object changes could then be identified based on the layer features, such as mean or standard deviation of the single image bands, shape and texture features (Fig. 8.6).

A good image segmentation should result in the creation of homogeneous objects, thereby enabling an object classification using strict thresholds. The aim is to find a set of feature parameters (spectral as well as textural) which enable the identification of changed areas. Creating ratios by dividing object values from t_1 by those of t_0 (spectral values and texture calculations) proves to result in a set of object parameters highly valuable for the identification of changed areas.

8.4.3 Change Detection Within the geoland2 Project

Within the geoland2 project the developments regarding the local component were threefold:

1. The creation of an update protocol to guarantee a common mapping approach
2. The development of an automated change detection algorithm
3. The implementation of the developed methodology for the updating of the Urban Atlas

At first an Urban Atlas update protocol was developed based on the Corine Land Cover change mapping approach (2000–2006). This protocol identifies the different change possibilities (e.g. new features, disappearance of features, fragmented change, etc.) and gives a specific solution for each change type.

This mapping protocol is necessary to implement the automatically detected changes in a uniform way. In the end the aim within the geoland2 project was to create a "full" Urban Atlas update for a set of European cities. In order to do so, the detected land cover changes need to be interpreted by a photo-interpreter to create a database of land use changes and ultimately come to an updated Urban Atlas map.

Different change mapping approaches have been developed, including a standard pixel-based differencing, a comparison of object texture, and a change identification based on the Pearsons' index. In total the Urban Atlas for seven European cities has been updated using this semi-automated change detection approach.

In contrast to this "full" update whereby an image-to-image comparison is performed, the "quick update" procedure compares the existing Urban Atlas (2006) map with the updated HR layer Imperviousness (2009). This change detection method considers the monitoring of urban sprawl, thereby focusing on the built-up of formerly non-built up areas.

Within the objects of the Urban Atlas classified as non-built-up (agricultural areas, forests, land without current use, green urban areas, . . .), the updated sealing layer is screened for the presence of sealed areas. In the next step, these highlighted areas are segmented based on the new HR image (2009).

This methodology offers a quick procedure for the monitoring of the urban expansion using existing datasets. Drawback is that it is only focused on certain land cover classes, thereby neglecting other possible land cover changes.

8.5 Conclusions

Automated change detection techniques prove to be applicable for detecting changes in the urban environment. More specific, the effort needed for identifying land use changes in order to update an Urban Atlas map is significantly reduced when using automatic change detection techniques for the detection of land cover changes. These change/no-change layers can be used as a direct input for the CAPI-based map updating.

Multiple methodologies have been studied, thereby comparing pixel- with object-based procedures. The object-oriented methods prove to be best suited for the task in mind.

Future research should focus on an improvement and validation (reduction of commission and omission errors) of the tested workflows, thereby also focusing on the classification of changes.

References

Baatz M, Schäpe A (2000) Multiresolution segmentation – an optimization approach for high quality multi-scale image segmentation. In: Strobl J et al (ed) Angewandte Geographische Informationsverarbeitung XII, AGIT symposium, Salzburg, Germany

Chavez PS Jr (1996) Image-based atmospheric corrections – revisited and improved. Photogramm Eng Remote Sens 62(9):1025–1036

Eliens A (2000) Principles of object oriented software development. Addison Wesley, Reading

Eugene YK, Johnston RG (1996) The ineffectiveness of the correlation coefficient for image comparisons. Technical report LA-UR- 96–2474, Los Alamos

Jensen JR (1996) Introductory digital image processing: a remote sensing perspective, 2nd edn. Prentice Hall, Upper Saddle River

Liu Y, Nishiyama S, Yano T (2004) Analysis of four change detection algorithms in bi-temporal space with a case study. Int J Remote Sens 25(11):2121–2139

Lu D, Mausel P, Brondizios E, Moran E (2004) Change detection techniques. Int J Remote Sens 25:2365–2407

Niemeyer I (1999) Object-based change detection – an unsupervised approach, OBIA06, J.F. Mas, Monitoring land-cover changes: a comparison of change detection techniques. Int J Remote Sens 20(1):139–152

Niemeyer I, Marpu PR, Nussbaum S (2008) Change detection using object features. In: Blaschke T, Lang S, Hay G (eds) Object-based image analysis. Springer, Berlin

Otsu N (1978) A threshold selection method from gray-level histogram. IEEE Trans Syst Man Cybern 8:62–66

Parthasarathy S, Anbazhagan N (2006) Analyzing the software quality metrics for object oriented technology. Inf Technol J 5:1053–1057

Rogerson PA (2002) Change detection threshold for remotely sensed images. J Geogr Syst 4:85–97

Van Wolvelaer J, Santos C, Deflorio AM, Garzon A (2010) Urban Atlas updating by semi-automatic change. In: Proceedings 30th EARSeL symposium: remote sensing for science, education, and natural and cultural heritage, Paris, France, 31 May–3 June 2010, pp 147–154

Part III
State of the Art Mapping Methods

Chapter 9
A Review of Modern Approaches to Classification of Remote Sensing Data

Lorenzo Bruzzone and Begüm Demir

9.1 Introduction

Remote sensing (RS) images are a rich information source for monitoring the Earth surface, e.g., for climate change analysis, urban area studies, forestry applications, risk and damage assessment, water quality assessment, crop monitoring. Due to the recent developments in both passive multispectral and hyperspectral sensors, as well as Synthetic Aperture Radar (SAR) active instruments, the role of this technology is becoming more and more important for environmental monitoring and anthropic applications. The new generation of satellite multispectral sensors characterized by very high geometrical resolution (VHR) (e.g., Ikonos, Quickbird, Spot-5, and Geoeye-1) can acquire images with a metric or sub-metric resolution. Hyperspectral sensors can acquire images characterized by a high spectral resolution that usually results in hundreds of observation channels. The acquisition of VHR SAR images from satellite platforms has also become possible thanks to recent TerraSAR-X, TanDEM-X and Cosmo-Skymed missions.

In this chapter we focus our attention on automatic classification methodologies for passive RS images aimed at generating land-cover maps. Land-cover maps are usually obtained from RS images by using supervised classification techniques, which require a set of labeled samples for training the classification algorithm. The information contained in the last generation of satellite RS data allows a classification of land-cover types with improved accuracy and robustness. However, for an efficient and effective analysis of the huge amount of data available, it is necessary to define automatic techniques that can address several critical problems related to the very high spectral and geometrical resolutions of the data.

High spectral resolution data (i.e., hyperspectral images) result in the possibility to discriminate land-cover classes with very similar spectral signatures with a high

L. Bruzzone (✉) • B. Demir
Department of Information Engineering and Computer Science,
University of Trento, Via Sommarive, 14, I-38123 Trento, Italy
e-mail: lorenzo.bruzzone@ing.unitn.it

I. Manakos and M. Braun (eds.), *Land Use and Land Cover Mapping in Europe: Practices & Trends*, Remote Sensing and Digital Image Processing 18,
DOI 10.1007/978-94-007-7969-3_9,

accuracy. However, supervised classification of hyperspectral images is a very complex task due to: (1) the non-stationary behavior of the spectral signatures of land-cover classes in the spatial domain of the scene, which is due to physical factors related to ground (e.g., different soil moisture or composition), vegetation, and atmospheric conditions; and (2) the small ratio between the number of training samples and the number of available spectral channels, which results in the Hughes phenomenon (i.e. the classification accuracy decreases by increasing the number of features given as input to the classifier over a given threshold, which depends on the number of training samples and the kind of classifier adopted (Hughes 1968)). All the aforementioned issues result in decreasing the robustness, the generalization capability, and the overall accuracy of classification systems used to generate the land-cover maps.

In the case of VHR data, the significant amount of geometrical details present in a very high-resolution scene completely changes the perspective of data analysis compared with high and moderate-resolution images provided by previous-generation satellite sensors (such as the multispectral Thematic Mapper and Enhanced Thematic Mapper Plus scanners). In particular, the improvement in spatial resolution simplifies the problem of mixed pixels present in standard images, while it increases the internal spectral variability (intraclass variability) of each land-cover class and decreases the spectral variability between different classes (interclass variability). Thus, on the one hand, the resulting high intraclass and low interclass variabilities lead to a reduction in the statistical separability of the different land-cover classes in the spectral domain, which in turn involves high classification errors (Carleer et al. 2004; Binaghi et al. 2003). In addition, the limited spectral resolution of VHR sensors, which depends on technological constraints, further increases the complexity of the classification problem (Carleer et al. 2004; Schowengerdt 2002). On the other hand, due to the high spatial resolution of the images, the geometrical information of the scene can also be considered in the classification process according to proper feature-extraction methodologies. Thus, the properties of VHR data require the development of specific methods for data analysis.

The aim of this chapter is to review different techniques proposed in the literature for the classification of multispectral and hyperspectral RS images. We focus our attention on the most recent methodological developments related to the classification of RS images acquired by new generation RS sensors. The rest of this chapter is organized as follows. Sections 9.2 and 9.3 review the techniques proposed in the literature for the classification of hyperspectral and VHR RS images, respectively. Section 9.4 introduces the most recent classification approaches presented in the literature. Finally, Sect. 9.5 draws the conclusion of this chapter.

9.2 Classification Techniques for Hyperspectral Images

Statistical and machine learning techniques have been widely used in the past decades for the analysis and classification of multispectral RS data. Many techniques have been presented in the literature including maximum likelihood (ML) classifier (Duda et al. 2001), spectral angle mapper (SAM) classifier (Kruse et al. 1993), k-nearest neighbor classifier (Samaniego et al. 2008), decision trees (van de Vlag and Stein 2007), genetic algorithms based classifiers (Bandyopadhyay and Pal 2001) and ant colony algorithms (Liu et al. 2008). In the past, many machine learning algorithms based on artificial neural networks (Haykin 1999) have also been used for classification of multispectral data (e.g., multilayer perceptrons (Bischof and Leona 1998; Yang et al. 1999), radial basis function neural networks (Bruzzone and Fernández-Prieto 1999; Giacinto and Bruzzone 2000)). Multispectral sensors proved to be effective for discriminating among different land-cover classes. However, they cannot discriminate classes with very similar spectral signatures. To deal with this problem, hyperspectral RS images can be used. However, the above-mentioned classification techniques decrease their effectiveness when applied to hyperspectral images. This depends on their sensitivity to both the large spatial variability of the signatures of land-cover classes and the Hughes phenomenon.

In order to handle the aforementioned problems, in the literature, different promising approaches have been proposed to hyperspectral image classification. In the recent years, kernel-based RS image classification techniques have become very popular, since these approaches are robust to the Hughes phenomenon and can provide reasonably high classification accuracy also in critical conditions. Four different kernel-based methods, i.e., regularized radial basis function neural networks, support vector machine (SVM) (Boser et al. 1992; Schölkopf and Smola 2001), Fisher discriminant analysis, and regularized AdaBoost have been analyzed and compared both theoretically and experimentally in the context of classification of hyperspectral data in Camps-Valls and Bruzzone (2005). The comparison has been carried out in terms of robustness to high dimension of the input space, low number of training samples, and noisy data. As a result, SVM is found to be most effective, yielding higher accuracies than the other kernel-based methods considered, and at a much lower computational cost. The success of SVMs in classification of hyperspectral data is justified by three main reasons (Camps-Valls and Bruzzone 2005; Gualtieri et al. 1999; Huang et al. 2002; Camps-Valls et al. 2004; Melgani and Bruzzone 2004): (i) their intrinsic effectiveness, with respect to traditional classifiers, which results in high classification accuracies and very good generalization capabilities (low sensitivity to the Hughes phenomenon); (ii) the convexity of the objective function used in the learning of the classifier, which results in a unique solution (i.e., the system cannot fall into sub-optimal solutions associated with local minima); (iii) the possibility of representing the convex optimization problem in a dual formulation, where only non-zero Lagrange multipliers are necessary for defining the separation hyperplane (which is a very important advantage in the case of large datasets) that is related to the sparsity of the solutions (Cristianini and Shawe-Taylor 2000). Another interesting and effective method introduced in the RS literature is the kernel Fisher discriminant (KFD) analysis (Mika et al. 1999; Murat Dundar and Landgrebe 2004). This method takes advantage of the same concept of kernel used in SVMs to obtain nonlinear solutions; however, it minimizes a different functional, and thus the solution is expressed in a different way. Another kernel-based classification technique is relevance vector machine (RVM), which has been presented in Demir and Ertürk (2007) for the classification of hyperspectral images. The most important advantage of RVMs compared to SVMs is that it can provide similar classification accuracy with a significantly reduced number of relevance vectors with respect to that of support vectors. Thus RVM is more suitable for applications that require low complexity and, possibly, real-time classification (Demir and Ertürk 2007).

9.3 Classification Techniques for Very High Geometrical Resolution Images

With the availability of VHR images it is possible to acquire detailed information on the shape and geometry of the objects that are present on the ground. This detailed information can be exploited by automatic classification systems to generate land-cover maps that exhibit a high degree of geometrical details. The precision that the classification system can afford in the characterization of the geometrical properties of the objects that are present on the ground is particularly relevant in many practical applications. In the recent literature, many papers have addressed the development of novel techniques for the classification of VHR RS images. In Camps-Valls et al. (2006), composite kernels defined in the context of SVMs are used to combine spatial and spectral information to obtain higher accuracy compared to that yielded using spectral information only. To this purpose, spatial feature vectors are obtained using either a simple mean or the mean and the standard deviation together of a certain neighborhood window of the corresponding spectral feature vector. Then, it is proposed to compute kernel matrices corresponding to spatial and spectral feature vectors separately and merging them using different combination approaches. In Pesaresi and Benediktsson (2001) and Tuia et al. (2009a) it is proposed to use morphological profiles (MPs) for the classification of VHR images. In these works, MPs are constructed by applying opening and closing operations to the image features. The characterization of spatial information with MPs is mostly suitable for representing the multiscale variability of the structures in the image, whereas it is not sufficient to model other geometrical features. To overcome this limitation, in Dalla Mura et al. (2010) it is proposed to use morphological attribute filters instead of the conventional MPs for better modeling the spatial information. The combination of spectral and spatial information is achieved based on Markov random field in Farag et al. (2005), and majority voting is used within regions obtained by the watershed segmentation algorithm in Tarabalka et al. (2008). In Unsalan and Boyer (2004), a technique for the identification of land developments across large-scale regions is presented. The proposed technique uses straight lines, statistical measures (length, orientation, and periodicity of straight line), and a spatial coherence constraint to identify three classes, namely: (1) urban; (2) residential; and (3) rural. In Shackelford and Davis (2003), an object-based approach is proposed to classification of dense urban areas from pan-sharpened multispectral Ikonos imagery. This approach exploits a cascade combination of a fuzzy pixel-based classifier and a fuzzy object-based classifier. The fuzzy pixel-based classifier uses spectral and simple spatial features to discriminate between roads and buildings, which are spectrally similar. Subsequently, a segmented image is used to model the spectral and spatial heterogeneities and to improve the overall accuracy of the pixel-based thematic map. Shape features and other spatial features (extracted from the segmented image) as well as the previously generated fuzzy classification map are used as inputs to an object-based fuzzy classifier. In Mott et al. (2002), an algorithm based on selective region growing is proposed to classify a high-resolution image. In the first step, the image is classified by taking into account only spectral information. In the second step, a classification procedure is applied to the previous map by taking into account not only spectral information but also a pixel distance condition to aggregate neighbor pixels. By reiteration, neighbor pixels that belong to the same class grow in a selective way, obtaining a final classification map. A pixel-based approach to the classification of VHR images is presented in Bruzzone and Carlin (2006). The proposed approach is defined on the basis of two modules: (1) a feature-extraction module that exploits an adaptive, multilevel, and complete hierarchical representation of the spatial context of each pixel in the scene under investigation; and (2) a classification module based on SVMs. A graph-based filter model for object-oriented classification in VHR imagery, namely locally weighted discriminating projection, is proposed in Chen et al. (2011). Thanks to this model, the traditional graph-based linear criteria can be extended to facilitate nonlinear mapping for linear nonseparable sample set.

9.4 Advanced Techniques for Classification of Remote Sensing Images

One of the main critical issues in the application of supervised classification methods to the analysis of RS images is the definition of a proper training set for the learning phase of the classification algorithm. In many RS applications, due to the practical impossibility to obtain a sufficient number of representative training samples for a reliable estimation of classifier parameters (particularly when many information classes are considered), supervised classification approaches may result in poor accuracies. Indeed, the amount and the quality of the available reference samples play an important role to obtain accurate land-cover maps. However, the collection of a sufficient number of high quality labeled samples is time consuming and costly. The quality of samples is related to different factors: (i) the high spatial correlation that exists when training patterns are collected in neighboring locations on the ground (thus violating the expected independence assumption) that results in reduction of the information conveyed by training samples with respect to the case of independent samples; (ii) the intrinsic variability of the spectral signatures of land-cover classes in the images (which have a non-stationary spatial behavior) that is related to the acquisition and the ground conditions (this is particularly critical in hyperspectral images where the spectral resolution of the sensor results in an intrinsic high sensitivity to small variation in these conditions). This results in a limitation of the capability to properly model the classification problem in the training set and thus in defining an effective classification model with supervised classifiers.

To deal with these problems, semisupervised learning (SSL) and active learning (AL) methods have been recently presented in the literature in the context of classification of RS images. Semisupervised classification methods exploit both training data and unlabeled samples in the learning phase of the classifier in order to obtain a general decision function that can take into account both the information present in the training set and the structure of all data in the feature space. This is particularly useful in ill-posed problems where the training set does not completely represent the real distribution of the data in the feature space. AL is introduced as an alternative to passive learning. Passive learning is the standard approach adopted for the definition of a training set in RS, which is based on the application of statistical sampling procedures that exploit the knowledge of the application domain for extracting ground reference samples and do not consider any interaction with the adopted supervised classifier. In AL, the idea is to optimize the definition of the training set by interacting with the criterion adopted by the classification technique for the decision process. The concept is that starting from an initial small number of training samples it is possible to exploit the classifier for defining a training set that completely describes the classification problem according to the mechanism adopted by the classifier for defining the discriminant functions. In this way, it is expected that the classifier drives the definition of the training set and that the expensive phase of collection of labels for irrelevant (i.e. redundant or non-useful) samples is avoided.

Another conceptually different approach to improve the statistic in the learning of a classifier is domain adaptation (DA). DA methods aim at classifying an image for which no prior information is available (target domain) by exploiting the ground reference samples already available for another similar image (source domain). Accordingly, the need and effort to recollect labeled samples is significantly reduced by reusing the already available information on the source domain to classify the target domain. DA (also known as partially supervised/unsupervised learning) methods define strategies that use the information available on the source domain to classify the target domain, assuming that the two domains may have different (but related) distributions.

The next subsections provide a discussion and a review on the use of SSL, AL and DA, for the classification of RS images. Table 9.1 summarizes the main characteristics of all the aforementioned learning methods.

Table 9.1 Main characteristics of the considered learning problems

Type of learning	Hypotheses	Objective
Supervised learning (Duda et al. 2001; Kruse et al. 1993; Samaniego et al. 2008; van de Vlag and Stein 2007; Bandyopadhyay and Pal 2001; Liu et al. 2008; Haykin 1999; Bischof and Leona 1998; Yang et al. 1999; Bruzzone and Fernández-Prieto 1999; Giacinto and Bruzzone 2000; Boser et al. 1992; Schölkopf and Smola 2001; Camps-Valls and Bruzzone 2005; Gualtieri et al. 1999; Huang et al. 2002; Camps-Valls et al. 2004; Melgani and Bruzzone 2004; Cristianini and Shawe-Taylor 2000; Mika et al. 1999; Murat Dundar and Landgrebe 2004; Demir and Ertürk 2007; Camps-Valls et al. 2006; Pesaresi and Benediktsson 2001; Tuia et al. 2009a; Dalla Mura et al. 2010; Farag et al. 2005; Tarabalka et al. 2008; Unsalan and Boyer 2004; Shackelford and Davis 2003; Mott et al. 2002; Bruzzone and Carlin 2006; Chen et al. 2011)	Labeled training data are available Samples to be classified are drawn from the same domain of training data	Exploit available labeled samples for the learning of the classification algorithm
Semisupervised learning (Bennett and Demiriz 1998; Bruzzone et al. 2006; Dundar and Landgrebe 2004; Chi and Bruzzone 2007; Camps-Valls et al. 2007; Zhou et al. 2004; Gómez et al. 2008; Belkin et al. 2006; Chi and Bruzzone 2005, 2006; Bruzzone and Persello 2009; Li et al. 2011, 2012)	Small number of labeled training data is available The samples to be classified are drawn from the same domain of training data	Exploit both labeled and unlabeled samples for the learning of the classification algorithm
Active learning (Mitra et al. 2004; Rajan et al. 2008; Tuia et al. 2009b; Demir et al. 2011a; Patra and Bruzzone 2011; Liu et al. 2009a, b)	Small number of labeled training data is available The samples to be classified are drawn from the same domain of training data	Define an optimized training set according to an iterative procedure based on interactions between the classifier and supervisor
Learning under domain adaptation (Bruzzone and Fernandez Prieto 2001; Bruzzone and Fernandez Prieto 2002; Bruzzone and Cossu 2002; Bruzzone and Marconcini 2010; Bahirat et al. 2010; Persello and Bruzzone 2011; Demir et al. 2011b)	Labeled training data are available for the source domain The samples to be classified are drawn from the target domain (target domain$\neq$source domain)	Adapt the classifier trained on the source domain for the target domain by using SSL or AL methods

9.5 Semisupervised Learning Techniques for Classification of Remote Sensing Images

The semisupervised paradigm represents an alternative to the traditional supervised classification methods, which entirely base the design of the classifier decision rule on the available training data, assuming that the training data completely represent the classification problem according to the rule adopted by the classifier. However, this may be a strong assumption in RS, where often we have ill-posed problems in which both the quantity of training data available and their quality are critical issues. Unlike the supervised approaches, semisupervised methods exploit both training data and unlabeled samples in the learning phase of the classifier, and thus, under proper assumptions, outperform standard supervised techniques especially when few training samples are available.

Recently, in the machine learning community, a growing attention has been focused on semisupervised approaches implemented under the cluster assumption: this assumption states that each cluster of samples is assumed to belong to one data class; thus, the decision boundary between classes is defined between clusters, i.e., in low-density regions of the feature space. In this context, transductive SVMs (TSVMs) (Vapnik 1998, 1999) and semisupervised SVMs (S^3VMs) (Bennett and Demiriz 1998) proved particularly effective in several applications. In particular, TSVMs and S^3VMs exploit specific iterative algorithms based on SVMs which gradually search a reliable separation hyperplane (in the kernel space) through a learning process that incorporates both labeled and unlabeled samples in the training phase. In Bruzzone et al. (2006) the authors proposed a semisupervised classifier specifically designed for addressing ill-posed problems in the context of RS. In particular, this technique has the following properties: (1) it is based on a transductive procedure that exploits a weighting strategy for unlabeled patterns on the basis of a time-dependent criterion; (2) it is able to mitigate the effects of suboptimal model selection (which is unavoidable in the presence of small size training sets); and (3) it can address multiclass problems. A similar method that exploits, in place of SVMs, a modified version of the Kernel Fisher's discriminant using labeled and unlabeled data has been presented in Dundar and Landgrebe (2004). In particular, the proposed technique is obtained through an optimization of a quadratic programming problem that minimizes the total cost of misclassified labeled data while maximizing the Rayleigh coefficient in the kernel space. Other effective approaches have been presented in Chi and Bruzzone (2007), where the authors introduced in RS two different S^3VM algorithms for the classification of hyperspectral data implemented and optimized in the primal formulation. In this case the constraints of the labeled and unlabeled samples are directly included in the cost function in order to obtain an unconstrained optimization problem. The first presented primal S^3VM optimizes the unconstrained objective function by the gradient descent technique, leading to the formulation of ∇S^3VMs. The second algorithm combines ∇S^3VMs with a graph-based kernel matrix.

Another class of promising methods includes graph-based semisupervised algorithms, which define a graph where the nodes are labeled and unlabeled patterns and edges reflect the similarity of samples. Each sample spreads its label information to its neighbors until a global stable state is achieved on the whole data set. Graph-based approaches aim at estimating an objective function on the graph which generally consists of a loss term and a regularizer. In this context, an interesting approach has been proposed in Camps-Valls et al. (2007) where a graph-based classifier is presented for hyperspectral images, which is based on the algorithm described in Zhou et al. (2004). This algorithm takes advantage of both the high number of unlabeled samples present in the image and the integration of contextual information. Another technique has been proposed in Gómez et al. (2008). In particular, the authors extended to the RS domain the Laplacian SVM technique proposed in Belkin et al. (2006), which introduces an additional regularization term on the geometry of both labeled and unlabeled samples by using the graph Laplacian. This method follows a noniterative optimization procedure in contrast to most transductive learning methods and provides out-of-sample predictions in contrast to graph-based approaches.

In order to increase the reliability of the semisupervised learning process, systems based on ensemble methods have also been devised. As an example, in Chi and Bruzzone (2005, 2006), the employment of semilabeled-sample-driven bagging techniques is proposed. In Bruzzone and Persello (2009) a context-sensitive semisupervised SVM (CS^4VM) classifier is presented. This method addresses classification problems where the available training set is not fully reliable, i.e., some labeled samples may be associated to the wrong information class (mislabeled patterns). A soft sparse multinomial logistic regression model is presented in Li et al. (2011) for the semisupervised classification of hyperspectral images. This model exploits both hard and soft labels corresponding to labeled and unlabeled training samples, respectively. Thanks to the soft labels obtained by subspace-based multinomial logistic regression algorithm (Li et al. 2012), this strategy effectively models the phenomenon of mixed pixels present in hyperspectral images.

It is worth nothing that the effectiveness of the semisupervised methods depends on the two main assumptions: (i) the considered data should follow the cluster assumption (i.e., samples of different classes belong to different clusters in the feature space); and (ii) the initial training samples should not be too far from the correct representation of the considered classification problem. It is very intuitive to understand that if the first assumption is not verified the semisupervised method cannot work properly. The second assumption is critical in the sense that the initial training samples pose a bias on the part of the feature space where looking for low density. If these samples are too far from the real decision hyperplane, the risk is to find a correct solution from the theoretical viewpoint that does not match the property of the addressed problem. For these reasons, the use of semisupervised methods should be done after a careful evaluation of the properties of the considered problem and when the available training set is incomplete, yet not too biased.

9.6 Active Learning Techniques for Classification of Remote Sensing Images

A different approach to both enrich the information given as input to the supervised classifier and improve the statistic of the classes is AL. AL techniques iteratively expand the size of an initial labeled training set selecting the most informative samples from a pool of unlabeled samples for manual labeling. At each iteration, the most informative unlabeled samples (for a given classifier) are selected based on a query function, labeled by a supervisor and added to the current training set. Finally, the supervised classifier is retrained with the samples moved from the pool to the training set. It is worth noting that the initial training set requires few labeled samples for the first training of the classifier and then is enriched iteratively by including the most informative samples selected from a pool. At the convergence, the training set is made up of a minimum number of samples "optimal" for the considered classifier. When the AL process is completed, the classifier is trained once again and the classification of the image under investigation is carried out. The selection of the most informative samples from a pool to be included in the training set on the basis of AL offers three main advantages: (i) the labeling cost is reduced due to the avoidance of redundant samples, (ii) the computational complexity of the learning phase is reduced due to the selection of an optimal subset of training samples (i.e., a set with a small number of most representative samples), and (iii) accurate classification accuracy can be obtained due to the improved class models estimated on a high quality training set on the basis of the classification rule used from the considered classifier. The supervisor is usually a human expert who gives the true class labels to the selected samples. For RS classification problems, the labeling of both the initial training set and of queried samples can be obtained by: (1) in situ ground surveys, (2) image photointerpretation, or (3) hybrid solutions (both photointerpretation and ground surveys).

In the RS literature some AL techniques have been recently presented to optimize the training set. In Mitra et al. (2004), the unlabeled sample that is closest to the classification boundary (i.e., classification margin) of each binary SVM in a One-Against-All (OAA) multiclass architecture is considered as the most informative and therefore included in the current training set at each iteration of the AL process. An AL technique that selects the unlabeled sample that maximizes the information gain is presented in Rajan et al. (2008). To estimate the information gain, the Kullback–Leibler (KL) divergence is calculated between the posterior probability distribution of the current training set and the training set obtained by including each unlabeled sample into the training set. Two different AL techniques for multiclass RS classification problems are presented in Tuia et al. (2009b). In the first technique, the unlabeled samples that both have the smallest distance to the decision hyperplane of each binary SVM in an OAA multiclass architecture and do not share the same closest support vector are selected as uncertain and added to the training set. The second technique assesses the uncertainty on decisions of a committee of classifiers, i.e., uncertain samples are those having maximum disagreement between a committee of classifiers. Disagreement among the classifiers is measured by the entropy in the distribution of the labels provided by the committee members for each sample. In Demir et al. (2011a), different AL techniques proposed in the machine learning literature are investigated for the multiclass SVM classification problems, and also a novel AL method is proposed. The latter firstly selects the most informative unlabeled samples by the Multiclass-Level Uncertainty strategy. Then it analyzes their distribution by using the k-means clustering in the kernel space. Finally, the most informative (i.e., most uncertain) sample of each cluster is added to the training set at each iteration of AL. In Patra and Bruzzone (2011), a cluster assumption based AL method is presented for addressing critical problems where significantly biased initial training sets are available. Label acquisition costs sensitive AL techniques are proposed in Liu et al. (2009a, b). Differently from the standard AL techniques proposed in the RS literature, these techniques assume that the collection of the label of each sample has a non-uniform cost that depends on the label acquisition process. The cost of labeling each sample is modeled only according to the overall traveling distance for collecting the labels of all the samples, ignoring the accessibility of the samples.

From a general viewpoint, the use of AL techniques has not the drawbacks mentioned for the semisupervised techniques. AL generally works properly in any classification problem. If the initial training set is biased, the effect that we can obtain is to reach the convergence of the learning slowly, i.e. with a higher number of iterations and thus requiring the labeling of a higher number of training samples (empirical results reported in many papers point out that in any case the correct convergence can be generally achieved with a smaller number of samples than those collected in traditional passive learning). The main drawbacks of AL are related to the need to implement a training data collection strategy that includes the classifier in the process. This is relatively easy when the training set can be defined via photointerpretation. It becomes apparently more challenging when label collection is performed on the ground. In this case the strategy for collecting samples should be driven by the classifier. This can be achieved by connecting the people operating on the ground and their GPS coordinates with the computer running the classifier, which can be either a remote server connected via radio link or a laptop that can be used on the ground.

9.7 Domain Adaptation Techniques for Classification of Remote Sensing Images

Another important problem in the classification of RS data is to update land-cover maps by the analysis of the RS images. Land-cover maps updating is important due to the availability of increased numbers of images regularly acquired by satellite-borne sensors on the same areas at different times (i.e., time series of remotely sensed images). Because of the new policies related to free availability of data (e.g., Landsat archive, future ESA Sentinel missions) this issue is becoming more and more strategic as time series are accessible to each potential users in a systematic way. Land-cover maps can be updated by direct supervised classification of each image in the time series. However, such an approach requires reliable ground reference data for all the available temporal images in order to properly train the classifier. In operational scenarios, as mentioned before, gathering a sufficient number of labeled training samples for each single image to be classified is not realistic due to the high cost and the related time consuming process of this task. Moreover, although the images in the time series refer to the same area, ground reference samples available for one of the images may not follow the same distribution in other acquired images due to several reasons, such as differences in the atmospheric conditions at the image acquisition dates, different acquisition system state, different levels of soil moisture, changes occurred on the ground, etc. In these situations, exploiting the classifier trained on the image for which training data are available may result in poor classification performance, and therefore recollection of labeled samples is necessary. To reduce the need and effort to recollect labeled samples, it is desirable to reuse the already available information on images acquired on the same area of interest (source domain) to classify new images acquired on the same area (target domain).

To deal with this problem, transfer learning (TL) techniques, and more in detail DA methods in TL, have been recently introduced in the RS literature (Bruzzone and Fernandez Prieto 2001; Demir et al. 2011b). DA is known also as partially supervised/unsupervised learning and is addressed with SSL or AL methods. On the one hand, SSL applies a classifier trained on the source domain to the target domain after tuning the parameters according to unlabeled data from the target domain (Bruzzone and Fernandez Prieto 2001; Bahirat et al. 2010). In other words, the information of reference training samples from the source domain is improved by costless unlabeled samples from the target domain to obtain a reliable classifier for the target domain. On the other hand, AL methods aim at improving (from the target domain point of view) the information of the source domain reference samples by iteratively adding samples selected from the target domain (Rajan et al. 2008; Persello and Bruzzone 2011; Demir et al. 2011b). Before inclusion in the training set, these samples should be manually labeled by a human expert, thus these methods have associated a cost that SSL techniques do not have. However AL methods try to reduce it by labeling the smallest possible number of unlabeled samples. This is achieved by selecting for labeling those samples that are the most informative from the target domain viewpoint, thus avoiding the huge cost of collecting large amount of labeled samples. In the literature examples of DA methods based on both SSL and AL are available. For example, in Bruzzone and Fernandez Prieto (2001) DA problems are addressed with SSL by updating on the basis of the distribution of the target domain the parameters of a parametric ML classifier already trained on the source domain. This method has been generalized in the context of the Bayes rule for cascade classification in order to exploit the temporal correlation between domains (Bruzzone and Fernandez Prieto 2002). Further improvements of this method are presented in Bruzzone and Cossu

(2002) and Bruzzone and Marconcini (2010). A multiple cascade classifier system is proposed in Bruzzone and Cossu (2002) including ML and neural-network classifiers, whereas differences on the set of land-cover classes between the domains are addressed by the joint use of an unsupervised change-detection method and of the Jeffreys-Matusita (JM) statistical distance measure in Bruzzone and Marconcini (2010). Another DA method with SSL based on an SVM classifier is described in Bahirat et al. (2010). In this work, the discriminant function for the target domain is initialized by the labeled samples of the source domain. Then the unlabeled patterns of the target domain that have a high probability to be correctly classified are iteratively included in the training set, whereas the labeled samples of the source domain are gradually removed.

In Rajan et al. (2008), Persello and Bruzzone (2011), and Demir et al. (2011b) DA problems are addressed with AL and thus, unlike (Bruzzone and Fernandez Prieto 2001; Demir et al. 2011b), a small number of labeled training samples is collected from the target domain together with the labeled samples of the source domain. In Rajan et al. (2008), the classifier parameters are initialized by the distributions estimated on the labeled samples of the source domain. Then the unlabeled samples of the target domain that have the maximum information gain (measured by the KL divergence) are included in the training set of the target domain after manual labeling. In Persello and Bruzzone (2011), the statistical parameters of a ML classifier are initialized by exploiting the labeled samples of the source domain, and then the most informative samples are iteratively selected from the target domain by AL to be added to the training set after manual labeling. In this method, during the AL process, the source domain samples that do not fit with the distribution of the classes in the target domain are removed. An AL technique developed in the context of the cascade classification of multitemporal RS images for updating land-cover maps is presented in Demir et al. (2011b). The proposed AL technique is based on the selection of unlabeled samples that have maximum uncertainty on their labels assigned by cascade classification, and explicitly exploits temporal correlation between multitemporal images by conditional entropy.

9.8 Discussion and Conclusion

In this chapter, we presented a literature survey on the most recent classification techniques for the analysis of multispectral and hyperspectral RS images. We observed that differences in both the type of classification problem and the type of RS data require the development of specific data analysis methods.

In the context of classification of hyperspectral RS data, techniques that are both robust to the Hughes phenomenon and to the nonstationary behavior of the spectral signatures of land-cover classes are required. In order to address the above-mentioned problems, kernel-based methods, particularly SVMs and RVMs, are found very effective in the RS literature. The most appealing properties of SVMs and RVMs for the classification of hyperspectral RS data are their high generalization capability and robustness to the Hughes affect (which allow SVMs and

RVMs to operate in large dimensional feature spaces with few training samples), and their sparse solutions (i.e., the solution is expressed as a function only of the most critical training samples in the distribution).

In the context of classification of VHR RS data, the exploitation of spatial information together with the spectral information is found to be very beneficial. To this end, several methods defined by exploiting composite kernels, morphological filters, Markov random fields, segmentation algorithms and region growing methods have been presented in the RS literature for the joint use of spectral and spatial information.

We also discussed the recent trends for the automatic classification of RS images. In particular, we focused on semisupervised learning (SSL), active learning (AL) and domain adaptation (DA). These approaches allow one to address classification problems in the critical conditions where the available labeled training samples are limited. These operational conditions are very common in RS data classification problems, due to the high cost and time related to the collection of labeled samples. We observed that semisupervised techniques can improve the classification accuracies obtained by supervised classifiers (thanks to using both labeled and unlabeled samples in the learning of the classification algorithm) depending on both: (1) the quality of the initial training set; and (2) the use of the data for which the cluster assumption is hold. We also stated that, unlike the semisupervised techniques, AL techniques can converge to good classification accuracies starting from any initial training set and without requiring the cluster assumption, thus resulting in a very useful tool for the collection of informative labeled samples in real RS problems. We also observed that DA methods are very promising for low-cost updating land-cover maps by using RS images periodically acquired on the same investigated area. These technique are developed under two assumptions: (i) the set of land-cover classes that characterizes the target domain should be the same as those included in the source domain, and (ii) the land-cover class statistical distributions should be sufficiently correlated (but not necessarily identical) between the domains.

Despite the promising developments discussed in this chapter, it is still necessary to develop more advanced classification methods that can efficiently extract information from the complex data acquired by the last generation of RS sensors. Furthermore, the above-mentioned recent developments, such as AL and DA, have some critical limitations, and thus need to be improved for operational RS scenarios. For example, most of the AL methods presented in the RS literature assume that the costs of labeling samples are equal to each other (i.e., uniform cost). However, the labeling cost of each sample in the case of ground surveys depends on both the accessibility of the sample and the traveling distance to reach the related locations. To deal with this problem, it is necessary to define labeling cost-sensitive AL methods. Furthermore, as mentioned before, DA methods are defined on the basis of two critical assumptions. However, in some real RS classification problems these assumptions could not be satisfied due to (i) the possible appearance and/or disappearance of the land-cover classes during time, and (ii) the possible high differences on the class statistical distributions in the image time series. Thus, it is necessary to study DA methods that can mitigate the aforementioned problems.

References

Bahirat K, Bovolo F, Bruzzone L, Chaudhuri S (2010) A novel domain adaptation maximum likelihood classifier for updating land-cover maps in complex scenarios. In: Proceedings of the SPIE conference on image and signal processing for remote sensing, Toulouse, France

Bandyopadhyay S, Pal SK (2001) Pixel classification using variable string genetic algorithms with chromosome differentiation. IEEE Trans Geosci Remote Sens 39(2):303–308

Belkin M, Niyogi P, Sindhwani V (2006) Manifold regularization: a geometric framework for learning from labeled and unlabeled examples. J Mach Learn Res 7:2399–2434

Bennett KP, Demiriz A (1998) Semi-supervised support vector machines. Proc Adv Neural Inform Process Syst 10:368–374

Binaghi E, Gallo I, Pepe M (2003) A neural adaptive model for feature extraction and recognition in high resolution remote sensing imagery. Int J Remote Sens 24(20):3947–3959

Bischof H, Leona A (1998) Finding optimal neural networks for land use classification. IEEE Trans Geosci Remote Sens 36(1):337–341

Boser BE, Guyon IM, Vapnik VN (1992) A training algorithm for optimal margin classifiers. In: Haussler D (ed) Proceedings of the in 5th annual ACM workshop on COLT, Pittsburgh, PA, pp 144–152

Bruzzone L, Carlin L (2006) A multilevel context-based system for classification of very high spatial resolution images. IEEE Trans Geosci Remote Sens 44(9):2587–2600

Bruzzone L, Cossu R (2002) A multiple cascade-classifier system for a robust a partially unsupervised updating of land-cover maps. IEEE Trans Geosci Remote Sens 40(9):1984–1996

Bruzzone L, Fernández-Prieto D (1999) A technique for the selection of kernel-function parameters in RBF neural networks for classification of remote-sensing images. IEEE Trans Geosci Remote Sens 37(2):1179–1184

Bruzzone L, Fernandez Prieto D (2001) Unsupervised retraining of a maximum-likelihood classifier for the analysis of multitemporal remote-sensing images. IEEE Trans Geosci Remote Sens 39(2):456–460

Bruzzone L, Fernandez Prieto D (2002) A partially unsupervised cascade classifier for the analysis of multitemporal remote-sensing images. Pattern Recogn Lett 23(9):1063–1071

Bruzzone L, Marconcini M (2010) Domain adaptation problems: a DASVM classification technique and a circular validation strategy. IEEE Trans Pattern Anal Mach Intell 32(5):770–787

Bruzzone L, Persello C (2009) A novel context-sensitive semisupervised SVM classifier robust to mislabeled training samples. IEEE Trans Geosci Remote Sens 47(7):2142–2154

Bruzzone L, Chi M, Marconcini M (2006) A novel transductive SVM for semisupervised classification of remote-sensing images. IEEE Trans Geosci Remote Sens 44(11):3363–3373

Camps-Valls G, Bruzzone L (2005) Kernel-based methods for hyperspectral image classification. IEEE Trans Geosci Remote Sens 43(6):1351–1362

Camps-Valls G, Gómez-Chova L, Calpe J, Soria E, Martín JD, Alonso L, Moreno J (2004) Robust support vector method for hyperspectral data classification and knowledge discovery. IEEE Trans Geosci Remote Sens 42(7):1530–1542

Camps-Valls G, Gomez-Chova L, Munoz-Mari J, Vila-Frances J, Calpe-Maravilla J (2006) Composite kernels for hyperspectral image classification. IEEE Geosci Remote Sens Lett 3 (1):93–97

Camps-Valls G, Bandos T, Zhou D (2007) Semi-supervised graph-based hyperspectral image classification. IEEE Trans Geosci Remote Sens 45(10):3044–3054

Carleer A, Debeir O, Wolff E (2004) Comparison of very high spatial resolution satellite image segmentation. In: Proceedings of SPIE conference image and signal processing remote sensing, vol 5238. Bellingham, Washington, USA, IX, pp 532–542

Chen X, Fang T, Huo H, Li D (2011) Graph-based feature selection for object-oriented classification in VHR airborne imagery. IEEE Trans Geosci Remote Sens 49(1):353–365

Chi M, Bruzzone L (2005) A semilabeled-sample-driven bagging technique for ill-posed classification problems. IEEE Geosci Remote Sens Lett 2(1):69–73

Chi M, Bruzzone L (2006) An ensemble-driven k-NN approach to ill-posed classification problems. Pattern Recognit Lett 27(4):301–307
Chi M, Bruzzone L (2007) Semisupervised classification of hyperspectral images by SVMs optimized in the primal. IEEE Trans Geosci Remote Sens 45(6):1870–1880
Cristianini N, Shawe-Taylor J (2000) An introduction to support vector machines. Cambridge University Press, Cambridge
Dalla Mura M, Benediktsson JA, Waske B, Bruzzone L (2010) Morphological attribute profiles for the analysis of very high resolution images. IEEE Trans Geosci Remote Sens 48(10):3747–3762
Demir B, Ertürk S (2007) Hyperspectral image classification using relevance vector machines. IEEE Geosci Remote Sens Lett 4(4):586–590
Demir B, Persello C, Bruzzone L (2011a) Batch mode active learning methods for the interactive classification of remote sensing images. IEEE Trans Geosci Remote Sens 49(3):1014–1031
Demir B, Bovolo F, Bruzzone L (2011b) Active-learning based cascade classification of multitemporal images for updating land-cover maps. In: Proceedings of the 6th international workshop on the analysis of multi-temporal remote sensing images, Trento, Italy, pp 57–60, July 2011
Duda RO, Hart PE, Stork DG (2001) Pattern classification, 2nd edn. Wiley, New York
Dundar MM, Landgrebe DA (2004) A cost-effective semisupervised classifier approach with kernels. IEEE Trans Geosci Remote Sens 42(1):264–270
Farag A, Mohamed R, El-Baz A (2005) A unified framework for map estimation in remote sensing image segmentation. IEEE Trans Geosci Remote Sens 43(7):1617–1634
Giacinto G, Bruzzone L (2000) Combination of neural and statistical algorithms for supervised classification of remote-sensing images. Pattern Recognit Lett 21(5):399–405
Gómez L, Camps-Valls G, Muñoz-Marí J, Calpe-Maravilla J (2008) Semisupervised image classification with Laplacian support vector machines. IEEE Geosci Remote Sens Lett 5(3):336–340
Gualtieri JA, Chettri SR, Cromp RF, Johnson LF (1999) Support vector machine classifiers as applied to AVIRIS data. In: Summaries 8th JPL airborne earth science workshop, JPL Publication. Pasadena, California, USA, 99–17, pp 217–227
Haykin S (1999) Neural networks: a comprehensive foundation. Prentice Hall, Englewood Cliffs
Huang C, Davis LS, Townshend JRG (2002) An assessment of support vector machines for land cover classification. Int J Remote Sens 23(4):725–749
Hughes GF (1968) On the mean accuracy of statistical pattern recognizers. IEEE Trans Inf Theory IT-14(1):55–63
Kruse FA, Lefkoff AB, Boardman JB, Heidebrecht KB, Shapiro AT, Barloon PJ, Goetz AFH (1993) The spectral image processing system (SIPS)—interactive visualization and analysis of imaging spectrometer data. Remote Sens Environ 44(2/3):145–163
Li J, Bioucas-Dias JM, Plaza A (2011) Semi-supervised hyperspectral image classification using a new (soft) sparse multinomial logistic regression model In: Proceedings of the 3rd workshop on hyperspectral image and signal processing: evolution in remote sensing, Lisbon
Li J, Bioucas-Dias J, Plaza A (2012) Spectral-spatial hyperspectral image segmentation using subspace multinomial logistic regression and Markov random fields. IEEE Trans Geosci Remote Sens 50(3):809–823
Liu X, Li X, Liu L, He J (2008) An innovative method to classify remote sensing image using ant colony optimization. IEEE Trans Geosci Remote Sens 46(12):4198–4208
Liu A, Jun G, Ghosh J (2009a) Spatially cost-sensitive active learning. In: SIAM International Conference on Data Mining (SDM), Sparks, Nevada, USA, pp 814–825
Liu A, Jun G, Ghosh J (2009b) Active learning of hyperspectral data with spatially dependent label acquisition costs. In: Proceedings of the IEEE international geoscience and remote sensing symposium, Cape Town, South Africa, pp V-256–V-259
Melgani F, Bruzzone L (2004) Classification of hyperspectral remote sensing images with support vector machines. IEEE Trans Geosci Remote Sens 42(8):1778–1790

Mika S, Rätsch G, Schölkopf B, Smola A, Weston J, Müller K-R (1999) Invariant feature extraction and classification in kernel spaces. In: Advances in neural information processing systems, vol 12. MIT Press, Cambridge, MA

Mitra P, Shankar BU, Pal SK (2004) Segmentation of multispectral remote sensing images using active support vector machines. Pattern Recognit Lett 25(9):1067–1074

Mott C, Andresen T, Zimmermann S, Schneider T, Ammer U (2002) Selective region growing-an approach based on object oriented classification routine. Proc Int Geosci Remote Sens Symp 3:1612–1614

Murat Dundar M, Landgrebe A (2004) A cost-effective semisupervised classifier approach with kernels. IEEE Trans Geosci Remote Sens 42(1):264–270

Patra S, Bruzzone L (2011) A fast cluster-based active learning technique for classification of remote sensing images. IEEE Trans Geosci Remote Sens 49(5):1617–1626

Persello C, Bruzzone L (2011) A novel active learning strategy for domain adaptation in the classification of remote sensing images. In: Proceedings of the IEEE international geoscience and remote sensing symposium, Vancouver, Canada, pp 3720–3723

Pesaresi M, Benediktsson JA (2001) A new approach for the morphological segmentation of high-resolution satellite imagery. IEEE Trans Geosci Remote Sens 39(2):309–320

Rajan S, Ghosh J, Crawford MM (2008) An active learning approach to hyperspectral data classification. IEEE Trans Geosci Remote Sens 46(4):1231–1242

Samaniego L, Bárdossy A, Schulz K (2008) Supervised classification of remotely sensed imagery using a modified k-NN technique. IEEE Trans Geosci Remote Sens 46(7):2112–2125

Schölkopf B, Smola A (2001) Learning With kernels—support vector machines, regularization, optimization and beyond. MIT Press, Cambridge, MA

Schowengerdt RA (2002) Remote sensing. Models and methods for image processing, 2nd edn. Academic, Norwell

Shackelford AK, Davis CH (2003) A combined fuzzy pixel-based and object-based approach for classification of high-resolution multispectral data over urban areas. IEEE Trans Geosci Remote Sens 41(10):2354–2363

Tarabalka Y, Chanussot J, Benediktsson JA, Angulo J, Fauvel M (2008) Segmentation and classification of hyperspectral data using watershed. In: Proceedings of the IGARSS, Boston, MA, pp III-652–III-655

Tuia D, Pacifici F, Kanevski M, Emery W (2009a) Classification of very high spatial resolution imagery using mathematical morphology and support vector machines. IEEE Trans Geosci Remote Sens 47(11):3866–3879

Tuia D, Ratle F, Pacifici F, Kanevski M, Emery WJ (2009b) Active learning methods for remote sensing image classification. IEEE Trans Geosci Remote Sens 47(7):2218–2232

Unsalan C, Boyer KL (2004) Classifying land development in high resolution panchromatic satellite images using straight-line statistics. IEEE Trans Geosci Remote Sens 42(4):907–919

van de Vlag DE, Stein A (2007) Incorporating uncertainty via hierarchical classification using fuzzy decision trees. IEEE Trans Geosci Remote Sens 45(1):237–245

Vapnik VN (1998) Statistical learning theory. Wiley, New York

Vapnik VN (1999) The nature of statistical learning theory, 2nd edn. Springer, Berlin

Yang H, van der Meer F, Bakker W, Tan ZJ (1999) A back-propagation neural network for mineralogical mapping from AVIRIS data. Int J Remote Sens 20(1):97–110

Zhou D, Huang J, Schölkopf B (2004) Learning with local and global consistency. In: Thrun S, Thrun S, Saul L, Schölkopf B (eds) Advances in neural information processing system, vol 16. MIT Press, Cambridge, MA, pp 321–328

Chapter 10
Recent Advances in Remote Sensing Change Detection – A Review

Antje Hecheltjen, Frank Thonfeld, and Gunter Menz

10.1 Introduction

In one of the most cited reviews, change detection is defined as "… the process of identifying differences in the state of an object or phenomenon by observing it at different times" (Singh 1989). Change detection can be seen as a processing chain encompassing several partly interlinked and overlapping steps: pre-processing, change extraction (CE), thresholding, change labeling, and accuracy assessment (Fig. 10.1). Selected input data (described here in Sect. 10.2) are first pre-processed; typically encompassing radiometric as well as geometric preparation (Sect. 10.3). Change extraction describes the algorithm applied to detect changes in the data sets (Sect. 10.4). The separation of changed and unchanged pixels is frequently achieved by applying thresholds to the change extraction result. The term "change labeling" was first used by Johnson and Kasischke in 1998 to distinguish it from "change detection" as a functionally and procedurally distinct process. Current prevalent needs extent beyond knowing simply that change has occurred; information is required regarding what has changed and how, and this has driven important developments in the process of change labeling in recent years.

Coppin et al. 2004 differentiate between change extraction (i.e. the change detection algorithm) and change labeling. Equating change labeling with change classification routines, they decided to limit their review to change detection algorithms only. Kennedy et al. (2009) presented a broader perspective of the change labeling process. They distinguished between two general methods of deriving labeled change maps, both employing differencing functions. The first method applied a mapping or classification function over the two input images separately; both of the classifications or continuous estimates were then compared through differencing, resulting in a change map. This is part of what we term

A. Hecheltjen (✉) • F. Thonfeld • G. Menz
Center for Remote Sensing of Land Surfaces (ZFL), University of Bonn,
Walter-Flex-Street 3, 53113, Bonn, Germany
e-mail: a.hecheltjen@uni-bonn.de; frank.thonfeld@uni-bonn.de; g.menz@uni-bonn.de

I. Manakos and M. Braun (eds.), *Land Use and Land Cover Mapping in Europe: Practices & Trends*, Remote Sensing and Digital Image Processing 18,
DOI 10.1007/978-94-007-7969-3_10,

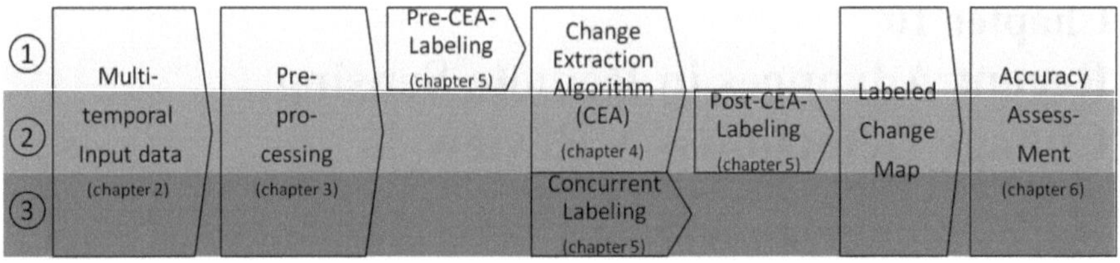

Fig. 10.1 The change detection processing chain – three general options to derive labeled change maps (*1*: pre-CE-labeling; *2*: post-CE-labeling; *3*: concurrent labeling)

"*pre-CE-labeling*". The second approach involved spectral differencing of the two input images directly; labeling is performed through applying a function to the difference image. This is part of what we call "*post-CE-labeling*". The authors mention three exceptions to this general scheme: (1) maintaining the existing classification for the unchanged pixels and applying classification models to label the changed pixels only; (2) direct change classification on a bi-temporal layer stack, and; (3) time series analysis.

Our review follows a similar broad understanding of change labeling. We define change labeling as the process of identifying and naming various types of changes. In its simplest form, the change extraction algorithm (CEA) results in a binary discrimination of the change data in which changed and unchanged pixels are identified. More complex procedures result in maps of land-use/land-cover conversions shown as from/to classes, or derive land-use/land-cover modifications indicated as increases or decreases of an index value. We distinguish the different change labeling operations according to their position in the change detection processing chain, referring to these operations respectively as pre-CE-labeling (Fig. 10.1, first row), post-CE-labeling (Fig. 10.1, second row), or concurrent labeling (Fig. 10.1, third row). Change labeling and its linkages with CEA are described in Sect. 10.5. Result of the change labeling step is a change detection map, the accuracy of which must then be assessed (Sect. 10.6).

Few of the existing change detection reviews attempt to consider SAR and optical data together (Milne 1988; Singh 1989; Mouat et al. 1993; Rignot and Van Zyl 1993; Polidori et al. 1995; Coppin et al. 2004; Lu et al. 2004; Radke et al. 2005). Thonfeld et al. (2010) is an exception. SAR sensor data have unique utility in environmental monitoring purposes due to their independence from solar illumination and ability to penetrate clouds. SAR data, however, have not been equally considered or utilized for change detection applications. Reasons for this include limited data availability and increased data costs, or limited processing capabilities. Although SAR data typically include a lower number of sensor bands (corresponding to the number of polarizations), multitemporal SAR data and derived information are increasingly being used in two principal ways: (1) to improve the description of temporally variable classes for classification analyses (Bruzzone et al. 2004; Quegan et al. 2000), and; (2) to assess the evolution of dynamic environments without necessarily classifying an image (e.g. Shepherd et al. 2001).

Typical applications of SAR including the use of differential interferometry (InSAR) to detect ground deformation due to earthquakes (Massonnet et al. 1993), subsidence (Strozzi et al. 2003) or slope instabilities (Strozzi et al. 2005), or for velocity measurements of glaciers and ice caps are usually not considered as change detection techniques. This paradigm seems to be generally accepted in the field of land cover change detection. These analyses are, however, used to detect changes on the earth's surface and are sometimes applied to land cover change detection. It therefore seems reasonable to integrate SAR methods in this change detection review.

10.2 Input Data

In many cases the purpose of a study determines the choice of input data. If long term monitoring is the objective, only few data sources are available; Landsat, Spot, AVHRR are the most widely used. If satellite imagery is required for short term analyses (e.g. emergency response), the user has to take advantage of all available information and select the most appropriate and useful contemporary high spatial and temporal resolution data. The choice of an appropriate image spectral or polarization band mix also depends on the application. Thus, input data sets – spectral bands or indices – and the selection of adequate, appropriate analytic methodologies are driven by the particular application. Since many radar systems provide only one (e.g. ERS-1/2) or two (e.g. ENVISAT ASAR) polarizations, SAR methods are frequently limited to very few input bands. However, for some algorithms the user can decide which band(s) to select, and functions such as image differencing, ratioing, regression or texture can be calculated for each band. In many studies, only one optimal band is selected (Ridd and Liu 1998). Principal Components Analysis (PCA) may be applied on all change bands to reduce data dimensionality (Jha and Unni 1994). Other change detection methods such as Change Vector Analysis (CVA) (Malila 1980), Iteratively Re-Weighted Multivariate Alteration Detection (IR-MAD) (Canty and Nielsen 2008), or PCA are usually applied to the whole data layer stack.

10.3 Preprocessing: Radiometric and Geometric Requirements

10.3.1 Radiometric Requirements

Many change detection and classification approaches do not require absolute atmospheric correction (Song et al. 2001). However, the data should represent similar atmospheric conditions and this can be achieved by relative atmospheric correction or radiometric normalization. Early relative atmospheric correction methods used pseudo-invariant features (PIF) identified in both images (Schott et al. 1988). State-of-the-art algorithms such as IR-MAD function completely automatically and do not require any knowledge of the acquisition conditions or of the study area. Several studies have analyzed the influence of radiometric correction on change detection accuracy (Collins and Woodcock 1996; Song et al. 2001; Schroeder et al. 2006) and all recommend applying radiometric correction. However, they also conclude that relative atmospheric correction performs as well as absolute correction and is usually easier to implement. Thus, although absolute atmospheric correction is not required in many cases, it is required if spectral libraries or field spectra are included in the analysis, and it is necessary to analyze spectral reflectances. But sun angle effects such as differing shadow lengths and incident illumination characteristics as well as additional variables such as different atmospheric conditions and sensor performance have to be corrected by adjusting the target images to a reference image, i.e. relative radiometric normalization. Vicente-Serrano et al. (2008) assessed different preprocessing methods for land cover change detection and suggested implementation of a sequence including cross-sensor calibration, absolute radiometric correction, topographic correction and automatic relative radiometric normalization when vegetation trends are analyzed. Relative normalization methods have to be applied following absolute methods since non-surface related noise must be harmonized. For the analysis of abrupt vegetation changes the use of Top of Atmosphere (TOA) reflectances and accurate relative radiometric normalization are sufficient to achieve useful results (Vicente-Serrano et al. 2008). However, these researchers did not include IR-MAD in their study. Since IR-MAD is based on linear regression, cross-sensor calibration using empirically derived slope and intercept parameters is redundant and may be omitted (Vogelmann et al. 2001). In recent change detection studies based on multiple images, one image is atmospherically corrected using an absolute procedure (e.g., Second Simulation of the Satellite Signal in the Solar Spectrum, or "6S") and all the other images are subsequently normalized to that master image using a relative method (e.g. IR-MAD) (Coops et al. 2010; Griffiths et al. 2012; Powell et al. 2008). For time series analysis, it is more important to have consistent surface reflectance measurements rather than true surface reflectance (Schroeder et al. 2006). This consistency is referred to as "common radiometric scale" (Song et al. 2001).

Cloud cover remains a challenge for change detection studies based on optical imagery. Cloud free scenes must be selected for bi-temporal change detection and trajectory analysis. Analysts sometimes have to employ images out of season. These data do not display the same acquisition conditions of sun angles or vegetation phenology, and additional and mostly irrelevant variations in the data are introduced. In a typical time series analysis, virtually every image can be used because in many cases the entire area of interest is not obscured by clouds, and those areas with cloud cover are likely to be unobscured on nearby images. For many regions in the world, it is nearly impossible to construct cloud free time series, and clouds and cloud shadows must be masked in each scene. Consequently, each pixel has its own time series, with all cloud and cloud shadow pixels as well as any missing values (e.g. gaps in Landsat ETM+ SLC-off images) being masked out (Zhu et al. 2012). Assuming a perfect masking function, cloud cover, cloud shadows and missing values are minor problems in time series analysis. Two powerful software packages have been recently developed for the systematic and automatic radiometric preprocessing of TM and ETM+ Landsat images: (1) the Landsat Ecosystem Disturbance Adaptive Processing System (LEDAPS) (Masek et al. 2006) performs image calibration to TOA reflectance, cloud-screening, and atmospheric correction to surface reflectance using the MODIS/6S approach, and; (2) Fmask (Function of mask), a tool for cloud and cloud shadow detection (Zhu and Woodcock 2012). Other cloud detection methods include the Automated Cloud Cover Assessment (ACCA) (Irish 2000; Irish et al. 2006) and the method by (Huang et al. 2010b).

10.3.2 Geometric Requirements

Precise geocoding and registration of images is mandatory to achieve reliable change detection results (Townshend et al. 1992). Even small inaccuracies may result in pseudo-changes – changes which are due to image misalignment and other artifacts rather than real changes. Thus, methods have been developed to reduce the effects of registration noise in remote sensing change detection (Gong et al. 1992; Bruzzone and Cossu 2003; Stow 1999). Assuming perfect co-registration of images, there may be pseudo-changes caused by differences in acquisition geometries; different sun angles may cause different shadow proportions or different viewing angles may result in distortions. Sun angle effects may be reduced by selecting images acquired under similar sun angle conditions; distortions due to viewing angle variation can be reduced by considering the pixel neighborhood during the change detection process (Castilla et al. 2009). The latter effects are more pronounced in high spatial resolution imagery acquired by sensors with off-nadir capabilities (e.g. Ikonos, Quickbird, Formosat-2, and RapidEye) and in areas of distinct relief such as urban areas.

Misregistration and differences in viewing geometries also affect SAR images (Gamba et al. 2006), although modern SAR satellite systems such as TerraSAR-X achieve subpixel accuracies, corresponding to errors of less than a meter (Yoon et al. 2009). Several approaches have been developed to mitigate geometric constraints; Gamba et al. (2006) describe a combined pixel and feature based change detection procedure.

10.4 Change Extraction Algorithms

Many change extraction algorithms were developed in the 1970s and 1980s (Table 10.1). With increasing computing power, growing image archives and the emergence of new sensors the number of new methods increased as well, and the

Table 10.1 Genealogy of change detection algorithms. (Green: algorithms for optical data; yellow: algorithms for SAR data; blue: algorithms for both, optical and SAR data). Only major innovations are shown

	1970s	1980s	1990s	2000s	2010s
Classification	Post-classification comparison (PCC)		Artificial neural networks (ANN)	Object-based methods	CMFDA
			Spectral mixture analysis (SMA)	Multisource PCC	
				Machine learning (e.g. SVM)	
				Change-restricted PCC	
Transformation		Kauth-Thomas Tassled Cap	Multivariate alteration detection/ maximum autocorrelation factor (MAD/MAF)	Kernel methods	Fast Fourier transformation
		Principal component analysis (PCA)	Iterative PCA	Object-based methods	Curvelets
			Iterative MAD	Wavelets	
			GrammSchmidt		
Algebra	Differencing	Vegetation index differencing	Coherence	Object-based methods	Robust image differencing (LCM)
	Ratioing	Regression	Differential SAR Interferometry (DInSAR)	Kernel methods	
		Change vector analysis (CVA)	Speckle tracking	Persistent scatterer interferometry	
		Differential radargrammetry	Speckle decorrelation	Multilevel CVA	
				Texture-based algebra	
				Offset tracking/ feature tracking	
Time series analysis			Trajectory analysis	Time series segmentation	BFAST

first comprehensive technical reviews soon followed (Milne 1988; Singh 1989). Newer sensors such as IKONOS, MERIS or TerraSAR-X have completely different spatial, spectral, radiometric and temporal resolution characteristics. New change extraction methods were required to take advantage of the improved sensor data and updated comprehensive reviews have been carried out to summarize major findings within the field of remote sensing change detection (Coppin et al. 2004; Lu et al. 2004; Radke et al. 2005).

A suite of well-known and frequently used change extraction algorithms was developed rather early in remote sensing history and many enhancements and further developments of existing methods were generated. The aim of Sect. 10.4 is to review the major change extraction techniques irrespective of application and data requirements. We have adopted the nomenclature for this discussion in part from Lu et al. (2004), in which each change extraction algorithm is assigned to one of four categories (Table 10.1):

1. *Image algebra*, which leads to change extraction based on spectral values, backscatter values, indices, texture features and related properties.
2. *Transformation* based change extraction uses transformed images properties such as principal components (PC).
3. *Classification* based approaches – there are several different strategies to use classification for change extraction, but the results of all of these CE techniques is a classified image. We review classification based CE strategies in Sect. 10.4.3.

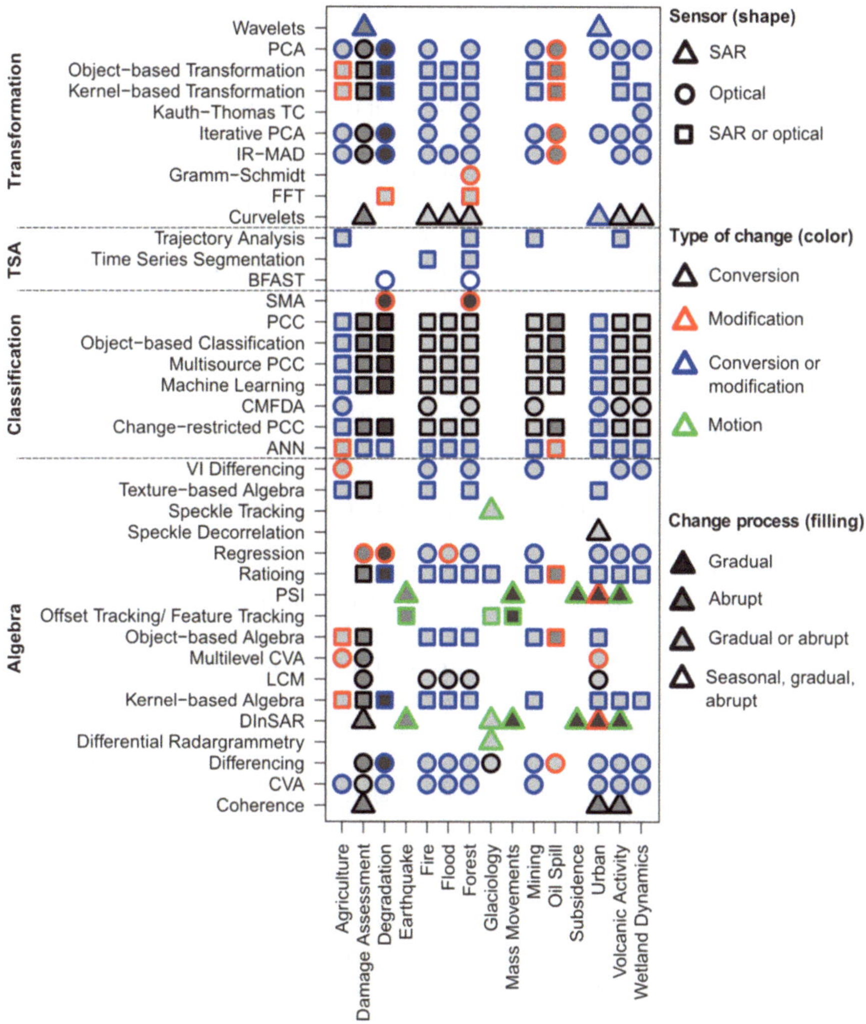

Fig. 10.2 Matrix of change detection algorithms (Shape of symbols refers to sensor, i.e. *triangle*: SAR, *circle*: optical, *square*: SAR and optical; outline color of symbols refers to type of change, i.e. *black*: conversion, *red*: modification, *blue*: conversion or modification, *green*: motion; fill color refers to change process, i.e. *very dark grey*: gradual, *dark grey*: abrupt, *light grey*: gradual or abrupt; *white*: seasonal, gradual and abrupt)

4. *Time series analysis*, which can be defined as CE process using more than two time steps. Clear distinctions are sometimes complicated, but we normally refer to time series analysis when trends are analyzed, frequently in conjunction with the detection of seasonal and abrupt changes.

In Fig. 10.2 change detection algorithms and selected fields of application are shown. Some of the algorithms are suitable for use with either SAR or optical image data only; others can be applied to any kind of input data. In the same manner, certain algorithms are suited for certain change types, i.e. conversion, modification, conversion or modification, or motion. Some of the algorithms are able to capture the character of a change process, thereby separating abrupt from gradual changes. Others are also able to detect seasonal processes. Finally, each algorithm can be assigned to one of the categories of algebra, classification, time series analysis (TSA) or transformation. The transformation categorization is not important for the understanding of change. Advantages and disadvantages as well as performance of each CEA are dependent on a number of factors including research goal, region of interest, and data availability. These factors are not shown in this figure.

10.4.1 Algebra Methods

10.4.1.1 Image Differencing and Vegetation Index Differencing

Image differencing was introduced in the 1970s (Weismiller et al. 1977) and is one of the earliest change detection algorithms. It is sometimes referred to as delta change detection (Weismiller et al. 1977). The technique includes the subtraction of a date one image from a coregistered second date image. Image differencing is frequently used as it is easy to apply and interpretation of results is straightforward. The image data layers may consist of original reflectance or radiance values, spectral indices or texture values. Vegetation index (VI) differencing (Nelson 1983) is one of the most frequently used change detection algorithms (Thonfeld et al. 2010) and can be seen as image differencing applied on vegetation indices. Due to their sensitivity to vegetation condition and abundance, VI differencing is usually applied to detect changes related to vegetation. Due to data gamma-distribution, ratioing techniques rather than image differencing should be applied to SAR data (Rignot and Van Zyl 1993). An effective way to visualize the results of an image differencing analysis is a RGB composite with the difference image in the red channel, the t_2 image in green and the t_1 image in blue (Castilla et al. 2009).

10.4.1.2 Ratioing

Although image ratioing is one of the earliest change detection algorithms, it is rarely used for optical images (Howarth and Wickware 1981). However, ratioing is better adapted to the statistical properties of SAR images than image differencing (Rignot and Van Zyl 1993). In ratio images, no-change areas are represented by values close to one whereas changes have either high values or values close to zero. Due to the non-Gaussian distribution of the ratio image (resulting in inappropriate standard deviation metrics, for example) thresholding techniques must be employed carefully (Chen 2007).

10.4.1.3 Normalized Difference Change Detection (NDCD)

The normalized difference change detection (NDCD) procedure developed by Gianinetto and Villa (2011) is a derivation of the image differencing approach that was originally described in Coppin and Bauer (1994). It is calculated in the same way as the Normalized Difference Vegetation Index (NDVI) (Tucker 1979) with the $t_1$and t_2 variables representing the pre- and after-event images, respectively:

$$\text{NDCD} = (t_2 - t_1)/(t_2 + t_1)$$

This method produces a normalized difference image for each image band with output values ranging between −1 (representing maximum reflectance decrease) and +1 (maximum reflectance increase) with a 0 value indicating no change. The method was initially developed to assess flooded areas in an urban environment but may be applied to a wide variety of environmental changes.

10.4.1.4 Regression

The multitemporal image regression method assumes a linear relationship between the images of two dates (Ridd and Liu 1998). The regression image is determined from the residuals of the least square regression for each reflectance band. Using the regression equations, a "no-change" image can be predicted from unchanged pixels. The difference between the predicted image and the 2nd real time-image denotes changes in each spectral band (Jha and Unni 1994). Although this method reduces errors caused by atmospheric distortions and haze, it is considered to be somewhat outdated since it requires user interaction and the selection of no-change pixels. In addition, performance improvements are negligible compared to image differencing techniques (Ridd and Liu 1998).

10.4.1.5 Change Vector Analysis (CVA)

Change Vector Analysis (CVA) is a bi-temporal change detection method which calculates change magnitude and direction (Malila 1980). The advantages of the method are its capability of using all spectral input information and the provision of directional information which facilitates the interpretation of occurring changes. Although CVA is not limited to selected bands, it is typically advantageous to use uncorrelated bands or reduce the number of input bands to meaningful spectral ranges (Bovolo and Bruzzone 2007). It is also possible to include textural features in the CVA calculation (He et al. 2011). The change magnitude is usually expressed as a Euclidian distance between two spectra, i.e. the square root of the sum of the squares of the differences for each input band. The direction can be calculated in four different ways: (1) sector-coding (Michalek et al. 1993); (2) vector direction cosines (Chen et al. 2003); (3) PCA in multitemporal space (Lambin and Strahler 1994); and, (4) direction in polar domain (Allen and Kupfer 2000). Although the additional direction information is one of the strengths of CVA, it is sometimes ignored and only change magnitude is calculated (Xian et al. 2009; Xian and Homer 2010).

10.4.1.6 Phase-Related SAR Change Extraction Methods

One key characteristic of SAR images is that the detected signals are composed of a real and an imaginary component which can be expressed as amplitude, i.e. the absolute value of the complex number, and phase. Consequently, SAR-based change detection methods can be based on phase or on intensity values (the square of amplitude).

Persistent Scatterer Interferometry (PSI)

The persistent scatterer interferometry (PSI) technique (Ferretti et al. 2007) makes use of image features which are stable over long periods of time such as buildings. Using PSI allows for the detection of ground deformation in the range of millimeters and centimeters per year. As such, PSI indicates neither environmental modification nor conversion. Rather, it denotes ground changes due to subsidence or deformation caused by natural forces such as earthquakes, by mining (Wegmüller et al. 2010) or groundwater extraction (Bell et al. 2008).

Coherence Change

Coherence is a measure of the phase correlation of two images. If changes occur between two image acquisitions, the phase is decorrelated. Stable regions show high coherence. Most vegetation surfaces decorrelate within very short times (seconds to minutes) due to the motion of branches, leaves, and stems. High coherence, on the other hand, is often related to non-vegetated and man-made surfaces. Similar imaging geometry for both images is required.

For change detection of man-made objects it is more appropriate to use the coherence difference approach. For this technique, two image pairs are required – the first pair can be termed the pre-image pair, the second pair as the after-image pair. Coherence is then calculated for both the pre- and the after-image pairs. Subsequently, the coherence of the pre-image pair is subtracted from the after-image pair coherence (Liao et al. 2008). If vegetation has been replaced by an urban object, then the low coherence of the pre-image pair has turned into high coherence of the after-image pair. However, if a forest has been cleared and reforested, it will not be apparent in the coherence difference image since both coherence images will show low values. Therefore, coherence differencing only identifies changes between coherent and incoherent surfaces. Thus, the application of coherence difference techniques is recognized as more efficient at detecting changes as well as better able to discriminate changes within the context of multi-temporal features (Liao et al. 2008; Thonfeld and Menz 2011).

Differential SAR Interferometry (DInSAR)

Interferometry uses the phase difference of at least two SAR images that have similar acquisition geometries but that were acquired at different times or with different orbital positions (Bamler and Hartl 1998). While interferometry is a powerful technique to generate digital elevation models, differential interferometry can also be used to quantify various types of surface displacements, including ground motion caused by tectonic processes, subsidence due to mining, gas, water, and oil withdrawal, and glacier and ice sheet movement. Such processes can be identified over large temporal and spatial scales at centimeter or even millimeter precision. Bamler and Hartl (1998) and Rosen et al. (2000) present extensive introductions to the principles of radar interferometry. Interferometry is considered a powerful technique for the detection of changes (Rignot and Van Zyl 1993; Polidori et al. 1995) and is one of the most frequently applied methods in radar remote sensing.

10.4.1.7 SAR Backscatter-Related Change Extraction Methods

In addition to methods such as ratioing that can be applied equally to optical and SAR data, a number of change extraction methods specific to SAR have been developed based on SAR backscatter analysis.

Differential Radargrammetry

Deriving height information from SAR amplitude images with different viewing geometries is called radargrammetry. Converting the parallaxes to surface displacements instead of heights is the process of differential radargrammetry (Polidori et al. 1995). Although this methodology is somewhat dated, however, it has been applied to map sea ice (Leberl 1983).

Speckle Decorrelation

The occurrence of speckle is a characteristic of SAR data. Speckle is caused by distributed ground targets which contribute to the signal of a resolution cell leading to multiplicative noise and a "salt and pepper" effect in the image. Intensity images can be correlated in a predefined sliding window as can phase images. Speckle decorrelation is supposed to be caused by changes in backscatter elements. Thus, the decorrelation of intensity images should be identical to the loss of coherence (Zebker and Villasenor 1992). However, in some cases the loss of correlation in the intensity images may also be caused by other phenomena such as the cardinal effect in urban areas (Yonezawa and Takeuchi 2001). Combining both coherence and speckle decorrelation may reveal complementary information regarding changes on the ground.

Offset Tracking

DInSAR has been introduced as powerful technique to quantify surface motion at large spatial and temporal scales. However, this technique fails if the phase information is decorrelated. In some cases it is possible to select stable point scatterers, which can then be used for PSI. When no stable point scatterers can be detected even PSI must be excluded. Feature tracking or offset tracking is an alternative method based on backscatter intensity (Strozzi et al. 2002). Two intensity images acquired at different times are first co-registered based on the motion-free image elements. A spatial neighborhood for a pixel of the first image is then defined, and the position of the most similar pixel cluster is identified in the second image. Offset tracking is frequently applied to quantify glacier motion since glacial surfaces often show phase decorrelation in the acquisition interval of current SAR satellite sensors (Fallourd et al. 2011; Pritchard and Vaughan 2007). SAR offset tracking is similar to feature tracking in optical images but has some advantages, especially in the field of ice motion detection since it is not affected by cloud cover, darkness or sensor saturation (Pritchard and Vaughan 2007).

Feature tracking in optical images can be accomplished using the COSI-Corr tool which co-registers and correlates satellite or aerial images (Leprince et al. 2007). Areas of application for COSI-Corr include the identification of ground deformation due to severe earthquakes (Wei et al. 2011), monitoring glacier motion (Herman et al. 2011), and monitoring sand dunes (Necsoiu et al. 2009).

Speckle Tracking

In contrast to offset tracking, speckle tracking is used to analyze speckle rather than robust physical features (Pritchard and Vaughan 2007). The use of amplitude rather than phase characteristics (i.e. interferometry) is typically less accurate but has other advantages: amplitude data are more robust to decorrelation and are not only sensitive to displacement in range (Michel et al. 1999).

10.4.2 Transformation-Based Methods

Several methods are based on the transformation of the input data in a higher dimensional feature space (e.g. PCA, Gram-Schmidt transformation, Kauth-Thomas

Tasseled Cap transformation (TC), IR-MAD, Minimum Noise Fraction transformation (MNF), Maximum Autocorrelation Factor transformation (MAF), or Chi^2 transformation).

10.4.2.1 Principal Component Analysis (PCA)

PCA can be applied to data arrays containing all spectral bands of two acquisition dates (Ingebritsen and Lyon 1985; Byrne et al. 1980). Unchanged pixels are oriented along the first PC axis as a consequence of the assumed linear relation between unchanged pixels. The second PC contains change information (Coppin and Bauer 1994). Change information will be oriented along the second PC axis which is orthogonal to the first (uncorrelated) axis (Wiemker 1997). PCs are calculated from eigenvectors based on all pixels, including the changed pixels. Thus, the first PC (the "no-change" axis) contains erroneous information introduced by the change pixels. To minimize the influence of the change pixels, Wiemker (1997) developed an iterative approach which calculates the covariance matrix incorporating a weighting coefficient for all pixels that quantifies the probability of each to be a no-change pixel. This method greatly improves the identification of no-change pixels and consequently improves the change information. This technique has been adapted to other approaches such as IR-MAD.

Although PCA is rarely employed as a change detection methodology, the technique is often used to reduce the dimensionality of change extraction results derived by other means, e.g. image differencing applied to all layers in an image data stack (Jha and Unni 1994).

10.4.2.2 Gramm-Schmidt Transformation

Another transform used for change detection is the Gramm-Schmidt orthogonalization. Collins and Woodcock (1994) applied this technique to Landsat images in order to detect tree mortality. In a related study, Collins and Woodcock (1996) compared the utility of three linear transformation methods; PCA, TC, and Gramm-Schmidt. They concluded that PCA and TC showed better results than Gramm-Schmidt; they also specifically recommended the use of the TC transform since the TC components are easier to interpret. The Gramm-Schmidt transform is currently not frequently used.

10.4.2.3 Iteratively-Reweighted Alteration Detection (IR-MAD)

A robust and automatic CE method is the iteratively re-weighted multivariate alteration detection (IR-MAD) (Canty and Nielsen 2008). This method is an extension of the multivariate alteration detection (MAD) developed by Nielsen and Conradsen (1997). IR-MAD is designed to identify unchanged pixels which can subsequently be used to define a regression equation for radiometric normalization of multispectral images. Since one output is a Chi^2 image representing the change probability of each pixel, the method can be used for change extraction. MAD components are also computed as a second output. As they are uncorrelated, every MAD component can be referenced to a different type of change. No radiometric preprocessing is required. In many instances, IR-MAD (or its MAD forerunner) are used for automatic radiometric normalization (Griffiths et al. 2012; Schroeder et al. 2006, 2011).

10.4.2.4 Other Transformations

Numerous other transformation-based methods have been developed. These include Chi^2 analysis (Ridd and Liu 1998), Minimum Noise Fraction (MNF) methods (Gianinetto and Villa 2007) or Fast Fourier Transformation (FFT). These transforms are rarely used today as better results can often be obtained through use of simpler algorithms that produce more easily interpreted results.

An innovative recent development in the field of change extraction is the use of kernel transformation methods, including the kPCA (Nielsen and Canty 2008) and kMAF techniques (Nielsen et al. 2010). These methods allow for the solution of non-linear problems and thus often show improved change extraction results.

10.4.2.5 Wavelets, Curvelets and Other – Lets

Bovolo and Bruzzone (2005) utilized the wavelet representation of a logarithmic scaled ratio image to suppress speckle effects in change extraction results. Final results are generated by an adaptive scale-driven fusion. A multi-resolution representation of the change image derived using wavelets has also been applied to optical data in order to minimize misregistration effects (Carvalho et al. 2001).

A recent advance in SAR change detection is the use of curvelets (Schmitt et al. 2010). This method transforms two co-registered logarithmic scaled SAR amplitude images in the curvelet domain and differentiates the curvelet coefficients. The results are then transformed back to the spatial domain. Curvelets have been shown to provide a better representation than wavelets, surfacelets and Laplacian pyramids (Schmitt et al. 2010). Significant advantages of the curvelet method are its wide range of application (including flood, forest fire, and earthquake monitoring, ship detection, or detection of new buildings), and its capability to process single and multi-polarized data, even data acquired at differing incidence angles (Schmitt et al. 2010).

10.4.3 Classification-Based Methods

10.4.3.1 Post-Classification Comparison (PCC)

There are numerous classification-based approaches used to detect and identify changes. The oldest and presumably most widely used of all such methods is post-classification comparison (PCC) (Jensen et al. 1987). PCC is based on the separate classification of two images taken at different times. Following image classification, change matrices are derived which can then be interpreted as from/to classes. From those matrices change statistics are calculated. The primary advantage of this method is that it operates independent of input data. Thus, classification results derived from SAR and optical data or other data can be compared. No radiometric preprocessing or adjustment between images is required. However, since none of the single classifications are of perfect accuracy, the errors of input classifications compound and decrease the overall accuracy of the final change extraction error matrix. In general, the PCC technique tends to produce more inclusion errors ("false alarms") than methods based on spectral or textural features.

A further development of the PCC approach is one that we term "change-restricted PCC". This hybrid process operates in the following way. Multidate image data are first acquired and classified. Change extraction is applied to t_1 and t_2 imagery using a reflectance-based method such as CVA. A conservative threshold is then applied to differentiate between changed and unchanged areas. Finally, the changed areas are reclassified using the class descriptions of the unchanged areas (Xian et al. 2009; Xian and Homer 2010). It is important that all classes are described since the t_2 image may include classes that are not present in t_1 imagery.

10.4.3.2 Continuous Monitoring of Forest Disturbance Algorithm (CMFDA)

Continuous Monitoring of Forest Disturbance Algorithm (CMFDA) is a method recently developed for use with high temporal resolution satellite image time series data sets (Zhu et al. 2012). CMFDA exploits the temporal/spectral trajectory of each individual pixel (its "history") thereby defining a multitemporal class description. Thus, any date of the satellite image can be predicted and subtracted from a real acquisition. Assuming perfect masking of clouds and cloud shadows within the imagery, the remnant differences can be attributed to land cover change. Applying this method allows for the immediate detection of changes and the results become more robust the longer the data time series that is available before and after the change event.

10.4.3.3 Multitemporal Spectral Mixture Analysis (SMA)

Based upon the definition of pure end-members, high-spectral dimension data sets can be used for bi-temporal change detection. Since even fine changes can be captured with multitemporal spectral mixture analysis (SMA) the procedure is sensitive for intra-class changes such as forest degradation. Usually, four end-members (green vegetation, non-photosynthetically active vegetation, soil, and shadow) are defined based on spectral libraries, *in situ* field measurements or image properties. For comparability, end-members are derived from or converted to reflectances, i.e. atmospherically corrected spectra. It is assumed that the spectra in the image are composed of a linear combination of pure spectra. Changes in the proportions of the end-members of each pixel between acquisitions can be referred to ecosystem changes. Adams et al. (1995) used SMA to compare end-member fractions within a multi-year Landsat TM image data set covering a tropical environment. To better capture fractional degradation patterns, the Normalized Difference Fraction Index (NDFI) was developed and shown to be more sensitive for changes than the individual end-member fractions (Souza Jr. et al. 2005). Thus, very fine year-to-year changes can be monitored. Other examples of the usage of SMA for change detection can be found in dry regions. Dawelbait and Morari (2012) applied SMA to assess land degradation in Sudan. Several studies exist which tried to assess gradual changes based on multitemporal SMA. Yang et al. (2012) calculated end-members for each image of a Landsat time series and regressed them against tree cover to monitor long-term tree cover dynamics in semi-arid woodlands. They compared the results with those of vegetation indices and concluded that they depend greatly upon specific climatic conditions (i.e. wet vs. dry years). Since vegetation indices are heavily affected by background signals generated by soils and other factors, SMA may be the preferred approach for use in sparsely vegetated dry lands (Hostert et al. 2003).

10.4.4 Time Series Analysis (TSA)

Most published change detection studies analyze two temporal steps, and may be understood as bi-temporal before/after analyses. Such analyses are inherently limited, however, as many ecosystem processes cannot be detected when only two dates of images are analyzed. Remote sensing based time series analyses of change detection were developed in part to overcome this limitation. This approach is not new – such analyses have been performed for many years. To date, however, the use of time series analyses for environmental change detection has not been the subject of a systematic review. Most long term time series have been neglected in reviews because of their rather coarse spatial resolution. NOAA-AVHRR data are an example. Indeed, these data have not been considered for use in classical change detection, a term which typically refers to bi-temporal change analyses. A recent development in change detection research has been the implementation of time series methods utilizing higher spatial resolution data sets such as Landsat. Despite the Landsat Long Term Acquisition Plan (Arvidson et al. 2001) and systematic, long-term acquisitions (Markham et al. 2004; Williams et al. 2006) gaps are present in the various data time series for most locales, reducing the number of available scenes, sometimes to only a few per year. The number of usable data sets is further reduced by the presence of cloud cover within the imagery (Ju and Roy 2008). Instead of seeking only cloud-free images, all available data should be considered when compiling a time series (Zhu et al. 2012).

Promoted by the opening of the comprehensive Landsat archive, and novel data distribution policies which make data sets available at no cost, the development of time series methods has accelerated and many studies using Landsat time series have been published (Griffiths et al. 2012; Huang et al. 2010a; Kennedy et al. 2010, 2012; Pflugmacher et al. 2012; Powell et al. 2010; Schroeder et al. 2011; Vogelmann et al. 2011; Wulder et al. 2012; Zhu et al. 2012).

Time series reveal more information about the nature of changes that occur within the environment. There are only few surfaces in the environment which are not subject to changes. Even rather stable surfaces such as manmade objects like roofs, roads and other artificial objects change their condition over time. Thus, change is a function of time, and the suitability of data to detect changes is greatly dependent upon their temporal dimension. Processes on Earth's surface exhibit different temporal behavior. Seasonal changes, trends, abrupt changes, and cyclic events often are all components of a single reflectance signal. Often such underlying processes can only be separated by employing dedicated time series methods in combination with an adequate data time series. In recent change detection studies, compositing techniques have been developed to reduce seasonal effects in time series of annual images (Griffiths et al. 2012). Remote sensing time series analysis focuses mainly on the description of environmental processes and their modeling. Recent studies also make use of the data to model past processes in order to predict future ones. If the temporal behavior of each pixel is known, an image of any point of time can be constructed, i.e. predicted (Zhu et al. 2012). Given this information, each new acquisition can be compared with the predicted one and differences may then be referenced directly to changes on the ground.

Bi-temporal change detection methods may be able to detect gradual and abrupt changes as well as trends in the environment, but likely will not be adequate to describe the temporal characteristics of the environmental processes underlying these phenomena. Thus, time series analysis should be favored to reveal such processes when adequate data are available.

10.4.4.1 Trajectory Analysis

Trajectory analyses may be viewed as an intermediate methodology. In contrast to time series, which represent continuous temporal information, temporal trajectories are capable of indicating information in longer temporal increments – typically in years – but are generally unable to depict underlying environmental processes. Many approaches aim at removing or minimizing seasonal variations from long term response signals. Schroeder et al. (2007) applied a time-invariant regression method to construct a fitted trajectory curve for analysis of stand recovery in a forest environment. Huang et al. (2010a) used a z-score method to normalize seasonal variations in forest analysis. Kennedy et al. (2010) took advantage of a de-spiking algorithm for analysis of forest disturbance and recovery.

Time series analyses may be used to identify spatial and temporal changes while simultaneously characterizing changes as either abrupt changes, disturbances, or longer term trends. However, due to data set gaps caused by limited availability or cloud cover, complete long-term time series of data from similar sensors are rare. Thus, annual data are selected to perform analyses on trajectories. The major advantage of annual time series is the potential to separate abrupt from gradual changes. Examples are given for change differentiation from fires and grazing pressure in rangelands (Hostert et al. 2003; Röder et al. 2008b; Sonnenschein et al. 2011; Stellmes et al. 2010) and changes in forests (Röder et al. 2008a; Viedma et al. 1997).

Lawrence and Ripple (1999) published one of the first studies describing the application of Landsat-like data that utilized more than two time steps. They analyzed vegetation regeneration in southwestern Washington State following the Mount St. Helens volcanic eruption. Their analysis was based on unsupervised clustering and subsequent change curve extraction. The change curves were calculated as polynomial curves fitted to the cluster means. The Vegetation Change Tracker (VCT) described by Huang et al. (2010a) is a similar algorithm. VCT uses spectral indices to monitor forest changes and fire impact (Vogelmann et al. 2011). Sen et al. (2012) used trajectories of several indices derived from Landsat images to monitor mining revegetation.

10.4.4.2 Time Series Segmentation

The Landsat-based detection of Trends in disturbance and Recovery (LandTrendR) is a sophisticated method of fitting temporal trajectories to spectral curves using both regression and point-to-point fitting (Kennedy et al. 2010). This technique allows for the discrimination of fast and slow change processes. Like most time series methods, it is based on spectral indices. It is frequently used for forest monitoring, including mapping the effects of insect driven disturbance on tree mortality and surface fuels (Meigs et al. 2011).

10.4.4.3 Breaks for Additive Seasonal and Trend (BFAST)

Breaks For Additive Seasonal and Trend (BFAST) is a method which was originally developed for use with MODIS time series data (Verbesselt et al. 2010a, b). As the name indicates it separates the observed signal – an NDVI time series, as an example – into a seasonal component, a trend component and residuals which can be interpreted as random fluctuations. The algorithm allows for the detection of long-term phenological changes (Verbesselt et al. 2010a, b). The strength of the algorithm lies in its ability to detect abrupt changes in the trend component and is particularly effective for analysis of natural ecosystems such as forests and savannahs. The approach has been extended to allow for near real-time detection of changes (Verbesselt et al. 2012).

10.4.5 Object-Based Approaches

The availability of high spatial resolution imagery from the Ikonos, Quickbird, Spot 5, Kompsat-2, Formosat-2 sensor systems has driven the development of analytic methods appropriate for the new data. Object-based methods have been used successfully in numerous applications, including mapping shrub encroachment (Laliberte et al. 2004) and forest change (Castilla et al. 2009; Desclée et al. 2006). Object-based approaches generally consist of deriving homogeneous image objects (which ideally correspond to real world objects); then calculating object statistics and spatial indices; and finally applying existing change detection algorithms on the previously derived object features. Following this procedure, object-based image differencing (Desclée et al. 2004), MAD (Listner and Niemeyer 2011; Niemeyer et al. 2007), and PCC (Laliberte et al. 2004) have been successfully applied. All of the object-based methods shown above achieve accurate results (which also often correspond to human perception). Difficulties remain, however, in defining an appropriate number of segmentation levels and in integrating the information to one change extraction result. Change objects are sometimes derived from the change extraction result rather than from the spectral input data (Castilla et al. 2009). That means an arbitrary change detection method is applied on a per-pixel basis and the output is then subjected to the segmentation. This can be an advantage when large proportions of spurious changes are identified due to geometric misalignment between two images. Also, since only change objects are segmented, it is not necessary to define several segmentation levels. Small areas of false changes can be suppressed by applying a minimum mapping unit limit (Castilla et al. 2009). An exhaustive introduction to the field of object-based changed detection is presented by Chen et al. (2012).

10.5 Change Labeling

10.5.1 Categorization of Change Extraction Algorithms Regarding Their Suitability for Change Labeling

Within the context of change labeling, the CEA discussed in Sect. 10.4 can be divided into three groups:

1. Algorithms that return the *intensity or probability of change*, such as image differencing or image ratioing, both of which are types of algebraic algorithms. Such algorithms are useful to identify change "hot spots" and to derive change masks; however they are of limited use in deriving change labels.
2. CEA that produce *images containing information about kinds of changes*, including bi-temporal PCA as well as other transformation algorithms. Some algorithms also return information on both change intensities and the type of

change. Examples of such algorithms include CVA which derives change magnitudes and change directions, and IR-MAD with its Chi^2 image and MAD components. Such algorithms can be used for further change analysis in order to derive labeled change maps.
3. CEA that return *labeled change maps directly without further analyses*, including PCC and Support Vector Machines (SVM) change detection. Both of these techniques are considered as classification algorithms and as such do not require any additional analysis for change labeling.

10.5.2 Change Labeling Approaches

Depending on which CEA (Sect. 10.4) is implemented, change labeling happens either before, concurrent with or after the application of the CEA. Pre-CE-labeling assigns mono-temporal informational classes or indices to the input data. This may be through the calculation of an index like NDVI, a transformation like TC (separating information on greenness, brightness, and wetness), or a classification. The CEA then performs the analysis of changes within or between those informational classes, employing, as an example, an image algebra operator such as image differencing (Sect. 10.4.1). Concurrent labeling means that change labeling is inherent in the CE, as in classification approaches like direct multi-date classification. Post-CE-labeling assigns information on the kinds of changes to the results of the CEA. In the case of a transformation type CEA (Sect. 10.4.2), this might be the correlation of PCs with input bands or simply the visual interpretation of the CE results using the input data in combination with ancillary information. In the case of an algebra based CEA, post-CE-labeling is done by implementing a thresholding function to separate change from no-change or positive change from negative change.

10.5.2.1 Pre-CE-Labeling: Setting the Ground

The results of a CEA are typically more easily interpreted if the input data have undergone some transformation procedure(s), particularly if the CEA involves image differencing. Differencing two pixel values of earth observation image data indicates only that the pixel value has become larger or smaller, and cannot provide any additional information about the specific meaning of these changes in pixel values. In some cases, if the test site is very well known and the spatial subset of the scene contains only known change types, the data analyst may be able to interpret changes in pixel values and attribute accurate additional meaning to them. An example is a forest environment where changes can only be attributed to logging and regrowth. In other cases where such interpretation is difficult, a data transformation prior to the CEA can aid the process of adding meaning to the CE results. CVA may be thought of as a more sophisticated form of image differencing

(Sect. 10.4.1.5). Every environmental change (regardless of direction) is a variant combination of spectral response, recorded as increasing or decreasing brightness values in the image input bands. Pre-CE-labeling approaches are thus also proposed for CVA. In his original 1980 paper on CVA, Malila described spectral transformations like the TC transformation (Kauth and Thomas 1976) as optional processing steps prior to performing the actual CVA. The TC transformation was developed to transform Landsat MSS data to better derive information on vegetation greenness, brightness and wetness. It was later enhanced to make full use of Landsat TM bands (not including the thermal band) by Crist and Cicone (1984). These techniques are transferable to other sensors only if these sensors cover the same spectral bands; such sensor systems include Landsat ETM+, MODIS or the planned Sentinel-2 system. Fung (1990) analyzed the applicability of the Kauth-Thomas TC brightness, greenness and wetness components for change detection. Relatively few bi-temporal studies make use of TC components. These are, however, more frequently used for time series analysis within forested environments (Griffiths et al. 2012; Hais et al. 2009; Powell et al. 2010).

The Pre-CE-Labeling procedure also includes the calculation of *indices*; the result of image differencing or CVA is thus not merely a change in digital numbers or reflectances within the data, but may be interpreted as an increase or decrease in a specific index. Two examples of widely-used indices are the NDVI for green vegetation and the NDWI for surface water (McFeeters 1996). Special indices have also been developed for SAR data, including the SWI soil water index (Zhao et al. 2008). An extensive list of available indices and formulas, along with information regarding their interoperability among different sensors (as well as additional information) can be found in the Index Database (IDB) at: http://www.indexdatabase.de/ (Henrich et al. 2009).

For more specific change labeling tasks, when the change classes of interest cannot be characterized using generic functions such as TC or by available indices, application-specific *spectral features* can be developed and implemented. Johnson and Kasischke (1998) used multivariate discriminant analysis to examine the spectral responses of and to develop spectral signatures for three change classes of interest: new clay topsoil deposits, new asphalt, and new vegetation. A full-dimensional CVA change image containing all possible change directions was the starting point for this analysis. *A priori* knowledge regarding the relevance of identified changes was utilized to assess individual change areas within the study site and to develop spectral features. This method can be utilized to develop spectral features for any change classes of interest; limited only by the spectral discrimination among change classes.

Image *texture* is frequently employed to improve classification accuracy and to support change detection and monitoring studies (Del Frate et al. 2008). Briggs and Nellis (1991) utilized image texture characteristics in the monitoring of heterogeneity within Prairie ecosystems. Texture data are typically used in combination with spectral information. With the availability of new high spatial resolution sensors such as Quickbird and Ikonos, image texture information is also used to detect new buildings in informal settlements and refugee camps (Klonus et al. 2012), as well as to assess damage in populated areas following natural disasters. Several texture measures of the grey level co-occurrence matrix (GLCM) have been used for change extraction (Haralick et al. 1973; Klonus et al. 2012). Changes are calculated as the difference of two texture features derived from image data acquired at different times.

Numerous *multitemporal features* have been derived using SAR image time series with the goal of improving the characterization of distinct classes present within a study area, thereby improving overall classification performance. These features include, among others:

- the mean annual variation (MVA) described by Quegan et al. (2000)
- standard deviation
- normalized standard deviation
- logarithmic measure based on normalized standard deviation
- saturation
- standard deviation of decibel (dB) values
- maximum-minimum ratio in dB (Bruzzone et al. 2004)

Several studies have been published describing the use of multitemporal features in combination with SAR backscatter intensity for assessing flood dynamics (Martinez and Le Toan 2007) and urban change detection (Liao et al. 2008; Thonfeld and Menz 2011). It must be noted that multitemporal features may also be applicable to change detection of optical imagery, but to date have not been applied in this way.

Further, for SAR data, especially *polarimetric* SAR data, *parameters* such as backscatter coefficients, correlation coefficients and phase differences can be applied singly or in combination. Such parameters can be used to describe selected change classes and to enhance the interpretability of the change extraction results. Conradsen et al. (2003) use this capability in combination with test statistics in the complex Wishart distribution for change detection of agricultural fields. For some applications, especially those involving high spatial resolution imagery, it can be useful to apply *image segmentation* and *object feature* extraction before applying the CEA. This procedure has been implemented for the IR-MAD algorithm; corresponding plug-ins for the Definiens/eCognition software are readily available (Listner and Niemeyer 2011).

In a strict sense, PCC (Sect. 10.4.3.1) is image differencing with pre-CE-labeling. The separate classification of two input images can therefore also be considered as Pre-CE-Labeling. The advantage is that change labels are easily interpretable and change detection utilizing data from multiple sensors is possible. Error propagation through the separate classifications can be a disadvantage of this approach, however. Table 10.2 provides an overview of pre-CE-labeling techniques.

10.5.2.2 Post-CE-Labeling: Labeling Change Types and Change Intensities and Interpretation of Change Extraction Results

The results of CEA performed on satellite image data, rather than on spectral features or indices, are typically difficult to interpret accurately. Post-CE-labeling approaches have been developed to address this issue.

Table 10.2 Pre-CE-labeling techniques

Pre-CE-labeling	Associated CEA	Sensors	Applications/ classes	References
Tasseled Cap Transformation	Image differencing; CVA	Landsat	Greenness, brightness, wetness	Kauth and Thomas (1976) and Malila (1980)
Indices, e.g. NDVI, NDWI, SWI	Image differencing; CVA	Mainly optical but also scatterometer (e.g. SWI); cf. index database (IDB)	Any; e.g. vegetation, surface water, soil moisture	Henrich et al. (2009), Tucker (1979), McFeeters (1996), and Zhao et al. (2008)
Spectral features	Image differencing; CVA	Optical	Any; e.g. clay, asphalt, vegetation	Johnson and Kasischke (1998)
Multitemporal features		SAR, partly transferable to optical	e.g. flood, urban change	Bruzzone et al. (2004), Liao et al. (2008), Martinez and Le Toan (2007), Quegan et al. (2000), and Thonfeld and Menz (2011)
Polarimetric parameters, e.g. backscatter coefficients, correlation coefficient, phase difference	Test statistics	Polarimetric SAR, e.g. ALOS-Palsar, Radarsat-2, TerraSAR-X, ENVISAT-ASAR	e.g. crops	Conradsen et al. (2003)
Image segmentation and extraction of object features	IR-MAD	Optical	Any	Listner and Niemeyer (2011)
Classification	Image differencing (=PCC)	Any (optical and SAR)	Any	Singh (1989)

The most common approach to aid interpretation of CE-results is the use of *thresholds* to separate change and no-change-areas. Apart from a manual, iterative search for optimal thresholds, a number of additional automated methods are in use. A popular method of change thresholding is the use of a fixed value threshold, – two standard deviations, as an example (Singh 1989). More sophisticated automated approaches include Gaussian fitting (Bazi et al. 2005), the Expectation-Maximization (EM) algorithm (Bazi et al. 2009), Markov Random Fields (MRF) (Bruzzone and Prieto 2000, 2002), and the histogram corner method for unimodal thresholding (Castilla et al. 2009; Rosin 2001). The extent and location of change pixels identified within an image can differ depending upon the thresholding method implemented. It is often useful, therefore, to consider applying several different thresholding methods and selecting the results that best fit the criteria for the particular application.

A relatively simple approach to identifying and labeling CVA change directions involves examining the spectral response increases and decreases in specific spectral bands that are known to indicate certain types of change. As an example, consider a simple CVA using two Landsat images: decreases in green vegetation may be identified by an increased response in band 3 and decrease in band 4; all such change directions could be coded by the analyst in red. Increases in green vegetation would be indicated by a decrease in band 3 response and an increase in band 4, and all such change directions could be coded in green. Johnson and Kasischke (1998) refer to this technique as *"change space sectors"*.

Given a *classification* of each of the two input images, CE results may be easily interpreted. This approach can operate as follows. CE is performed on two satellite images resulting in a full-dimensional change image. Applying a threshold, change areas are identified. The changes are labeled according to both classifications. If the classes for a specific change pixel are different in both classifications, the pixel is labeled as a conversion class. If the classes are identical, the change pixel is labeled as modification within that particular class. Johnson and Kasischke (1998) employ this method in combination with CVA. This procedure has two advantages over the PCC method: (1) modifications can be accurately identified, and (2) pseudo-changes due to classification errors are not misidentified as modifications.

CE results, and in particular the types of changes identified based on PCA are generally difficult to interpret. PCs are orthogonal and uncorrelated, and it can therefore be expected that changes represented by one PC are not included in another PC. However, due to the nature of the transformation – which is data driven – it cannot be generalized which components contain which kind of changes. Collins and Woodcock (1996) performed PCA on a bi-temporal Landsat TM image stack. To label the PCs they applied *stepwise regression*, linking the results of PCA CE with forest mortality and choosing the components significant for forest mortality from the transformation matrix.

The *Maximum Autocorrelation Factor algorithm (MAF)* (Nielsen and Canty 2009) or the *kernelized (kMAF)* version (Nielsen et al. 2010) can be applied as a post-processing routine to MAD or PCA CE. This procedure cannot directly apply change classes to label each pixel; rather, it assists the post-CE-labeling process. MAF makes use of spatial autocorrelation and preserves the spatial context, which is desirable since objects of change generally occupy several contiguous pixels in a scene. Post-processing with MAF/kMAF tends to focus on extreme changes with high spatial autocorrelation.

To label the MAF/MAD variates, Canty and Nielsen (2006) proposed a clustering approach called *Fuzzy Maximum Likelihood Estimation (FMLE)* (Gath and Geva 1989). FMLE is similar to the EM algorithm. This approach has also been transferred to object based change detection for high resolution imagery (Niemeyer et al. 2007). An overview of post-CE-labeling techniques is provided in Table 10.3.

Table 10.3 Post-CE-labeling techniques

Post-CE-labeling	Associated CEA	Sensors	Applications/classes	References
Thresholding	Any, e.g. image differencing, CVA	Any (optical and SAR)	Change – no change	Bazi et al. (2005), (2009), Bruzzone and Prieto (2000), (2002), Castilla et al. (2009), and Rosin (2001)
Change space sectors	CVA	Optical	e.g. vegetation	Johnson and Kasischke (1998)
Classification	Any, e.g. CVA	Any (optical and SAR)	Any	Johnson and Kasischke (1998)
Stepwise regression	PCA	Optical, e.g. Landsat	e.g. forest mortality	Collins and Woodcock (1996)
MAF, kMAF	MAD, PCA	Optical	Change – no change; specific classes, e.g. mining	Nielsen and Canty (2009) and Nielsen et al. (2010)
Clustering, e.g. FMLE	MAD, MAF/MAD	Optical, high resolution optical	e.g. new buildings	Canty and Nielsen (2006) and Niemeyer et al. (2007)

Table 10.4 Concurrent labeling techniques

Concurrent-labeling	Associated CEA	Sensors	Applications/classes	References
Direct multidate classification (DMC)	None	Any	Any	Hecheltjen et al. (2010) and Kuemmerle et al. (2008)
Compound classification	None	Any	Any	Demir et al. (2012)

10.5.2.3 Concurrent Labeling: Labeling Change Classes

Classification-based CEAs simultaneously detect and label changes (Sect. 10.4.3), we therefore summarize them under the term *'concurrent labeling'* (Table 10.4). Concurrent labeling methods incorporate the advantages of PCC-CE (Sects. 10.4.3.1 and 10.5.2.1), principally the compilation of maps showing actual change classes. The concurrent labeling approach incorporates these advantages while overcoming the principal disadvantage of PCC: the propagation of thematic error throughout the derived datasets due to the integration and subsequent subtraction of two separate classifications. These methods are considered supervised classification algorithms which require the development and use of training data.

Rather than classifying both input images separately, *direct multidate classification (DMC)* performs change classification directly on bi-temporal data layer stacks. The difficulty in this approach is defining an accurate set of training data for the change classes. Kuemmerle et al. (2008) applied SVM for direct multidate classification to assess farmland abandonment in the Carpathians. Hecheltjen et al. (2010) employed DMC change classification with SAR and optical data to assess land cover changes due to mining.

Demir et al. (2012) conclude that *compound classification* is less critical with regard to the definition of training data sets when compared with DMC. They apply active learning approaches to model temporal dependence between the images.

10.6 Accuracy Assessment

Accuracy assessment remains a challenge in CE analyses, principally due to the limited availability of complete timely *in situ* ground information. Incomplete or imperfect ground data have substantial impact on the accuracy estimation from confusion matrices (Foody 2009, 2010). Further difficulties arise in natural environments characterized by phenological changes or seasonal and long term environmental variations. Depending upon the aim of the individual study, it is important to differentiate between targeted changes and natural fluctuations; thus, accuracy assessment may in some cases be subjective. However, strategies have been developed to minimize mislabeling; limiting map updating to changed areas only, is an example. Li and Zhou (2009); Morisette and Khorram (2000); Van Oort (2007) have all conducted studies dedicated to change detection accuracy assessment.

10.7 Conclusion

In this review we have shown that a wide variety of strategies have been developed and applied to conduct change detection studies. There exists, however, no single CE method which is best for all applications or all sensors. Sensors must be carefully chosen to provide relevant spectral and temporal information for a given application. The core of the CE process – the labeling of changes – can be performed in essentially three different ways: pre-CE-labeling, post-CE-labeling, and concurrent labeling. The CEA itself can be selected from a variety of algorithms. An increasing number of studies take advantage of satellite spectral data time series. Early work revealed the temporal SAR backscatter behavior of different land cover types over time periods on the order of one year. To date, however, there is no study based on the application of a time series of SAR data to discriminate trends from abrupt changes. Calculated scene-by-scene difference imaging has helped to identify substantial changes between data acquisitions. As summarized previously, multitemporal SAR data have been frequently used for land cover classification, as well as for monitoring ground deformation, glacier movements or other mass movements. However, currently only preliminary studies have been performed evaluating the use of a limited SAR time series data set for land cover change detection. Object-based approaches, time series analysis and increasing use of SAR data can be seen as major innovations in change detection research during the past decade – mainly driven by progress in data policy and data availability. There are many CEAs to extract changes. The labeling of the changes, however, provides the essential information about the nature of change.

Acknowledgements This work was carried out within the ENVILAND 2 research project funded by the Space Agency of the German Aerospace Center (DLR) with federal funds of the German Federal Ministry of Economics and Technology on the basis of legislation by the German Parliament grant no. 50EE0844 – 50EE0847. We acknowledge fruitful discussions with Mort Canty, Matthias Braun and the ZFL team.

References

Adams JB, Sabol DE, Kapos V, Almeida Filho R, Roberts DA, Smith MO, Gillespie AR (1995) Classification of multispectral images based on fractions of endmembers: application to land-cover change in the Brazilian Amazon. Remote Sens Environ 52(2):137–154

Allen TR, Kupfer JA (2000) Application of spherical statistics to change vector analysis of Landsat data: southern Appalachian spruce–fir forests. Remote Sens Environ 74(3):482–493

Arvidson T, Gasch J, Goward SN (2001) Landsat 7's long-term acquisition plan — an innovative approach to building a global imagery archive. Remote Sens Environ 78(1–2):13–26

Bamler R, Hartl P (1998) Synthetic aperture radar interferometry. Inverse Probl 14(4):R1–R54

Bazi Y, Bruzzone L, Melgani F (2005) An unsupervised approach based on the generalized Gaussian model to automatic change detection in multitemporal SAR images. IEEE Trans Geosci Remote Sens 43(4):874–887

Bazi Y, Melgani F, Bruzzone L, Vernazza G (2009) A genetic expectation-maximization method for unsupervised change detection in multitemporal SAR imagery. Int J Remote Sens 30(24):6591–6610

Bell JW, Amelung F, Ferretti A, Bianchi M, Novali F (2008) Permanent scatterer InSAR reveals seasonal and long-term aquifer-system response to groundwater pumping and artificial recharge. Water Resour Res 44:1–18

Bovolo F, Bruzzone L (2005) A detail-preserving scale-driven approach to change detection in multitemporal SAR images. IEEE Trans Geosci Remote Sens 43(12):2963–2972

Bovolo F, Bruzzone L (2007) A theoretical framework for unsupervised change detection based on change vector analysis in the polar domain. IEEE Trans Geosci Remote Sens 45(1):218–236

Briggs JM, Nellis MD (1991) Seasonal variation of heterogeneity in the tallgrass prairie: a quantitative measure using remote sensing. Photogramm Eng Remote Sens 57(4):407–411

Bruzzone L, Cossu R (2003) An adaptive approach to reducing registration noise effects in unsupervised change detection. IEEE Trans Geosci Remote Sens 41(11):2455–2465

Bruzzone L, Prieto DF (2000) Automatic analysis of the difference image for unsupervised change detection. IEEE Trans Geosci Remote Sens 38(3):1171–1182

Bruzzone L, Prieto DF (2002) An adaptive semiparametric and context-based approach to unsupervised change detection in multitemporal remote-sensing images. IEEE Trans Image Process 11(4):452–466

Bruzzone L, Marconcini M, Wegmuller U, Wiesmann A (2004) An advanced system for the automatic classification of multitemporal SAR images. IEEE Trans Geosci Remote Sens 42 (6):1321–1334

Byrne GF, Crapper PF, Mayo KK (1980) Monitoring land-cover change by principal component analysis of multitemporal Landsat data. Remote Sens Environ 10(3):175–184

Canty MJ, Nielsen AA (2006) Visualization and unsupervised classification of changes in multispectral satellite imagery. Int J Remote Sens 27(18): 3961–3975

Canty MJ, Nielsen AA (2008) Automatic radiometric normalization of multitemporal satellite imagery with the iteratively re-weighted MAD transformation. Remote Sens Environ 112 (3):1025–1036

Carvalho LMT, Fonseca LMG, Murtagh F, Clevers JGPW (2001) Digital change detection with the aid of multiresolution wavelet analysis. Int J Remote Sens 22(18):3871–3876

Castilla G, Guthrie RH, Hay GJ (2009) The Land-cover Change Mapper (LCM) and its application to timber harvest monitoring in Western Canada. Photogramm Eng Remote Sens 75(8):941–950

Chen CH (ed) (2007) Image processing for remote sensing. CRC Press, Boca Raton, 400 pp

Chen J, Gong P, Chunyang H, Pu R, Shi P (2003) Land-use/land-cover change detection using improved change-vector analysis. Photogramm Eng Remote Sens 69(4):369–379

Chen G, Hay GJ, Carvalho LMT, Wulder MA (2012) Object-based change detection. Int J Remote Sens 33(14):4434–4457

Collins JB, Woodcock CE (1994) Change detection using the Gramm-Schmidt transformation applied to mapping forest mortality. Remote Sens Environ 50(3):267–279

Collins JB, Woodcock CE (1996) An assessment of several linear change detection techniques for mapping forest mortality using multitemporal Landsat TM data. Remote Sens Environ 56 (1):66–77

Conradsen K, Nielsen AA, Schou J, Skriver H (2003) A test statistic in the complex wishart distribution and its application to change detection in polarimetric SAR data. IEEE Trans Geosci Remote Sens 41(1):4–19

Coops NC, Gillanders SN, Wulder MA, Gergel SE, Nelson T, Goodwin NR (2010) Assessing changes in forest fragmentation following infestation using time series Landsat imagery. For Ecol Manag 259(12):2355–2365

Coppin PR, Bauer ME (1994) Processing of multitemporal Landsat TM imagery to optimize extraction of forest cover change features. IEEE Trans Geosci Remote Sens 32(4):918–927

Coppin P, Jonckheere I, Nackaerts K, Muys B, Lambin E (2004) Digital change detection methods in ecosystem monitoring: a review. Int J Remote Sens 25(9):1565–1596

Crist EP, Cicone RC (1984) A physically-based transformation of thematic mapper data—the TM tasseled cap. IEEE Trans Geosci Remote Sens GE-22(3):256–263

Dawelbait M, Morari F (2012) Monitoring desertification in a savannah region in Sudan using Landsat images and spectral mixture analysis. J Arid Environ 80:45–55

Del Frate F, Pacifici F, Solimini D (2008) Monitoring urban land cover in Rome, Italy, and its changes by single-polarization multitemporal SAR images. IEEE J Sel Top Appl Earth Obs Remote Sens 1(2):87–97

Demir B, Bovolo F, Bruzzone L (2012) Detection of land-cover transitions in multitemporal remote sensing images with active-learning-based compound classification. IEEE Trans Geosci Remote Sens 50(5):1930–1941

Desclée B, Bogaert P, Defourny P (2004) Object-based method for automatic forest change detection. In: IEEE international geoscience and remote sensing symposium. IGARSS'04, Anchorage Alaska, vol 5, pp 3383–3386

Desclée B, Bogaert P, Defourny P (2006) Forest change detection by statistical object-based method. Remote Sens Environ 102(1–2):1–11

Fallourd R, Harant O, Trouve E, Nicolas J-M, Gay M, Walpersdorf A, Mugnier J-L et al (2011) Monitoring temperate glacier displacement by multi-temporal TerraSAR-X images and continuous GPS measurements. IEEE J Sel Top Appl Earth Obs Remote Sens 4(2):372–386

Ferretti A, Savio G, Barzaghi R, Borghi A, Musazzi S, Novali F, Prati C, Rocca F (2007) Submillimeter accuracy of InSAR time series: experimental validation. IEEE Trans Geosci Remote Sens 45(5):1142–1153

Foody GM (2009) The impact of imperfect ground reference data on the accuracy of land cover change estimation. Int J Remote Sens 30(12): 3275–3281

Foody GM (2009. 2010) Assessing the accuracy of land cover change with imperfect ground reference data. Remote Sens Environ 114: 2271–2285

Fung T (1990) An assessment of TM imagery for land-cover change detection. IEEE Trans Geosci Remote Sens 28(4):681–684

Gamba P, Dell'Acqua F, Lisini G (2006) Change detection of multitemporal SAR data in urban areas combining feature-based and pixel-based techniques. IEEE Trans Geosci Remote Sens 44(10):2820–2827

Gath I, Geva AB (1989) Unsupervised optimal fuzzy clustering. IEEE Trans Pattern Anal Mach Intell 11(7):773–781

Gianinetto M, Villa P (2007) Rapid response flood assessment using minimum noise fraction and composed spline interpolation. IEEE Trans Geosci Remote Sens 45(10):3204–3211

Gianinetto M, Villa P (2011) Mapping hurricane Katrina's widespread destruction in New Orleans using multisensor data and the Normalized Difference Change Detection (NDCD) technique. Int J Remote Sens 32(7):1961–1982

Gong P, Ledrew EF, Miller JR (1992) Registration-noise reduction in difference images for change detection. Int J Remote Sens 13(4):773–779

Griffiths P, Kuemmerle T, Kennedy RE, Abrudan IV, Knorn J, Hostert P (2012) Using annual time-series of Landsat images to assess the effects of forest restitution in post-socialist Romania. Remote Sens Environ 118:199–214

Hais M, Jonášová M, Langhammer J, Kučera T (2009) Comparison of Two types of forest disturbance using multitemporal Landsat TM/ETM+ imagery and field vegetation data. Remote Sens Environ 113(4):835–845

Haralick RM, Shanmugam K, Dinstein I (1973) Textural features for image classification. IEEE Trans Syst Man Cybern SMC-3(6):610–621

He C, Wei A, Shi P, Zhang Q, Zhao Y (2011) Detecting land-use/land-cover change in rural–urban fringe areas using extended change-vector analysis. Int J Appl Earth Obs Geoinfor 13(4):572–585

Hecheltjen A, Waske B, Thonfeld F, Braun M, Menz G (2010) Support vector machines for multitemporal and multisensor change detection. In: ESA SP-686, European Space Agency, Bergen

Henrich V, Götze E, Jung A, Sandow C, Thürkow D, Gläßer C (2009) Development of an Online indices-database: motivation, concept and implementation. In: EARSeL proceedings, EARSeL, Tel Aviv 3 pp. http://www.earsel6th.tau.ac.il/~earsel6/CD/PDF/earsel-PROCEED INGS/3064%20Henrich.pdf

Herman F, Anderson B, Leprince S (2011) Mountain glacier velocity variation during a retreat/advance cycle quantified using sub-pixel analysis of ASTER images. J Glaciol 57(202):197–207

Hostert P, Röder A, Hill J (2003) Coupling spectral unmixing and trend analysis for monitoring of long-term vegetation dynamics in Mediterranean rangelands. Remote Sens Environ 87(2–3):183–197

Howarth PJ, Wickware GM (1981) Procedures for change detection using Landsat digital data. Int J Remote Sens 2(3):277–291

Huang C, Goward SN, Masek JG, Thomas N, Zhu Z, Vogelmann JE (2010a) An automated approach for reconstructing recent forest disturbance history using dense Landsat time series stacks. Remote Sens Environ 114(1):183–198

Huang C, Thomas N, Goward SN, Masek JG, Zhu Z, Townshend JRG, Vogelmann JE (2010b) Automated masking of cloud and cloud shadow for forest change analysis using Landsat images. Int J Remote Sens 31(20):5449–5464

Ingebritsen SE, Lyon RJP (1985) Principal components analysis of multitemporal image pairs. Int J Remote Sens 6(5):687–696
Irish RR (2000) Landsat 7 automatic cloud cover assessment. Algorithm Multispectr Hyperspectr Ultraspect Imag VI:348–355. SPIE Proceedings 4049
Irish RR, Barker JL, Goward SN, Arvidson T (2006) Characterization of the Landsat-7 ETM Automated Cloud-Cover Assessment (ACCA) algorithm. Photogramm Eng Remote Sens 72(10):1179–1188
Jensen JR, Ramsey EW, Mackey HE, Christensen EJ, Sharitz RR (1987) Inland wetland change detection using aircraft MSS data. Photogramm Eng Remote Sens 53(5):521–529
Jha CS, Unni NVM (1994) Digital change detection of forest conversion of a dry tropical Indian forest region. Int J Remote Sens 15(13):2543–2552
Johnson RD, Kasischke ES (1998) Change vector analysis: a technique for the multispectral monitoring of land cover and condition. Int J Remote Sens 19(3):411–426
Ju J, Roy DP (2008) The availability of cloud-free Landsat ETM+ data over the conterminous United States and globally. Remote Sens Environ 112(3):1196–1211
Kauth R, Thomas G (1976) The tasselled cap – a graphic description of the spectral-temporal development of agricultural crops as seen by LANDSAT. In: LARS symposium on machine processing of remotely sensed data, vol Paper 159. Purdue. http://docs.lib.purdue.edu/lars_symp/159
Kennedy RE, Townsend PA, Gross JE, Cohen WB, Bolstad P, Wang YQ, Adams P (2009) Remote sensing change detection tools for natural resource managers: understanding concepts and tradeoffs in the design of landscape monitoring projects. Remote Sens Environ 113(7):1382–1396
Kennedy RE, Yang Z, Cohen WB (2010) Detecting trends in forest disturbance and recovery using yearly Landsat time series: 1. LandTrendr–temporal segmentation algorithms. Remote Sens Environ 114:2897–2910
Kennedy RE, Yang Z, Cohen WB, Pfaff E, Braaten J, Nelson P (2012) Spatial and temporal patterns of forest disturbance and regrowth within the area of the northwest forest plan. Remote Sens Environ 122:117–133
Klonus S, Tomowski D, Ehlers M, Reinartz P, Michel U (2012) Combined edge segment texture analysis for the detection of damaged buildings in crisis areas. IEEE J Sel Top Appl Earth Obs Remote Sens 5(4):1118–1128
Kuemmerle T, Hostert P, Radeloff V, van der Linden S, Perzanowski K, Kruhlov I (2008) Cross-border comparison of post-socialist farmland abandonment in the Carpathians. Ecosystems 11(4):614–628
Laliberte AS, Rango A, Havstad KM, Paris JF, Beck RF, Mcneely R, Gonzalez AL (2004) Object-oriented image analysis for mapping shrub encroachment from 1937 to 2003 in southern New Mexico. Remote Sens Environ 93(1–2):198–210
Lambin EF, Strahler AH (1994) Change-vector analysis in multitemporal space: a tool to detect and categorize land-cover change processes using high temporal-resolution satellite data. Remote Sens Environ 48(2):231–244
Lawrence RL, Ripple WJ (1999) Calculating change curves for multitemporal satellite imagery: Mount St. Helens 1980–1995. Remote Sens Environ 67(3):309–319
Leberl FW (1983) Photogrammetric aspects of remote sensing with imaging radar. Remote Sens Rev 1(1):71–158
Leprince S, Barbot S, Ayoub F, Avouac J-P (2007) Automatic and precise orthorectification, coregistration, and subpixel correlation of satellite images, application to ground deformation measurements. IEEE Trans Geosci Remote Sens 45(6):1529–1558
Li B, Zhou Q (2009) Accuracy assessment on multi-temporal land-cover change detection using a trajectory error matrix. Int J Remote Sens 30(5):1283–1296
Liao M, Jiang L, Lin H, Huang B, Gong J (2008) Urban change detection based on coherence and intensity characteristics of SAR imagery. Photogramm Eng Remote Sens 74(8):999–1006
Listner C, Niemeyer I (2011) Object-based change detection. Photogrammetrie Fernerkundung Geoinfor 4:233–245

Lu D, Mausel P, Brondizio E, Moran E (2004) Change detection techniques. Int J Remote Sens 25(12):2365–2401

Malila WA (1980) Change vector analysis: an approach for detecting forest changes with Landsat. In: Proceedings of the 6th annual symposium on machine processing of remotely sensed data, Purdue University, Indiana, pp 329–335

Markham BL, Storey JC, Williams DL, Irons JR (2004) Landsat sensor performance: history and current status. IEEE Trans Geosci Remote Sens 42(12):2691–2694

Martinez J-M, Le Toan T (2007) Mapping of flood dynamics and spatial distribution of vegetation in the Amazon floodplain using multitemporal SAR data. Remote Sens Environ 108(3):209–223

Masek JG, Vermote EF, Saleous NE, Wolfe R, Hall FG, Huemmrich KF, Gao F, Kutler J, Lim T-K (2006) A Landsat surface reflectance dataset for North America, 1990–2000. IEEE Geosci Remote Sens Lett 3(1):68–72

Massonnet D, Rossi M, Carmona C, Adragna F, Peltzer G, Feigl K, Rabaute T (1993) The displacement field of the Landers earthquake mapped by radar interferometry. Nature 364:138–142

McFeeters SK (1996) The use of the Normalized Difference Water Index (NDWI) in the delineation of open water features. Int J Remote Sens 17(7):1425–1432

Meigs GW, Kennedy RE, Cohen WB (2011) A Landsat time series approach to characterize bark beetle and defoliator impacts on tree mortality and surface fuels in conifer forests. Remote Sens Environ 115(12):3707–3718

Michalek JL, Wagner TW, Luczkovich JJ, Stoffle RW (1993) Multispectral change vector analysis for monitoring coastal marine environments. Photogramm Eng Remote Sens 59(3):381–384

Michel R, Avouac J-P, Taboury J (1999) Measuring ground displacements from SAR amplitude images: application to the Landers Earthquake. Geophys Res Lett 26(7):875–878

Milne AK (1988) Change direction analysis using Landsat imagery: a review of methodology. In: IEEE international geoscience and remote sensing symposium. IGARSS'88. Remote sensing: moving toward the 21st century, vol 1, pp 541–544

Morisette JT, Khorram S (2000) Accuracy assessment curves for satellite-based change detection. Photogramm Eng Remote Sens 66(7):875–880

Mouat D, Mahin G, Lancaster J (1993) Remote sensing techniques in the analysis of change detection. Geocarto Int 8(2):39–50

Necsoiu M, Leprince S, Hooper DM, Dinwiddie CL, McGinnis RN, Walter GR (2009) Monitoring migration rates of an active subarctic dune field using optical imagery. Remote Sens Environ 113(11):2441–2447

Nelson RF (1983) Detecting forest canopy change due to insect activity using Landsat MSS. Photogramm Eng Remote Sens 49(9):1303–1314

Nielsen AA, Conradsen K (1997) Multivariate Alteration Detection (MAD) in multispectral, Bi-temporal image data: a new approach to change detection studies. Technical report IMM-REP-1997-11. IMM, Department of Mathematical Modelling, Technical University of Denmark, Denmark

Nielsen AA, Canty MJ (2008) Kernel principal component analysis for change detection. In: Image and signal processing for remote sensing XIV, 7109:71090T–1–10. SPIE, Cardiff, Wales, UK

Nielsen AA, Canty MJ (2009) Kernel principal component and maximum autocorrelation factor analyses for change detection. In: Image and signal processing for remote sensing XV, 7477:74770T–1–6. SPIE

Nielsen AA, Hecheltjen A, Thonfeld F, Canty MJ (2010) Automatic change detection in RapidEye data using the combined MAD and kernel MAF methods. In: IEEE International Geoscience and Remote Sensing Symposium (IGARSS 2010), Honolulu, pp 3078–3081

Niemeyer I, Marpu PR, Nussbaum S (2007) Change detection using the object features. In: IEEE International Geoscience and Remote Sensing Symposium, IGARSS 2007, pp 2374–2377, Barcelona

Pflugmacher D, Cohen WB, Kennedy RE (2012) Using Landsat-derived disturbance history (1972–2010) to predict current forest structure. Remote Sens Environ 122:146–165
Polidori L, Caillault S, Canaud J-L (1995) Change detection in radar images: methods and operational constraints. In: IEEE International Geoscience and Remote Sensing Symposium. IGARSS'95. Quantitative remote sensing for science and applications, vol 2. Firenze Italy, 1529–1531
Powell SL, Cohen WB, Yang Z, Pierce JD, Alberti M (2008) Quantification of impervious surface in the Snohomish water resources inventory area of Western Washington from 1972–2006. Remote Sens Environ 112(4):1895–1908
Powell SL, Cohen WB, Healey SP, Kennedy RE, Moisen GG, Pierce KB, Ohmann JL (2010) Quantification of live aboveground forest biomass dynamics with Landsat time-series and field inventory data: a comparison of empirical modeling approaches. Remote Sens Environ 114(5):1053–1068
Pritchard HD, Vaughan DG (2007) Widespread acceleration of tidewater glaciers on the Antarctic Peninsula. J Geophys Res 112(F3):F03S29
Quegan S, Le Toan T, Yu JJ, Ribbes F, Floury N (2000) Multitemporal ERS SAR analysis applied to forest mapping. IEEE Trans Geosci Remote Sens 38(2):741–753
Radke RJ, Andra S, Al-Kofahi O, Roysam B (2005) Image change detection algorithms: a systematic survey. IEEE Trans Image Process 14(3):294–307
Ridd MK, Liu J (1998) A comparison of four algorithms for change detection in an urban environment. Remote Sens Environ 63(2):95–100
Rignot EJM, Van Zyl JJ (1993) Change detection techniques for ERS-1 SAR data. IEEE Trans Geosci Remote Sens 31(4):896–906
Röder A, Hill J, Duguy B, Alloza JA, Vallejo R (2008a) Using long time series of Landsat data to monitor fire events and post-fire dynamics and identify driving factors. A case study in the Ayora Region (Eastern Spain). Remote Sens Environ 112(1):259–273
Röder A, Udelhoven T, Hill J, del Barrio G, Tsiourlis G (2008b) Trend analysis of Landsat-TM and -ETM+ imagery to monitor grazing impact in a rangeland ecosystem in Northern Greece. Remote Sens Environ 112(6):2863–2875
Rosen PA, Hensley S, Joughin IR, Li FK, Madsen SN, Rodriguez E, Goldstein RM (2000) Synthetic aperture radar interferometry. Proc IEEE 88(3):333–382
Rosin PL (2001) Unimodal thresholding. Pattern Recogn 34(11):2083–2096
Schmitt A, Wessel B, Roth A (2010) Curvelet-based change detection on SAR images for natural disaster mapping. Photogrammetrie Fernerkundung Geoinformation 2010(6):463–474
Schott JR, Salvaggio C, Volchok WJ (1988) Radiometric scene normalization using pseudoinvariant features. Remote Sens Environ 26(1):1–16
Schroeder TA, Cohen WB, Song C, Canty MJ, Yang Z (2006) Radiometric correction of multi-temporal Landsat data for characterization of early successional forest patterns in Western Oregon. Remote Sens Environ 103(1):16–26
Schroeder TA, Cohen WB, Yang Z (2007) Patterns of forest regrowth following clearcutting in Western Oregon as determined from a Landsat time-series. For Ecol Manag 243(2–3):259–273
Schroeder TA, Wulder MA, Healey SP, Moisen GG (2011) Mapping wildfire and clearcut harvest disturbances in boreal forests with Landsat time series data. Remote Sens Environ 115(6):1421–1433
Sen S, Zipper CE, Wynne RH, Donovan P (2012) Identifying revegetated mines as disturbance/recovery trajectories using an interannual Landsat chronosequence. Photogramm Eng Remote Sens 78(3):223–235
Shepherd A, Wingham DJ, Mansley JAD, Corr HFJ (2001) Inland thinning of pine island glacier, West Antarctica. Science 291(5505):862–864
Singh A (1989) Review article digital change detection techniques using remotely-sensed data. Int J Remote Sens 10(6):989–1003

Song C, Woodcock CE, Seto KC, Lenney MP, Macomber SA (2001) Classification and change detection using Landsat TM data: when and how to correct atmospheric effects? Remote Sens Environ 75(2):230–244
Sonnenschein R, Kuemmerle T, Udelhoven T, Stellmes M, Hostert P (2011) Differences in Landsat-based trend analyses in drylands Due to the choice of vegetation estimate. Remote Sens Environ 115(6):1408–1420
Souza CM Jr, Roberts DA, Cochrane MA (2005) Combining spectral and spatial information to Map canopy damage from selective logging and forest fires. Remote Sens Environ 98(2–3):329–343
Stellmes M, Udelhoven T, Röder A, Sonnenschein R, Hill J (2010) Dryland observation at local and regional scale — comparison of Landsat TM/ETM+ and NOAA AVHRR time series. Remote Sens Environ 114(10):2111–2125
Stow DA (1999) Reducing the effects of misregistration on pixel-level change detection. Int J Remote Sens 20(12):2477–2483
Strozzi T, Luckman A, Murray T, Wegmuller U, Werner CL (2002) Glacier motion estimation using SAR offset-tracking procedures. IEEE Trans Geosci Remote Sens 40(11):2384–2391
Strozzi T, Wegmuller U, Werner CL, Wiesmann A, Spreckels V (2003) JERS SAR interferometry for land subsidence monitoring. IEEE Trans Geosci Remote Sens 41(7):1702–1708
Strozzi T, Farina P, Corsini A, Ambrosi C, Thüring M, Zilger J, Wiesmann A, Wegmüller U, Werner C (2005) Survey and monitoring of landslide displacements by means of L-band satellite SAR interferometry. Landslides 2(3):193–201
Thonfeld F, Menz G (2011) In: Coherence and multitemporal intensity metrics of high resolution SAR images for urban change detection. DLR, Oberpfaffenhofen, pp 1–9. http://sss.terrasar-x.dlr.de/papers_sci_meet_4/oral/LAN0125_thonfeld.pdf
Thonfeld F, Hecheltjen A, Braun M, Menz G (2010) From algorithms to processing chains: a review of land cover and land use change detection methodologies. In Proceedings of the ESA living planet symposium, vol ESA SP-686. European Space Agency, Bergen, Norway
Townshend JRG, Justice CO, Gurney C, McManus J (1992) The impact of misregistration on change detection. IEEE Trans Geosci Remote Sens 30(5):1054–1060
Tucker CJ (1979) Red and photographic infrared linear combinations for monitoring vegetation. Remote Sens Environ 8(2):127–150
Van Oort PAJ (2007) Interpreting the change detection error matrix. Remote Sens Environ 108(1):1–8
Verbesselt J, Hyndman R, Newnham G, Culvenor D (2010a) Detecting trend and seasonal changes in satellite image time series. Remote Sens Environ 114(1):106–115
Verbesselt J, Hyndman R, Zeileis A, Culvenor D (2010b) Phenological change detection while accounting for abrupt and gradual trends in satellite image time series. Remote Sens Environ 114(12):2970–2980
Verbesselt J, Zeileis A, Herold M (2012) Near real-time disturbance detection using satellite image time series. Remote Sens Environ 123:98–108
Vicente-Serrano SM, Pérez-Cabello F, Lasanta T (2008) Assessment of radiometric correction techniques in analyzing vegetation variability and change using time series of Landsat images. Remote Sens Environ 112(10):3916–3934
Viedma O, Meliá J, Segarra D, Garcia-Haro J (1997) Modeling rates of ecosystem recovery after fires by using Landsat TM data. Remote Sens Environ 61(3):383–398
Vogelmann JE, Helder D, Morfitt R, Choate MJ, Merchant JW, Bulley H (2001) Effects of landsat 5 thematic mapper and Landsat 7 enhanced thematic mapper plus radiometric and geometric calibrations and corrections on Landscape characterization. Remote Sens Environ 78(1–2):55–70
Vogelmann JE, Kost JR, Tolk B, Howard S, Short K, Chen X, Huang C, Pabst K, Rollins MG (2011) Monitoring landscape change for LANDFIRE using multi-temporal satellite imagery and ancillary data. IEEE J SelTop Appl Earth Obs Remote Sens 4(2):252–264

Wegmüller U, Walter D, Spreckels V, Werner CL (2010) Nonuniform ground motion monitoring with TerraSAR-X persistent scatterer interferometry. IEEE Trans Geosci Remote Sens 48(2):895–904

Wei S, Fielding E, Leprince S, Sladen A, Avouac J-P, Helmberger D, Hauksson E et al (2011) Superficial simplicity of the 2010 El Mayor-Cucapah Earthquake of Baja California in Mexico. Nat Geosci 4(9):615–618

Weismiller RA, Kristof SJ, Scholz DK, Anuta PE, Momin SM (1977) Change detection in coastal zone environments. Photogramm Eng Remote Sens 43(12):1533–1539

Wiemker R (1997) An iterative spectral-spatial Bayesian labeling approach for unsupervised robust change detection on remotely sensed multispectral imagery. Lect Notes Comput Sci 263–270

Williams DL, Goward S, Arvidson T (2006) Landsat: yesterday, today, and tomorrow. Photogramm Eng Remote Sens 72(10):1171–1178

Wulder MA, Masek JG, Cohen WB, Loveland TR, Woodcock CE (2012) Opening the archive: how free data has enabled the science and monitoring promise of Landsat. Remote Sens Environ 122:2–10

Xian G, Homer C (2010) Updating the 2001 national land cover database impervious surface products to 2006 using Landsat imagery change detection methods. Remote Sens Environ 114(8):1676–1686

Xian G, Homer C, Fry J (2009) Updating the 2001 national land cover database land cover classification to 2006 by using Landsat imagery change detection methods. Remote Sens Environ 113(6):1133–1147

Yang J, Weisberg PJ, Bristow NA (2012) Landsat remote sensing approaches for monitoring long-term tree cover dynamics in semi-arid woodlands: comparison of vegetation indices and spectral mixture analysis. Remote Sens Environ 119:62–71

Yonezawa C, Takeuchi S (2001) Decorrelation of SAR data by urban damages caused by the 1995 Hyogoken-nanbu Earthquake. Int J Remote Sens 22(8):1585–1600

Yoon YT, Eineder M, Yague-Martinez N, Montenbruck O (2009) TerraSAR-X precise trajectory estimation and quality assessment. IEEE Trans Geosci Remote Sens 47(6):1859–1868

Zebker HA, Villasenor J (1992) Decorrelation in interferometric radar echoes. IEEE Trans Geosci Remote Sens 30(5):950–959

Zhao D, Kuenzer C, Fu C, Wagner W (2008) Evaluation of the ERS scatterometer-derived soil water index to monitor water availability and precipitation distribution at three different scales in China. J Hydrometeorol 9(3):549–562

Zhu Z, Woodcock CE (2012) Object-based cloud and cloud shadow detection in Landsat imagery. Remote Sens Environ 118:83–94

Zhu Z, Woodcock CE, Olofsson P (2012) Continuous monitoring of forest disturbance using all available Landsat imagery. Remote Sens Environ 122:75–91

Chapter 11
Synergies from SAR-Optical Data Fusion for LULC Mapping

Björn Waske

11.1 Introduction

Nowadays Earth Observation (EO) systems play a major role in supporting environmental programs and monitoring compliances, such as the European Global Monitoring for Environment and Security (GMES – Copernicus) program. Copernicus aims on the provision of reliable and current information on our planet and its environmental state in three Earth system domains "Land", "Atmosphere", and "Marine". Moreover, the corresponding services support the management of humanitarian crises, natural disasters and man-made crisis (Aschbacher and Milagro-Pérez 2012). Products in the Land-domain, comprise accurate and cross-border harmonized information on land cover and land cover change, including information on seasonal and annual changes, the vegetation state and the monitoring of the water cycle. Overall these products will support decision-making and various monitoring applications in context of land use and land cover change, water quality, spatial planning, and global food security (Copernicus 2012).

In context of Copernicus as well as other environmental treaties, land cover mapping is the most commonly used remote sensing application. Consequently, the development of adequate classification methods is an ongoing research field. However, map accuracies are driven, among others, by factors such as input images, training data, study site, and classification method (e.g., Smith et al. 2003; Waske and Benediktsson 2007).

Many land cover classifications aim on areas that are characterized by great temporal variability and typical spatial patterns of highly frequent land cover changes between different crops. As a matter of fact, multitemporal imagery can significantly improve the classification accuracy when compared to the accuracy achieved by single-date applications (Brisco and Brown 1995; Blaes et al. 2005;

B. Waske (✉)
Institute of Geographical Sciences, Freie Universität Berlin, Germany
e-mail: bjoern.waske@fu-berlin.de

I. Manakos and M. Braun (eds.), *Land Use and Land Cover Mapping in Europe: Practices & Trends*, Remote Sensing and Digital Image Processing 18,
DOI 10.1007/978-94-007-7969-3_11,

McNairn et al. 2009). Nevertheless, the potential and availability optical images may be limited by weather conditions, particularly in regions of Central Europe. Multitemporal data sets within one growing season can be produced reliably by synthetic aperture radar (SAR). Moreover, optical and SAR data provide different, often synergetic information on land cover. As a matter of fact the classification accuracies can be increased by combining optical and SAR data (e.g., Blaes et al. 2005; Waske and Benediktsson 2007; Waske and van der Linden 2008; McNairn et al. 2009). Regarding recently launched EO missions with increased revisit times and better spatial resolutions like the German TerraSAR-X and RapidEye multisensor applications become even more attractive. In addition, upcoming missions like ESAs satellite constellation Sentinel will further increase the availability of multisensor data sets, e.g. due to Sentinel-1 and Sentinel-2.

Overall, users can choose between various EO data sets. Nevertheless, the number of studies and applications that are based on multisensor data sets is still limited. One reason might be the demand for adequate classification strategies. Frequently used methods as the maximum likelihood classifier are often limited when classifying multisensor and multitemporal imagery. Although some multisensor applications are based on conventional classification methods, more sophisticated approaches from the field of machine learning, such as support vector machines and classifier ensemble as Random Forests, have been shown to perform more accurate results (e.g., Briem et al. 2002; Gislason et al. 2006; Waske and van der Linden 2008; Waske and Braun 2009).

Unfortunately implementations of such methods often may require complex frameworks and often expert knowledge and/or programming skills. Moreover, they do not necessarily support the direct use of remote sensing imagery. In contrast to this, widely used (commercial) remote sensing software packages typically do not include more recent classifier developments (Waske et al. 2012).

The major aim of this study is to foster the operational use of Random Forests (RF) classification approach as well as the use of multisensor remote sensing data, by using a freely available and platform independent tool for remote sensing image classification, called imageRF (Waske et al. 2012).

The specific objective of this study is the classification of multisensor data (Enivsat ASAR and SPOT) from agricultural areas. To demonstrate the positive impact of different sensors, classifications with different input data sets are performed. Moreover, the study demonstrates the general potential of RF in this context. The impact of the two parameters on the accuracy and stability of ensembles in terms of the classification result is investigated.

11.2 Background

11.2.1 Random Forests

The concept of classifier ensembles, e.g., Random Forests, is based on the idea of combining outputs from more than one classifier to enhance classification accuracy.

In many cases a self-learning decision tree (DT) is used, even though the use of more complex classification methods was investigated (e.g., Kim et al. 2003; Zortea et al. 2008; Waske et al. 2010). However, decision trees are computationally light and simple to handle.

RF (Breiman 2001) use a set of k DT classifiers. Each tree within the ensemble is trained with a randomly generated subset of the original training samples. Moreover, at each split node of the tree a subset of m randomly selected input features is used, with $m < D$ and D as the dimension of the input image. The number of randomly selected features is user defined and often set to the square root of the total number of input features, i.e., D the dimension of the input data set. Another common default value is $m = \log(D)$.

The final classification, i.e., the assignment of each pixel to an information class (i.e., the specified land-cover classes), is based on a majority vote over all k classifier decisions. The proportion of votes for a certain class $\omega_j \in \Omega = \{\omega_1, \ldots, \omega_c\}$ over all k trees can be also interpreted as the (pseudo)probability of occurrence of that class:

$$P_x(\omega_j) = \frac{n_j}{k}$$

with n_j as the number of trees classifying pixel x as class ω_j. A pixel is finally classified to that class with the highest likelihood (i.e., using a simple majortiy vote).

Beside the classification output, the RF algorithm provides additional parameters. The out-of-bag (OOB) error seems particularly interesting, because it enables an independent assessment of the classification accuracy, without providing additional test data. Each individual tree in the ensemble is based on a randomly chosen 2/3 of the training data, while 1/3 is left out for each specific tree. These out of bag samples are classified by that particular tree. The OOB-error is derived by the classification error of all left-out samples, averaged over the total number of trees (Breiman 2001). The variable importance, which utilizes the OBB measure, provide information in the impact of each band on the classification accuracy.

RF have been widely used for ecological applications (e.g., Prasad et al. 2006; Cutler et al. 2007) and were successfully introduced in context of land cover classification of remote sensing data (e.g., Gislason et al. 2006; Chan and Paelinckx 2008; Waske and van der Linden 2008; Waske and Braun 2009; Loosvelt et al. 2012). These results show that RF can perform similar or even outperform other methods in terms of accuracy.

11.2.2 imageRF Classification Software

The training and classification were performed using *imageRF* (Waske et al. 2012), which is a freely available IDL-based tool for RF classification and regression analysis of remote sensing imagery. In this study *imageRF* was run in the

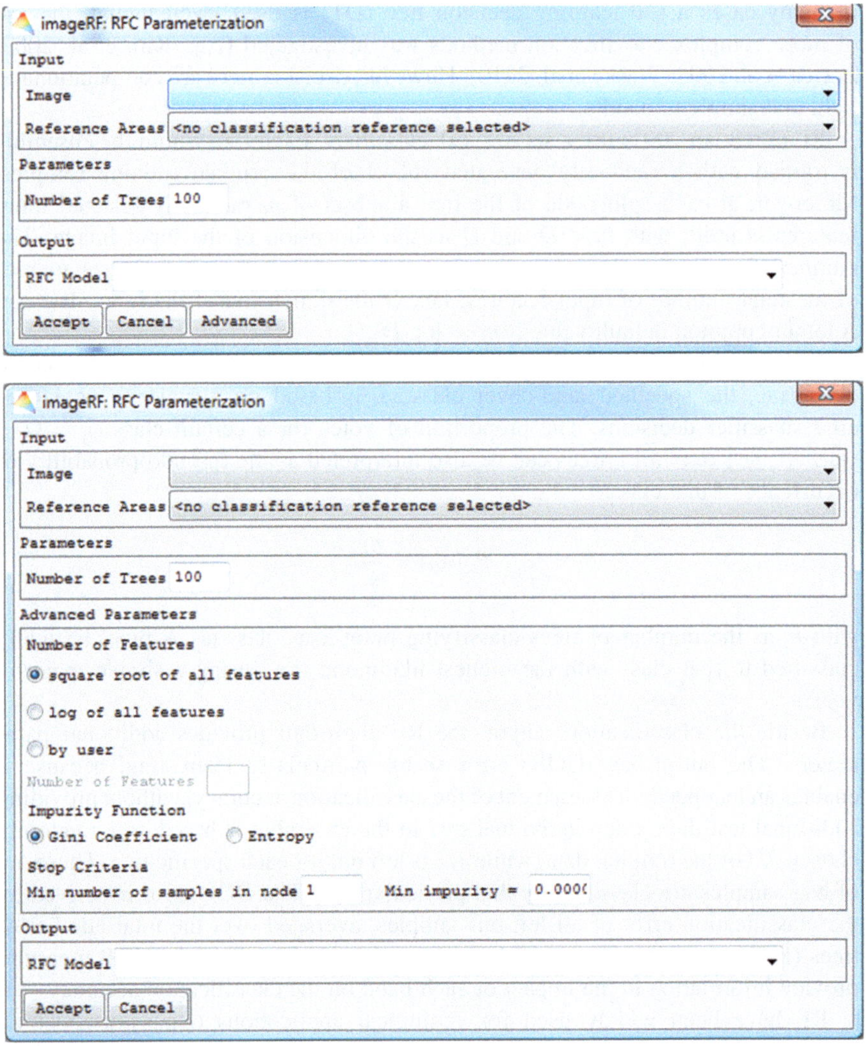

Fig. 11.1 Basic (*top*) and advanced (*bottom*) parameterization menu for RF classification, using imageRF in the EnMAP-Box

EnMAP-Box[1] a freely available, license-free and platform-independent processing environment for remote sensing imagery. However, *imageRF* can be fully integrated into the commercially available IDL/ENVI environment. *imageRF* includes among others (i) generic image files for the input images, reference data and results, (ii) separate training and application of the RF model, i.e., a trained model is saved and can be applied several times, and (iii) a user-friendly GUI (Fig. 11.1) for the

[1] http://www.geographie.hu-berlin.de/abteilungen/geomatik/projects/enmapbox
http://indus.caf.dlr.de/forum/

definition of the following parameters: number of iterations, number of randomly selected features, stopping and split criterion. Although these parameters can be freely defined by the user, reliable default parameters are provided. Regarding other remote sensing applications, which usually report the classification accuracy, and contrary to the original RF implementation the OOB accuracy (i.e., 100 – obb-error) is provided.

11.3 Study Site and Data Set

The study site is located near the city of Bonn in Western Germany, in the federal state North Rhine-Westphalia at an average altitude of approximately 170 m ASL (Fig. 11.2). The almost flat area is dominated by agriculture and thus characterized by the typical spatial patterns that result from.

Fig. 11.2 Location of the study site (indicated by the *square*)

The classification was conducted on a set of multitemporal SAR and multispectral images. Ten Envisat ASAR alternating polarization (HH/HV) and image mode (VV) images were acquired between 16. March and 13. September, 2007. The multispectral data set consist of two SPOT-4 images from 24. May and 4. August, 2007. The SAR data was calibrated to backscatter intensity following common procedures. The SAR inherent noise was reduced by multitemporal speckle filter. Finally, the SAR images were orthorectified with a spatial accuracy of approximately one pixel, using a digital elevation model, orbit parameters and a multispectral reference image.

11.4 Results and Discussion

11.4.1 Experimental Setup

To underline the impact of the input data on the classification accuracy, RF was applied several times, using (i) the SAR data, (ii) the SPOT data and (iii) the multisensor data set.

For the generation of independent training and validation sets a stratified random sampling approach was chosen, using the corresponding ground truth data as a priori knowledge. To investigate the impact of the number of training samples on the classification accuracy, different sizes of the training sets were generated containing 50, 100, 150 and 200 samples per land cover class. Each sample set size was generated ten times and the final classification accuracies were averaged and the standard deviation was derived. The same independent validation set was used for the evaluation of all classification results. Accuracy assessment was performed using overall accuracies and confusion matrices in order to derivate the producer's and user's accuracies.

The classifications in this study are performed following common default values (i.e., square root of the number of features). Moreover, various classifications were generated by systematically increasing the number of randomly selected features (i.e., m) as well as the number of trees that are combined for the final class decision (i.e., k). In doing so, the impact of the two parameters on the accuracy and stability of ensembles in terms of the classification result is investigated.

11.4.2 Classification Results, Using Single-Source Data Sets

The experiments in this paper were conducted on the multisensor data sets consisting of SAR and multispectral imagery using different number of training samples. RF was applied three times: on the multitemporal SAR, the multispectral and the multisensor data set.

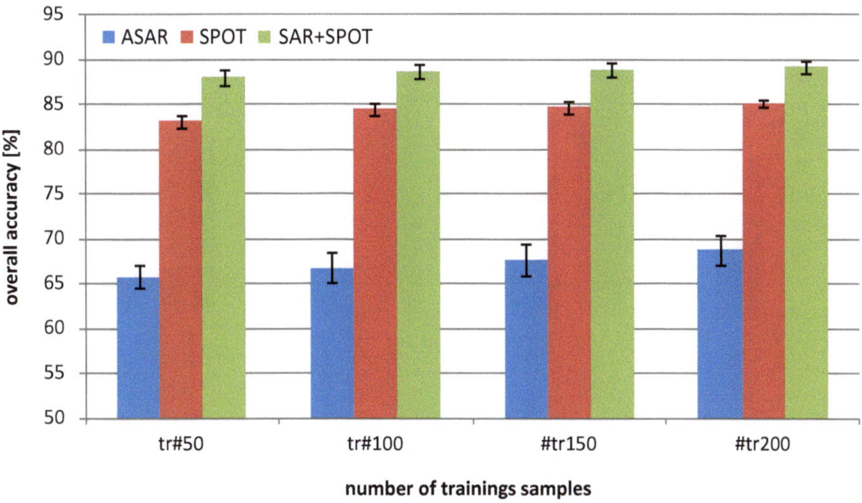

Fig. 11.3 Overall accuracies, using different data sets and number of training samplers. *Error bars* indicate the standard deviation over 10 trails

As expected, the experimental results clearly show the potential of Random Forest in terms of accuracy and stability of the results (Fig. 11.3). Although the accuracies are slightly increase, with an increasing number of training samples, RF performs well with small training sample sets. Moreover, the accuracies show a small standard deviation, i.e., the impact of different training samples on the classification accuracy is relatively low. These findings are in accordance with previous studies (e.g., Waske and Braun 2009).

The class accuracies, provided by the SAR and multispectral imagery, confirm the different nature of these data sets (Fig. 11.4). While some classes are most accurately classified by the SPOT data set, the ASAR data is more adequate for the separation of other classes. Comparing the different data sets, it can be assessed that the 2-month SPOT data set outperforms the multitemporal ASAR imagery in terms of accuracies. Nevertheless, the classification of ASAR data results in higher producer's accuracies for wheat and potatoes and higher user's accuracy for barley. Moreover, the accuracies are generally increased by a multisensor data analysis (see Sect. 11.4.3).

11.4.3 Classification Results Using Multisensor Data Sets

Comparing the multisensor-based results, the total accuracy increased by approximately 4 % compared to the classification results achieved on the SPOT data and up to 2 % when compared to the accuracy achieved with the ASAR data (Fig. 11.3). The positive impact from the multisensor data set on the classification accuracy is independent from the number of training samples.

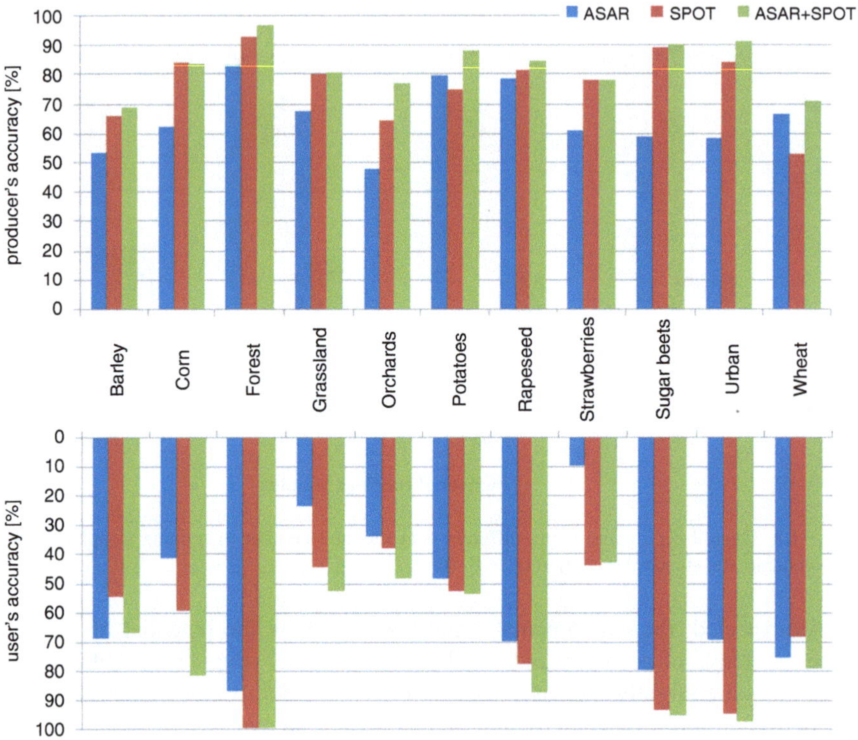

Fig. 11.4 Producer's and user's accuracies, using accuracies, using different data sets and training sample set tr#200

These findings are confirmed by the producer's and user's accuracies. The producer's accuracies for orchards, potatoes and wheat as well as user's accuracies of corn, grassland and orchards are increased by multisensor image analysis. In addition, the producer's and user's accuracies are more balanced, when compared to the single-sensor results (Fig. 11.4).

Nevertheless classification accuracies of some land cover classes (e.g., potatoes, strawberries, and orchards) are relatively low, irrespectively of the used input data set. Reasons for this could be the nature of these classes and the study site itself. Many potatoes and strawberries fields are of relatively small size and thus, the classification accuracy is decreased. This is in accordance to other studies that discussed the negative effect of landscape heterogeneity and decreasing plot sizes on the classification accuracies (e.g., Smith et al. 2003).

Moreover, classes as strawberries are characterized by temporal and spatial variability, e.g., due to agricultural management (e.g., temporary covers by plastic foil and straw). The ground beneath orchards, which are characterized by a relative sparse canopy cover, is often covered by grassland. As a result, orchards can appear as a mixture between grassland and forest (Waske and Benediktsson 2007).

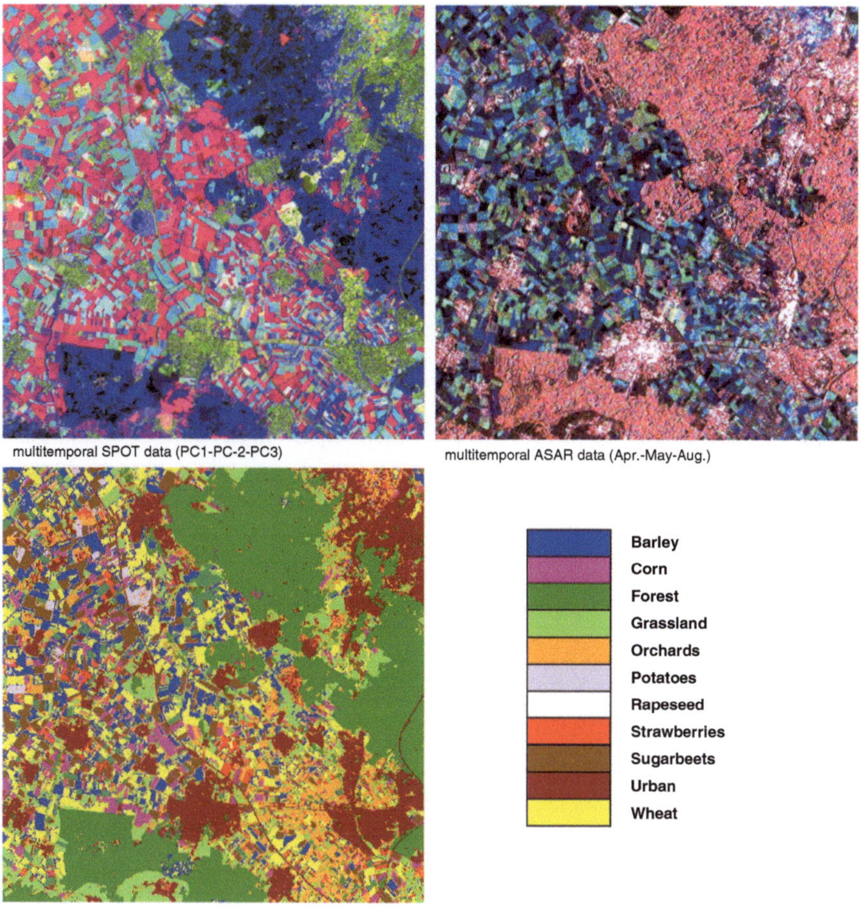

Fig. 11.5 Land cover map, using RF with tr'200 and multitemporal SPOT (PCs were only used for visualization) and Envisat ASAR data

The visible assessment of the classification results confirms the good performance of a multisensor image analysis, using RF (Fig. 11.5). The SAR inherent noise, which is often obvious in SAR-based land cover maps, is clearly reduced. Although some noise exists, most fields are clearly assigned to a specific land cover class. Edges along different field plots are clearly indentified, even along relatively small objects.

11.4.4 Impact of RF Parameters

Figure 11.6 shows the OOB accuracy, depending on the number of trees within the RF (i.e., *k*). Irrespective of the data set, the generation of very small RF was

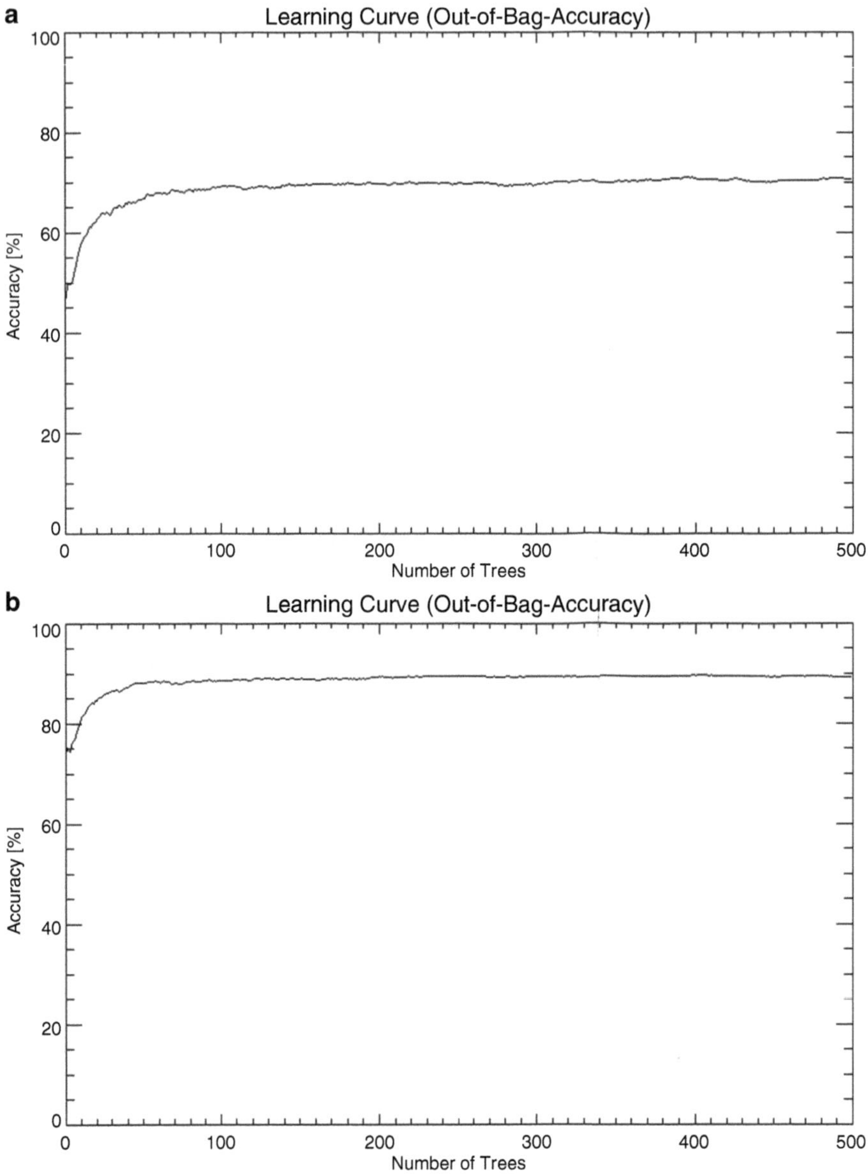

Fig. 11.6 Learning curve (out-of-bag-accuracy) provided by imageRF, using SAR (**a**) and multisensor (**b**) data

ineffective in terms of accuracies. The maximum OOB accuracy is achieved by generating RF with approximately 50–100 iterations, while the use of additional trees did not further improve the OOB accuracy. This is also confirmed by the classification accuracy, derived from the independent test data set. When varying

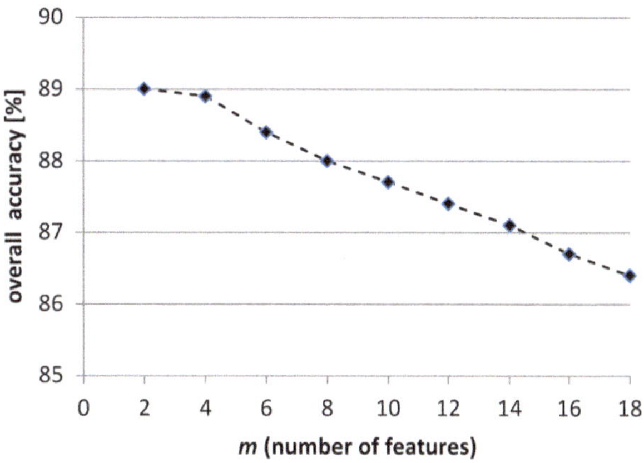

Fig. 11.7 Overall accuracy achieved on the multisensor data, using tr#200 and different number of features (i.e., ASAR acquisitions and SPOT bands)

the number of iterations (e.g., $k = 25, 50, \ldots 500$), overall accuracies achieved on the multisensor data set vary by ~2 % (not shown in detail).

The number of randomly selected features (i.e., m) was systematically set to 2, 4, …, 18, to investigate the impact of this parameter on the classification accuracy (Fig. 11.7). The results confirm that the widely used default recommendation, m is set to the square root of the number of bands (Gislason et al. 2006), is useful, while a larger value results in a reduction of the overall accuracy. However, in this experiment smaller values provide similar accuracies. Thus, one may set m to the log of the number of bands, which is another known default value.

11.5 Conclusion

Overall, RF appears very well suited for classifying multitsensor data. The results show clearly that a multisensor image analysis is superior to standard single-source classification methods.

The good results for separating agricultural classes, which are often difficult to separate, particularly with mono-temporal data sets, underline the value of a multisensor multitemporal image analyses. As already discuss in other studies RF performs well, even with a small number of training samples and without any prior feature selection. Another advantage of RF is computational very light and very simple to handle, because it mainly depends on two user defined values k (i.e., number of iterations) and m (i.e., number of randomly selected features). Moreover, the impact of these parameters on the classification accuracy is relatively low and reliable default recommendations can be given, respectively.

The RF implementation *imageRF* that was used in this study further simplifies the use of RF. The user-friendly implementation is freely available, can handle common remote sensing file formats, and provides reliable default parameterization, which can be flexible modified in a GUI. Behind these facts, the use of

RF as well as multisensor image analysis is advanced and seems particularly interesting for user-oriented applications and operational monitoring services, e.g. in context of GMES and ESAs upcoming Sentinel missions.

Acknowledgements Remote sensing and field data were acquired within the ENVILAND research project (FKZ 50EE0404), funded by the German Aerospace Center (DLR) and the Federal Ministry of Economics and Technology (BMWi). SAR data were made available by ESA under CAT 1 (C1P 3115). The SPOT image is provided by the European OASIS program (OASIS 58 – CE 6324).

References

Aschbacher J, Milagro-Pérez MP (2012) The European Earth monitoring (GMES) programme: status and perspectives. Remote Sens Environ 120:3–8

Blaes X, Vanhalle L, Defourny P (2005) Efficiency of crop identification based on optical and SAR image time series. Remote Sens Environ 96(3–4):352–365

Breiman L (2001) Random forests. Mach Learn 45(1):5–32

Briem GJ, Benediktsson JA, Sveinsson JR (2002) Multiple classifiers applied to multisource remote sensing data. IEEE Trans Geosci Remote Sens 40(10):2291–2299

Brisco B, Brown RJ (1995) Multidate SAR/TM synergism for crop classification in Western Canada. Photogramm Eng Remote Sens 61(8):1009–1014

Chan JCW, Paelinckx D (2008) Evaluation of Random Forest and Adaboost tree-based ensemble classification and spectral band selection for ecotope mapping using airborne hyperspectral imagery. Remote Sens Environ 112:2999–3011

Copernicus (2012) http://www.copernicus.eu/pages-principales/overview/. Last access: January 2014

Cutler DR, Edwards TC, Beard KH, Cutler A, Hess KT, Gibson J, Lawler JJ (2007) Random forests for classification in ecology. Ecology 88(11):2783–2792

Gislason PO, Benediktsson JA, Sveinsson JR (2006) Random forests for land cover classification. Pattern Recognit Lett 27(4):294–300

Kim HC, Pang S, Je HM, Kim D, Yang Bang S (2003) Constructing support vector machine ensemble. Pattern Recognit 36(12):2757–2767

Loosvelt L, Peters J, Skriver H, De Baets B, Verhoest NEC (2012) Impact of reducing polarimetric SAR input on the uncertainty of crop classifications based on the random forests algorithm. IEEE Trans Geosci Remote Sens 50(10):4185–4200. doi:10.1109/TGRS.2012.2189012

McNairn H, Champagne C, Shang J, Holmstrom D, Reichert G (2009) Integration of optical and Synthetic Aperture Radar (SAR) imagery for delivering operational annual crop inventories. ISPRS J Photogramm Remote Sens 64(5):434–449

Prasad AM, Iverson LR, Liaw A (2006) Newer classification and regression tree techniques: bagging and random forests for ecological prediction. Ecosystems 9(2):181–199

Smith JH, Stehman SV, Wickham JD, Yang L (2003) Effects of landscape characteristics on land-cover class accuracy. Remote Sens Environ 84:342–349

Waske B, Benediktsson JA (2007) Fusion of support vector machines for classification of multisensor data. IEEE Trans Geosci Remote Sens 45:3858–3866

Waske B, Braun M (2009) Classifier ensembles for land cover mapping using multitemporal SAR imagery. ISPRS J Photogramm Remote Sens 64:450–457

Waske B, van der Linden S (2008) Classifying multilevel imagery from SAR and optical sensors by decision fusion. IEEE Trans Geosci Remote Sens 46:1457–1466

Waske B, van der Linden A, Benediktsson JA, Rabe A, Hostert P (2010) Sensitivity of support vector machines to random feature selection in classification of hyperspectral data. IEEE Trans Geosci Remote Sens 48(7):2880–2889

Waske B, van der Linden S, Oldenburg C, Rabe A, Hostert P (2012) ImageRF – a user-oriented implementation for classifying remote sensing images by random forests. Environ Model Softw 35:192–193

Zortea M, De Martino M, Serpico S (2008) A SVM ensemble approach for spectral-contextual classification of optical high spatial resolution imagery. In: Proceedings of the IGARSS, Boston, MA, pp 1489–1492

Chapter 12
Application of an Object-Oriented Method for Classification of VHR Satellite Images Using a Rule-Based Approach and Texture Measures

Stanislaw Lewinski, Zbigniew Bochenek, and Konrad Turlej

12.1 Introduction

Since the launch of commercial software for object-oriented data analysis numerous research and application works were undertaken, in order to apply this concept and elaborate semi-automatic methods for land cover classification based on satellite images. The research works were concentrated on two main aspects of object-oriented approaches: multi-resolution segmentation to adjust objects to terrain elements in an optimal way and on classification methods, exploiting comprehensively spectral, spatial and textural features of image objects, as well as their mutual relationships. Applications range from studies using multi-resolution satellite data (Whiteside 2005) to very high-resolution images (QuickBird, Ikonos), which enabled more effective analysis of texture and shape features (Wei et al. 2005; Kressler et al. 2005; De Kok and Wężyk 2008; de Kok et al. 2008). A common classification approach was based on applying training areas for particular land cover classes and a Standard Nearest Neighbour Classifier to assign objects to land cover categories (Yuan and Bauer 2006; Hajek 2005; Elmqvist et al. 2008). Alternative approaches for the classification process comprise the use of parametric values of spectral and texture type as well as hierarchical classification workflows, based on a decision tree

S. Lewinski (✉)
Institute of Geodesy and Cartography, 27 Modzelewskiego St., 02-679 Warsaw, Poland

Space Research Centre of the Polish Academy of Sciences, Earth Observation Group, Bartycka 18A, 00-716 Warsaw, Poland
e-mail: stlewinski@cbk.waw.pl

Z. Bochenek • K. Turlej
Institute of Geodesy and Cartography, Remote Sensing Department, 27 Modzelewskiego St., 02-679 Warsaw, Poland
e-mail: zbigniew.bochenek@igik.edu.pl; konrad.turlej@igik.edu.pl

I. Manakos and M. Braun (eds.), *Land Use and Land Cover Mapping in Europe: Practices & Trends*, Remote Sensing and Digital Image Processing 18,
DOI 10.1007/978-94-007-7969-3_12,

method (Lewinski and Bochenek 2008; Lucas et al. 2007; Su Wei et al. 2008). The presented work on an object based classification approach emerged from the needs formulated within the Geoland 2 SATChMo Core Mapping Service.

12.2 Satellite Data and Study Areas

Very high-resolution satellite images acquired by the Korean Multi-Purpose Satellite KOMPSAT-2 formed the study base. The main characteristics of the KOMPSAT 2 images are as follows: four spectral bands (blue, green, red and infrared) plus one panchromatic channel; 4-m spatial resolution in multispectral mode and 1-m spatial resolution in panchromatic mode; 15 × 15 km scene size. The images were collected by the European Space Agency for Geoland 2 Project in August–September 2009 and pre-processed at ESA facilities prior to their delivery for further research works. Satellite data were converted to the level 1R, including geometric correction. Unfortunately, no atmospheric correction was performed and it was a major challenge in the construction of automatic classification algorithms.

Four images located in Poland, Sweden (south-east), Great Britain and Spain (close to Madrid) were available for the study. The main test area was located in central-western Poland, in Wielkopolska region. The area encompasses all main land cover classes; agricultural land is predominant in this region, while also mixed forests, small cities, rural settlements and grassland patches occur. The method was developed on the Polish study site and its applicability and transferability was tested and adapted using the three other areas.

12.3 Methodical Approach

12.3.1 General Assumptions

The general idea in this work was to prepare a semi-automatic method for land cover classification that could be used for operational mapping throughout Europe for the nine main land cover categories related to CORINE Land Cover level 1 and 2 nomenclature (CORINE 1993):

- Urban/artificial
- Bare non-cultivated ground
- Agricultural areas
- Forest/woodland/trees
- Sparse woody vegetation
- Grassland
- Other vegetation (moorland, etc.)
- Water
- Snow and ice

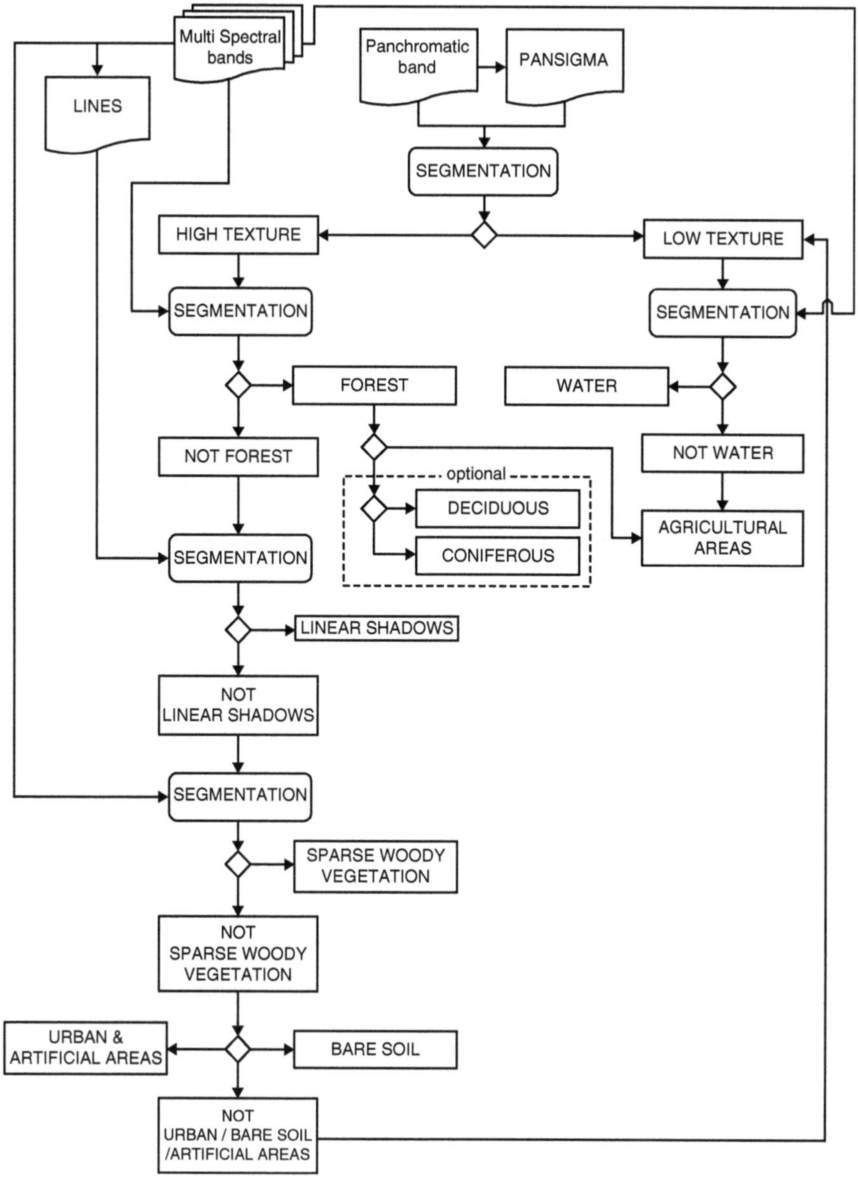

Fig. 12.1 Object-oriented classification workflow using an iterative decision-tree approach

The approach is based on a decision tree that sequentially separates land cover classes. The main difficulty is hence the identification of appropriate object features and a criteria for their classification. The approach follows a stepwise approach (Fig. 12.1); individual steps are described in more detail in Sect. 12.3.2:

- Stage 1 – division of the study area into two object groups, the first group characterized by high texture values, the second by low values respectively;

- Stage 2 – classification of the high texture group, including urban/artificial class, forests/woodland/trees, sparse woody vegetation and bare non-cultivated ground;
- Stage 3 – classification of the low-texture group, comprising agricultural areas, grasslands, snow and ice (if existing) as well as water;
- Stage 4 – re-classification of existing classes in order to refine the classification output.

12.3.2 *Hierarchical Classification of Land Cover Categories*

The panchromatic image of the study area was used to generate/divide high- and low texture areas using texture filters. Two sigma filters were used for this purpose, which emphasize the inner and outer object boundaries, respectively. The sum of two filtered images formed the resultant PAN SIGMA image, which was used at the next stages of analysis for deriving texture information.

The panchromatic image was used in conjunction with original one for a first segmentation, i.e. division of the study area into homogeneous objects. The scale parameter in this segmentation was chosen to be quite high in order to keep the resultant objects as large as possible. The PAN SIGMA image was divided into high- and low texture areas by using quantiles, following a solution proposed by de Kok and Wężyk (2008). This measure allows to split the analyzed distribution (range of PAN Sigma values in this case) into two sub-distributions, which e.g. can represent two groups of land cover, characterized by different textures. The crucial point is to determine the quantile value, which could make such a division with high thematic accuracy. This can be done by an iterative process through analysis of division into high and low texture, while applying various thresholds (quantile values) within the test area. In theory such a quantile value should be fixed for various satellite scenes, but in practice, it can change depending on character and differentiation of the land cover. As a result of this step two groups of objects were created; a high texture group, including urban/artificial class, forests/woodland/trees, sparse woody vegetation and bare non-cultivated ground, as well as a low texture group, comprising agricultural areas, grasslands and water. The high texture areas were assumed to be a fixed property of urban/forest (spectral discontinuity) while spectral low variability (spectral continuity) was assumed to be a characteristic property of agriculture/water and large grass fields.

In parallel to the discrimination of high and low texture areas the extraction of lines, represented mainly by linear shadows, was performed. That step was done on the basis of the infrared channel analyzing the orientation of linear features within the IR image. The resulting layer with lines was used at the later stages of the classification to refine the final classification product.

Once the high texture image was created, it was segmented again, including the multispectral channels (green, red, infrared) and a lower scale factor in order to obtain smaller objects. After this segmentation the division of high texture areas into the forest/woodland tree class and non-woodland class was done. As a criterion of this division a spectral feature called RatioR was utilized. This customized feature is defined by the formula (de Kok and Wężyk 2008):

$$\mathrm{RatioR} = \mathrm{R}/(\mathrm{R} + \mathrm{G} + \mathrm{B} + \mathrm{IR} + \mathrm{PAN})$$

where R, G, B, IR, PAN denote the reflectance values in respective spectral channels.

The threshold value of the RatioR feature was also determined by quantiles. In addition, the Normalized Difference Vegetation Index (NDVI)

$$\mathrm{NDVI} = (\mathrm{IR} - \mathrm{R})/(\mathrm{IR} + \mathrm{R})$$

was included in the processing of high texture areas together with the RatioR in order to classify forest/woodland/trees objects. In the presented classification workflow this class could be optionally sub-divided into two forest categories: coniferous, and deciduous forests. In this case, a sampling approach was utilized: samples for coniferous and deciduous forests were collected and a standard nearest neighbour classifier was applied.

The stage of delineating the forests/woodland category is equipped with additional procedures, allowing to extract from this class areas, which despite of high texture belong to agricultural category. Two types of extraction can be done: agricultural fields characterized by high texture and openings located within forest areas. For the first type of extraction threshold values in the infrared band and texture homogeneity measure in this channel were utilized, while for the second one brightness threshold values and homogeneity PAN measure were applied.

Once classification of forest/woodland/trees category had been completed, the analysis of the remaining area within high texture category was started. As a first step of this analysis a segmentation of this area was done with a low scale factor, using the line layer, prepared in the previous step. Subsequently, the classification of shadows was performed through the application of a threshold value to the line layer. At the final step of this stage a new segmentation of the remaining high texture area was performed with a low scale factor using the three spectral bands (red, green, infrared).

The resulting area was in turn classified, in order to distinguish the next land cover category – sparse woody vegetation. Two feature attributes were used for this purpose: NDVI and RatioR, with a priority put to the NDVI threshold and complementary utilization of the RatioR threshold. The remaining high texture area was classified in the next step with the aim to discriminate the two land cover categories: urban/artificial class and non-cultivated bare ground. In case of the urban/artificial class two features were used for discrimination:

- a customized feature – standard deviation in red channel divided by the square root of the area of the object
- a texture homogeneity measure in the PAN image

In case of non-cultivated bare ground the threshold for the brightness index was applied in conjunction with a texture homogeneity measure in the panchromatic band. Three basic land cover classes exist in the low texture group – water, grasslands and agricultural areas. They were considered in this third step. At the

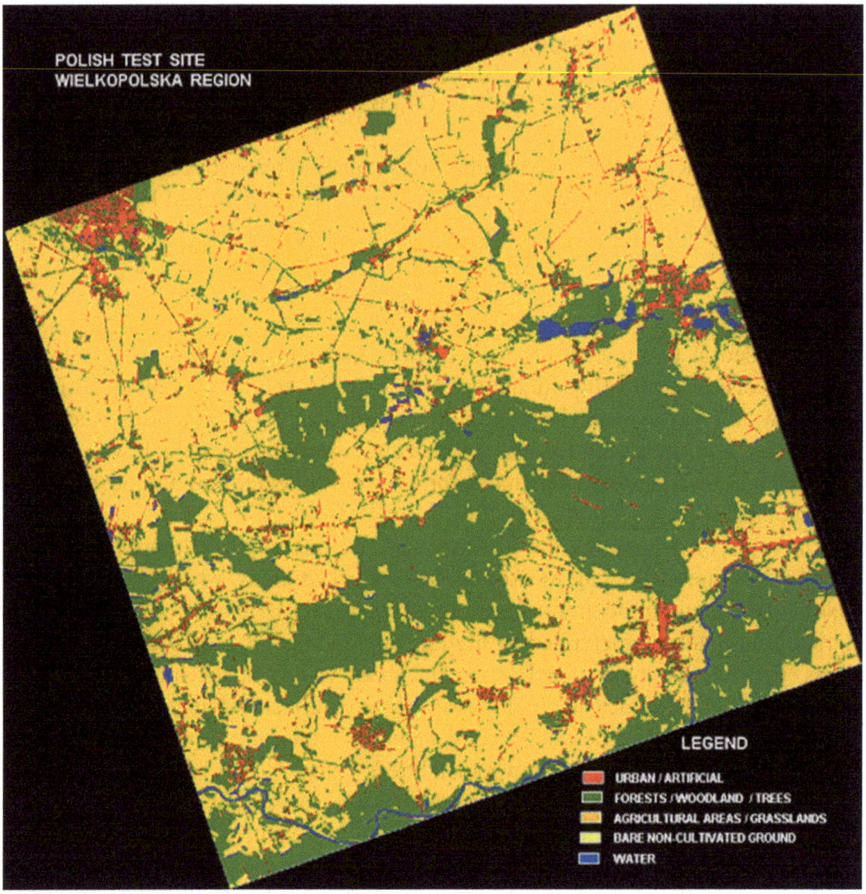

Fig. 12.2 Classification image of Polish test site – Wielkopolska region

initial phase an additional segmentation of the low texture regions was performed, with a possibility to refine the results through the use of a second segmentation based on spectral differences between objects in the infrared band. After the completion of the segmentation procedure water bodies were classified applying a threshold in IR reflectance. The remaining classes in the low texture regions were assumed to represent agricultural areas and grassland. No division was finally made between these two land cover categories, due to unfavourable dates of acquisition of the satellite images (late summer/early autumn). In order to discriminate these classes with high accuracy a multi-temporal approach would be more appropriate.

At the fourth stage of the classification workflow re-classification of some classes was done. This concerned especially the class of linear shadows as well as some areas representing bare non-cultivated ground. Latter were shifted to urban/artificial class, when neighboring this class to a high degree. At the end a generalization process was conducted with the requirement to preserve objects, which fulfill the condition of a Minimum Mapping Unit >0.25 ha. The final classification image for the Polish study area is presented in Fig. 12.2.

Table 12.1 Accuracy assessment of classification of Kompsat 2 image

	Reference data					
Classified data	Forests/ woodland	Urban/ artificial	Bare ground	Water	Agriculture/ grassland	Total
Forests	**176**	6	0	0	12	**194**
Urban	4	**24**	2	3	6	**39**
Bare ground	1	2	**8**	0	11	**22**
Water	0	0	0	**23**	2	**25**
Agri & Grassland	3	0	1	0	**216**	**220**
Total	**184**	**32**	**11**	**26**	**247**	**500**
Accuracy (%)						
Producer's	95.7	75.0	72.7	88.5	87.4	
User's	90.7	61.5	36.4	92.0	98.2	
Overall accuracy **89.4 %**						

12.3.3 Accuracy Assessment

An accuracy assessment was conducted using 500 randomly-stratified control points. Stratification was done by land cover classes (sparse woody vegetation was not classified since it was not present in this study area). For each control point with its surroundings the land cover was interpreted visually based on original image color composites. Results of the accuracy assessment are presented in the form of error matrix in Table 12.1.

12.4 Results and Discussion

The overall accuracy of the classification reached 89.4 % fulfilling the requirements for generic land use maps prepared within GMES Land Services (85 %). The obtained accuracy is supported by a high Kappa coefficient (0.83).

The best results were achieved for three land cover categories: forests, water and agricultural areas/grasslands. Both assessments – from point of view of producer's and user's accuracy – proved to have a high level of accuracy. In each category results of accuracy assessment exceeded the pre-determined level of 85 %. However, one must remember, that agricultural areas/grassland class was a broad land cover category, so while separating it into two independent classes the accuracy level could be slightly lower.

The urban/artificial class and bare non-cultivated ground revealed moderate, although acceptable producer's accuracies. Precision of discrimination of urban/ artificial class mainly depended on its detailed assigning to the high-texture group. Bare non-cultivated ground was mixed with agricultural bare soil, while having low texture characteristics, at the same time decreasing its level of recognition, especially from the user's point of view.

The applicability of the presented method was tested on the other three test sites spread out in various regions of Europe: Sweden, Great Britain and Spain.

The same accuracy assessment procedure has been applied for each test site, with 500 points randomly distributed across the study area. The following values of overall accuracy were obtained: 72.8 %, 72.8 % and 82.4 %, respectively. They are lower than in the case of the Polish scene, but having regard to the geographical differences of the analyzed area they can be considered as satisfactory. In operational phase of SATChMo project results of automatic classification were corrected manually.

12.5 Conclusions

The presented work demonstrates an object-oriented, rule-based classification method, utilizing threshold values for spectral, texture and context features in a decision tree work flow. The advantage of the proposed method against approaches requiring reference samples is its simplicity – each land category is described by one or two features, which allows the operator to adjust this features quite easily in order to achieve acceptable result of class delineation. Moreover, the whole process of classification can be highly automated and easily modified through adding new features or classes when needed.

Application of multi-resolution segmentation at different stages of the classification process is another characteristic element of the proposed method. It enables to adjust size and shape of objects to particular land cover classes more precisely than in a one-step segmentation approach. An innovation of the proposed approach is also the use of a textural image created through filtering the very high-resolution panchromatic image in order to separate land cover categories characterized by high or low texture at the initial stage of classification.

The presented approach was verified through its application to three other test sites, located in Sweden, Great Britain and Spain. The subsequent results obtained in the course of SATChMo project indicate especially, that the proposed method can be effectively applied for land cover mapping in a temperate zone of Europe, with minor adjustments of features used for class separation. In case of the Mediterranean zone, characterized by specific types of vegetation and agriculture, the proposed approach requires some additional modifications, but its general idea is fully applicable. Further research work on improving the proposed method are still required in order to increase accuracy of the classes urban/artificial as well as bare non-cultivated ground. An efficient solution for grassland delineation, based on multi-temporal approach is also required.

Acknowledgements The presented work has been done within FP7 Geoland-2 project, financed by the European Commission and supported by the Ministry of Higher Science and Education. The method was tested using KOMPSAT-2 satellite images collected by ESA (European Space Agency) for Geoland 2 project – Application SATChMo Core Mapping Service.

References

CORINE (1993) Land cover. Technical guide

De Kok R, Wężyk P (2008) Principles of full autonomy in image interpretation. The basic architectural design for a sequential process with image objects. In: Blaschke Th, Lang S, Hay GJ (eds) Object-based image analysis. Series: lecture notes in geoinformation and cartography. Springer, Berlin/Heidelberg, ISSN: 1863–2246, pp 697–710

De Kok R, Wezyk P, Weidenbach M (2008) The role of edge objects in full autonomous image interpretation. In: Proceedings of GEOBIA 2008 – pixels, objects, intelligence, GEOgraphic object based image analysis for the 21st century, Calgary, Alberta, Canada. http://www.isprs.org/proceedings/XXXVIII/4-C1/Sessions/Session1/6717_DeKok_Proc_pap.pdf

Elmqvist B, Ardo J, Olsson L (2008) Land use studies in drylands: an evaluation of object-oriented classification of very high resolution panchromatic imagery. Int J Remote Sens 29(24):7129–7140

Hajek F (2005) Object-oriented classification of remote sensing data for the identification of tree species composition. In: Proceedings of ForestSat 2005 conference, May 31–June 3, 2005, Boras, Sweden. http://www.ecognition.com/sites/default/files/229_forestsat2005_20__20filip_20hajek.pdf

Kressler FP, Steinnocher K, Kim YS (2005) Enhanced semi-automatic image classification of high-resolution data. In: Proceedings of the IGARSS 2005 symposium. Seoul, Korea. July 25–29, 2005. http://www.ecognition.com/sites/default/files/240_igarsskressler2.pdf

Lewinski St, Bochenek Z (2008) Rule-based classification of SPOT imagery using object-oriented approach for detailed land cover mapping. In: Proceedings of 28th EARSeL symposium "Remote sensing for a changing Europe", Istanbul, Turkey, 2–5 June 2008, pp 197–204

Lucas R, Rowlands A, Brown A, Keyworth S, Bunting P (2007) Rule-based classification of multi-temporal satellite imagery for habitat and agricultural land cover mapping. ISPRS J Photogramm Remote Sens 62(3):165–185

Su W, Li J, Chen Y, Liu Z, Zhang J, Low TM, Suppiah I, Hashim SAM (2008) Textural and local statistic for the object-oriented classification of urban areas using high-resolution imagery. Int J Remote Sens 29(1):3105–3117

Wei W, Chen X, Ma A (2005) Object-oriented information extraction and application in high-resolution remote sensing image. In: Proceedings of the IGARSS 2005 symposium. Seoul, Korea. July 25–29, 2005, pp 3803–3806

Whiteside T (2005) A multi-scale object-oriented approach to the classification of multi-sensor imagery for mapping land cover in the top end. In: Proceedings of NARGIS 2005 – application in tropical spatial science. 4th–7th July 2005 Charles Darwin University, Darwin, NT, Australia. http://www.ecognition.com/sites/default/files/272_18.1_20whiteside__20tim.pdf

Yuan F, Bauer ME (2006) Mapping impervious surface area using high resolution imagery: a comparison of object-based and per pixel classification. In: Proceedings of ASPRS 2006 annual conference, Reno, Nevada; May 1–5, 2006. http://www.ecognition.com/sites/default/files/184_asprs2006_0178.pdf

Chapter 13
Remote Sensing of Vegetation for Nature Conservation

Sebastian Schmidtlein, Ulrike Faude, Stefanie Stenzel, and Hannes Feilhauer

13.1 Introduction

A rapidly changing environment with land use and climate as the most dynamic components causes new challenges for nature conservation and management of protected areas. Dealing with these changes requires a systematic monitoring. To date, such monitoring programs are mostly backed by expert guess or permanent observation plots. Both have their merits but the plot-based approach is certainly more objective. However, even in the case of appropriate sampling, plots provide only punctual information and changes in the area between plots are easily missed. This gap can be closed by remote sensing.

Remote sensing applications may help experts by substantiating their guess and by assessing the spatial relevance of changes. Today, remote sensing methods in nature conservation and management include a broad range of approaches from visual interpretation to numerically based techniques. Similarly broad is the origin of data, which includes aerial photography, airborne and satellite-borne multispectral and hyperspectral optical data as well as data from active remote sensing using LIDAR or radar sensors.

Despite this broad range of possibilities, visual interpretation of images continues to be the most widespread technique used in nature conservation (Gross et al. 2009). High-resolution digital and analogous imagery is readily available at

S. Schmidtlein (✉) • S. Stenzel
Institute of Geography and Geoecology, KIT Karlsruhe,
Kaiserstr. 12, 76128 Karlsruhe, Germany
e-mail: schmidtlein@kit.edu

U. Faude
EFTAS Münster, Oststr. 2-18, 48145 Münster, Germany

H. Feilhauer
Institute of Geography, FAU Erlangen-Nürnberg, Kochstr. 4/4, 91054 Erlangen, Germany

I. Manakos and M. Braun (eds.), *Land Use and Land Cover Mapping in Europe: Practices & Trends*, Remote Sensing and Digital Image Processing 18,
DOI 10.1007/978-94-007-7969-3_13,

no or low cost and can be handled with simple approaches. Trained interpreters are able to gain a great amount of information from these data. Yet, there remains a lot of unused potential. Several reasons have been discussed, such as limited data access, the complexity of more advanced approaches in terms of comprehensibility and the need of established methods (Van den Borre et al. 2011). Practitioners are often not inclined to support studies that are (or appear to be) experimental or they are frustrated by overly enthusiastic promises of remote sensing scientists. In addition, high costs for specialized software may have put a break to the implementation of advanced approaches.

Fueled by the hope for cost-efficient vegetation mapping, the discussion about the potential and possible implementations of remote sensing continues. Compared to a few years ago, remote sensing data have become more affordable and some powerful software products for their analysis are available even free of cost (e.g., grass GIS, RGui, EnMap-Toolbox). The fact that remote sensing is able to provide information that could not be derived by conventional approaches plays a minor role in the discussion about a possible application in conservation and management. In our opinion budgeting issues should not be the only reason to opt for or to decide against the use of remote sensing. Instead, possible extra gains should be taken into account. For instance, a combination of remote sensing and field work may derive detailed information about continuous gradients in species composition or ecosystem functioning that cannot be derived from field-based data alone (Schmidtlein and Sassin 2004; Feilhauer et al. 2011; Schmidtlein et al. 2012). In general, semi-automated approaches in remote sensing are reproducible, which is a desirable property especially in monitoring programs. In addition, these approaches are indeed often cost efficient for what they can deliver. This is especially true in case of remote or large areas. A prominent example for such applications are the mapping tasks in the context of the European habitat directive and the related Natura 2000 network that requires monitoring of huge areas in 6 year intervals (Council of the European Communities 1992).

Such demands raise the question which data and techniques are useful in conservation and management. We are of the opinion that until further notice and with the exception of simply structured targets, only training of semi-automated methods based on samples taken from the image itself provide the amount of detail that is needed. The reason is that signatures of species and vegetation types are highly variable in space and time. Thus, spectral libraries or physically based approaches are not in the focus of this chapter. We aim to give a brief overview on the promising approaches but will also point to the difficulties. The chapter focuses on the detection of individual plant species, the mapping of distinct habitats, plant communities and diversity patterns, as well as on approaches that take into consideration gradual transitions. We put an emphasis on applications that are relevant to Europe. Other review articles dealing with these or similar topics have been written by Kennedy et al. (2009), Newton et al. (2009), Van den Borre et al. (2011), Wiens et al. (2009), as well as Xie et al. (2008) to name just a few.

13.1.1 Which Remote Sensing Data Is Useful?

Despite successful applications based on multispectral data (see, e.g., Förster et al. 2008; Somodi et al. 2012, or Feilhauer et al. 2013 for an assessment of the potential), hyperspectral data are currently on the tip of everyone's tongue when talking about the future of optical remote sensing in nature conservation. With the detailed spectral resolution of these sensors, the detection of spectral differences between different habitats or species becomes more likely. Even though hyperspectral satellites (like EnMAP) are planned to be launched over the next years, hyperspectral data will continue to be delivered by airborne sensors for a while. Accordingly, spatial and temporal coverage is typically limited. Sensors like the upcoming Sentinel-2 with 13 spectral bands may prove a valuable replacement for the time being. When using hyperspectral data it should also be kept in mind that specialized methods are required that can make use of the rich spectral information and that are robust towards highly inter-correlated spectral bands. In particular if 3-d patterns of the canopy are relevant, LIDAR and radar data are another option or a good add-on (Wulder et al. 2007). A special strength of radar is the ability to penetrate surfaces and to address water content in soils and vegetation. This is potentially useful for identifying wetlands or monitoring wetland degradation. Further, these data may allow for the detection of changes in the 3-d habitat structure that are an urgent problem for the function of habitats (Nagendra et al. 2012). For example, increases of shrub and tree cover affect the occurrence of light-demanding plant species in fallow farmland and most birds react instantaneously to changes in vertical patterns. Fusions of optical remote sensing data with LIDAR or Synthetic Aperture Radar (SAR) may improve the results in this context. For example, the use of very high resolution imagery and aerial photographs (Laliberte et al. 2004; Walker and Briggs 2007), involving texture analysis, object-based approaches or stereoscopy, can complement the information on 3-d structure (Waser et al. 2008; Nagendra et al. 2013; Hantson et al. 2012). In some cases, hyperspectral data has been used for this purpose (Harris et al. 2003; Spanhove et al. 2012). While bush encroachment is an obvious field of application for this kind of analysis, few studies try to address grassland structure. For example, recent attempts aim to develop a method for detecting cutting dates from multitemporal TerraSAR-X data (Schuster et al. 2011). The frequency and date of mowing can be useful for estimating the conservation value of grassland.

Three important questions should be answered before choosing data or starting a flight campaign: (1) what spatial coverage do I need?, (2) what spectral coverage makes sense?, and (3) what spatial resolution (ground sampling distance, GSD) is necessary? Especially concerning flight campaigns, it is often opted for the highest spatial resolution in order to preserve all possibilities. Yet, especially the decision on spatial resolution may be of great consequence: typically, the signal-to-noise ratio decreases with increasing spatial and spectral resolution for a given sensor sensitivity. This may lead to technical issues that exceed the benefits of a higher resolution. Also, the higher the spatial resolution, the more flight strips or tiles are needed. This comes with higher processing efforts and sometimes interfering illumination effects.

It is thus preferable to select a spatial resolution for which it can be assumed that the targets, be it species or habitats, are well depicted. For example, to characterize a plant community, a pixel should contain a mixture of species. The best size for determining individual plants avoids such mixtures and returns a pure species signal. Yet, species identification becomes difficult if the area on ground covered by the pixel is so small that the reflection of one individual or of a mono-stand becomes too heterogeneous (see Nagendra 2001; Wulder et al. 2004). Considering the spectral resolution, a large number of bands is not always necessary. Tests with simulated data show, that in certain cases a broad spectral coverage may be more important than high spectral resolution (Feilhauer et al. 2013). Still, a high spectral resolution offers a maximum flexibility in the selection of spectral features. This can be important since, with our incomplete knowledge about radiative transfer and due to the influence of unknown surface components, surprises regarding which spectral bands are relevant are not uncommon.

Remote sensing of vegetation will not always lead to satisfactory results. The success is not only highly dependent on data and methods, but also on actual reflectance differences between vegetation units or species across the study area.

13.1.2 On the Uniqueness of Plant Species Appearances

When aiming to map certain plant species, species assemblages, or habitats, we assume that their respective reflectance is unique (Castro-Esau et al. 2006). Yet, even if plants feature a multitude of anatomical structures, the basic elements remain the same (Sorby 1873). All plants (but a few) are basically green most of the time (Kumar et al. 2001) and their reflectance may not be as unique as we hope for. Usually the vegetation optimum, i.e., the peak in plant development, is the best time to map vegetation patterns since differences in species traits are fully developed (Feilhauer et al. 2010; Feilhauer and Schmidtlein 2011; Laba et al. 2005). The contrary may apply if species traits are less clear. Often, litter and soil are related to species occurrences and variation in reflectance that is caused by these surface components may serve as a proxy for species occurrences. In these cases, good results may be achieved outside the vegetation optimum (Schmidt and Skidmore 2003; Feilhauer and Schmidtlein 2011; Schmidtlein et al. 2012). This applies to empirical approaches with a calibration based on samples within the area of an image at the time of acquisition. Approaches that rely on collections of reflectance taken elsewhere or at another time hardly profit from such correlations.

Variables that differ between individual plants, like state of growth or turgor pressure, exert a decisive impact on reflectance. Spatial differences in plant development due to health, stress, and disturbance (e.g., land use) add to mapping uncertainties (Carter 1993; Carter and Knapp 2001; Gausmann 1984; Sanger 1971; Sorby 1873). In fact, Castro-Esau et al. (2006) showed that intra-species variability can be higher than inter-species variability, especially across large areas, and thus blur if not offset possible relations between species and reflectance. In such

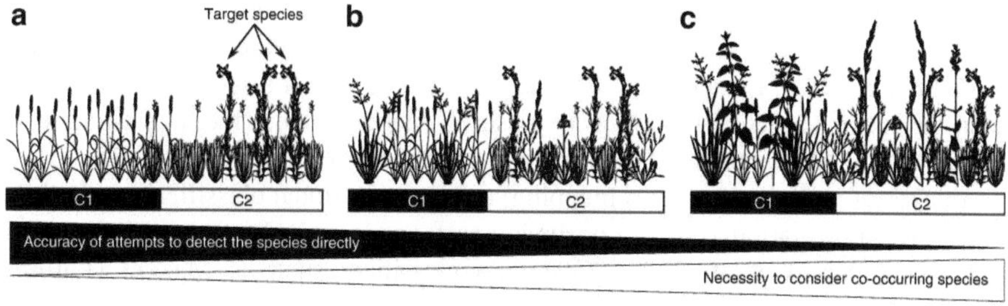

Fig. 13.1 Detecting invasive species in a mixed stand of two communities C1 (not invaded) and C2 (invaded). Increased mixing with other species (from **a–c**) affects the accuracy of the direct detection of individual species. However, the canopy reflectance of an entire plant community may serve as a proxy for characteristic species

cases the use of multiseasonal remote sensing (which makes use of differences in growth over the year) may be helpful but can also even worsen the situation, especially if non-taxonomically caused differences are spatially variable across seasons. Multiseasonal remote sensing has often improved the mapping results (see, e.g., Townsend and Walsh 2001 for a discussion on the value of multiseasonal multispectral data and Noujdina and Ustin 2008 for a comparison of monotemporal and multiseasonal hyperspectral data), yet does not guarantee success since noise in the data is also accumulated (Feilhauer et al. 2013; Ghioca-Robrecht et al. 2008; Langley et al. 2001).

13.1.3 Species Mapping – Looking for a Needle in the Haystack?

With respect to individual species, practitioners in conservation and management focus on two sorts of species worth mapping and monitoring – either invasive species that threaten an ecosystem or target species that are rare, endangered, or act as umbrella species (i.e., indicator species for threatened habitats). Since endangered species are often small and unobtrusive, their direct detection is mostly not feasible with remote sensing methods. Yet, this does not mean that they cannot be detected indirectly, for example via co-occurring species. In contrast, invasive species often feature remote sensing friendly properties such as prominent flowers, high spatial coverage, and the formation of mono-stands.

Direct mapping of species that are present in the upper canopy is possible if individuals of a species are sufficiently large or if the species forms sufficiently large mono-stands. Besides this uniqueness of reflectance it is also important whether the species is surrounded by a homogeneous background or mixed with other species (Fig. 13.1). The importance of the complexity of the environment for mapping success (Andrew and Ustin 2008) is closely linked to issues caused by mixed pixels and thus to an appropriate spatial resolution. For direct mapping of individual species, the appropriate selection of distinct phenological stages for data acquisition or multiseasonal remote sensing are likely to contribute to a high mapping accuracy (Everitt et al. 1992; Laba et al. 2005; Langley et al. 2001). If a species grows higher than the species in its surroundings (e.g., shrub encroachment), the incorporation of LIDAR or SAR data can lead to another improvement (Hantson et al. 2012).

In case of individual species mapping, most conventional classifiers are of little help since information about all occurring surface elements in the area is required (e.g., Underwood et al. 2003). One-class classifiers such as Maxent (Phillips et al. 2006), Support Vector Data Description (SVDD) (Tax and Duin 2004), and One-Class Support Vector Machines (OCSVM) (Schölkopf et al. 2001) are more adapted. These classifiers require only spatial data on the occurrence of the target species and hence enable an efficient acquisition of ground-truth data. By learning from the spectral properties of species occurrences, one-class classifiers aim to distinguish this class from the landscape background (i.e., the image) and result in fuzzy predictions of class membership probabilities.

Understory plants can hardly be addressed directly via their spectral signature. Only in special cases a characteristic phenological development like early greening allows for a direct detection (e.g., Tuanmu et al. 2010). The situation for species with small or few individuals that vanish in a mixture of other species is similar. This hampers, for example, the detection of invasive species in an early stage of invasion. However, the characteristic reflectance of plant communities is often a good proxy for an indirect mapping of hidden plant species. This means that it might be worth the try to address these species based on their ecological requirements and co-occurring species that are triggering canopy reflectance. If this procedure is followed it should be kept in mind that the result can be based on actual as well as on potential occurrences – a fact that may pose methodical issues. Remote sensing depicts the current condition of the Earth surface and if we derive maps of potential habitat the outcomes will always be biased by current land use and thus land cover. Bradley et al. (2012) criticize that this has often led to difficulties in the interpretation of results. The aim of a mapping project thus needs to be considered carefully and the usefulness of remote sensing data must be scrutinized.

13.1.4 Mapping Vegetation Types and Ecotones

Considering the issues in detecting species, mapping the distribution of vegetation types may seem easier. However, this task comes along with other challenges: Have you ever stood in the field and asked yourself where one vegetation type stops and another begins? This may often be a difficult and frustrating task because the species composition of vegetation often changes gradually. The continuous change is referred to as ecotone or floristic gradient. Apart from the continuum in space, another continuum exists in floristic similarity because even if patches of vegetation are well distinct in the field they are hardly ever identical in terms of their species composition. For these reasons, the use of discrete classes for vegetation mapping is

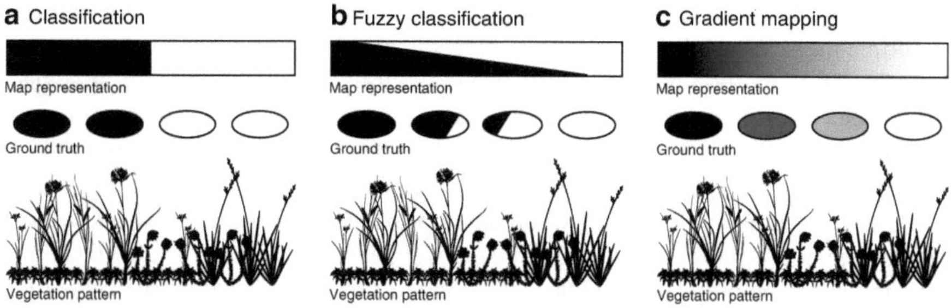

Fig. 13.2 Addressing patterns in natural vegetation with continuously changing species composition. The patterns can be described as either hard classes (**a**), fuzzy classes (**b**), or floristic gradients (**c**)

affected by two issues. First, the spatial delineation of vegetation types or plant communities (Fig. 13.2a) is prone to uncertainties and varies between observers or algorithms. Second, "typical" stands are difficult to identify in the field but such "typical" stands are a fundamental requirement to train and validate classifications. This issue is even more complicated if the artificial nature of the pixel is taken into consideration where mixed stands are the rule. Taken together, it may be easier and closer to ecological thinking (Foody and Trodd 1993; Foody 1996) to refrain from juggling with different classification systems and to aim to preserve the vegetation continuum. Several approaches to map the vegetation continuum have been developed to date. In general, these approaches can be classified in two categories, either based on fuzzy classification (Fig. 13.2b) or gradient mapping (Fig. 13.2c). Both approaches have successfully been used to map natural ecosystems.

A fuzzy classification can be achieved with various algorithms designed for conventional 'hard' classifications. These classifiers generate much information beyond the mere assignment of a pixel to a class (Wang 1990). This information, for example the probability of a pixel to belong to a certain class, can be used to generate a fuzzy classification that represents vegetation patterns closer to ecologic reality. Several algorithms are suitable for fuzzy classification and have been used to map the distribution of various heathland types (see, e.g., Foody 1992, 1996), grassland (e.g., Oldeland et al. 2010), and transitional gradients from heath to forest ecosystems (e.g., Wood and Foody 1989). Similarly, spectral unmixing techniques offer great potential for fuzzy vegetation mapping. Instead of probability maps this approach results in predictions of pixel-based cover fractions of vegetation types (e.g., Roberts et al. 1998). As with individual species, habitats and communities can be mapped with one-class classifiers and most of them have a fuzzy mode. For example, SVDD has proved to be able to detect single vegetation types (Sanchez-Hernandez et al. 2007a, b) in a landscape matrix. Maximum Entropy modeling does so too (Amici 2011) and even performed better in comparison to OCSVM (Li and Guo 2010).

Despite all benefits, the application of fuzzy classification is hampered by two issues. First, fuzzy approaches generate a separate map for each class. In particular for heterogeneous and complex ecosystems, this leads to a bulky result consisting of multiple maps. Second, fuzzy classification still requires a priori defined vegetation classes for calibration and validation (Foody 1999) – certainly a source of subjectivity and error. The identification of what may be considered a 'pure' representation of a vegetation type is a topic that drove to despair generations of phytosociologists. It is unlikely that remote sensing resolves the issue.

Gradient mapping does not require any a priori defined classes. Instead, it takes advantage of ordination techniques adopted from ecology to describe the vegetation continuum as an interval-scaled floristic gradient based on a quantitative vegetation sample. Ordination techniques are very common in ecology and are basically a dimensionality reduction of the vegetation data. Various numerical algorithms (including the well-known Principal Component Analysis) may be used for this task. The resulting floristic gradients or similarity spaces are subsequently regressed against the gray values corresponding to the vegetation sample. To cope with the inter-correlation inherent to the spectral data, sophisticated regression techniques such as Partial Least Squares regression (Wold et al. 2001), Support Vector regression (Smola and Schölkopf 2004), or Random Forest regression (Breiman 2001) are required.

13.1.5 Diversity and Different Ways to Tackle It

Biodiversity belongs to the ecosystem properties that are most frequently addressed with remote sensing. Still, the outcome of the related studies differs considerably. These differences may be due to the fact that there is little agreement about the meaning of the term 'biodiversity'. Biodiversity includes – but is not limited to – measures of species richness or evenness of species shares within a defined area (i.e., Whittaker's Alpha diversity shown in Fig. 13.3) (Whittaker 1960), the change rate in species composition related to a certain distance (i.e., Whittaker's Beta diversity), the diversity of functional attributes as well as genetical, phylogenetical, structural, or chemical diversity. In addition, these measures are generally restricted to certain parts of a system (e.g., to vascular plants).

Biodiversity patterns are ecosystem properties on a meta scale and there is only little reason to believe that they are globally related to spectral features. Meta analyses of remotely-sensed maps showing the vegetation properties of interest (e.g., species composition, biochemical composition, or structures) using moving windows or measures of spatial variability are a viable way to assess diversity patterns coarser than the pixel resolution. Similarly, the spectral heterogeneity of a vegetation stand is sometimes linked to compositional changes and may thus be used as a proxy for species richness (Gould 2000; Hall et al. 2012; Rocchini et al. 2004). The opposite conclusion (homogeneous reflectance indicates low diversity) is, however, more likely to be wrong (Schmidtlein and Sassin 2004). If species feature a similar spectral signature or if diversity patterns are located at sub-pixel scales, the observed heterogeneity may underestimate the actual species

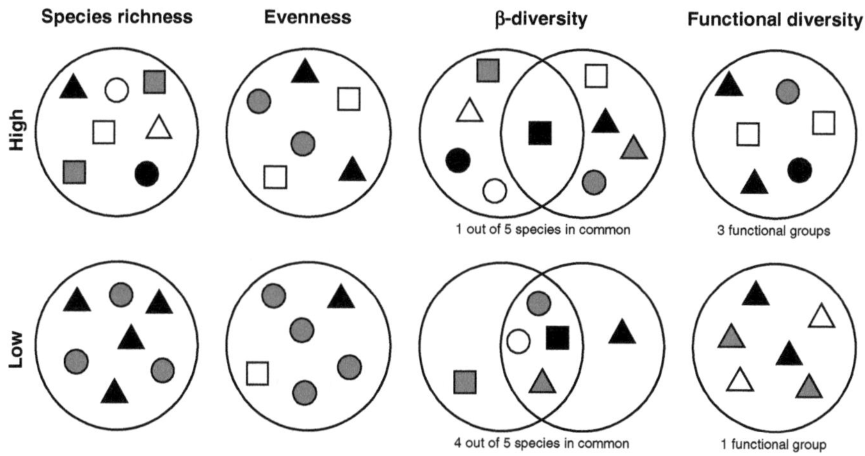

Fig. 13.3 Different measures and concepts of biodiversity. Each symbol represents a species, the shape of the symbols illustrate functional groups

richness. The spatial resolution of the imagery dictates the spatial scale at which diversity is quantified and needs to be chosen with care.

Although the diversity within a pixel is more difficult to address following causal relationships, simple proxies often allow for a local estimation of parameters such as species richness. For example, mere canopy height may indicate species richness in forest ecosystems (Goetz et al. 2007). Also, some vegetation types generally feature a higher diversity than others. Vegetation maps showing the distribution of these types may hence allow for a rough assessment of biodiversity (e.g., Hernandez-Stefanoni and Ponce-Hernandez 2004). Such proxies are, however, only valid for the ecosystem under investigation, hardly transferable to other systems and are hence no viable solution for a general assessment of biodiversity.

The capability of remote sensing to address functional diversity depends on the functions of interest. Specific aspects, such as plant strategies, can be described comprehensively (Schmidtlein et al. 2012). Since these characteristics are often closely linked to plant optical properties, modeling and mapping of functional diversity should be easier than assessments of species richness. This applies even more to the chemical diversity of vegetation canopies, which is related to both species richness and functional diversity (Townsend et al. 2008), and can be scaled from the leaf to the landscape level (Asner and Martin 2008). Because many biochemicals are also responsible for optical properties of vegetation stands, the biochemical variation can in general be assessed with high accuracy using hyperspectral data. Problems arise, however, if biochemistry varies locally due to plant stress or leaf exposure to light. The approach has to date only been tested for tropical forest systems.

13.2 Synthesis

Remote sensing for nature conservation is a field with numerous applications that are only partly exploited. Depending on how close a mapping target is linked to optical or structural properties of vegetation, success and failure are close by. The potential for a complete automation of most mapping procedures mentioned in this chapter is very low and field work will always be needed to calibrate new models. This is bad news for the hard science party in remote sensing. Good news is that the products are not cheap replacements of field work but valuable add-ons that provide detailed spatial information on target variables with only little additional effort compared to field-based studies. These add-ons may help to fulfill the existing demand for reliable and objective information in conservation planning and may further have the potential to take monitoring attempts to the next level. However, this has not been fully realized in the conservation community. Constant interaction, communication, and understanding of ecologists and remote sensing scientists are required to take full advantage of the synergies between both fields.

Acknowledgments The author's activities in remote sensing for nature conservation have been or are partly funded by the German Ministry of Economics and Technology, German Space Agency (MSAVE project, DLR FKZ 50EE1032) and by the European Community's Seventh Framework Programme (MS-MONINA project, grant 263479).

References

Amici V (2011) Dealing with vagueness in complex forest landscapes: a soft classification approach through a niche-based distribution model. Ecol Inform 6:371–383

Andrew ME, Ustin SL (2008) The role of environmental context in mapping invasive plants with hyperspectral image data. Remote Sens Environ 112:4301–4317

Asner GP, Martin RE (2008) Spectral and chemical analysis of tropical forests: scaling from leaf to canopy levels. Remote Sens Environ 112:3958–3970

Bradley BA, Olsson AD, Wang O, Dickson BG, Pelech L, Sesnie SE, Zachmann LJ (2012) Species detection vs. habitat suitability: are we biasing habitat suitability models with remotely sensed data? Ecol Model 244:57–64

Breiman L (2001) Random forests. Mach Learn 45:5–32

Carter GA (1993) Responses of leaf spectral reflectance to plant stress. Am J Bot 80:239–243

Carter GA, Knapp AK (2001) Leaf optical properties in higher plants: linking spectral characteristics to stress and chlorophyll concentration. Am J Bot 88:677–684

Castro-Esau KL, Sánchez-Azofeifa GA, Rivard B, Wright SJ, Quesada M (2006) Variability in leaf optical properties of mesoamerican trees and the potential for species classification. Am J Bot 93:517–530

Council of the European Communities (1992) Council directive 92/43/EEC of 21 May 1992 on the conservation of natural habitats and of wild fauna and flora

Everitt JH, Alaniz MA, Escobar DE, Davis ME (1992) Using remote sensing to distinguish common (*Isocoma coronopifolia*) and Drummond goldenweed (*Isocoma drummondii*). Weed Sci 40:621–628

Feilhauer H, Schmidtlein S (2011) On variable relations between vegetation patterns and canopy reflectance. Ecol Inform 6:83–92

Feilhauer H, Oerke EC, Schmidtlein S (2010) Quantifying empirical relations between planted species mixtures and canopy reflectance with PROTEST. Remote Sens Environ 114:1513–1521

Feilhauer H, Faude U, Schmidtlein S (2011) Combining Isomap ordination and imaging spectroscopy to map continuous floristic gradients in a heterogeneous landscape. Remote Sens Environ 115:2513–2524

Feilhauer H, Thonfeld F, Faude U, He KS, Rocchini D, Schmidtlein S (2013) Assessing floristic composition with multispectral sensors – a comparison based on monotemporal and multiseasonal field spectra. Int J Appl Earth Obs Geoinf 21:218–229

Foody GM (1992) A fuzzy sets approach to the representation of vegetation continua from remotely sensed data: an example from lowland heath. Photogramm Eng Remote Sens 58:221–225

Foody GM (1996) Fuzzy modelling of vegetation from remotely sensed imagery. Ecol Model 85:3–12

Foody GM (1999) The continuum of classification fuzziness in thematic mapping. Photogramm Eng Remote Sens 65:443–451

Foody GM, Trodd NM (1993) Non-classificatory analysis and representation of heathland vegetation from remotely sensed imagery. GeoJournal 29:343–350

Förster M, Frick A, Walentowski H, Kleinschmit B (2008) Approaches for utilising QuickBird data for the monitoring of NATURA 2000 habitats. Community Ecol 9:155–168

Gausmann HW (1984) Evaluation of factors causing reflectance differences between sun and shade leaves. Remote Sens Environ 15:177–181

Ghioca-Robrecht DM, Johnston CA, Tulbure MG (2008) Assessing the use of multiseasonal Quickbird imagery for mapping invasive species in a Lake Erie coastal marsh. Wetlands 28:1028–1039

Goetz S, Steinberg D, Dubayah R, Blair B (2007) Laser remote sensing of canopy habitat heterogeneity as a predictor of bird species richness in an eastern temperate forest, USA. Remote Sens Environ 108:254–263

Gould W (2000) Remote sensing of vegetation, plant species richness and regional biodiversity hotspots. Ecol Appl 10:1861–1870

Gross JE, Goetz SJ, Cihlar J (2009) Application of remote sensing to parks and protected area monitoring: introduction to the special issue. Remote Sens Environ 113:1343–1345

Hall K, Reitalu T, Sykes MT, Prentice HC (2012) Spectral heterogeneity of QuickBird satellite data is related to fine-scale plant species spatial turnover in semi-natural grasslands. Appl Veg Sci 15:145–157

Hantson W, Kooistra L, Slim PA (2012) Mapping invasive woody species in coastal dunes in the Netherlands: a remote sensing approach using LIDAR and high-resolution aerial photographs. Appl Veg Sci 15:536–547

Harris AT, Asner GP, Miller ME (2003) Changes in vegetation structure after long-term grazing in pinyon–juniper ecosystems: integrating imaging spectroscopy and field studies. Ecosystems 6:368–383

Hernandez-Stefanoni JL, Ponce-Hernandez R (2004) Mapping the spatial distribution of plant diversity indices in a tropical forest using multi-spectral satellite image classification and field measurements. Biodivers Conserv 13:2599–2621

Kennedy RE, Townsend PA, Gross JE, Cohen WB, Bolstad P, Wang YQ, Adams P (2009) Remote sensing change detection tools for natural resource managers: understanding concepts and tradeoffs in the design of landscape monitoring projects. Remote Sens Environ 113:1382–1396

Kumar L, Schmidt K, Dury S, Skidmore A (2001) Imaging spectrometry and vegetation science. In: van der Meer FD, de Jong SM (eds) Imaging spectroscopy: basic principles and prospective applications. Kluwer Academic Publishers, Dordrecht, pp 111–155

Laba M, Tsai F, Ogurcak D, Smith S, Richmond ME (2005) Field determination of optimal dates for the discrimination of invasive wetland plant species using derivative spectral analysis. Photogramm Eng Remote Sens 71:603–611

Laliberte AS, Rango A, Havstad KM, Paris JF, Beck RF, McNeely R, Gonzalez AL (2004) Object-oriented image analysis for mapping shrub encroachment from 1937–2003 in southern New Mexico. Remote Sens Environ 93:198–210
Langley SK, Cheshire HM, Humes KS (2001) A comparison of single date and multitemporal satellite image classifications in a semi-arid grassland. J Arid Environ 49:401–411
Li W, Guo Q (2010) A maximum entropy approach to one-class classification of remote sensing imagery. Int J Remote Sens 31:2227–2235
Nagendra A (2001) Using remote sensing to assess biodiversity. Int J Remote Sens 22:2377–2400
Nagendra H, Lucas R, Honrado J, Jongman R, Tarantino C, Adamo M, Mairota P (2013) Remote sensing for conservation monitoring: assessing protected areas, habitat extent, habitat condition, species diversity, and threats. Ecol Indic 33:45–59
Newton AC, Hill RA, Echeverria C, Golicher D, Rey Benayas JM, Cayuela L, Hinsley SA (2009) Remote sensing and the future of landscape ecology. Prog Phys Geogr 33:528–546
Noujdina NV, Ustin SL (2008) Mapping Downy Brome (*Bromus tectorum*) using multidate AVIRIS data. Weed Sci 56:173–179
Oldeland J, Dorigo W, Lieckfeld L, Lucier A, Jürgens N (2010) Combining vegetation indices, constrained ordination and fuzzy classification for mapping semi-natural vegetation units from hyperspectral imagery. Remote Sens Environ 114:1155–1166
Phillips SJ, Anderson RP, Schapire RE (2006) Maximum entropy modeling of species geographic distributions. Ecol Model 190:231–259
Roberts DA, Gardner M, Church R, Ustin S, Scheer G, Green RO (1998) Mapping Chaparral in the Santa Monica Mountains using multiple endmember spectral mixture models. Remote Sens Environ 65:267–279
Rocchini D, Chiarucci A, Loiselle SA (2004) Testing the spectral variation hypothesis by using satellite multispectral images. Acta Oecologica 26:117–120
Sanchez-Hernandez C, Boyd DS, Foody GM (2007a) Mapping specific habitats from remotely sensed imagery: support vector machine and support vector data description based classification of coastal saltmarsh habitats. Ecol Inform 2:83–88
Sanchez-Hernandez C, Boyd DS, Foody GM (2007b) One-class classification of a specific land-cover class: SVDD classification of fenland. IEEE Trans Geosci Remote Sens 45:1061–1073
Sanger JE (1971) Quantitative investigation of leaf pigments from their inception in buds through autumn coloration to decomposition in falling leaves. Ecology 52:1075–1089
Schmidt KS, Skidmore AK (2003) Spectral discrimination of vegetation types in a coastal wetland. Remote Sens Environ 85:92–108
Schmidtlein S, Sassin J (2004) Mapping of continuous floristic gradients in grasslands using hyperspectral imagery. Remote Sens Environ 92:126–138
Schmidtlein S, Feilhauer H, Bruelheide H (2012) Mapping plant strategy types using remote sensing. J Veg Sci 23:395–405
Schölkopf B, Platt JC, Shawe-Taylor J, Smola AJ, Williamson RC (2001) Estimating the support of a high-dimensional distribution. Neural Comput 13:1443–1471
Schuster C, Ali I, Lohmann P, Frick A, Förster M, Kleinschmit B (2011) Towards detecting swath events in TerraSAR-X time series to establish NATURA 2000 grassland habitat swath management as monitoring parameter. Remote Sens 3:1308–1322
Smola AJ, Schölkopf B (2004) A tutorial on Support Vector Regression. Stat Comput 14:199–222
Somodi I, Čarni A, Ribeiro D, Podopnikar T (2012) Recognition of the invasive species *Robinia pseudoacacia* from combines remote sensing and GIS sources. Biol Conserv 150:59–67
Sorby HC (1873) On comparative vegetable chromatography. Proc R Soc Lond 21:442–483
Spanhove T, Van den Borre J, Delalieux S, Haest B, Paelinckx D (2012) Can remote sensing estimate fine-scale quality indicators of natural habitats? Ecol Indic 18:403–412
Tax DMJ, Duin RPW (2004) Support vector data description. Mach Learn 54:45–66
Townsend PA, Walsh SJ (2001) Remote sensing of forested wetlands: application of multitemporal and multispectral satellite imagery to determine plant community composition and structure in southeastern USA. Plant Ecol 157:129–149

Townsend AR, Asner GP, Cleveland CC (2008) The biochemical heterogeneity of tropical forests. Trends Ecol Evolut 23:424–431
Tuanmu MN, Vina A, Bearer S, Xu W, Ouyang Z, Zhang H, Liu J (2010) Mapping understory vegetation using phenological characteristics derived from remotely sensed data. Remote Sens Environ 114:1833–1844
Underwood E, Ustin S, DiPietro D (2003) Mapping nonnative plants using hyperspectral imagery. Remote Sens Environ 86:150–161
Van den Borre J, Paelinckx D, Mücher CA, Kooistra L, Haest B, De Blust G, Schmidt AM (2011) Integrating remote sensing in Natura 2000 habitat monitoring: prospects on the way forward. J Nat Conserv 19:116–125
Walker JS, Briggs JM (2007) An object-oriented approach to urban forest mapping in phoenix. Photogramm Eng Remote Sens 73:577–583
Wang F (1990) Improving remote sensing image analysis through fuzzy information representation. Photogramm Eng Remote Sens 56:1163–1169
Waser LT, Baltsavias E, Ecker K, Eisenbeiss H, Feldmeyer-Christe E, Ginzler C, Küchler M, Thee P, Zhang L (2008) Assessing changes of forest area and shrub encroachment in a mire ecosystem using digital surface models and CIR-aerial images. Remote Sens Environ 112:1956–1968
Whittaker RH (1960) Vegetation of the Siksiyou Mountains, Oregon and California. Ecol Monogr 30:279–338
Wiens J, Sutter R, Anderson M, Blanchard J, Barnett A, Aguilar-Amuchastegui N, Avery C, Laine S (2009) Selecting and conserving lands for biodiversity: the role of remote sensing. Remote Sens Environ 113:1370–1381
Wold S, Sjöström M, Eriksson L (2001) PLS-regression. A basic tool of chemometrics. Chemom Intell Lab Syst 58:109–130
Wood TF, Foody GM (1989) Analysis and representation of vegetation continua from Landsat Thematic Mapper data for lowland heaths. Int J Remote Sens 10:181–191
Wulder M, Hall RJ, Coops NC, Franklin SE (2004) High spatial resolution remotely sensed data for ecosystem characterization. BioScience 5:511–521
Wulder MA, Han T, White JC, Sweda T, Tsuzuki H (2007) Integrating profiling LIDAR with Landsat data for regional boreal forest canopy attribute estimation and change characterization. Remote Sens Environ 110:123–137
Xie Y, Sha Z, Yu M (2008) Remote sensing imagery in vegetation mapping: a review. J Plant Ecol 1:9–23

Chapter 14
Modeling Urban Sprawl

Roland Goetzke

14.1 Introduction

Urban populations are increasing worldwide. The percentage of people in Europe living in cities increased from 51 % in 1950 to 70 % in 2000 and is expected to reach 84 % in 2050 (United Nations Population Division 2010). The growth of cities and the social and environmental consequences associated with this phenomenon are subjects of intensive debate. Especially suburban sprawl within the Anglo-American context has been well documented and is a major target of contemporary criticism (Blais 2010; Kunstler 1994; Warner 1972). It is viewed as an expression of wealth and individual freedom (Gillham and MacLean 2002), as well as a manifestation of chaos requiring restraint through planning (Batty 2008). The challenges of urban sprawl raised the need for new concepts in urban planning and sustainable growth policies, like smart growth or new urbanism (Freilich et al. 2010).

Uncontrolled urban sprawl has diverse environmental, economic, social and aesthetic consequences. The conversion of land-use evolving from urban growth lead to alterations of biogeochemical cycles, hydrologic systems, and biodiversity in the city and – because of a city's "ecological footprint" – well beyond its boundaries (Grimm et al. 2008). Urban sprawl is not only a phenomenon of Anglo-American cities, where it is associated with low-density expansion of private residential housing. In Europe cities have historically developed a dense core; but since the 1950s urban sprawl has also become a challenge for urban planning and policy (EEA 2006).

Cities belong to the most complex land-use systems and their sprawling nature often seems unplanned, even chaotic. But being complex does not necessarily mean disordered. Cities are "...the example par excellence of complex systems (Batty 2008)": They show emergence, non-linearity, feedback, and path-dependence.

R. Goetzke (✉)
Department of Geography, University of Bonn,
Meckenheimer Allee 166, 53115, Bonn, Germany
e-mail: goetzke@uni-bonn.de

I. Manakos and M. Braun (eds.), *Land Use and Land Cover Mapping in Europe: Practices & Trends*, Remote Sensing and Digital Image Processing 18,
DOI 10.1007/978-94-007-7969-3_14,

Cities are self-maintaining and their physical expansion is a function of the sum of the manifold decisions made by their constituent parts, i.e. residents, planners, and economic interests. This makes urban growth an appropriate and fascinating subject for scientists engaged in modeling (Berling-Wolf and Wu 2004). There is also an increasing demand by policy makers and city planners for models that can support their decision-making process by forecasting future states of cities or future scenarios resulting from different planning assumptions (Geertman and Stillwell 2004). In the last four decades a great number of urban growth models have been developed and incorporated into Spatial Decision Support Systems (SDSS) and Planning Support Systems (PSS) (Geertman and Stillwell 2009). Despite this development much criticism remains concerning the gap between planning practice and scientific research (Couclelis 2005).

This paper provides a comprehensive overview of the processes and problems of urban sprawl with a focus on European urban areas. A summary of common urban growth models and modeling techniques is given and the role of remote sensing in these processes is examined. An example application illustrates model coupling as one part of ongoing research activities.

14.2 Urban Sprawl

Despite considerable research on the subject of urban sprawl there is no agreement on a comprehensive definition of urban sprawl. According to Chin (2002) four elements have to be considered: urban form, land-use, impacts, and densities. Urban sprawl and urban growth are both terms that describe spatial processes that may result in a similar land-use pattern. While *urban growth* can take place in a coordinated way, *urban sprawl* is often associated with an uncoordinated growth that is larger than reasonable. A third term in this context is *urbanization*, which describes an a-spatial process referring to the complex changes of life style resulting from the impact of cities on society (Clark 1982). The terms urban growth and urban sprawl are used in the following interchangeably, because the focus of this paper is on the physical expansion of urban structures and not on the processes leading to these observable patterns.

14.2.1 Patterns, Processes, Problems, and Policies

Urban sprawl accelerated in Northern America and Europe since the second half of the twentieth century. With an increasing population in the cities, a lack of affordable housing in the centers, and higher levels of income the consumer demand for housing changed. At the same time an increased mobility and changes in the mode of private and public transport enabled access to areas at further distances from the city leading to a spreading suburban development. It is this suburban development that has been principally identified as urban sprawl.

Most widely cited as urban sprawl are low density expanses of single-use development leading to uniform suburban patterns. Additionally, there is "leap-frog" or scattered development, referring to discontinuous growth away from central cores resulting in a fragmented land-use pattern (Gillham and MacLean 2002). Another form is "commercial strip development", which typically follows arterial roads lined with shopping centers and commercial areas (Gillham and MacLean 2002).

The environmental and economic impacts of urban sprawl are numerous and have triggered the debates about this phenomenon. These impacts include the consumption of valuable agricultural land, natural habitat fragmentation, and in recent years the loss of biodiversity (Grimm et al. 2008); increased air pollution, greenhouse gas emission and traffic congestion (Gillham and MacLean 2002); high infrastructure and development costs and exhaustive energy consumption (Burchell 2005); the loss of landscape heritage values and the evolution of displeasing urban patterns (Antrop 2004). A key indicator for unsustainable land-use changes in urbanized areas is the increase of impervious land-cover, which directly affects surface water runoff and infiltration, soil functions and local climate (Arnold and Gibbons 1996).

The continuing loss of arable land due to urban growth has caused the German federal government to implement a reduction of the daily land conversion (or "take") in the national sustainability strategy, from 129 ha in the year 2000 to 30 ha in 2030. In 2009/2010 the daily land take for settlement and infrastructure areas in Germany was 77 ha (Hoymann et al. 2012). In addition to the above listed specific problems, spatial planning has to deal with urban sprawl under the conditions of globalization, demographic change and the need for developing and implementing strategies for climate change adaptation (EEA 2012).

14.2.2 Trends of Urban Sprawl in European Cities

European cities are often idealized as "compact cities" with higher densities and shorter distances than North-American cities (Dieleman and Wegener 2004). But since the 1950s Europe's cities have expanded by 78 % while their population grew by only 38 % (EEA 2006). The development of European cities is diverse. Kasanko et al. (2006) found that Southern European cities are still the most compact. In Eastern Europe and parts of Central Europe most cities are characterized by discontinuous residential structures and low population densities, while in Western Europe and parts of Central Europe cities show a diverse development. This could only partially be confirmed by Schwarz (2010), who also identified low density cities in Italy and Spain as well as compact cities in the UK.

The driving forces of urban sprawl in Europe are manifold and vary between countries, and within regions and cities. Economic drivers range from macro-economic factors like the European Integration to micro-economic factors such as land prizes. Population growth is, of course, an important factor, but urban sprawl can also be observed in areas where population is declining (Couch et al. 2005). Social factors include new housing preferences and pull-factors resulting from inner-city problems. Transportation factors like private car ownership or improved public transport provide greater freedom to the localization of people. Policy interventions play a major role in urban sprawl, but are of all the relevant factors the most difficult to quantify. Weak land-use planning or a lack of coordination can lead to undesirable urban patterns. In German cities it can be observed that in recent years, due to inter-communal competition for residents and tax revenue, the urban development changed from the demand side to the supply side, resulting in the construction of new housing estate areas and business parks (Mainz 2005; Siedentop et al. 2009).

The areas in Europe, where urban sprawl is most obvious, are the regions with high population density and economic activity, notably the Netherlands, Belgium, Luxembourg, Western and Southern Germany, Northern Italy, and the Paris and London regions (EEA 2006). But there are new "hot spots" and corridors with higher growth rates than in these already highly urbanized areas. EU Structure and Cohesion funds have had significant implications on urban sprawl. Regions such as Ireland, Portugal, Eastern Germany, and several areas in Spain, which have benefited from EU regional policies, have also recently shown high growth rates (EEA 2006). These countries have invested high amounts of these funds to improve and expand their transportation network. Such direct infrastructure investments affect land development by improving accessibility to and availability of land, which attracts transport-related industries, followed by jobs and residents (Christiansen and Loftsgarden 2011).

Today, urban sprawl in Europe affects not only urban areas, but the rural countryside as well (Antrop 2004). As a result, new urban structures emerge which have urban as well as rural characteristics, resulting in an "urban–rural continuum" (Sieverts 2005). Current trends of urban growth in Germany show an increasing polarization (Hoymann et al. 2012) where most areas show both urban growth and a declining population.

14.3 Models of Urban Growth and Urban Dynamics

In the early 1960s the computer became an essential tool for geographers and urban planners and the dynamic modeling of recurring urban patterns became a new field of research (Hägerstrand 1967). Already in that early stage critics fundamentally questioned the purpose of urban simulations in the planning context (Lee 1973) – and the same worries about the loss of contact with policy problems, high processing power requirements, and confusing methodologies are still discussed today (Couclelis 2005).

14.3.1 Scope and Objectives of Urban Growth Models

Operational, spatially explicit urban growth models did not become available until the 1990s. Urban models developed in the last two decades are valuable tools for three groups of users: (1) researchers developing modeling tools and investigating processes in the urban environment, (2) land-use planners needing support in realizing visions for future city development, and (3) citizens who benefit from participation in their community and use model results as illustrations of future projections. The three groups have by nature very different backgrounds and expectations related to urban models. Couclelis (2005) offers an in-depth view into the trade-offs concerning models in research and planning.

Traditionally, computer simulations of urban growth have been particularly developed in the context of community planning where population growth, economic trends and transport-related effects on economic growth were the main concerns. Currently, the consequences of land-use and environmental impacts of growing cities are becoming increasingly important.

14.3.2 Theories and Modeling Techniques

A diverse suite of modeling approaches has evolved during the last decade, some of these deal explicitly with processes leading to urban growth, others consider urban growth as one type of land-use change amongst others. A selection of these approaches is presented in the following paragraphs. Recent surveys of operational land-use change models offer deep insights into elementary model concepts and characteristics, including aspects of urban growth (Agarwal et al. 2002; Briassoulis 2000; Koomen and Stillwell 2007; Verburg et al. 2004b). Reviews of urban growth models in particular cover the applicability of models in a planning context (US EPA 2000) and a differentiation of models arising out of transportation, economic, and environmental sciences (Berling-Wolf and Wu 2004). Other reviews examine urban models based on specific modeling techniques, such as Cellular Automata (CA) and Multi-Agent Systems (MAS) (Benenson and Torrens 2004; Haase and Schwarz 2009). In addition to CA and MAS other modeling techniques commonly used in urban growth models include Statistical Regression (SR), Spatial Optimization (SO), and Machine Learning (ML). Table 14.1 provides an overview of well documented land-use change models used in the context of urban growth as well as the analytic techniques applied in these models.[1]

[1] The techniques are not evaluated or ranked in terms of performance. All presented models use a combination of modeling techniques in their algorithms.

Table 14.1 Methodological background of several models of urban growth

Model	CA	MAS	SR	SO	ML	References
California Urban Futures (CUF II, CURBA)	–	–	+	+	–	Landis (2001)
CLUE (Clue-S, Dyna-Clue, EU-ClueScanner)	+	–	+	+	–	Batisani and Yarnal (2009), Verburg et al. (2002) and Lavalle et al. (2011)
Environment Explorer	+	–	+	+	–	de Nijs et al. (2004) and Engelen et al. (2003)
GEOMOD	+	–	+	–	+	Poelmans and Van Rompaey (2009) and Pontius et al. (2001)
ILUTE	–	+	+	–	–	Miller et al. (2004)
Land Transformation Model	+	–	–	–	+	Pijanowski et al. (2002)
Land Use Scanner	–	–	+	+	–	Koomen et al. (2011)
MOLAND	+	–	–	+	–	Lavalle et al. (2004)
PUMA	–	+	+	–	–	Ettema et al. (2007)
SLEUTH	+	–	+	–	+	Clarke et al. (1997) and Silva and Clarke (2005)
Urban SIM	–	+	+	–	–	Waddell (2002)
What If?	–	–	–	+	–	Klosterman (2008)

CA cellular automata, *MAS* multi agent systems, *SR* statistical regression, *SO* spatial optimization, *ML* machine learning

14.3.2.1 Cellular Automata (CA)

CA are related to the behavior of complex, self-reproducing systems (Tobler 1979). All CA use the same assumptions and have the same basic structure:

- CA have a uniform cellular space,
- every cell has one state out of a finite number of states,
- every cell has a finite neighborhood,
- the transition of a cell from one state to another is possible and is determined by local transition rules that can be deterministic or stochastic, and
- the cell states are updated in a discrete sequence of time steps.

Tobler (1979) envisaged the concept of CA for geographical applications. They became increasingly important in urban growth modeling (Batty and Xie 1994; Clarke et al. 1997; Couclelis 1985; Torrens and O'Sullivan 2001; White and Engelen 1997). CA are quite simple systems, which can be used to model complex dynamic systems like cities, which are characterized by emergence, self-organization, and non-linearity (Barredo et al. 2003). Assigning the transition rules is a crucial part in creating realistic simulations with CA, and often SR or ML modeling techniques are used for this function. Established urban growth models based on CA are SLEUTH – also known as the Clarke Urban Growth Model (Clarke et al. 1997) –, the Environment Explorer (Engelen et al. 2003), and MOLAND (Lavalle et al. 2004).

14.3.2.2 Multi Agent Systems (MAS)

While CA focus on the discrete cell space representing individual land-use cells or land-parcels, MAS deal with the decision processes of the key actors – or "agents" – in the land-use system and their impact on the land-use. Agents interact with each other, they are autonomous, they share their environment with other agents, they communicate, and they make land-use decisions. Agents are entities, like e.g. individuals, households, cities, or political structures. MAS are actually non-spatial, but in urban growth models cells are considered as containers for agents. Information can be distributed from one agent to another through neighborhood effects in the raster. In contrast to CA agents in MAS are able to move along cells.

MAS are highly complex and require intense parameterization and computing power. For these reasons, many MAS are based on simplified hypothetical landscapes (Verburg et al. 2004b). Hybrid MAS/CA models are often employed, where CA represent the landscape and the land-use changes that are induced by agents in the MAS (Torrens 2001). While CA have advantages in simulating changes in urban extent and infrastructure, MAS better model the processes resulting in these changes, such as population dynamics or changes in transport flows.

14.3.2.3 Statistical Regression (SR)

SR is a modeling technique utilized for detecting logic correlations between land-use changes and underlying driving forces. Multiple linear regression is a common SR technique. Logistic regression must be employed rather than linear regression when SR is applied to raster data, where land-use is represented in the form of discrete classes. Examples for the application of SR for the explanation of urban land-use changes can be found in Batisani and Yarnal (2009), Goetzke (2011) and Verburg et al (2004a).

14.3.2.4 Spatial Optimization

Spatial Optimization (SO) is a technique which is highly focused on the applicability of models in a planning context, delivering stakeholders a set of alternative scenarios instead of a single solution. The land-use allocation that is performed by a model has to meet *ex ante* assumptions about the environment, the economic activity of agents, and policy impacts. Models implementing SO attempt to minimize parameter deviations, such as those of predefined land-use demands or claims. The resultant alternatives should then be treated by stakeholders as propositions for further analysis and not as ultimate solutions (Ligmann-Zielinska et al. 2008).

14.3.2.5 Machine Learning (ML)

ML is used in bottom-up modeling approaches, where either expected land-use changes cannot be predicted or no linear relationship between land-uses and driving forces can be presumed. With ML, a model is fitted during calibration until it best fits reference data; ML is thus employed in many CA-based models. A well documented example of a CA model using ML during calibration is the SLEUTH model (Dietzel and Clarke 2007). Other ML approaches used to improve urban growth models are Artificial Neural Networks (Li and Yeh 2002) and Support Vector Machines (SVM) (Huang et al. 2010; Yang et al. 2008).

14.3.3 Remote Sensing and Models of Urban Growth

Cellular space is well suited to represent real world phenomena. Thus, raster information derived from remote sensing imagery is a major data source for urban growth models. Extensive research has conclusively shown remote sensing techniques to be valuable in mapping urban areas and urban change (Bhatta 2010; Gamba and Herold 2009; Weng 2012). Increasingly available suites of remote sensing data in different resolutions from new instruments (including radar and hyperspectral data) reveal new insights into urban systems. Typically, urban models utilize discrete land-cover classes or types of urban land-use (e.g. residential and industrial/commercial uses). Residential uses can be derived directly from remote sensing data; industrial/commercial uses, however, require interpretation and inference of socioeconomic information.

Many urban growth models are implemented at local or regional scales, where individual data sets are collected using medium to high resolution satellite data or topographic maps. For urban growth models to be implemented at national or trans-border scales, comprehensive data sets are required that permit proper comparison between regions. Recent activities have generated land-use datasets that synoptically characterize European urban areas; these datasets have been produced based on remote sensing data and can serve as a base data for urban models. These include CORINE Land Cover and the GMES/Copernicus Urban Atlas (see also Chaps. 2, 3, 4 and 5). Several international initiatives have also developed global land-use/-cover products that can be used in large-scale urban modeling applications. The GlobCover and GLC2000 datasets show global urban areas at comparatively coarse resolutions of 300 and 1,000 m, respectively. Although these datasets employ the same classification nomenclature, they are not fully comparable due to different classification mechanisms.

14.4 Example: Combined Use of Two Land-Use Change Models

A priority research topic within the field of urban growth modeling that requires additional attention is the integration of different modeling concepts in order to address the heterogeneity and multi-scale characteristics of urban land-use systems. This can be expressed in the combining of bottom-up and top-down modeling methodologies (Rounsevell et al. 2012). In the following section examples of a combined use of different modeling approaches are presented with the aim of expanding established concepts in this field.

Conceptually, the Clue-S model (Verburg et al. 2002) relies on the competition among land-uses. The model allocates regional demands for land-use changes based on local land-use suitability maps that have been calculated using logistic regression. Thus, the model is capable of simulating changes in an entire land-use system and to provide insights into the driving forces leading to these changes. Clue-S may also be used to analyze scale dependencies by bringing regional ("top-down") demands and local suitability together. In this model urban areas are treated as one land-use class amongst others. If the demand for urban land is high and the areas best suited for urban land have already been "urbanized", urban growth will compete in Clue-S with other land-use types having a lower demand (e.g. forest or agriculture), but a higher local suitability in the respective areas.

The SLEUTH model (Clarke et al. 1997) has been widely used in urban growth simulations. The model has a CA component simulating the development of urban areas and a land-cover component that simulates land-cover changes other than urban. The SLEUTH urban growth model follows a "bottom-up" approach. The model parameters that determine the form of urban growth are defined during a calibration process where the model performs brute force Monte Carlo iterations and compares the model results with reference data to find the optimal parameter combination. The model simulates four different kinds of urban growth: spontaneous growth, new spreading centers, organic (edge growth) and road influenced growth. In addition to maps of existing urban extent, primary model input data include terrain slope maps, transportation network maps and an exclusion layer containing development restrictions.

In this study a modified version of the CA component of SLEUTH is used. It differs significantly from the original model. First, in the original model a set of Pearson's regression statistics is used to determine the growth coefficients. These are calculated by comparing different landscape metrics produced as model output with reference data. This procedure has been replaced by the multiple resolution validation (MRV) method (Pontius et al. 2004). Second, the number of reference datasets used during calibration is reduced from 4 to 2. The original algorithm needs besides an initial dataset at least three control datasets in order to calculate the regression scores. Four subsequent datasets (and ideally a fifth) are also required in order to validate the model; in practice these are often difficult to acquire.

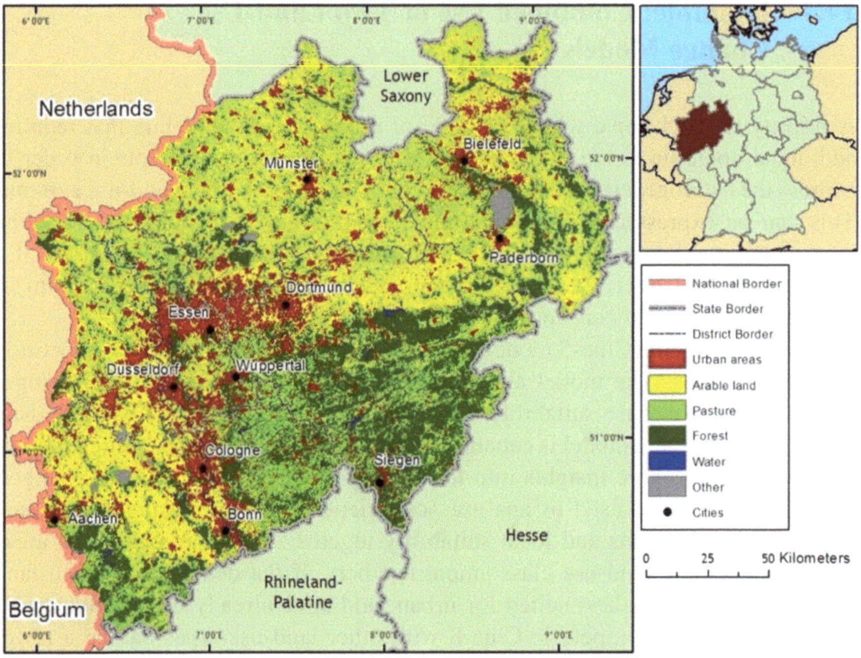

Fig. 14.1 Map of the study area North Rhine-Westphalia (NRW) based on classified Landsat data of 2005

We therefore utilize the MRV to compare the simulated built-up areas with a single control map during the process of model calibration. This reduction is reasonable when the time-domain of the simulation covers only a few decades during which approximately linear growth can be assumed. The original calibration methodology should be implemented in cases where the entire history of urban development is to be modeled, resulting in the characteristic s-shaped growth rate.

Both the Clue-S and the modified Urban Growth Model (UGM) component of SLEUTH have been implemented in the JAVA-based modeling platform XULU (eXtandable Unified Land Use Modeling Platform) (Schmitz et al. 2007). The core program of XULU contains management functions, import/export routines, as well as a GUI. The models are implemented as XULU plug-ins, requiring that the model algorithms be translated into JAVA programming language. XULU stores all data in a data pool, which offers the opportunity for model coupling. The output of one model can serve as input for a second model, with both models running simultaneously in XULU. A coupling routine has been developed that copies the urban extent simulated by the UGM into the Clue-S model. In that way, all land-use classes except the "urban" class are modeled by Clue-S and the urban areas are modeled by UGM (Goetzke 2011).

Both models have been applied separately as well as coupled to a study area covering the federal state of North Rhine-Westphalia (NRW) in Germany (Fig. 14.1). The study area comprises the highly urbanized areas along the Rhine river and the Ruhr area, as well as rural areas. These rural areas are dominated by

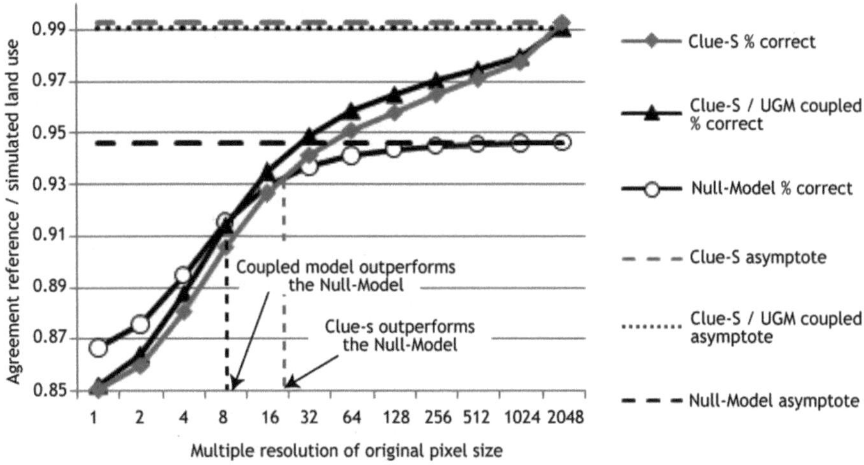

Fig. 14.2 Agreement between a reference map of the year 2001 and three model results: Null-Model, coupled version of Clue-S and UGM, Clue-S. Clue-S performs better than the Null-Model at a resolution 16× pixel size and the coupled model at 8× pixel size

agricultural land-use in the lowlands and by pasture and forests in the upland regions. Despite a declining population since 2000, the average land consumption in NRW remains 15 ha per day. Reference land-use data were available from classified Landsat data for 1975, 1984, 2001, and 2005, which were spatially resampled at a resolution of 100 × 100 m (Goetzke 2011). All land-use maps were classified with an overall accuracy of above 85 %. Land-use classes included urban, agricultural land, pasture, forest, water, and other (mining and military training areas). The land-use data reveal an increase in urban area of 63 %, or approximately 187,000 ha (5.4 % of the total land area) during the last 30 years within the study area.

The models have been calibrated by simulating land-use changes from 1984 to 2001. The final simulation covered the 2001–2025 period. Two scenarios were calculated: one followed a "business-as-usual" form of urban growth and the second alternative simulated urban growth under the achievement of sustainability goals. Data from 2005 were utilized to validate the models during the simulation. Because of the relatively short 2001–2005 time period and the expected small amount of land-use change during that period, another validation has been performed by simulating land-use changes during 1975–1984. All model results during calibration and validation were compared to a "Null-Model", where no land-use changes are assumed during the relevant time period. As large elements of the landscape within the study area are persistent, the Null-Model often delivers better results than a land-use model (Pontius et al. 2004). Goodness-of-fit statistics for each model are measured by the average agreement at all resolutions calculated by the MRV (Pontius et al. 2004) (see Table 14.2). For the calibrated model, the MRV shows that Clue-S outperforms a Null-Model assumption at 16× original pixel size (1.6 km), while the coupled model delivers better results than a Null-Model at 8× original pixel size, or 800 m (Fig. 14.2).

Table 14.2 Performance of the calibrated models Clue-S, UGM, and a coupled version of Clue-S and UGM for simulating land-use changes in NRW for the years 1984–2001 (validation results are shown in brackets: *1975–1984; 2001–2005*)

Model	MRV agreement for all land-use classes	MRV agreement for urban land-use
Clue-S	0.934 (*0.926; 0.975*)	0.978 (*0.976; 0.994*)
UGM	–	0.984 (*0.977; 0.996*)
Coupled Clue-S/UGM	**0.939** (*0.932; 0.979*)	**0.984** (*0.977; 0.996*)[a]
Null-Model	0.924 (*0.927; 0.986*)	0.972 (*0.969; 0.998*)

The values indicate the overall weighted agreement between the simulated and observed maps at different resolutions by the multiple resolution validation method described in Pontius et al. (2004)
[a]Urban areas identical with UGM

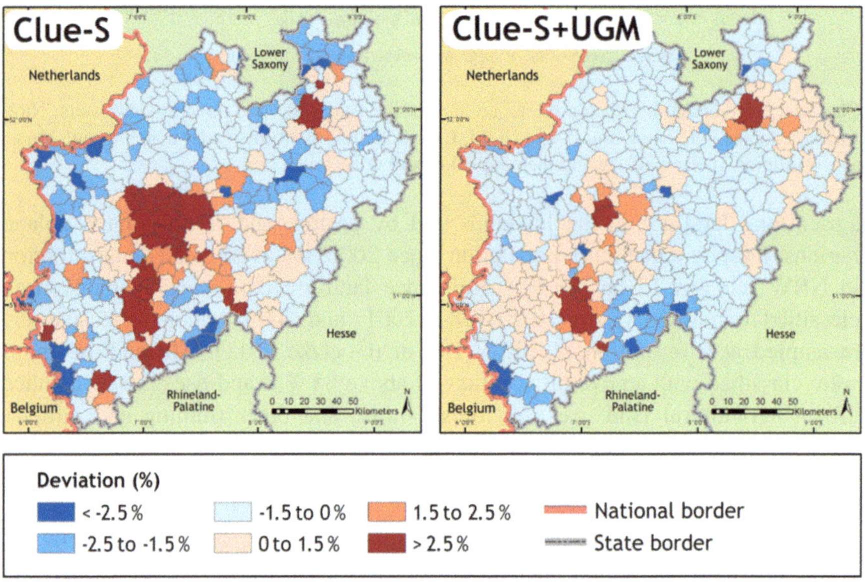

Fig. 14.3 Deviation between simulated and observed urban growth with the Clue-S model (*left*) and the coupled Clue-S/UGM model (*right*) for the reference year 2001

Table 14.2 shows the significantly better performance of the coupled Clue-S/UGM model, which uses UGM to simulate urban growth. The Null-Model performed better only for the 2005 validation year, due to the high degree of persistence in the landscape between 2001 and 2005 as determined by the high level of agreement between the Null-Model and the reference map.

When examining the results of this test, it is useful to consider both the average agreement for the entire study area as well as a regional comparison of model performances. The Clue-S model tended to overestimate urban growth in existing highly urbanized areas. According to the calculated probability maps, pixels with the highest suitability for urban land-use are located close to existing urban areas. In agglomerations like the Ruhr area, less urban growth appeared than was estimated by the model. The deviation between observed and simulated growth was much lower in the coupled version of the model (Fig. 14.3).

14.5 Perspectives in Urban Growth Modeling

With a very diverse methodological background and the increasing availability of land-use/land-cover data thanks to progress in remote sensing, the prerequisites for modeling urban dynamics are available. However, many established models still do not address the process of simultaneous urban growth and contraction (Haase et al. 2012). Also, the differing pathways and interactions typically present in urban and rural areas still need more attention (Verburg et al. 2004b). Especially when human decision-making is considered, urban growth models – MAS is an example – tend to become increasingly complex, limiting their application and transferability. Another challenge is the integration of political planning measures in urban models; implementing strategies for climate change adaption and mitigation is an important example (Verburg et al. 2012). Many urban growth models still lack adequate validation methodologies, limiting the use of model results as a basis for discussion (Verburg et al. 2004b). Significant new findings can also be expected by combining modeling approaches from different disciplines. Examples are the EU-ClueScanner developed for the European Commission DG Environment (Koomen et al. 2010) and the Land Use Scanner applied by the German Federal Institute for Research on Building, Urban Affairs and Spatial Development (Goetzke et al. 2012). Both models are programmed in the GeoDMS environment and are based on the modeling approach taken from the Dyna-CLUE model using the numerical algorithm of the well-established Land Use Scanner model.

Under the regime of climate change along with new challenges resulting from changing demographical and environmental conditions, urban models are becoming increasingly important and they are becoming more valuable for planners, researchers, and ultimately the public.

Acknowledgements This study was carried out in the Remote Sensing Research Group (RSRG, Department of Geography, University of Bonn) with the support of Gunter Menz (RSRG). I am also grateful to Andreas Rienow (University of Bonn) for his competent technical and linguistic assistance. The land-use data used in this study has been created at the Center for Remote Sensing of Land Surfaces (ZFL) at the University of Bonn, Germany, within the project "Visualisierung von Landnutzung und Flächenverbrauch in Nordrhein-Westfalen mittels Satelliten- und Luftbildern" funded by the Ministry of Environment and Nature Conservation, Agriculture and Consumer Protection of the federal state of NRW, Germany.

References

Agarwal C, Green GM, Grove JM, Evans TP, Schweik CM (2002) A review and assessment of land-use change models: dynamics of space, time, and human choice. USDA Forest Service, Newton Square

Antrop M (2004) Rural–urban conflicts and opportunities. In: Jongman RHG (ed) The new dimensions of the European landscape. Springer, Dordrecht, pp 83–91

Arnold CL, Gibbons CJ (1996) Impervious surface coverage: the emergence of a key environmental indicator. J Am Plan Assoc 62:243–258

Barredo JI, Kasanko M, McCormick N, Lavalle C (2003) Modelling dynamic spatial processes: simulation of urban future scenarios through cellular automata. Landsc Urban Plan 64:145–160
Batisani N, Yarnal B (2009) Urban expansion in centre county, Pennsylvania: spatial dynamics and landscape transformations. Appl Geogr 29:235–249
Batty M (2008) The size, scale, and shape of cities. Science 319:769–771
Batty M, Xie Y (1994) From cells to cities. Environ Plan B Plan Des 21:31–48
Benenson I, Torrens PM (2004) Geosimulation: object-based modeling of urban phenomena. Comput Environ Urban Syst 28:1–8
Berling-Wolf S, Wu J (2004) Modeling urban landscape dynamics: a review. Ecol Res 19:119–129
Bhatta B (2010) Analysis of urban growth and sprawl from remote sensing data. Springer, Berlin
Blais P (2010) Perverse cities: hidden subsidies, wonky policy, and urban sprawl. UBC Press, Vancouver
Briassoulis H (2000) Analysis of land use change: theoretical and modeling approaches. Regional Research Institute, University of West Virginia, Morgantown
Burchell RW (2005) Sprawl costs: economic impacts of unchecked development. Island Press, Washington, DC
Chin N (2002) Unearthing the roots of urban sprawl: a critical analysis of form, function and methodology, vol 47. CASA, London
Christiansen P, Loftsgarden T (2011) Drivers behind urban sprawl in Europe (No. 1136), TØI Report. Institute of Transport Economics. Norwegian Centre for Transport Research, Oslo
Clark D (1982) Urban geography: an introductory guide. Taylor & Francis, London
Clarke KC, Hoppen S, Gaydos LJ (1997) A self-modifying cellular automaton model of historical urbanization in the San Francisco Bay area. Environ Plan B Plan Des 24:247–261
Couch C, Karecha J, Nuissl H, Rink D (2005) Decline and sprawl: an evolving type of urban development – observed in Liverpool and Leipzig. Eur Plan Stud 13:117–136
Couclelis H (1985) Cellular worlds: a framework for modeling micro – macro dynamics. Environ Plan 17:585–596
Couclelis H (2005) “Where has the future gone?” Rethinking the role of integrated land-use models in spatial planning. Environ Plan 37:1353–1371
De Nijs TCM, de Niet R, Crommentuijn L (2004) Constructing land-use maps of the Netherlands in 2030. J Environ Manage 72:35–42
Dieleman F, Wegener M (2004) Compact city and urban sprawl. Built Environ 30:308–323
Dietzel C, Clarke KC (2007) Toward optimal calibration of the SLEUTH land use change model. Trans GIS 11:29–45
EEA (2006) Urban sprawl in Europe. The ignored challenge (Report No. 10/2006), EEA Report. European Environment Agency, Copenhagen
EEA (2012) Urban adaption to climate change in Europe. Challenges and opportunities for cities together with supportive national and European policies. (Report No. 2/2012), EEA Report. European Environment Agency, Copenhagen
Engelen G, White R, de Nijs T (2003) Environment explorer: spatial support system for the integrated assessment of socio-economic and environmental policies in the Netherlands. Integr Assess 4:97–105
Ettema D, de Jong K, Timmermans H, Bakema A (2007) PUMA: multi-agent modelling of urban systems. In: Koomen E, Stillwell J, Bakema A, Scholten HJ (eds) Modelling land-use change, The GeoJournal library. Springer, Dordrecht, pp 237–258
Freilich RH, Sitkowski RJ, Mennillo SD (2010) From sprawl to sustainability: smart growth, new urbanism, green development, and renewable energy. American Bar Association, Chicago
Gamba P, Herold M (eds) (2009) Global mapping of human settlement: experiences, datasets, and prospects, Har/Dvdr. ed. CRC Press, Boca Raton
Geertman S, Stillwell J (2004) Planning support systems: an inventory of current practice. Comput Environ Urban Syst 28:291–310
Geertman S, Stillwell JCH (2009) Planning support systems best practice and new methods. Springer, Dordrecht
Gillham O, MacLean AS (2002) The limitless city: a primer on the urban sprawl debate. Island Press, Washington, DC

Goetzke R (2011) Entwicklung eines fernerkundungsgestützten Modellverbundes zur Simulation des urban-ruralen Landnutzungswandels in Nordrhein-Westfalen. Dissertation, Rheinische Friedrich-Wilhelms-Universität Bonn, Bonn

Goetzke R, Dosch F, Beckmann G, Hoymann J, Distelkamp M (2012) Wie viel Fläche wird wo und wie verbraucht? Trends, Szenario 2030 und Bewertung. In: Meinel G, Schumacher J, Behnisch M (eds) Flächennutzungsmonitoring IV. Genauere Daten – Informierte Akteure – Praktisches Handeln, IÖR Schriften. RHOMBOS-Verlag, Berlin, pp 185–194

Grimm NB, Faeth SH, Golubiewski NE, Redman CL, Wu J, Bai X, Briggs JM (2008) Global change and the ecology of cities. Science 319:756–760

Haase D, Schwarz N (2009) Simulation models on human-nature interactions in urban landscapes: a review including spatial economics, system dynamics, cellular automata and agent-based approaches. Living Rev Landsc Res 3:1–45

Haase D, Haase A, Kabisch N, Kabisch S, Rink D (2012) Actors and factors in land-use simulation: the challenge of urban shrinkage. Environ Model Softw 35:92–103

Hägerstrand T (1967) The computer and the geographer. Trans Inst Br Geogr 42:1–19

Hoymann J, Dosch F, Beckmann G, Distelkamp M (2012) Trends der Siedlungsflächenentwicklung. Status quo und Projektion 2030 (No. 09/2012), BBSR-Analysen Kompakt. Bundesinstitut für Bau-, Stadt- und Raumforschung, Bonn

Huang B, Xie C, Tay R (2010) Support vector machines for urban growth modeling. GeoInformatica 14:83–99

Kasanko M, Barredo J, Lavalle C, McCormick N, Demicheli L, Sagris V, Brezger A (2006) Are European cities becoming dispersed? A comparative analysis of 15 European urban areas. Landsc Urban Plan 77:111–130

Klosterman RE (2008) A new tool for new planning. The what if? Planning support system. In: Brail RK (ed) Planning support systems for cities and regions. Lincoln Institute of Land Policy, Cambridge, pp 85–100

Koomen E, Stillwell J (2007) Theories and methods. In: Koomen E, Stillwell J, Bakema A, Scholten HJ (eds) Modelling land-use change, The GeoJournal library. Springer, Dordrecht, pp 1–21

Koomen E, Diogo V, Hilferink M, van der Beek M (2010) EU-ClueScanner100m; model description and validation results. VU University Amsterdam, Amsterdam

Koomen E, Hilferink M, Borsboom-van Beurden J (2011) Introducing Land Use Scanner. In: Koomen E, Borsboom-van Beurden J (eds) Land-use modelling in planning practice, The GeoJournal library. Springer, Dordrecht, pp 3–22

Kunstler JH (1994) The geography of nowhere: the rise and decline of America's man-made landscape. Touchstone, New York

Landis JD (2001) CUF, CUF II, and CURBA: a family of spatially explicit urban growth and land-use policy simulation models. In: Brail RK, Klosterman RE (eds) Planning support systems. Integrating geographic information systems, models, and visualization tools. ESRI Inc, Redlands, pp 157–200

Lavalle C, Barredo JI, McCormick N, Engelen G, White R, Uljee I (2004) The MOLAND model for urban and regional growth forecast. A tool for the definition of sustainable development paths (No. EUR 21480 EN). Joint Research Centre, Ispra

Lavalle C, Baranzelli C, Batista e Silva F, Mubareka S, Gomes C, Koomen E, Hilferink M (2011) A high resolution land use/cover modelling framework for Europe: introducing the EU-ClueScanner100 model. In: Murgante B, Gervasi O, Iglesias A, Taniar D, Apduhan B (eds) Computational science and its applications – ICCSA 2011, Lecture notes in computer science. Springer, Berlin/Heidelberg, pp 60–75

Lee DB (1973) Requiem for large-scale models. J Am Inst Plan 39:163–178

Li X, Yeh AG-O (2002) Neural-network-based cellular automata for simulating multiple land use changes using GIS. Int J Geogr Inf Sci 16:323–343

Ligmann-Zielinska A, Church R, Jankowski P (2008) Spatial optimization as a generative technique for sustainable multiobjective land-use allocation. Int J Geogr Inf Sci 22:601–622

Mainz M (2005) Ökonomische Bewertung der Siedlungsentwicklung, Beiträge zum Siedlungs- und Wohnungswesen. V&R Unipress, Göttingen
Miller EJ, Douglas Hunt J, Abraham JE, Salvini PA (2004) Microsimulating urban systems. Comput Environ Urban Syst 28:9–44
Pijanowski B, Brown DG, Shellito BA, Manik GA (2002) Using neural networks and GIS to forecast land use changes: a land transformation model. Comput Environ Urban Syst 26:553–575
Poelmans L, Van Rompaey A (2009) Detecting and modelling spatial patterns of urban sprawl in highly fragmented areas: a case study in the Flanders–Brussels region. Landsc Urban Plan 93:10–19
Pontius RG Jr, Cornell JD, Hall CAS (2001) Modeling the spatial pattern of land-use change with GEOMOD2: application and validation for Costa Rica. Agric Ecosyst Environ 85:191–203
Pontius RG Jr, Huffaker D, Denman K (2004) Useful techniques of validation for spatially explicit land-change models. Ecol Model 179:445–461
Rounsevell MDA, Pedroli B, Erb K-H, Gramberger M, Busck AG, Haberl H, Kristensen S, Kuemmerle T, Lavorel S, Lindner M, Lotze-Campen H, Metzger MJ, Murray-Rust D, Popp A, Pérez-Soba M, Reenberg A, Vadineanu A, Verburg PH, Wolfslehner B (2012) Challenges for land system science. Land Use Policy 29:899–910
Schmitz M, Bode T, Thamm HP, Cremers AB (2007) XULU – a generic JAVA-based platform to simulate land use and land cover change (LUCC). In: Oxley L, Kulasiri D (eds) Proceedings of MODSIM 2007 international congress on modelling and simulation. Modelling and Simulation Society of Australia and New Zealand, pp 2645–2649
Schwarz N (2010) Urban form revisited-selecting indicators for characterising European cities RID A-5409-2011. Landsc Urban Plan 96:29–47
Siedentop S, Junesch R, Straßer M, Zakrzewski P, Samaniego L, Weinert J (2009) Einflussfaktoren der Neuinanspruchnahme von Flächen, vol 139, Forschungen. BMVBS, BBSR, Bonn
Sieverts T (2005) Zwischenstadt. Zwischen Ort und Welt, Raum und Zeit, Stadt und Land, 3rd edn. Birkhäuser, Basel
Silva EA, Clarke KC (2005) Complexity, emergence and cellular urban models: lessons learned from applying Sleuth to two Portuguese metropolitan areas. Eur Plan Stud 13:93–115
Tobler WR (1979) Cellular geography. In: Gale S, Olson G (eds) Philosophy in geography. Reidel, Dordrecht, pp 379–386
Torrens PM (2001) Can geocomputation save urban simulation? Throw some agents into the mixture, simmer and wait . . . (Working Paper No. 32), UCL working paper series. UCL Centre for Advanced Spatial Analysis, London, UK
Torrens PM, O'Sullivan D (2001) Cellular automata and urban simulation: where do we go from here? Environ Plan B Plan Des 28:163–168
United Nations Population Division (2010) World urbanization prospects: the 2010 revision. United Nations, New York
US EPA (2000) Projecting land-use change. A summary of models for assessing the effects of community growth and change on land-use patterns. United States Environmental Protection Agency, Washington, DC
Verburg PH, Soepboer W, Veldkamp AT, Limpiada R, Espaldon V, Mastura SSA (2002) Modelling the spatial dynamics of regional land use: the CLUE-s model. Environ Manage 30:391–405
Verburg PH, Ritsema van Eck JR, de Nijs TCM, Dijst MJ, Schot PP (2004a) Determinants of land-use change patterns in the Netherlands. Environ Plan B Plan Des 31:125–150
Verburg PH, Schot PP, Dijst MJ, Veldkamp AT (2004b) Land use change modelling: current practice and research priorities. GeoJournal 61:309–324
Verburg P, Koomen E, Hilferink M, Pérez-Soba M, Lesschen J (2012) An assessment of the impact of climate adaptation measures to reduce flood risk on ecosystem services. Landsc Ecol 27:473–486

Waddell P (2002) UrbanSim: modeling urban development for land use, transportation, and environmental planning. J Am Plan Assoc 68:297–314

Warner SB (1972) The urban wilderness: a history of the American city. Harper & Row, New York

Weng Q (2012) Remote sensing of impervious surfaces in the urban areas: requirements, methods, and trends. Remote Sens Environ 117:34–49

White R, Engelen G (1997) Cellular automata as the basis of integrated dynamic regional modelling. Environ Plan B Plan Des 24:235–246

Yang Q, Li X, Shi X (2008) Cellular automata for simulating land use changes based on support vector machines. Comput Geosci 34:592–602

Part IV
National Practice Examples

Chapter 15
Land Information System Austria (LISA)

Gebhard Banko, Reinfried Mansberger, Heinz Gallaun, Roland Grillmayer, Rainer Prüller, Manfred Riedl, Wolfgang Stemberger, Klaus Steinnocher, and Andreas Walli

15.1 Current Status of Land Monitoring in Austria

These days different federal public organizations generate, maintain and provide land cover (LC) and land use (LU) information in Austria. In addition, all nine federal state authorities of Austria, several cities and municipalities, and providers of infrastructure (e.g. road and rail-network, telecommunications, energy, etc.) acquire regional or local land cover and/or land use datasets. However, most of these products have different properties concerning spatial and temporal resolution, thematic classes and indefinite LC or LU classifications, which lead to limited data interoperability and to high economic costs.

A pioneer work has been accomplished in the Austrian-Hungarian monarchy throughout the eighteenth and nineteenth century by revised mapping approaches under the governance of the Austrian Military Survey (Lego 1968). These first land monitoring approaches at scale 1:28,880 were carried out in parallel with the attempt

G. Banko (✉)
Umweltbundesamt GmbH (Environment Agency Austria), Spittelauer Lände 5, 1090 Vienna, Austria
e-mail: gebhard.banko@umweltbundesamt.at

R. Mansberger (✉)
University of Natural Resources and Life Sciences, Peter Jordan-Straße 82, 1190 Vienna, Austria
e-mail: reinfried.mansberger@boku.ac.at

H. Gallaun
Joanneum Research, Institute for digital remote sensing and geoinformation, Steyrergasse, 17, 8010 Graz
e-mail: heinz.gallaun@joanneum.at

R. Grillmayer
University of Applied Sciences Wiener Neustadt for Business and Engineering Ltd., Departement for Geoinformation, Johannes Gutenberg-Straße, 3, 2700 Wiener Neustadt
e-mail: roland.grillmayer@fhwn.ac.at

I. Manakos and M. Braun (eds.), *Land Use and Land Cover Mapping in Europe: Practices & Trends*, Remote Sensing and Digital Image Processing 18,
DOI 10.1007/978-94-007-7969-3_15,

to establish detailed landscape and property information for mainly taxation purposes through large-scale maps (1:2,880) in form of the stable cadaster[1] (1817–1861). Therefore information on basic land cover characteristics like arable land, permanent grassland, orchards, vineyards, forest, water area and built-up area is available back to the 2nd Austrian Military Survey (1807–1848, Abart et al. 2011).

The Federal office of Metrology and Surveying (BEV) has kept up this parallel production processes for the cadastral information and topographic map production. In addition to the geometric information on parcel boundaries the (digital) cadastral maps contain assorted information on land cover and land use classes. This information is updated at irregular intervals, object specific (e.g. building plan from building permits) or project specific (large area updates based on aerial photographs). Simultaneously, digital topographic maps (1:50,000) are updated in intervals of 5–10 years based on aerial photographs and field verification. Due to a recent regulation (BANU-V 2010) and the implementation of a digital production these two processes will be merged within the next years. They will be based on one common data model and make use of novel approaches for change detection from orthophotos.

For control and application of agricultural subsidies (IACS) a central registry has been built up in the last years (INVEKOS-GIS). This registry contains, beside others, the information on the geometrical boundary of different agricultural utilization that is digitized by farmers based on orthophotos. The digital cadastral map has played an important role as basic background mapping reference. Due to the "greening" of common agricultural policy (CAP), this system will be enhanced until 2014 by landscape elements digitized from orthophotos. Landscape elements consist predominately trees and bushes that are digitized either as point information (MMU[2] > 4 m^2) or as polygon information (MMU > 100 m^2).

Since the introduction of geographic information systems in the regional authorities most administrative information is created in digital format. This lead

[1] "Stable Cadaster" is a synonym for the "Franziszeischen Cadaster" (1768–1835). The term stable describes the invariability of the taxation over time due to improvements of soil (Abart et al. 2011).

[2] Minimum Mapping Unit.

R. Prüller
Technical University Graz, Institute for remote sensing and photogrammetry,
Steyrergasse 30/I, 8010 Graz
e-mail: rainer.prueller@tugraz.at

M. Riedl
Amt der Tiroler Landesregierung, Landesstatistik und tiris,
Heiliggeiststraße 7-9, 6020 Innsbruck
e-mail: manfred.riedl@tirol.gv.at

W. Stemberger • A. Walli
GeoVille Information Systems GmbH, Sparkassenplatz, 2, 6020 Innsbruck
e-mail: stemberger@geoville.com; walli@geoville.com

K. Steinnocher
Austrian Institute of Technology, Donau-City-Straße, 1, 1220 Vienna
e-mail: klaus.steinnocher@AIT.ac.at

in the nine federal states to the creation and usage of nine different land use maps for the main land use classes like settlements, traffic and forest areas.

The development of a common, intermodal traffic graph across Austria as a common transport reference system (Graphen Integrations Plattform – GIP) is currently under construction.

Besides these state-run mapping products more and more commercial and public domain products find their way into administrative applications. Administrative bodies have the growing need to display also the surrounding of their area of interest and not to display their area as isolated map. The commercial products that are widely available are "Google-Maps" and "Bing-Maps", whereas the public domain dataset "Open Street Map" shows a fast increase of both, i.e. completeness and level of detail – depending on the number of active mapping communities.

15.2 LISA – Development of a Countrywide Approach

15.2.1 Why LISA?

The availability of homogeneous LC and LU datasets for Austria is an indispensable public necessity, needed for political decisions, effective administration, successful corporate governance and personal usage of the citizens. In particular such geoinformation is required by departments of public administration at federal and regional level for the interests of regional planning, forestry and agriculture, water management, natural hazard management, environmental and nature conservation for the periodic monitoring of changes. Detailed data on land cover and land use are also required in the private sector in fields such as site planning and geo-marketing to name but a few. The land cover data and land use data currently available in Austria do no longer meet these requirements. Available datasets are either not homogeneous, outdated, not compatible with other European LC or LU datasets, simply not available for use or do not provide adequate coverage and/or resolution (Grillmayer and Schneider 2004). The users require not that much a one-off snapshot in time, but they prefer regularly updated information on land cover and land use changes over time.

The Global Monitoring for Environment and Security (GMES/Copernicus) program enabled at national level to carry out the project "Land Information System Austria" (LISA). The objective of LISA is to achieve a consensus on a new Austrian land cover data base and to apply cutting edge science, innovative technology and to provide cost efficiency by combining satellite with high resolution in-situ data, to achieve economy of scale and sustainability of funding through a shared effort across different administration units. Instead of producing specific land information at specific levels, a land information system has to fulfill a multi-user approach. It is compiled out of the "core" requirements that serve not only one but also many user levels (from local to regional to national to international) (Weichselbaum et al. 2010).

In LISA the major challenge was the integration of the different basic data sources for homogeneous area-wide land cover/use data production, also permitting for continuous monitoring and periodic updates. The drawback of established nomenclatures, like CORINE Land Cover, is their inherent mixture of land cover and land use classes. Although this mixture is quite useful for some specific practical applications, it is not useful for operational semi-automated production chains. Therefore the European Environment Agency (EEA) has established a European Action Group for Land Monitoring in Europe (EAGLE) that is currently developing a new data model for land monitoring recognizing the need to explicitly differentiate a standardized set of land cover and land use classes as well as a their characteristics in an object oriented approach (Arnold et al. 2013).

An important enabler for LISA was the paradigm change in photogrammetry within the last decade: digital cameras provided images with better radiometric and spectral resolution. Additionally, they allowed the assessment of at least four spectral bands. Due to improved sensors the geometric resolution is comparable with analogue photogrammetric cameras. GNSS (Global Navigation Satellite System) combined with IMU (Inertial Measuring Unit) accelerated the process of orientation by measuring directly the positions of projection centers and sensor planes (exterior orientation). The availability of digital images promoted the development of (semi-)automatic methods for the assessment of geometric and thematic information (e.g. generation of digital surface models, classifications). As the height of object is a valuable parameter for a classification processes, new methods for the derivation of height information from stereoscopic photographs are useful. Especially the development of innovative algorithms (e.g. semi-global matching, Hirschmüller 2008) enable the derivation of Digital Surface Models (DSM) and in combination with Airborne Laser Scanner Data (ALS) the assessment of heights of individual objects. Finally, the continuous improvement of hardware (e.g. faster processors, cloud computing, extended disk storage) was a significant ingredient for an automatized, large area segmentation and classification process, as implemented in the LISA-project.

15.2.2 National Spatial Data Infrastructure

Without an adequate National Spatial Data Infrastructure (NSDI) and agreed licensing and usage models, the development of an effective land monitoring system would not have been possible. The core remote sensing backbone for LISA are orthophotos. For the last few years, three different administrative bodies (BEV, federal provinces and ministry for agriculture) jointly finance and acquire aerial images in a 3 years interval and produce orthophotos with a ground resolution of $20*20\ cm^2$. Most of the aerial photographs are acquired with Vexcel UltraCam-X and the recent overlapping standards are 80 % within flight direction and 40 % across flight direction. These overlapping standards will enable the optimal computation of Digital Surface Models (DSM) from aerial images in the near future.

Airborne laser scanning (ALS) data has been acquired for almost the whole territory of Austria. However, it is expected that updates will cover rather small and dynamic areas, but not the entire country within the next years to come. Nevertheless ALS data provide information on the height of the terrain (DTM) and they are essential to derive a normalized difference surface model (nDSM). Current cost estimates underline the importance of the DSM generation from aerial images. The production of orthophotos is currently in the range of 25–35 €/km^2 and ALS data acquisition are mostly beyond 100 €/km^2. First draft cost estimates for DSM generation from aerial images are around 5–10 €/km^2. However, it can be expected that the production of a DSM will soon be integral part of orthophoto production even with the capability to produce "true orthophotos".

Whereas land cover can be directly derived from remote sensing data, land use aspects rely largely on thematic spatial data infrastructure. The building census as one example for in-situ data in Austria is generated as point information expressing the main usage of buildings. Due to data privacy protection of statistical data this information is only available in form of aggregated 250*250 m^2 grid cells that are due to the coarse resolution only of little help for the level of detail in LISA. In Austria land use zoning plans from the spatial planning authorities are with exceptions available in digital format. They are valuable data sources to improve the quality of land use mapping. Last, but not least environmental inventories on bogs and mires, wetlands, dry grasslands and riparian zones are available to be integrated in the land use data. Many environmental inventories can be quite outdated (e.g. the last complete dry grassland inventory of Austria dates back to 1986), but they are nevertheless valuable indicators for specific land cover and/or land use.

15.2.3 European Spatial Data Infrastructure

European Services in the frame of GMES/Copernicus land monitoring are integrated into the LISA concept. On the one side, they provide multi-temporal coverage for areas that can hardly be classified with a single image per year. On the other side, they provide information on potential hot-spot areas of LC/LU changes on a coarse scale. Namely, the change in levels of soil sealing and forest coverage can be expressed using the high resolution layers from the GMES services (EC 2011, 2010). Especially the upcoming next generation of operational European satellites under the SENTINEL program is of major interest being integrated into national land monitoring programs. They will enhance the 3-D information from national spatial data infrastructure (exact geometry of Orthophotos combined with object height information from digital surface models) with a fourth dimension: time. The SENTINEL-2 data with a geometric resolution of 10*10 m^2 and a revisiting capability of 2–3 days for Austria will be used to further describe and populate land monitoring objects according to their variability in time and phonological development, thus retrieving important information on ecosystem status.

15.2.4 Future Potential of Very High Resolution Satellite Imagery

A cost and acquisition efficient additional alternative is the vastly increasing number of very high resolution (VHR), multispectral satellites. With 10 satellites offering ground resolutions of 50 cm currently waiting for launch and 15–20 systems planned in the future, the investments of the industry and public in this technology is significant. These days very high resolution satellites offer approximate equally expensive and fast as well as agile acquisition capacities (no overflight permissions are needed, a much greater ground coverage capacity, and image acquisitions are automatically repeated until cloud free coverages of an area of interest are given), thereby providing a complementary image and DSM source to existing orthophoto coverages. Moreover, the satellite missions are often supported by national initiatives or funding, providing preferred data provision for applications of national and civil importance. In Europe, the Plèiades system, as the successor to the SPOT satellite series has taken advantage of existing multi-national cooperation on the Spot and Helios programmes, where Austria, Belgium, Spain and Sweden have agreed to share the costs and possible benefits from the programme. In return for its investment in the forthcoming Plèiades system, Austria has been assured an access to the resources in proportion to its investment. Thereby public administrations may complement their image data acquisition campaigns with VHR images and optimize investments into aerial imagery, airborne laser scanning and VHR images to profit from the individual advantages of the systems in order to improve environmental monitoring and maintain higher monitoring cycles.

15.2.5 Object-Orientated Data Model

The user demands were translated into data model specifications in an iterative and cooperative procedure between service providers, users and scientific advisory board. Most land cover nomenclatures rely on hierarchical nomenclatures. It has to be noted that modern land information systems have to provide flexible solutions for further improvements and amendments. Therefore, object-orientated data models (Egenhofer et al. 1992) – originally developed within computer sciences – are applied in national applications like in Spain (SIOSE) or Austria (LISA). In an object-orientated concept of data modeling, real world entities are represented by objects, namely called "geographic features". A geographic feature is characterized by its feature properties as well as methods for their changing (Mitášová et.al. 1996). Instead of using the hierarchically Data Model, where only the type of land cover class/land use class is stored in a feature property, it is advisable to use the object-orientated Data Model (OODM). In OODM, even more characteristics of the geographic feature (e.g. biomass density, height, wetness index, sealing degree) can be stored as feature property (so called "object attributes").

The Environment Agency Austria has developed two distinct data models on distinct levels of scale in LISA:

- Land Cover data model and a separated
- Land Use data model.

Table 15.1 List of land cover classes

class			MMU	MVU	accuracy
abiotic	built-up	1 building	25	50	98% object accuracy 95% n.G.
		2 other constructed area	25	100	95%
	non-built up	3 bare soil	50	500	90%
		4 screes	50	500	90%
		5 bare rock	50	500	85%
	water	6 surface water	50	200	95%
		7 snow	50	500	85%
		8 ice	50	500	85%
biotic	woody	9 trees	25	25 50 1.000	95% strata 1: 25-50m2, object accuracy Strata 2: 50-1.000 m2, Strata 3: >1.000 m2
		10 bushes	50	500	90%
		11 dwarf shrubs	50	500	85%
	herbaceous	12 herbaceous vegetation	50	500	95%
		13 reeds	50	500	95%
unclassified		14 shadow	500	500	

The Land Cover data model contains 13 pure land cover classes that are purely derived from remote sensing data (orthophotos, height information and satellite imagery) using automated segmentation and classification techniques in an approximate scale of 1:10,000. The minimum mapping units (MMU) varies between the different classes (25–50 m^2) (Table 15.1).

The Land Cover classes can be described using a subset of classification parameters from the FAO Land Cover Meta Language (LCML, DI Gregorio and Jansen 2000) that establishes the ISO 19144 standards. In addition a semantic transformation is made to the INSPIRE Land Cover Classes (ILCC) that were developed as part of the INSPIRE draft technical guidelines on land cover (INSPIRE 2013). In the ILCC the class "arable land" exists, which constitutes a problematic class, as it is rather a land use class than a land cover class, if classified from a mono-temporal image only. Therefore the LISA logic avoided this mixture between land cover and land use and defined the land cover on purely biophysical properties and context sensitive parameters enabling nevertheless an additional attribution of objects (e.g. according to temporal characteristics). The development of LISA is therefore in-line with the new approaches of the EAGLE data model that is currently developed for European applications (Arnold et al. 2013). As objects are derived using image segmentation techniques, the border between objects is kept even between objects assigned to the same class, if they can be further differentiated according to defined attributes (e.g. field structure in agriculture).

The Land Use classes are grouped according to the six main land use classes: settlements, traffic, agriculture, forestry, natural areas (vegetated and non vegetated), and water systems. Although from land cover information a range of land use classes can be directly derived, the complete data model relies on the integration of sectorial geospatial in-situ data like cadastral information, land use zoning plans, traffic infrastructure, etc. The challenge for this production process is to make all kind of sectorial data available and to agree on licensing permissions. The comparable scale of the land use maps is about 1:25,000 with MMUs starting from 1,000 m^2 (e.g. settlements, technical infrastructure, forest) up to 5,000 m^2 (e.g. alpine grassland, glaciers). The land use data model contains 25 classes and 72 attributes. Attributes are for example used to further differentiate the settlement area according to the dominant use (residential, industrial, commercial or mixed use) and the dominant type of buildings (single houses structure, closed building structure, semi-detached, large blocks, etc.).

15.2.6 Land Cover Mapping

The feasibility of land cover mapping was tested for 49 pilot areas, each covering 30 km^2. Criteria for the selection of the pilot areas were the representativeness of regions (federal provinces) and landscape diversity. The aim of this study was to optimize methods and processing chains for mapping land cover for meeting the needs and requirements of the users and to enable the cost estimation for a country wide rollout. The study was done in a five stage approach:

- Production of land cover maps for all pilot areas (iteration 1)
- Verification and validation (qualitative by users, quantitative by scientific staff)
- Improvement of algorithms and workflow based on results and feedback of validation team
- Production of final land cover maps for 17 pilot areas (iteration 2)
- Validation (quantitative).

Two different service providers (GeoVille Information Systems GmbH / Innsbruck and Joanneum Research Forschungsgesellschaft mbH/Graz) produced all land cover maps. Orthophotos with a resampled geometric ground resolution of 50 cm, normalized difference surface models (nDSM) and GMES satellite images (IMAGE 2006, IMAGE 2009) were used as input data for the classification. The minimal mapping unit – defined by the scientific staff in accordance with the users – was dependent on the specific object class (e.g. 25 m^2 for buildings and trees, 50 m^2 for herbaceous vegetation).

The mapping of land cover objects could be achieved with a high degree of automation, as the features for the differentiation of classes directly could be derived from the available remote sensing data. The mapping of the land cover object was carried out in segment-based method: Segments were identified by spectral, radiometric and texture parameters. Using a comprehensive set of rules the classification was done for the individual land cover classes in a multi-level approach. Subsequently, corrections and improvement of the automatically produced land cover were outlined manually.

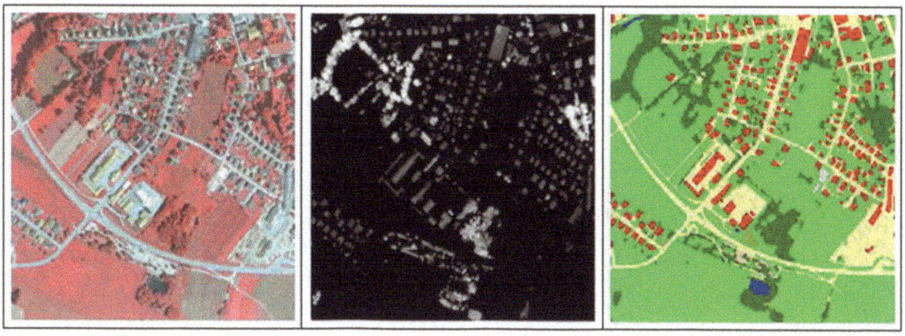

Fig. 15.1 Land cover map (Schrems, Austria) (*left*: orthophoto 2009, *middle*: nDSM 2006, *right*: land cover map)

The validation of the land cover maps produced in this first stage showed weaknesses in the differentiation between "buildings" and "other sealed areas" as well as between "stocked areas" (areas covered with trees) and "bushes". That was the reason to integrate Airborne Laser Scanning (ALS) data for the production of the final land cover maps (Fig. 15.1). The use of nDSM (normalized digital surface models) derived from ALS data led to improved results and reduced the time for manual post-processing. For future roll-out, digital surface models derived from the operationally acquired aerial images can replace the ALS derived digital surface models. Appropriate methods were tested successfully for an image block with 104 aerial images.

The accuracy of the prototypes of land cover maps was assessed in a statistical manner using a stratified random sampling method (Congalton and Green 2009). The number of samples was optimized for getting significant information (confidence interval 95 %, probability level 4 %) about the results for individual regions, landscape diversities, and for the specific producers. The achieved accuracy of the final map (95.5 % overall accuracy, based on 5,000 visually interpreted reference samples) was meeting the requirements of the users (both for commission and omission errors).

15.2.7 Land Use Mapping

The process chain for producing land use maps postulates the availability of land cover maps as the derivation of specific land use classes could be achieved – in a first approach – by aggregating the land cover objects (e.g. settlements are derived by aggregation of specific land cover types according to their spatial distribution). In addition to this aggregated information, geo-base data (Digital Cadastral Map) and other sectorial geodata (e.g. zoning plans, Integrated Agricultural Control System – IACS, building census data) were used as input to identify land use objects in the six main categories.

In a second stage the land use classes were refined using the available geoinformation with the result of in total 25 sub-land use classes and 72 attributes. For each of the sub-classes specific rules were implemented in a GIS (Geographic Information System). The complexity of the system can be seen in the high number

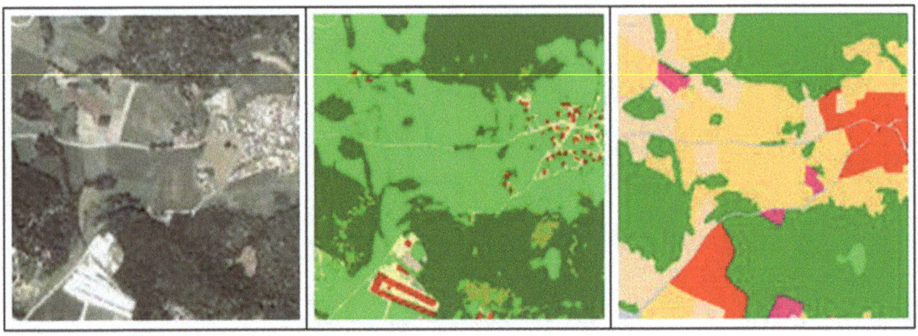

Fig. 15.2 Land use map (surrounding of City of Klagenfurt, Austria) (*left*: orthophoto 2009, *middle*: land cover map, *right*: land use map)

of single steps in the process chain (up to 80 sub-processes, Stemberger et al. 2012). The prototypes for land use maps were produced for 17 pilot areas in iteration 2. The challenge was the heterogeneity of the provided geodata, caused by the fact that according to the Austrian constitution the assessment of geodata lies in the responsibility of different levels of authorities. As an example, zoning plans are produced individually by the municipalities, environmental data are provided by the regional (provincial) governments, and cadastral data are under the responsibility of a federal institution (Federal Office of Surveying and Metrology) with the result of data sets with different geometric accuracies, different scales, different dates of assessment, and different thematic levels of detail (Fig. 15.2).

The validation of the prototypes of land use maps was performed for all 17 test sites. In contrast to the validation of land cover, the mapped polygons were used as sample units to enable an assessment of the thematic and geometric quality of the outlined objects. In total, 4,500 samples were selected (10 % of all polygons) in a stratified randomly approach to ensure that at least five samples per test area were evaluated for each class. Each sample was evaluated according to the geometric quality of the delineation (four geometric quality levels based on shape and size criteria: fit, largely fit, partly fitting, wrong), as well as according to the quality of the assigned classes (four thematic quality levels: correct, plausible, problematic, wrong).

Across all areas 90 % of all samples were rated within the two geometric criteria "fit" and "largely fit". The total thematic accuracy is 88 % (samples rated with the thematic criteria "correct" or "plausible"). The analysis of the results showed that some improvement in the quality could be achieved by sharpening the definition of land use blocks.

15.2.8 Change Mapping

LISA was designed to serve and fulfill common land monitoring needs, providing information on the status quo, and even more important on the changes occurring in

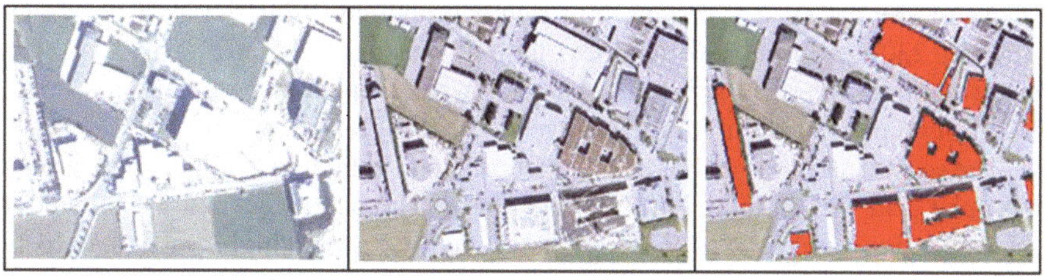

Fig. 15.3 Change detection (City of Innsbruck, Austria) (*left*: orthophoto 1999, *middle*: orthophoto 2008, *right*: recognized changes)

Austria's landscape. So the core functionality for detecting and implementing changes in the land cover database of LISA also was developed within the project. This covers the challenging synchronization of two technical processes, the development of the Earth Observation based method to automatically detect and map the changes and the implementation of the mapped geometric and thematic changes into the LISA database.

The concept developed for change detection strongly relies on object orientation, which enables the users to track back the changes for every single object during its lifetime. Thus, a new house is determined and the alterations done – e.g. in size – can be monitored using the spatial-temporal database. The challenge is to avoid technical changes (e.g. due to changes in improved methods or improved geometrical accuracy of the base data). Effects caused by different acquisition parameters of aerial photographs (e.g. tilts, shadows) had to be considered in the algorithm to avoid the detection of "false" changes. This could be achieved by using the land cover polygons of the initial state, freezing their geometry and by modifying this geometry (and topology) based significant changes in the image data of the later date. Therefore very small sliver polygons due to technical changes could be avoided. The definition of shape parameters, of neighborhood characteristics, potential changing paths, and other plausibility checks are features within the program to improve the results. The test for the detection of land cover changes was done for nine pilot areas each 30 km^2 sized. The results of the land cover interpretation as described in Sect. 15.2.6 were used as initial state. In consideration of the available recording times for three test areas the change detection was outlined for an up-dating (newer images available), for the remaining six test areas the interpreted changes are based on down-dating (images of a previous photo flight available). The selection of the nine test areas was representative for different landscape types (e.g. settlements or alpine forests). For all test sites the nDSM (representing object heights) were only available for the classification of the initial state (Fig. 15.3).

The validation of change detection of land cover requires a more complex analysis than validating single-date land cover classifications. The number of potential correct results extends the number of classes by the number of changes between classes. The maximum number of potential results is therefore n*n, leading to a confusion matrix of (n*n)*(n*n), which is difficult to interpret. In addition the sampling design requires a large number of sample points.

As an alternative, a two steps approach was applied (Prüller et al 2011). In the first step 2300 representative samples out of all "change polygons" (polygons of all test sites that were classified as changes) were compared with reference data. Ninety-two percentage of all the changes were recognized correctly.

The second step was a simple change/no change validation for the whole area in a regular grid with in total 2,400 sample points. The overall accuracy estimated in the validation process was 97 %. This seems to be very high, but it has to be considered that this value is influenced to a great extent by the high number of sample points (2,100) classified with "no changes[3]". In this case the figures on values of producer (84 %) and user accuracy (83 %) are more representative. Maucha and Büttner (2008) have pointed out that for rare observations (<5 % of area percentage) the differences between commission errors and omission errors are substantial and have implications on the statistical design. Potential for improvements are given by the availability of nDSMs or at least DSM (Digital Surface Models) as input for all land cover interpretations of the time series.

15.3 LISA – Selected Applications

The results of LISA – land cover and land use maps – opens new fields of land monitoring applications. Some examples, which had been tested in the LISA project, are characterized in the following chapters.

15.3.1 Upscaling of LISA-Data Sets to CORINE Land Cover

The development of transformation methods for the LISA database is proposed to ensure future flexibility and compatibility with different national LC/LU monitoring initiatives within Europe (e.g. DML-DE, SIOSE) and to ensure that LISA can serve as the Austrian input layer for European level LC/LU datasets (e.g. CORINE). The functionality of the method for the spatial upscaling and aggregation of classes to the specifications of a target model is demonstrated in Figs. 15.4 and 15.5.

The transformation consists of a semantic transformation (translation of classes) and a geometric transformation. The geometric transformation (upscaling geometrical properties of polygons) is performed as iterative process in four steps of MMU upscaling: from 1,000 to 3,000 m^2 and 1, 5 and 25 ha. The typical CORINE Land Cover priority rules (EEA 2000) for aggregation are not strictly adapted, but modified according to the dominant landscape type (e.g. within a forest dominated landscape agricultural parcels are aggregated on cost of forest polygons, whereas in agricultural landscapes the order is reversed and forest polygons are aggregated on cost of agricultural polygons).

[3] The interpretation of "no changes" was necessary, as the samples were randomly selected over the whole test area to avoid omission errors.

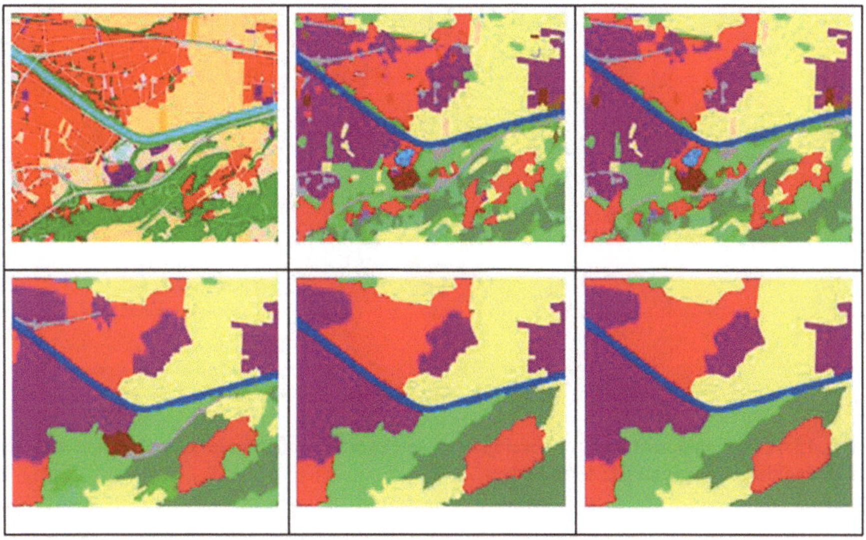

Fig. 15.4 Semantic and geometric transformation of LISA land use classes into CORINE Land Cover Classes *Upper row*: *left*: original LISA land use classes with 1,000 m^2 MMU, *middle*: LISA LU with 3,000 m^2 MMU, *right*: LISA LU with 1 ha MMU; *Lower row*: *left*: LISA LU with 5 ha, *middle*: LISA LU with 25 ha, *right*: LISA LU with 25 ha and cartographic processing

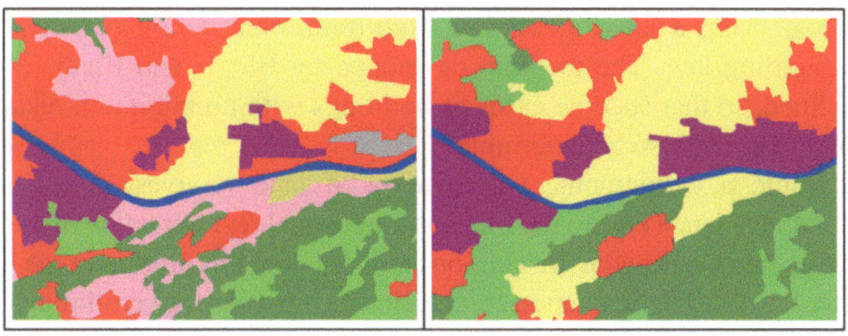

Fig. 15.5 Comparison between transformed LISA product and CLC 2006 (*left*: upscaled LISA data set with 25 ha MMU; *right*: CORINE Land Cover 2006)

15.3.2 Accounting of Land Reserves Available for Construction

On the basis of the Regional Planning Acts, the regional spatial planning departments are required to report on the development of built-up areas and land re-serves available for building at community level. The results of these evaluations provide information on the location and extent of existing land reserves, densification and reserves of arable land at intervals of at least 5 years.

Compared to the land cover employed in previous assessments, LISA can supply substantially higher number of classes, more appropriate geometric properties and

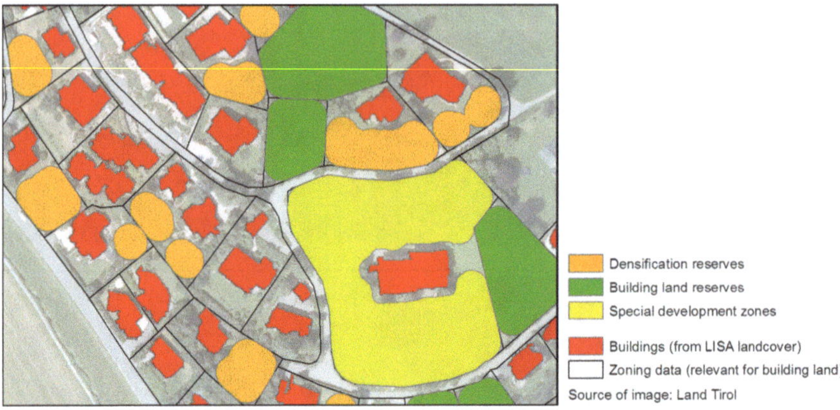

Fig. 15.6 Designation of land for development, densification and reserves of arable land together with buildings of special LISA land cover

the potential for national roll-out. As part of the demonstration phase, the entire workflow was adapted, highly automated and applied to the province of Tyrol.

The analysis with single-owned land parcels, which can be distinguished as "developed" or "undeveloped". Undeveloped areas represent potential land reserves, while built-up areas have basic consolidated reserves. In a second step, all potential areas within a parcel which are potentially available for construction are analysed using a moving-window algorithm, considering parameters of the regional building law. The method, developed by Blome and Riedl (2007), was further refined, so that specific cases of buildings with no minimum distance from the arable areas can also be determined.

As a result, a cartographic visualization of land reserves, densification and reserves of arable land (Fig. 15.6), which is a statistical representation of the available land reserves for construction. In addition to automating this process in the form of an ArcGIS tool, a flexible adaptation to changes in input data given through different data situations is also possible.

15.3.3 Classification of Coverage Types in Settlements

Monitoring urban settlements represents an essential input to spatial planning tasks. The location of various building types, reserved areas, and their future demand in urban settlements are integral information for regional policy. The objective of this application is the identification of different coverage types within settlement areas based on LISA land cover data. Seven categories are selected for the classification: detached houses, semi-detached houses, terraced houses, apartment buildings, perimeter block development, large storage buildings (typical for industry/trade), and high-rise buildings. The methodology relies on integrating location, footprint,

Fig. 15.7 Coverage type map (City of Salzburg) (*yellow*: single houses; *red*: serial house, *brown*: block of buildings; *violet*: industrial building; *blue*: high-rise building; *grey*: not classified)

and height parameters from the land cover data with plot boundaries from the cadaster. Combination of these data sets enables identification of building structures, from open to closed coverage types. The differentiation of detached, semi-detached, and terraced houses can only be achieved by the integration of plot boundaries. Additional parameters are the number of buildings per plot, as well as the green-area ratio. Small building objects of less than 35 m^2 are eliminated from the data set (Steinnocher et al. 2011), as they in general do not representative buildings that are suitable for permanent housing in Austria.

The methodology has been implemented in ArcGIS 9.2 and tested on two pilot project sites, the City of Salzburg and the City of Innsbruck. Realization in GIS starts with an intersection of land cover data with the plot boundaries and proceeds with a spatial join in order to get information from both data sets. For each building parameters such as height and footprint, number of buildings per plot, green-area ratio, and plot boundaries are calculated. The final classification is based on predefined rules referring to a set of selected parameters for each building type.

The result of this investigation for the center of Salzburg is presented in Fig. 15.7. Approximately 90 % of all buildings could be classified automatically. A first qualitative analysis shows some problems in the interpretation due to wrong building heights (partly caused by different acquisition dates of aerial photographs and ALS-data) and due to complex building shapes.

15.4 From Project Status to Operational Roll-Out

The initial development of LISA was funded by resources of the Federal Ministry of Transport, Innovation and Technology within the Austrian Space Applications Program (ASAP) of the Austrian Research Promotion Agency (FFG). The development was implemented as a three-stage process: Concept Phase (2009–2010) – Completion Phase (2010–2012) – Roll-out Phase (from 2012).

The Concept phase is presented in the chapters “Land Cover Mapping” and “Land Use Mapping”. In this project stage stakeholders, users and the scientific/ technical community achieved a consensus about the specifications for the new Austrian land cover/use data model and land monitoring system. During an iterative process, a technically feasible requirement catalogue and data model have been established, processing chains developed, prototypes have been produced and verified, and the costs been determined. Beside the demonstration prototypes covering in total 49 test diverse test sites (>3,000 km^2), the main results were a user community defined, technical feasible and scientifically verified requirement catalogue, the architecture and data model to be used.

During the Completion Phase – described in the chapters “Change Mapping” and “LISA – Selected Applications” critical components enabling the operational roll out of LISA as a national and operational monitoring solution were developed. These included establishment of the data model in a physical database, the implementation of fully automated change detection and upscaling method (spatial aggregation to European level LC/LU datasets, such as CORINE) and the demonstration of LISA for the usage in spatial planning applications.

The Roll-out Phase is already initiated through single initiatives of regional governments which financed land cover/land use data collections according to LISA specifications for part of their territory. This represented a major achievement in making LISA a sustainable user accepted standard. Nevertheless, the ultimate vision is to initiate single funded, country-wide roll-out to fulfill the initial user requirements for a homogenous and operational land monitoring system for Austria.

For a roll-out phase the project has to be transformed into a sustainable administrative cooperation. A unique opportunity in time for a country wide LISA roll-out is currently opened up in the frame of the Austrian Spatial Development Concept 2011 (ÖROK 2011). The spatial development concept is a strategic steering instrument for overall spatial planning and development and is updated every 10 years. For the implementation phase so called “ÖREK partnerships” are created to raise the commitment of the ÖROK members. One of the partnerships deals with “Land monitoring and management” and will focus on administrative, organizational and financial framework for the roll-out of LISA. Up to now it has been agreed that the national mapping agency will take over the data model for land cover and extract this information according to the flight planning for the Orthophoto-production every 3 years starting first time in 2014.

A further roll-out is currently carried out within the frame of the project “Cadaster Environment” financed by the European Space Agency (ESA). It will provide land cover data according to the LISA data model for a total of 10,000 km^2

around the main urban regions in Austria by 2014 and an additional change alert layer for almost the entire country (84,000 km^2). The change alert layer contains all permanent land cover changes (e.g. agriculture to urban) with a MMU of 0.5 ha using multitemporal coarser satellite image observations (in the range of 10–30 m) e.g. LANDSAT data from 2006 to 2009 (as substitution for upcoming SENTINEL data). The change alert regions, identified by the LCC alert mapping, are used as focus area for a finer analysis of the type of change using VHR land cover mapping.

The results of the LISA and "Cadaster Environment" project can be found on and downloaded free from the project homepage: www.landinformationssystem.at.

15.5 Lessons Learnt

From the experiences gathered during the various project phases the following lessons learnt can be concluded that are of relevance for similar efforts in other European countries.

- A new land monitoring system can only be developed, if all major stakeholders (data providers, users, service providers, scientific advisory board) are involved and communication/discussion is (also a financial) key element of the project
- The national data infrastructure is recognized and a key asset together with the provision of satellite image services under the COPERNICUS program
- Results of LISA are developed with a core user group (governmental organizations on various levels) and their requirements have to be carefully balanced to financial constraints for acceptance of results
- The LISA land cover and land use data model represent the minimum standard ("lowest common denominator") of requirements of local, provincial and federal authorities, thus providing consistent and homogenous LC/LU information
- For European Integration and upscaling of data it is absolutely necessary to orient novel approaches in line with the newly developed EAGLE data model

Acknowledgement LISA was developed by various public and private institutions in close cooperation with numerous public agencies at state and at federal province level. LISA was funded and supported by the BMVIT on the FFG in ASAP VI (concept phase LISA I 06.2009-10.2010) and ASAP VII (completion phase LISA II, 11.2010-06.2012). The *Cadaster Environment* project is financed by ESA under contract number 4000106954/12/I-LG-1.

References

Abart G, Ernst J, Twaroch C (2011) Der Grenzkataster: Grundlagen, Verfahren und Anwendungen. Neuer Wissenschaftlicher Verlag, Wien/Graz. ISBN 978-3-7083-0770-1

Arnold S, Kosztra B, Banko G, Smith G, Hazeu G, Bock M, Valcarcel Sanz N (2013) The EAGLE concept – a vision of a future European Land Monitoring Framework. In: Lesaponara (ed) Earsel Symposium, Matera, Italy, June 2013 (in print)

Banu V (2010) Verordnung des Bundesministers für Wirtschaft, Familie und Jugend über die Angabe und Definition der Benützungsarten und Nutzungen im Grenzkataster (Benützungsarten-Nutzungen-Verordnung), BGBl. 116, 15 Apr 2010
Blome P, Riedl M (2007) Erstellung von Baulandbilanzen – Methodik und erste Ergebnisse. In: RO-Info. Tiroler Raumordnung und Regionalentwicklung, Juli 2007, Heft 33, S. 6–9
Congalton RG, Green K (2009) Assessing the accuracy of remotely sensed data: principles and practices, 2nd edn. CRC Press/Taylor & Francis Group, Boca Raton
Di Gregorio A, Jansen LJM (2000) Land-Cover Classification System (LCCS): classification concepts and user manual. FAO/UNEP/CooperazioneItaliana, Rome, 177 pp. (Incl. CD-ROM with application software)
EC-European Commission (2010) Regulation (EU) No 911/2010 of the European Parliament and of the Council of 22 September 2010 on the European Earth monitoring programme (GMES) and its initial operations (2011 to 2013)
EC-European Commission (2011) Communication 831 on the European Earth monitoring programme (GMES) and itsoperations (from 2014 onwards)
EEA (2000) CORINE Land cover technical guide – Addendum 2000, Technical report No. 40, European Environment Agency, Copenhagen
Egenhofer MJ, Frank AU (1992) Object-oriented modeling for GIS. URISA J 4(2):3–19. http://www.spatial.maine.edu/~max/oomodeling.pdf. Accessed 30 Apr 2012
Grillmayer R, Schneider W (Hrsg.) (2004) Tagungsband zum Workshop: Geodaten zur Landbedeckung in Österreich, Shaker Verlag GmbH, Aachen. http://www.rali.boku.ac.at/5763.html. Accessed 30 Apr 2012
Hirschmüller H (2008) Stereo processing by semi-global matching and mutual information. EEE Trans Pattern Anal Mach Intell 30(2):328–341
INSPIRE Thematic Working Group Land Cover (2013) INSPIRE Data Specification for the spatial data theme Land Cover, D2.8.II.2_v3.0rc3, 4.2.2013; Directive 2007/2/EC of the European Parliament and of the Council of 14 March 2007 establishing an Infrastructure for Spatial Information in the European Community
Lego K (1968) Geschichte des Österreichischen Grundkatasters. Bundesamt für Eich- und Vermessungswesen/Eigenverlag, Wien
Maucha G, Büttner G (2008) Recommendations Quantitative assessment high-resolution soil sealing layer, EEA-thematic report, Copenhagen
Mitášová I, Višňovcová J, Hájek M (1996) Approaches to data modeling in GIS. Department of Theoretical Geodesy of Slovak Technical University. http://e-collection.library.ethz.ch/eserv/eth:25176/eth-25176-01.pdf. Accessed 30 Apr 2012
ÖROK (2011) Austrian spatial development concept 2011, Office of the Austrian conference on spatial planning (ÖROK), Vienna. http://www.oerok.gv.at/fileadmin/Bilder/2.Reiter-Raum_u._Region/1.OEREK/OEREK_2011/Dokumente_OEREK_2011/OEREK_2011_EN_Download version.pdf. Accessed 26 Apr 2012
Prüller R, Stemberger W, Gallaun H, Grillmayer R, Krenn P, Mansberger R, Steinnocher K, Walli A (2011) Nutzen von innovativen Technologien für eine flächendeckende, flexible Landbeobachtung Österreichs. In: Strobl J, Blaschke T, Griesebner H (Hrsg.) Angewandte Geoinformatik 2011 – Beiträge zum 23. AGIT-symposium, Salzburg. Wichmann, Berlin/Offenbach, S. 239–244
Steinnocher K, Banko G, Weichselbaum J (2011) Planungsrelevante Datengrundlagen für Österreich: LISA – Land Information System Austria. In: Schrenk M, Popovich V, Zeile P (eds) Real Corp 2011. Proceedings of the 16th international conference on urban planning, Regional Development and Information Society, Essen, Germany, May 2011, pp 707–714
Stemberger W (ed) et al (2012) Land information system Austria –LISA II, Final report submitted to the FFG; Austrian Space Application Program (ASAP VII), Project No. 828302
Weichselbaum J (ed) et al (2010) Land information system Austria – LISA I, Final report submitted to the FFG; Austrian Space Application Program (ASAP VI), Project No. 819754

Chapter 16
Digital Land Cover Model for Germany – DLM-DE

Michael Hovenbitzer, Friederike Emig, Christine Wende, Stephan Arnold, Michael Bock, and Stefan Feigenspan

16.1 Introduction: Background and Motivation of the DLM-DE

16.1.1 Federal Structure and Land Surveying Responsibilities in Germany

The tasks of the German Federal Agency for Cartography and Geodesy (Bundesamt für Kartographie und Geodäsie – BKG) include, besides others, the provision of basic spatial and geodetic reference data mainly for the needs of the Federal Government, but also for the administrative, economic and scientific sectors and the citizens (BKG 2014). The land surveying authorities of the 16 Federal States

M. Hovenbitzer (✉) • F. Emig • C. Wende
Bundesamt für Kartographie und Geodäsie (BKG – Federal Agency for Cartography and Geodesy), Richard-Strauss-Allee 11, 60598 Frankfurt am Main, Germany
e-mail: michael.hovenbitzer@bkg.bund.de; friederike.emig@bkg.bund.de; christine.wende@bkg.bund.de

S. Arnold
Statistisches Bundesamt (Destatis – Federal Statistical Office), Gustav-Stresemann-Ring 11, 65189 Wiesbaden, Germany
e-mail: stephan.arnold@destatis.de

M. Bock
Deutsches Zentrum für Luft- und Raumfahrt (DLR – German Aerospace Center), Königswinterer Straße 522-524, 53227 Bonn, Germany
e-mail: michael.bock@dlr.de

S. Feigenspan
Umweltbundesamt (UBA – Federal Environment Agency), Wörlitzer Platz 1, 06844 Dessau-Roßlau, Germany
e-mail: stefan.feigenspan@uba.de

I. Manakos and M. Braun (eds.), *Land Use and Land Cover Mapping in Europe: Practices & Trends*, Remote Sensing and Digital Image Processing 18,
DOI 10.1007/978-94-007-7969-3_16,

together with BKG form the Working Committee of the Surveying Authorities of the Federal States of Germany (AdV). All topographic data produced by AdV members is called ATKIS (Authoritative Topographic-Cartographic Information System) (AdV 2009) and comprises several datasets with different scales of 1:10,000–1:25,000 (Basis-DLM), 1:50,000–1:100,000 (DLM50), 1:250,000–500,000 (DLM250) and 1:1 Mio (DLM1000). The Federal States produce large scaled data up to 1:100,000, while BKG produces small scaled data from 1:250,000 to 1 Mio. The Basis-DLM is the key dataset, which is used as the source for any other derived topographic dataset at smaller scale.

The continuous updating of the ATKIS ® Basis-DLM lies within the responsibility of the land surveying authorities of the States. It is done mainly based on airborne ortho-imagery. The average update cycle of aerial ortho-imagery in each Federal State is 3 years. Hence, landscape's topography is captured and mapped stepwise over a 3 years cycle, but not for the whole territory at once. Within these cycles, different levels of updating priority exist, depending on the topographic relevance of the captured feature type; infrastructure and built-up areas are mapped with a higher priority than vegetation, water bodies or wetlands. BKG collects the ATKIS ® Basis-DLM data from the States to derive its own federal products for the entire territory of Germany. Also, BKG is a contact point for the provision and reselling of the States' products to other authorities and to the public.

16.1.2 History of CORINE Land Cover in Germany

The first production of CORINE Land Cover (CLC) 1990 in Germany was organized and managed by the Federal Statistical Office (Destatis, formerly StBA), located in Wiesbaden. In accordance with the Council Regulation (EEC) No 1210/90 of 7 May 1990, the European Environment Agency (EEA) together with the European Environment Information and Observation Network (EIONET) were established in 1994. Due to this action, the Federal Environment Agency of Germany (Umweltbundesamt – UBA) was appointed as National Focal Point (NFP) in general, as well as National Reference Center (NRC) for land cover and land use in particular, and took over the responsibility for CLC, which is maintained at UBA through the subsequent CLC phases 2000, 2006, 2012 (Keil et al. 2005, 2009) till the present days. As the main source of information, satellite imagery (a mosaic named IMAGE1990, 2000, 2006) centrally acquired by the European Space Agency (ESA) was used for automated image analysis, visual interpretation and on-screen digitizing. Following the production of the first full coverage of CLC in 1990, a shift of work procedure has been made from "mapping first" to "change first", which means that based on the starting situation of the previous CLC dataset, the changes were captured first, before the full area coverage of CLC 2000 has been created by combining the changes with the previous dataset. After all, CLC production was executed mainly independent from other national basic topographic mapping and surveying activities.

16.1.3 Motivation for the Concept of DLM-DE

In principle, CLC is a well-established land cover dataset. In Germany many existing applications are using CLC data for land monitoring purposes on national and sub-national levels. However, nationwide users meanwhile require higher spatial and temporal resolutions on land cover data and land cover changes than conventional CLC data can offer.

The expectation of cost saving through a more efficient use of existing data and avoidance of redundant data production for both national and European purposes in parallel is another driving factor of DLM-DE's development. The integration of national topographic reference data into pan-European thematic datasets aims at a more consistent and harmonized interoperability between national data sources and geo-information data on the European level.

Aforementioned reasoning has led to the new methodological bottom-up approach: Creating a consistent and harmonized production chain from national to European datasets on the one hand, and on the other hand aiming at cost efficiency by using already existing data as input instead of redundant data production.

16.1.4 Initiation of DLM-DE

Based on a decision of the German Interministerial Committee for Geo-Information (IMAGI), which dates back to May 2007, a trilateral cooperative project between BKG, UBA, and the Federal Ministry for the Environment, Nature Conservation and Nuclear Safety (BMU) was initiated with the long term aim to create a high resolution vector dataset on land cover and land use information by using already existing national data in combination with remote sensing methods. The project was called "Digitales Landschaftsmodell für die Aufgaben und Zwecke des Bundes – DLM-Bund" to meet the requirements and purposes of the Federal Governmental Body, later renamed to "Digitales Landbedeckungsmodell für Deutschland (Digital Land Cover Model for Germany) – DLM-DE".

Since 2006 BKG has the responsibility for the production management and implementation part of DLM-DE (with subcontractors), while the role of UBA under BMU is still the NRC for land cover and responsible for the grant agreement and delivery of CLC to the EEA (see Fig. 16.1). A main target application of DLM-DE, besides other national and regional usages, is the derivation of CORINE Land Cover data from the federal dataset DLM-DE (Arnold 2009).

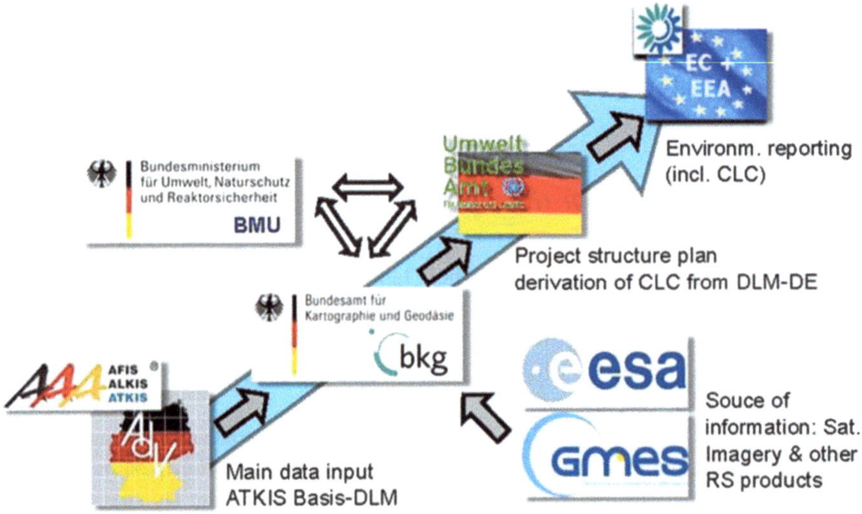

Fig. 16.1 Involved contributors in the German land cover bottom-up approach

16.2 Concept of DLM-DE

In this chapter the concept of DLM-DE shall be explained. Having started with a feasibility study in 2007, the DLM-DE2009 was the first dataset to cover the entire area of Germany. The enhanced concept of the subsequent DLM-DE2012 is characterized by several improvements in processing and production to create a higher flexibility and a broader scope of application.

16.2.1 DLM-DE Feasibility Study 2007

The BKG assigned to the German Remote Sensing Data Center (DFD) at the German Aerospace Center (DLR) a feasibility study according to BKG's and UBA's requirements in order to develop a first plan and concept for the DLM-DE and to estimate the expenses and the technical feasibility of a nation-wide production. The task was to verify and update national available topographic data and bring it with remote sensing methods to the recent state (2007 as reference year). Four tiles in size of a topographic map sheet 1:100,000 were selected as test sites.

16.2.1.1 Input Data and Source of Information

Following the principle of a bottom-up approach to integrate existing national data, the ATKIS ® Basis-DLM was used as input data. Besides this vector dataset, additional data were integrated. The main source of information was the

IMAGE2006 satellite mosaic from the conventional CLC2006 production phase. Other GMES products were also taken into account like the raster datasets Fast Track Service (FTS) Soil Sealing, and GSE Forest. Also, soil sealing results from the REFINA (DIFU 2012) project of the year 2000 were integrated in the study, where they were available.

16.2.1.2 Study Results

The overall result of this study was the affirmation of the technical feasibility of the intended project. The best practice outcomes of the study, e.g. which working steps are necessary, which processing can be automated, what is the estimated manual effort, were included in the subsequent planning process for the entire country-wide coverage of DLM-DE (Bock et al. 2008). Datasets like REFINA or GSE Forest were useful, where available. However, because of their limited regional extend by that time, they were not recommended for nation-wide integration into the workflow. Two main results among others were the fusion of the layer-structured vector data of the ATKIS ® Basis-DLM in a single layer form. This modification will in the following be mentioned as flattened structure. A second result was the additional assignment of preliminary CLC codes to its features due to a semantic transformation prior initiating the update and verification process itself.

16.2.2 DLM-DE2009

16.2.2.1 Input Data and Source of Information

The produced vector data from the ATKIS ® Basis-DLM (according to old data model, BKG 2011a) were used as the fundamental input dataset for DLM-DE2009 (BKG 2012; Arnold et al. 2010). The input data were taken as a "data freeze" from the GeoDataCentre (GDZ) of BKG by the date of April 2009 for further preparatory pre-processing. By that time, the data at GDZ were roughly between 6 months and several years old (depending on the last individual update activities and deliveries from the Federal States to BKG). Only CLC-relevant polygon feature types of the thematic categories 'settlement, transportation, vegetation and water bodies' were used from the ATKIS data; point and line features were not included. Due to the step-wise patchwork update cycle of the ATKIS ® Basis-DLM, it was necessary to refer to remote sensing satellite data as source of land cover information. By these, it is possible, to update practically the whole country for a single reference year.

The main source of information were multi-temporal satellite imagery of the sensor RapidEye with a ground sampling distance (GSD) of 5 m and additional imagery from the system Disaster Monitoring Constellation (DMC) with a GSD of 32 m. Due to various reasons, for some minor patches the acquisition of RapidEye imagery had to be postponed to 2010 (Hovenbitzer et al. 2011).

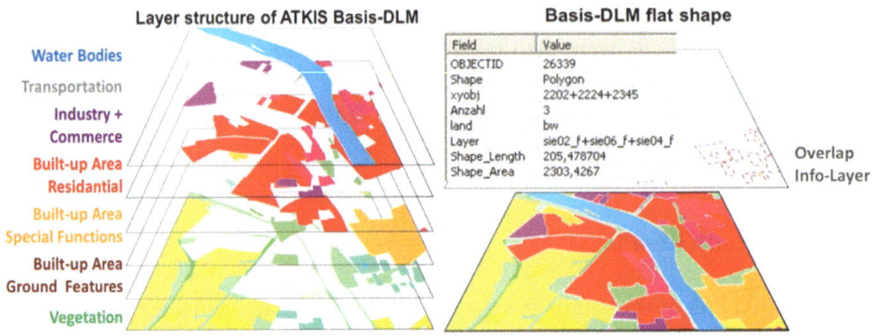

Fig. 16.2 Original layer structure of the ATKIS ® Basis-DLM flattened for the DLM-DE update process

Besides the RapidEye imagery, other auxiliary data were used, e.g. the GMES (now Copernicus) product Soil Sealing Layer 2009, the IMAGE2006 collection and authoritative digital orthophotos (DOP).

16.2.2.2 Semantic Transformation

In preparation of deriving the CLC data, the ATKIS feature type catalogue was compared with the CLC nomenclature by creating a semantic transformation table. Therefore the commonalities in the definitions of the ATKIS feature types and the 37 CLC-classes that are relevant in Germany (out of 44 classes in total for Europe) were analyzed. So all DLM-DE features received a preliminary CLC coding, which had to be verified and updated using remote sensing data.

16.2.2.3 Pre-processing of ATKIS ® Basis-DLM

The ATKIS data are provided as a set of different layers. To simplify the handling, the input layers are flattened by transferring it into a seamless and non-overlapping data structure. Using GIS-based workflow, all overlapping situations are eliminated, and only one single polygon feature remains in case of previous overlap of ATKIS layers (see Fig. 16.2). This flattening procedure follows an ordered sequence of the feature types dependent on their relevance for land cover modeling regarding the derivation of CLC. This pre-processing step brings the input data also to a format and shape similar to the CLC target dataset.

16.2.2.4 Production and Results

The verification and updating of DLM-DE itself as the main data production phase was outsourced through a call for tender to the geo-information industry.

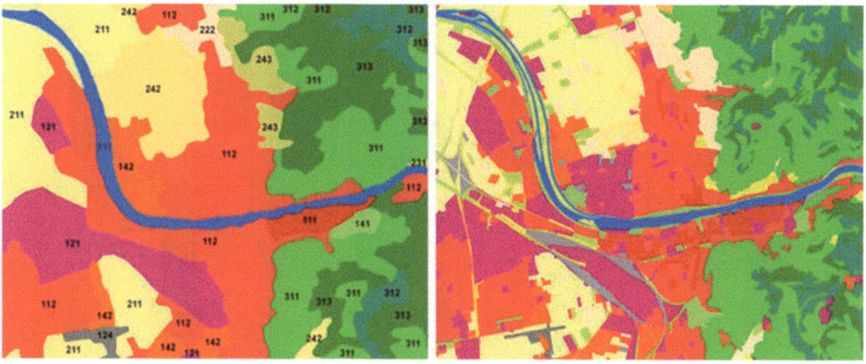

Fig. 16.3 Heidelberg, Germany. Comparison of conventional CLC2006 with 25 ha MMU (*left*) and DLM-D2009 with 1 ha MMU (*right*) dataset, both colored according to CLC legend (Source: *CLC2006* (*left side*), © *EEA 2006; DLM-DE2009* (*right side*), © *BKG/Geobasis-DE 2013*)

The preliminary (unique or multiple) CLC-codes, that were attached to the initial dataset, were then verified and updated through a semi-automated procedure based on remote sensing methods (supervised imagery classification, probability estimations, visual interpretation). The minimum mapping unit (MMU) was set to 1 ha for all features. The outcome of the DLM-DE project is a high resolution vector dataset on land cover and land use with a MMU of 1 ha for all features types. Figure 16.3 shows a snap shot of DLM-DE2009 results compared to the conventional CLC mapping result.

16.2.3 DLM-DE2012

Following DLM-DE2009 as the first mapping phase of the dataset, the year 2012 has been chosen to be the reference year for the second DLM-DE production. Regarding the availability of needed additional imagery the synchronization of the DLM-DE with the production phase of CLC2012 comprises a large benefit, as the ESA provides satellite data for the reference year 2012 to the member states. The different additional images allow a higher quality of multi-temporal analysis. Continuing the approach of DLM-DE2009 the procedure in 2012 has changed at a central point. As it was in 2009, the input data still is the ATKIS ® Basis-DLM (according to new AAA data model, BKG 2011b) and the MMU is again set to 1 ha. Instead of directly using the nomenclature of CLC (like in 2009), for the update of DLM-DE2012 a new class system separating land cover and land use was established, which is described in the following chapter, also to achieve better interoperability with the ATKIS data of the Federal States.

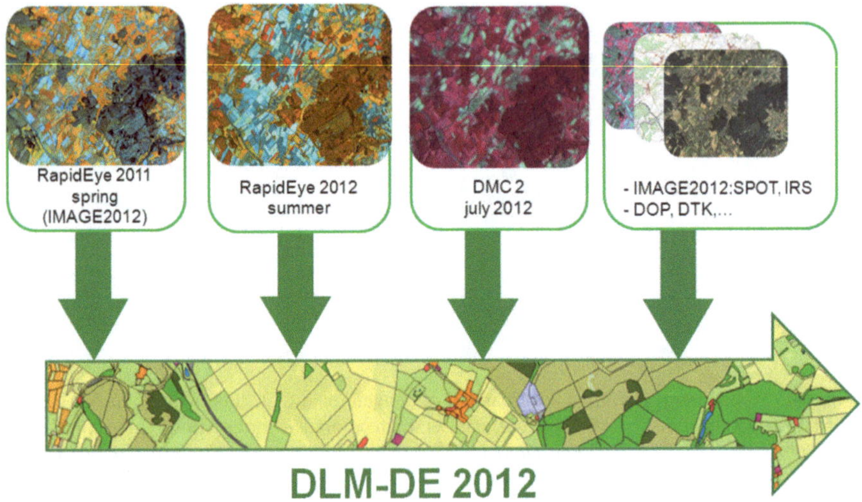

Fig. 16.4 Data used for production of DLM-DE2012 (Includes material © (2011, 2012) RapidEye AG, Germany. All rights reserved: © [2012] DMC International Imaging Limited of the United Kingdom; © BKG/Geobasis-DE 2013)

16.2.3.1 Input Data and Source of Information

Concordant with the previous 2009 production phase, the ATKIS ® Basis-DLM is used as the basic input data for the DLM-DE2012 production. In addition, selected results from the DLM-DE2009 mapping are integrated in the 2012 starting dataset.

Different satellite imageries are used as main source of land cover information. The primary source is one coverage from the RapidEye constellation for summer 2012. Furthermore and similar to 2009 one coverage of DMC is used for multi-temporal analysis, especially the distinction between grassland and cropland. The IMAGE2012 data is integrated in the production process as well, because it proved to be very useful for the classification of forest area types.

It was intended to make use of the land cover information provided by the High Resolution Layers from the Copernicus Land Monitoring Services as well, but these layers were not finished, when the production of DLM-DE2012 started. An overview of input and auxiliary data for DLM-DE2012 production shows Fig. 16.4.

16.2.3.2 Enhanced Classification: Separating Land Cover and Land Use

The new classification system applied for mapping takes place in between the CORINE nomenclature on one hand and the nomenclature of the ATKIS ® Basis-DLM on the other hand. For this purpose it describes the aspects of land use and land cover in two separate codes (LU and LC). CORINE Land Cover is

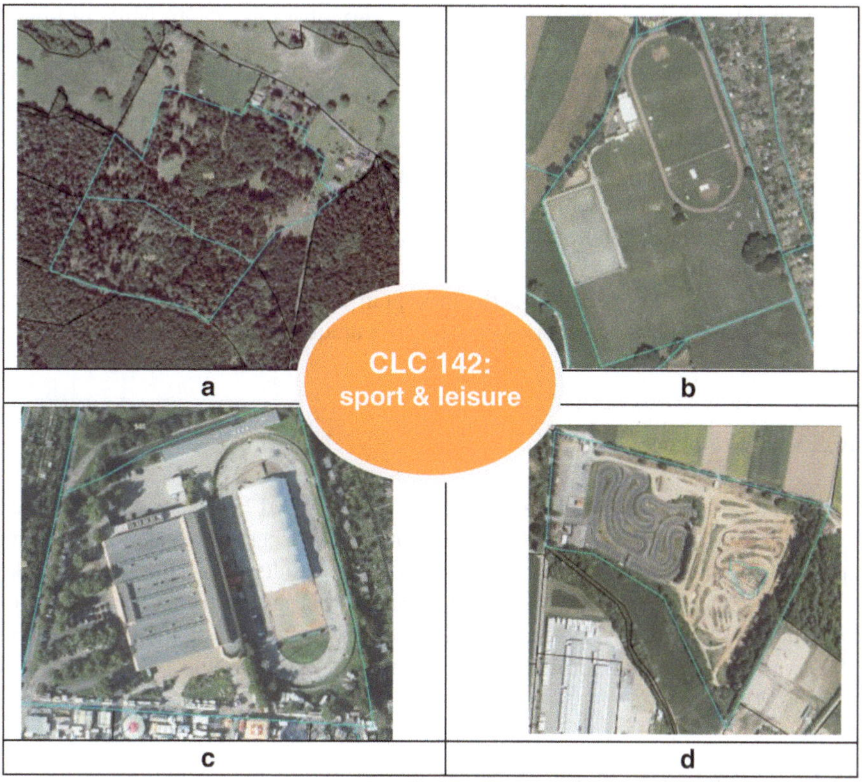

Fig. 16.5 Different kinds of leisure facilities (aerial ortho-imagery, © BKG/Geobasis-DE 2013). (**a**) Wildlife park. (**b**) Sports ground (grass). (**c**) Ice rink. (**d**) Go-cart and motorcycle race course

traditionally mapped on the source of satellite imagery and due to this historical development it focuses very strongly on the land cover aspect (although it mixes cover and functionality in case of built-up areas and infrastructure). In contrast the ATKIS ® Basis-DLM contains very detailed information on the land use. The separation of LC and LU manages not only the transformation from ATKIS ® Basis-DLM into the nomenclature of CORINE Land Cover, but also allows (partly) the way back to ATKIS. This opens up the possibility to support the Federal States with additional detailed information, especially in the field of natural classes (forests and semi-natural areas, wetlands, etc.). By separating the land cover and the land use aspects, the update is much easier for the interpreters. Instead of being forced to make a decision about the land use of features they can clearly classify the land cover as it appears in the imagery. The information about the usage is already contained in the ATKIS ® Basis-DLM and for the majority of features it is not possible to retrieve a proper classification of land use based on optical image data.

Figure 16.5 illustrates the benefits and needs of the distinction between land use and land cover. Each of the examples is classified as "142: sport & leisure facilities" according to the CLC nomenclature. But in fact these objects show very different types of land cover. Thus the new classification system helps to obtain more detailed information on the real land cover while supporting the derivation of CLC (in this case class "142") by referring to the land use information of the ATKIS ® Basis-DLM.

As a result of the pre-processing each DLM-DE polygon is given both a preliminary LC and a preliminary LU code that has to be verified and updated during the production process by (semi-automated) image interpretation. The final classes according to the CLC nomenclature will be created in a post-processing step after the update. Each CLC-class can be populated by combining one of the new codes for land cover and one for land use. As the excerpt in Table 16.1 shows, the design of the land cover and land use categories guarantees a unique CLC-assignment.

16.2.3.3 Pre-processing

The ATKIS ® Basis-DLM has been "frozen" again in 2012 in its status as provided by BKG's GeoDataCentre. All LC/LU relevant feature types are selected and brought again into a flattened structure. Additionally some results from DLM-D2009 (like forest area types) were integrated into the geometries of the input dataset of DLM-DE2012 through GIS operations. Several other pieces of information from 2009 were included, but not with their geometry, only as attribute.

16.2.3.4 Production Phase

All the pre-processing steps in preparation of the production phase (preliminary LC/LU-codes, integration of former results and geometric flattening) have been done by BKG. The actual update of DLM-DE2012 is again accomplished by a subcontractor after a call-for-tender procedure. The schedule foresees the finalization of DLM-DE2012 by the mid of 2014.

In parallel to the production, the developments for the derivation of the CLC change 2006–2012 and the CLC2012 dataset have started. These tasks lie again in the responsibility of BKG (in cooperation with UBA and DFD/DLR).

16.3 Transition from DLM-DE to CLC

The reporting duty from UBA towards the EEA does not only include a 25 ha CLC2012 dataset, but also the CLC change between 2006 and 2012. This results in two major tasks, fulfilled by BKG with support from DLR. The generalization from 1 ha CLC2012 to 25 ha CLC2012 and the creation of a 5 ha CLC change layer between 2006 and 2012.

Table 16.1 Excerpt of the cross-table that leads to final CLC-codes

	Land use								
Land cover	Housing	Manufacturing	Community	Harbour	Extraction site	...	Air traffic	Sport and leisure	...
Buildings (sealing >80 %)	111		111	123	131	...	124		...
Buildings (sealing 50–80 %)	112		121	123	131		124	142	
Buildings (sealing 30–50 %)	112		121	123	131		124	142	
Facilities		121	121	123	131		124	142	
Sealed areas w/o buildings		121	121	123	131		124	142	
...	...								
Broadleaves	141	121	121	123	131	...	124	142	...
Conifers	141	121	121	123	131		124	142	
Conifers and broadleaves (mixed)	141	121	121	123	131		124	142	
...	...								

Fig. 16.6 Gap between two lakes (DLM-DE/CLC 1 ha – *left side*, © BKG/Geobasis-DE) that has to be closed during generalization in comparison with CLC2006 (*right side*, © EEA 2006)

16.3.1 CLC2012 Derivation and Generalization

As the first step of deriving CLC, the updated land cover and land use codes of the DLM-DE2012 are transformed by combining them into CLC classes according to the cross-table shown in Table 16.1. This results in the first intermediate product, the CLC2012 1 ha dataset. This provides the basis for the subsequent generalization. During a first iteration another intermediate product with 5 ha is generated, before the final 25 ha CLC2012 dataset is created. For the basic part of this complex process a generalization tool prototype is used that has been developed in a cooperation between BKG and the IKG (Institute for Cartography and Geoinformation) of the University of Hannover (Thiemann et al. 2009). This tool is used for some basic operations, whereas further individual processing steps for regional or class-dependent improvements have been developed at BKG. This fine-tuning is necessary to handle objects that obviously belong together because of spatial vicinity (and therefore have to be merged during the generalization process), but do not share a common border. For lakes and settlements it is especially important to resolve and close these gaps. If their vicinity is not taken into account during the generalization, too many of those objects would be lost in the final dataset. Figure 16.6 shows an example that illustrates the importance of this vicinity aspect. Also, some testings on geometric simplification and generalization have been done at the DFD/DLR (Nieland 2009).

Another very important issue that has to be taken into account is the scale-dependency of some CLC classes. Especially the complex classes ("242 – complex cultivation patterns", "243 – land principally occupied by agriculture, with significant areas of natural vegetation" and "313 – mixed forest") have to be treated differently from the remaining classes. Candidate polygons that can contribute to these classes on 5 or 25 ha need to be identified already in the 1 ha dataset to

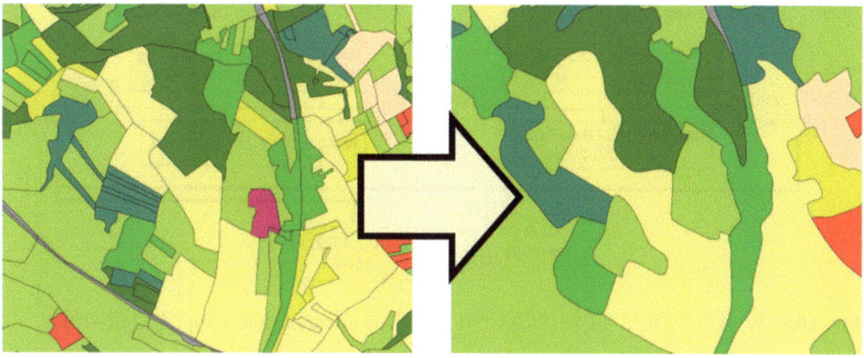

Fig. 16.7 Results of generalization from 1 to 5 ha (DLM-DE2012 – *left side* 1 ha. *right side* 5 ha, © BKG/Geobasis-DE)

preserve them grouped as clusters during the single generalization steps. If this aspect is not considered, all these features would be resolved into neighboring polygons according to the basic generalization rules like the longest common border or the thematically similarity. BKG has developed several algorithms to identify the proper candidate polygons in order to generate meaningful features for the complex classes on the level of 5 and 25 ha MMU.

The EEA's technical guidelines for CLC (Büttner et al. 2002, page 40) provide a priority table (similarity matrix) of the CLC classes for the traditional way of CLC production. This table brings all classes into relation, showing index values from 1 to 8 to represent the similarity between classes and give an indicator for the merging of neighboring polygons below the MMU during the mapping process. This table has been revised and adapted due to the needs of a fully automatic generalization process.

The algorithms for the last generalization from the intermediate 5 ha (see Fig. 16.7) to 25 ha CLC2012 are currently under the final development stage.

16.3.2 Land Cover Change 2006–2012

Besides the full dataset CLC2012, the derivation of the changes between CLC2006 and CLC2012 is an important part of the data delivery to EEA. Traditionally CLC is produced by mapping the land cover changes by interpreting satellite imagery ("change first"). With the now applied bottom-up approach to generate CLC2012 based on high resolution vector data, the conventional procedure is no longer feasible. In this case the change has to be derived in a subordinate process.

Because of the different production processes it is not possible to directly compare the (traditional) CLC2006 with the CLC2012 derived from DLM-DE2012. The diverging shape and structure of geometries would result in a large amount of pseudo-changes. DLM-DE2009 would be more appropriate because of the same

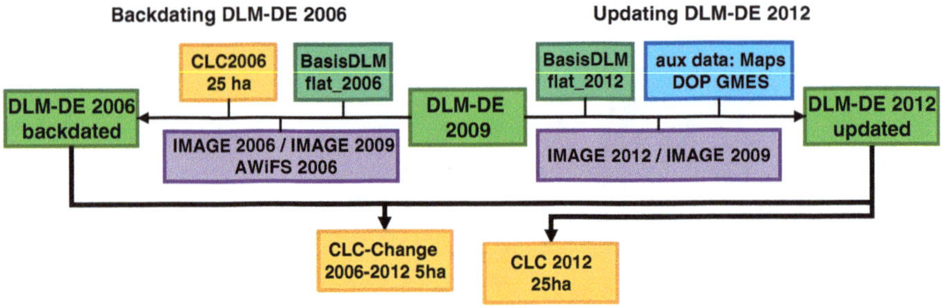

Fig. 16.8 Workflow DLM-DE2012/CLC2012 and land cover change mapping 2006–2012

geometrical basis, but the reference year is not suitable. To make this temporal comparison still possible, UBA has subcontracted the DLR to perform a backdating of DLM-DE2009 to 2006 (see Fig. 16.8).

In doing so, DLM-DE2009 is used as starting point and is reconstructed based on IMAGE2006, CLC2006 and archive ATKIS data from 2006 and brought back to the status of the reference year 2006 by using a backdating procedure (Keil et al. 2010). With the DLM-DE2012 being the future-projected counterpart, and the DLM-DE2006 backdated as the past-projected counterpart, the CLC 2006–2012 change dataset with 5 ha MMU can be generated.

The algorithms for the calculation of the change are as well developed at the BKG. Candidate features for changes are identified by comparing the 1 ha datasets DLM-DE2012 and DLM-DE2006 backdated. The relevant changes go through a generalization process and result in indicative change layer features that fulfill the MMU of 5 ha. Additionally, the geometry is improved and adapted to the CLC2012 5 ha dataset to guarantee full geometric consistency (snapping). Figure 16.9 shows an example for the derivation of CLC 2006–2012 changes.

16.3.3 Data Backflow Between National and Sub-national Level

The DLM-DE concept also foresees the possibility of a data flow back of the DLM-DE data to the land surveying institutions of the Federal States in a way that they can integrate the results in their own continuing updating process for ATKIS. On the long run, the DLM-DE concept comprises the vision of the iterative reduction of effort that is needed to derive European datasets from national sources, as European relevant LC/LU information to a certain extent can be integrated stepwise on the sub-national level through the data exchange between BKG and the AdV (see Fig. 16.10).

A test-wise attempt to integrate LC/LU information from DLM-DE2009 into ATKIS resulted in the understanding that LC/LU information described according to the CLC nomenclature can only be integrated selectively. In other words not

Fig. 16.9 Results of change derivation (*red* in the right image (CLC2012, 5 ha) based on the comparison between DLM-DE2012 1 ha (*above*) and DLM-DE2006 backdated 1 ha (*below*)). © BKG/Geobasis-DE

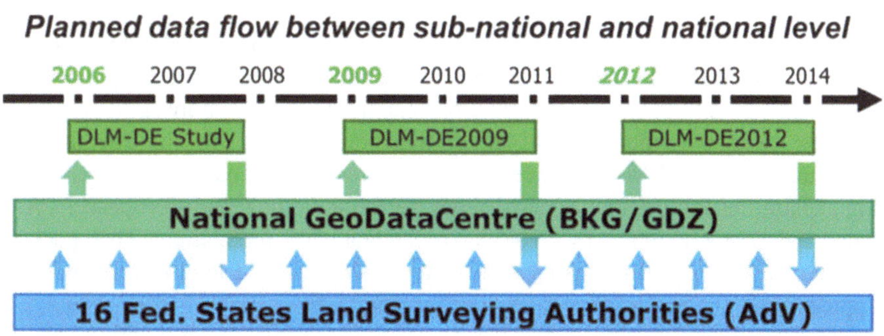

Fig. 16.10 Possible data flow of topographic and LC/LU data between national level (BKG) and sub-national level (AdV) along the time line

every CLC class is useful to support ATKIS feature types, because the definition of some CLC classes are semantically not congruent or too broad compared to the definition of ATKIS feature types. As a lesson learnt, the ATKIS feature types were not directly transformed into CLC classes and then update them due to the CLC nomenclature, but intermediate LC/LU components were used instead. This approach allows a broader use of the dataset besides the derivation of CLC. It offers also the possibility to give information on certain land cover features back to the Federal states and other interested institutions.

16.3.4 Integration of CLC-Relevant Attributes in ATKIS Feature Type Catalogue

As a reaction on the need for harmonization between national and European land surveying activities, it was achieved to add placeholders for certain CLC relevant land cover information according to its nomenclature into the German ATKIS feature type catalogue (FTC). This will help to make the derivation of some CLC classes based on national data easier, when it is implemented in the forthcoming next release of the FTC. This is a development that is independent from the above described DLM-DE specific set of LC/LU components. In particular this extension of the ATKIS feature type catalogue concerns the CLC classes "321-natural grassland", "324-transitional woodland-shrub", "421-salt marshes", "521-coastal lagoons" and "522–estuaries". In this context the first step is made on the way to the involvement of the Federal States' land surveying authorities in the European land monitoring bottom-up approach by handing over a part of the relevant data capture to their responsibility and competence, following the principle of subsidiarity.

16.4 Conclusion and Outlook

The transition of executive responsibility for land cover data production over the years from Destatis via UBA to the BKG as the national mapping agency manifests the changing significance of land monitoring activities from a more statistical point of view to a more holistic and integrative bottom-up approach. The targeted advantages of the DLM-DE concept are more precise information on LC/LU with high spatial resolution, higher consistency between national and European LC/LU data, and cost efficiency by avoiding redundant data production. This goes in direction of the pan-European vision of an object-oriented land cover modeling approach.

Besides Germany, also a number of other European member states already implemented also such a bottom-up approach deriving LC/LU data relevant for the European level out of national reference data. Two initiatives shall be mentioned in this context, an FP7-funded project called HELM (Harmonized European Land Monitoring) bringing together best practices, ideas, requirements and future visions, and also the EAGLE group (EIONET Action Group on Land Monitoring). The latter comprises of a principally voluntary cooperation of committed land monitoring experts throughout many European countries. They work on an enhanced data model for land monitoring that suits both European stakeholders' interests and national/sub-national land monitoring applications (EAGLE 2013; Arnold et al. 2013). The group is developing a new data model for land monitoring that is capable to store LC and LU information in a separate and more comprehensive way.

In relation with the developments in Germany in particular, the project "digital land cover model – DLM-DE" as a bottom-up approach will be continued. There are plans for the next update cycles in the years 2015 and 2018. BKG is strongly involved in the discussion process with the Federal states, which are interested in using the results and methods of the LC/LU production at BKG. Therefore, the goal is to integrate the land cover update process into the update workflow of topographic reference data of the Federal States.

References

AdV (2009) Dokumentation zur Modellierung der Geoinformationen des amtlichen Vermessungswesens (GeoInfoDok). Hauptdokument. Version 6.0.1, Stand: 02 July 2009. Arbeitsgemeinschaft der Vermessungsverwaltungen der Länder der Bundesrepublik Deutschland. English documentation (2006) can also be found on http://www.adv-online.de

Arnold S (2009) Integration von Fernerkundungsdaten in national und europäische Geodateninfrastrukturen – Ableitung von CORINE Land Cover–Daten aus dem DLM–DE. In: PFG Photogrammetrie – Fernerkundung – Geoinformation, Heft 2–2009, pp 123–135

Arnold S, Busch A, Grünreich D (2010) Das Projekt DLM-DE2009 Landbedeckung. In: Mitteilungen des Bundesamtes für Kartographie und Geodäsie, Band 45. Arbeitsgruppe Automation in Kartographie, Photogrammetrie und GIS (AgA). Tagung 2009, pp 9–22

Arnold S, Kosztra B, Banko G, Smith G, Hazeu G, Bock M, Valcarcel Sanz N (2013) The EAGLE concept – a vision of a future European Land Monitoring Framework. In: Lasaponara L, Masini N, Biscione M (eds) EARSeL symposium proceedings "Towards Horizon 2020", Matera, Italy, pp 551–568

BKG – Bundesamt für Kartographie und Geodäsie (2011a) Vektordaten BRD. Digitales Basis-Landschaftsmodell. Stand: 20 Sept 2011. http://www.geodatenzentrum.de/docpdf/basis-dlm.pdf

BKG – Bundesamt für Kartographie und Geodäsie (2011b) Vektordaten BRD. Digitales Basis-Landschaftsmodell (AAA-Modellierung). Stand: 29 Aug 2011. http://www.geodatenzentrum.de/docpdf/basis-dlm-aaa.pdf

BKG (2012) Produktbeschreibung "Digitales Landbedeckungsmodell für Deutschland DLM-DE2009". GeoBasis-DE Geodaten der Deutschen Landesvermessung – Bundesamt für Kartographie und Geodäsie, letzter Stand der Dokumentation: 30 Mar 2012, Frankfurt am Main. http://www.geodatenzentrum.de/docpdf/dlm-de2009.pdf

BKG (2014) www.bkg.bund.de/EN

Bock M, Keil M, Strunz G, Dietz A, Eisfelder C, Metz A, Rössig C (2008) Ergebnisse der Machbarkeitsstudie – Aktualisierung und Nutzung des DLM–DE für die Ableitung von CORINE Land Cover auf der Basis von Satellitendaten. Studie im Auftrag des BKG, German

Remote Sensing Data Center (DFD) at German Aerospace Center (DLR), Oberpfaffenhofen, unpublished

Büttner G, Feranec J, Jaffrain G (2002) Corine land cover update 2000 – technical guidelines. Technical report 89. EEA

DIFU – Deutsches Institut für Urbanistik (2012) Research for the reduction of land consumption and for sustainable land management (REFINA). Development and evaluation of remote sensing based area calculation for sustainable land management. http://www.refina-info.de/en/projekte/anzeige.phtml?id=3103

EAGLE (2013) EIONET Action Group on Land Monitoring in Europe. http://sia.eionet.europa.eu/EAGLE

Hovenbitzer M, Arnold S, Ru¨ckert F, Wende C (2011) Digitales Landschaftsmodell Deutschland DLM-DE. DGPF Tagungsband 20/2011

Keil M, Kiefl R, Strunz G (2005) CORINE land cover 2000 Germany – final report. German Remote Sensing Data Center (DFD) at German Aerospace Center (DLR), Oberpfaffenhofen

Keil M, Bock M, Esch T, Metz A, Nieland S (2009) CORINE land cover 2006 Germany – final report. German Remote Sensing Data Center (DFD) at German Aerospace Center (DLR), Oberpfaffenhofen

Keil M, Metz A, Bock M, Esch T, Nieland S, Feigenspan S (2010) Flächenerhebung und –statistik in CORINE Land Cover – Aktuelle Ergebnisse und Programmentwicklung. In: Meinel G, Schumacher U (eds) Flächennutzungsmonitoring II, Konzepte – Indikatoren – Statistik. IÖR Schriften Band 52. Rhombos Verlag, pp 93–107

Nieland S (2009) Entwicklung und Bewertung einer Methode zur automatisierten Generalisierung des DLM-DE auf den Zielmaßstab des CORINE Land Cover Datensatzes in Deutschland. Diplomarbeit Universität Innsbruck

Thiemann F, Sester M, Bobrich J (2009) Generalization of land-use data from topographic information. In: The international archives of the photogrammetry, remote sensing and spatial information sciences, vol 34, Part XXX

Chapter 17
Land Use & Land Cover Mapping in Europe: Examples from the UK

Geoffrey M. Smith

17.1 UK Philosophy

The United Kingdom (UK) has a long heritage in the development and application of land cover mapping approaches based on Earth Observation (EO) data. The production of UK national land cover products from EO data began as a technology driven proof-of-concept project in 1990, the Land Cover Map of Great Britain (LCMGB). Once in production LCMGB was brought under the banner of the Countryside Survey (CS), the national land monitoring program focusing on habitats. The next iteration, the Land Cover Map of 2000 (LCM2000), aimed to update LCMGB, but also upgraded the product with the adoption of new technology and respond to national policy drivers. The Land Cover Map of 2007 (LCM2007) continued this trend by integrating digital cartography and applying complex map generalization processes within its production.

The CS program included a series of scoping studies to assess user requirements and identify opportunities for the exploitation of new technology. These studies highlighted the conflicting requirements for national land cover mapping. There is a need for temporal consistency to assess change over time, but this is always alongside the need for improved quality/integration and reduced costs/production time while also responding to a broad and often changing set of policy requirements. There has also been a demand to deliver a product that services a broad and sometimes mutually exclusive set of user requirements. The UK LCMs could therefore be criticized for inconsistencies over time, but they have remained at the forefront of land cover mapping technology and continuously developed to address the needs of real users. The procedural developments in terms of parcel-based mapping, knowledge-based enhancements, parcel-level metadata and process operationalization have led the way in improving product accuracy and usability.

G.M. Smith (✉)
Specto Natura Ltd., College Road, Impington, Cambridge CB24 9PL, UK
e-mail: geoffsmith@specto-natura.co.uk

I. Manakos and M. Braun (eds.), *Land Use and Land Cover Mapping in Europe: Practices & Trends*, Remote Sensing and Digital Image Processing 18,
DOI 10.1007/978-94-007-7969-3_17,

The final overall accuracy of the LCMs have been comparable with the best performing land cover mapping exercises, including the more experimental approaches, of their time.

17.2 The Development

The origin and history of the UK land cover maps can be charted through a series of research and development projects, scoping studies, production activities and reviews.

17.2.1 Land Cover Map of Great Britain (LCMGB)

The LCMGB of 1990 was the first national land cover product for Great Britain derived from EO data. Land cover mapping was often cited as an application of EO, but few, if any, operational large extent projects had been completed at that time. The European CORINE project was under development, but it was in effect an extension of the visual interpretation approaches which had been applied to aerial photography for topographic mapping. LCMGB was a semi-automated classification of spectral data (Fuller et al. 1994a, b) from the Landsat Thematic Mapper (TM) instrument, mostly collected during 1988 and 1989. The final product contained 25 land cover classes (Table 17.1) recorded against a 25 m grid equivalent to the sampling scheme of the input Thematic Mapper (TM) data. The production included a set of rudimentary knowledge-based enhancements to cope with some classes out of context (Groom and Fuller 1996). The original design and purpose of LCMGB was based around a proof-of-concept approach with a specification which maximized the performance of the input data rather than addressing key user needs. Since production, LCMGB data (Fig. 17.1) have been licensed to over 500 users, including researchers, policy makers and commercial organizations, with wide ranging end uses.

17.2.2 Classification of Environment with Vector and Raster-Mapping (CLEVER-Mapping)

The LCMGB was successful, but the pixel-based approach gave an arbitrary grid structure to the landscape. The pixel-based approach also incorporated noise and unwanted natural variation into the classification process resulting in a somewhat speckled appearance with little, if any, information on landscape structure (Fuller et al. 1998). This situation and the continued development of geographic information and image analysis technology encouraged the development of object-based approaches which analyzed the EO data in units that were representative of 'real world' features.

Table 17.1 Comparison of the classification nomenclatures used for the LCMGB (1990), designed to maximize the capabilities of the data, and the LCM2000, aligned with an established nomenclature used in field surveys

LCMGB (1990)	LCM2000
Sea/estuary	Sea/estuary
Inland water	Water (inland)
Beach and coastal bare	Littoral rock
Saltmarsh	Littoral sediment
Upland bog	Saltmarsh
Lowland bog	Supra-littoral rock
Open shrub moor	Supra-littoral sediment
Dense shrub moor	Bog (deep peat)
Open shrub heath	Fen, marsh, swamp
Dense shrub heath	Dense dwarf shrub heath
Grass heath	Open dwarf shrub heath
Scrub/orchard	Montane habitats
Deciduous woodland	Broad-leaved/mixed woodland
Coniferous woodland	Coniferous woodland
Mown/grazed turf	Improved grassland
Meadow/verge meadow/semi-natural	Neutral grassland
Rough/marsh grass	Set-a-side grassland
Moorland grass	Calcareous grassland
Bracken	Acid grassland
Tilled land	Bracken
Ruderal weed	Arable cereals
Suburban/rural development	Arable horticulture
Continuous urban	Arable non-rotational
Inland bare ground	Suburban/rural developed
Felled forest	Continuous urban
	Inland bare ground

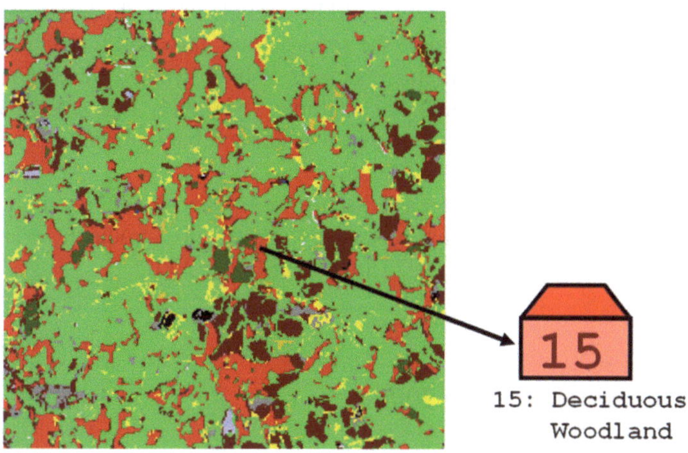

Fig. 17.1 The simple pixel-based classification of the 1990 Land Cover Map of Great Britain, using the standard UK nomenclature color table. In this case one pixel is selected which identifies itself as Deciduous Woodland (code 15, color *red*). (LCMGB)

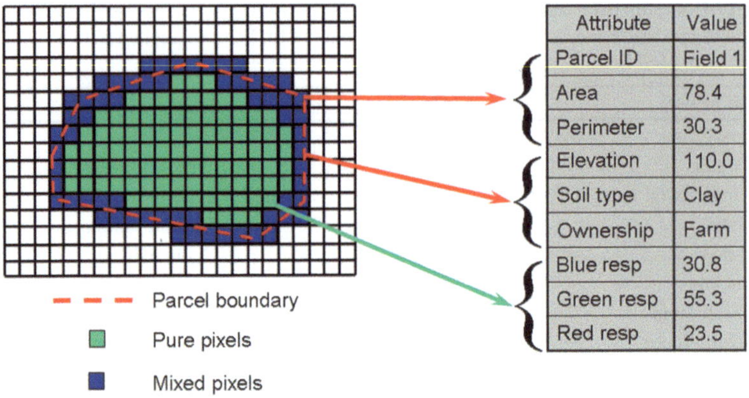

Fig. 17.2 The integration of vector, raster and ancillary data proposed by CLEVER-Mapping

The Classification of Environment with Vector and Raster-Mapping (CLEVER-Mapping) project developed an object-based classification procedure pre-empting the interest and growth of object-based image analysis (OBIA). This approach (Fig. 17.2) used a dataset of land parcel objects to control the analysis of the EO data thus avoiding the mixed pixels at the edge of each object, which were often misclassified in conventional per-pixel approaches due to their mixed spectral signatures for adjacent cover types (Dean and Smith 2003). The spectral response in each image band was averaged for the core pixels only within each object to minimize noise and unwanted natural variation. The averaged spectral responses for the core area were then applied to a standard maximum likelihood algorithm and the resulting classification attached to the object as a whole.

The object-based structure allowed different EO data types to be combined and a broad range of non-EO data to be included as attributes on the object. The later were used to perform complex knowledge-based enhanced. For instance, objects with elevations greater than a few meters could be excluded from the intertidal habitats and soil type could be used to refine the semi-natural grassland types that could be recorded. The topologically structured objects also allowed advanced spatial context enhancements to be applied. For example, it is likely that small patches of arable completely surrounded urban are incorrect and bare ground in a coniferous forest context is more likely to be felled forest.

The CLEVER-Mapping/OBIA approaches were tested in a relatively large scale object-based land cover mapping exercise by successfully producing a land cover map for the island of Jersey in 1997 (Smith and Fuller 2001).

17.2.3 Land Cover Map 2000

After the success of the mapping in Jersey, it was agreed that the LCM2000 (Fuller et al. 2002) would use the parcel-based processes and procedures developed within

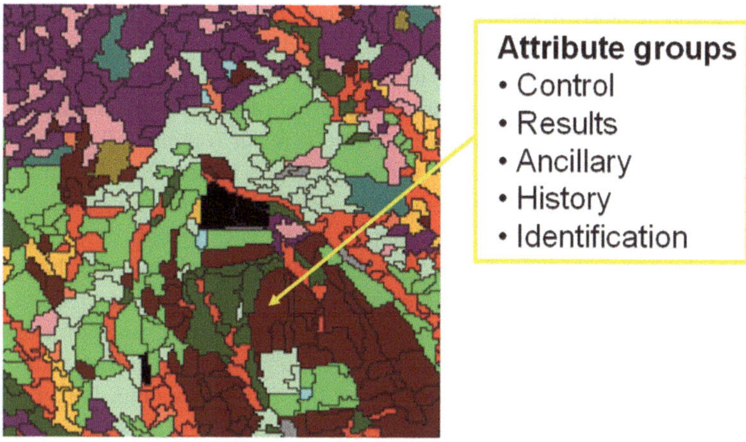

Fig. 17.3 Land Cover Map 2000 data showing the segment structure and the attribute groups on each object

the CLEVER-Mapping project. As well as improving the spatial structure of LCM2000 it was necessary to align the thematic content more closely with user requirements. The LCMGB nomenclature, although designed to maximize the performance of EO data, was not particularly policy relevant or aligned to existing classification schemes. The nomenclature for national monitoring as whole was under discussion and it was recommended that the Biodiversity Action Plan (BAP) Broad Habitats (Jackson 2000) should be selected and adopted as the LCM2000 nomenclature (Table 17.1).

LCM2000 essentially used the same type of input EO data as LCMGB, but applied to a variant of the CLEVER-Mapping procedure described earlier (Fuller et al. 2005). Unfortunately, a suitable vector land parcel data set was not available for the UK, therefore segmentation was used to group image pixels into areas broadly equivalent to land parcels. The segmentation results were generalized to give a minimum mapping unit (MMU) of 0.5 ha. The MMU was selected to capture relatively small features while still containing around 9 pixels, so that in a square object there would be at least one central pixel surrounded by 8 other. Segment-based approaches, in the same manner as parcel-based approaches, avoided mixed edge-pixels and used mean spectral responses for thematic identification. Ground reference data were used to identify image segments of known cover which would act as a sample of training areas from which to calculate the spectral reflectance statistics for each cover class. The classification used a maximum likelihood algorithm applied per-segment and recorded the most likely spectral subclass in statistical terms: in fact, it stored probabilities for the top five spectral subclasses, usually covering >90 % of the probability distribution. Knowledge-based enhancements were used to allocate an alternative class label, where more appropriate, for segments which were classified with low confidence or with classes out of their natural context.

As well as representing both the landscape structure and its thematic composition, the LCM2000 also recorded a rich set of object-level metadata (Fig. 17.3).

The LCMGB had only recorded a single land cover code against each pixel. Against each segment LCM2000 recorded a set of metadata grouped under the following headings: identification (e.g. unique identifier); context (non-remotely sensed ancillary information, e.g. elevation); control (information to guide the production, e.g., known (a priori) land cover type of a parcel); result (outputs of the production procedure, e.g. the predicted land cover type and a measure of its probability), remark (additional information associated with the classification procedure and results) and image information (e.g. number of pixels extracted).

In conclusion, LCM2000, for the first time, mapped land cover for the whole of the UK from satellite images. It offered a data structure which attempted to record the 'real' structure of the landscape and thus satisfied a wider range of user-needs. LCM2000 offered much more scope for use, integration and further exploitation than the conventional per-pixel products of the earlier mapping. It had a MMU which offered far better resolution than other available land cover datasets at that time and since.

17.2.4 Land Cover Map 2007

The LCM2000 was again a success with users, but technology had moved on and the use of GI data across the stakeholder community had increased thus improved integration was required. The most recent iteration, LCM2007, therefore used a similar parcel-based mapping approach to that used for LCM2000, but the spatial framework was based on existing national digital cartography from the Ordnance Survey's (OS) MasterMap (MM) product.

The use of digital cartography for the spatial structure more accurately reflects the true structure of the UK landscape (Smith 2008). OS MM is a topologically structured digital cartography layer for the whole country. The structured nature of the data provides a comprehensive cover of land parcel objects and their associated attribution allows the use of efficient, cost effective generalization to provide land parcels suitable for integration with EO data that has a spatial resolution of approximately 20–30 m (Smith et al. 2007). The half a billion objects within MM had to be reduced to around six million with a MMU of 0.5 ha and a minimum feature width (MFW) of 20 m. However, these objects did not represent all the required boundaries in the landscape and additional linework from agricultural surveys and image segmentation were required to complete the spatial framework (Smith and Morton 2010).

The objects were then labeled with land cover classes by a combination of object-based maximum likelihood classification of the EO data and knowledge-based enhancements driven by a range of ancillary data as in LCM2000. The final product (Fig. 17.4) was again an attribute rich land cover database but this time linked to national cartography designed to support a wider range of applications (Morton et al. 2011).

Fig. 17.4 The Land Cover Map 2007 showing the land parcel structure derived from existing digital cartography

17.3 Trends

The UK land cover products have always been at the forefront of technical and conceptual developments and their application in an operational domain.

17.3.1 New Technology

The key technical development has surrounded the use of object-based image analysis for the monitoring of 'real-world' objects which represent landscape structure. Even when the eCongition system was in its infancy, the LCM2000 had applied object-based approaches in an operational context exploiting a combination of research software, bespoke modules and commercial vector/raster OO database systems. The use of new technology continued in LCM2007 with development of an automated system for the generalization of detailed national cartography, which was deployed across a cluster processing system.

17.3.2 Integration

A major issue addressed by the UK land cover products has been the integration between the land cover maps, national field survey and other survey and monitoring activities. This has been achieved by the migration from pixels to segments and finally to the integration of national cartography within the process. This was also extended in LCM2007 by the use of agricultural land parcel (LPIS, IACS) data within the construction of the spatial framework. This has resulted in a land cover product which is more closely aligned with existing stakeholder datasets and their perceptions of the landscape.

17.3.3 Metadata

The complex production flowlines of LCM2000 and LCM2007 required and produced a large amount of information, which provided a rich source of additional data compared with the single land cover class that was recorded by LCMGB. The UK parcel-based production retained as much information as possible at the parcel-level (Smith and Fuller 2002). The thirty or so attributes attached to each land parcel were summarized into a set of parcel-level metadata. The utility of this information has been explored but the use made of it has been primarily academic as it is often still too advanced for operational users of land cover data.

17.3.4 Top-Down/Bottom-Up Processes

Recently some of the EO and supporting ancillary data have been made available through the GMES and other initiative and supplied by a European service in a consistent manner. The UK has then combined these data with local information and expertise to generate national products. All three UK LCMs have been generalized to form the UK contribution to the European CLC products (Fuller and Brown 1996; Smith et al. 2009). This demonstrates the benefits of a combination of top-down and bottom-up components when attempting to produce consistent and integrated land cover information at both the national and European levels.

17.3.5 Change

The issue of change has been considered throughout the development of the UK LCMs, but the implications of the changing methodology and nomenclature have meant that the generation of reliable change products was very challenging. As with other forms of survey, problems also exist in detecting what are often small degrees of change, against a background of uncertainty in the data. A range of technical and quality issues were identified which would need to be resolved to allow change mapping, so it was therefore recommended that the UK LCMs should go ahead without the requirement to deliver change information. Beneath this is a dilemma that faces any long-term longitudinal survey – whether to persist with what ultimately becomes an out-dated and sub-optimal technology to maintain consistency over time, or to adapt to new and improved methodologies as they become available.

17.3.6 Quality Assessment

Quantifying accuracy of land cover data is also inherently problematic, because of the difficulty in defining an objective and error-free reference against which results can be judged. At the outset, a target classification accuracy, when compared to independent field classifications, of 90 % was set for the UK LCMs. The main component of the CS was a field survey where in excess of 600 1-km squares were visited and mapped by field surveyors. These 1 km results were used to assess the quality of the UK LCMs and in practice, the results probably fall slightly short of the target at between 80 and 90 %, though this apparent error is to some extent inflated by uncertainties or inconsistencies in the field data.

17.4 Future

The national land cover maps in the UK have been seen to provide a valuable source of information, that was used not only to support countryside policy but also policy and research in many other areas. At present the UK is reviewing the CS program within which the LCMs sit, with a possible intention to rerun the surveys in the coming years. It is likely that the trends described above in technology and specification would continue in any future UK land cover product with the current needs for greater efficiency, integration and end user applications. A rolling program of land cover mapping has been suggested, but excluded in favor of attempts to more closely integrate the field survey and land cover mapping activities. At the European, national and regional level, the GMES program and its Land Monitoring Core Service (LMCS) will be an important force (and source of support) for future monitoring activities. The next UK LCM will undoubtedly be closely integrated with the LMCS to exploited centrally procured data and information products, validate them at the national level and contribute to European level reporting through initiatives such as CLC.

Acknowledgements The work reported in this chapter has been the results of a broad spectrum of activity under the banner of the Land Cover Maps within the UK Countryside Survey program. There has been a large team of people involved in these activities over the years from the Centre for Ecology and Hydrology (formerly the Institute for Terrestrial Ecology). The work was funded by a range of government departments and agencies led by the Department for Environment, Food and Rural Affairs, the Natural Environment Research Council and the UK Space Agency (formerly the British National Space Centre).

References

Dean AM, Smith GM (2003) An evaluation of per-parcel land cover mapping using maximum likelihood class probabilities. Int J Remote Sens 24:2905–2920

Fuller RM, Brown N (1996) A CORINE map of Great Britain by automated means. Techniques for automatic generalization of the Land Cover Map of Great Britain. Int J Geogr Inf Syst 8:937–953

Fuller RM, Groom GB, Jones AR (1994a) The Land Cover Map of Great Britain: an automated classification of Landsat Thematic Mapper data. Photogramm Eng Remote Sens 60:553–562

Fuller RM, Groom GB, Wallis SM (1994b) The availability of Landsat TM images for Great Britain. Int J Remote Sens 15:1357–1362

Fuller RM, Wyatt BK, Barr CJ (1998) Countryside survey from ground and space: different perspectives, complementary results. J Environ Manage 54:101–126

Fuller RM, Smith GM, Sanderson JM, Hill RA, Thomson AG, Cox R, Brown NJ, Clarke RT, Rothery P, Gerard FF (2002) The UK Land Cover Map 2000: construction of a parcel-based vector map from satellite images. Cartogr J 39:15–25

Fuller RM, Cox R, Clarke RT, Rothery P, Hill RA, Smith GM, Thomson AG, Brown NJ, Howard DC, Stott AP (2005) The UK Land Cover Map 2000: planning, construction and calibration of a remotely sensed, user-oriented map of broad habitats. Int J Appl Earth Obs Geoinf 7:202–216

Groom GB, Fuller RM (1996) Contextual correction: techniques for improving land cover mapping from remotely sensed images. Int J Remote Sens 17:69–89

Jackson DL (2000) JNCC Report No. 307 Guidance on the interpretation of the Biodiversity Broad Habitat Classification (terrestrial and freshwater types): definitions and the relationships with other habitat classifications. Joint Nature Conservation Committee, Peterborough

Morton D, Rowland C, Wood C, Meek L, Marston C, Smith G, Wadsworth R, Simpson IC (2011) Final report for LCM2007 – the new UK Land Cover Map, CS Technical Report No 11/07. Centre for Ecology and Hydrology

Smith GM (2008) The development of integrated object-based analysis of EO data within the UK national land cover products. In: Blaschke T, Lang S, Hay GJ (eds) Object-based image analysis: spatial concepts for knowledge-driven remote sensing applications. Springer, Berlin, pp 513–528

Smith GM, Fuller RM (2001) An integrated approach to land cover classification: an example in the Island of Jersey. Int J Remote Sens 22:3123–3142

Smith GM, Fuller RM (2002) Land Cover Map 2000 and meta-data at the land parcel level. In: Foody GM, Atkinson PM (eds) Uncertainty in remote sensing and GIS. Wiley, London, pp 143–153

Smith GM, Morton RD (2010) Real world objects in GEOBIA through the exploitation of existing digital cartography and image segmentation. Photogramm Eng Remote Sens 76:163–171

Smith G, Beare M, Boyd M, Downs T, Gregory M, Morton D, Brown N, Thomson A (2007) UK Land Cover Map production through the generalisation of OS MasterMap. Cartogr J 44:276–283

Smith GM, Morton RD, Gregory M (2009) Multi-scale land cover mapping in the UK driven by EO data and digital cartography. In: 3rd EARSeL Workshop Land Use and Land Cover, Bonn, Germany, 25–27 Nov 2009

Chapter 18
Operational Land Cover and Land Use Mapping in the Netherlands

Gerard W. Hazeu

18.1 Introduction

Land cover is changing rapidly in many parts of the world, particularly in areas with high population density like The Netherlands. The effect of these land cover changes becomes increasingly important in fields of work like spatial planning, water and environmental management. It is essential background information for environmental issues related to the land surface (Lambin et al. 1999; Kaufmann and Seto 2001; De Wit 2003; Chen and Wang 2010). In The Netherlands, remote sensing data combined with additional spatial data are recognised as an important source of information to characterise land cover and to detect its changes over time. At national scale the following four important databases are dealing with land cover:

1. Topographical database (Top10vector/Top10NL) (Kadaster 2003, 2007a, b)
2. Land Use database "Bestand BodemGebruik (BBG)" (CBS 2002)
3. National Land Use database "Landelijk Grondgebruiksbestand Nederland (LGN)" (De Wit 2003; Hazeu 2006; Hazeu et al. 2011)
4. CORINE Land Cover (Büttner et al. 2004; Feranec et al. 2007a; Hazeu et al. 2011)

All databases are covering the entire Netherlands, but they have different properties and are produced according to different specifications (updating cycle, thematic detail, data format etc.) (see Fig. 18.1 and section Data sources used). The LGN database will be dealt in more detail as it combines information from BBG and TOP10vector/Top10NL with other additional information, it has a clear land cover focus and is a well-established Dutch database that is regularly updated.

G.W. Hazeu (✉)
Alterra, Wageningen University and Research Centre, Droevendaalsesteeg 3,
6708 PB Wageningen, The Netherlands
e-mail: gerard.hazeu@wur.nl

I. Manakos and M. Braun (eds.), *Land Use and Land Cover Mapping in Europe: Practices & Trends*, Remote Sensing and Digital Image Processing 18,
DOI 10.1007/978-94-007-7969-3_18, © Springer Science+Business Media Dordrecht 2014

Fig. 18.1 An area in the Netherlands represented by four "land cover" databases showing the differences in thematic detail, format and appearance (Top10NLversion2011 (LL), BBG2008 (UL), LGN6 (UR) and CLC2006 (LR))

The first objective of this chapter is to present the Dutch LGN database. After a short historical background, the data sources used for the production of the last version of LGN (LGN6), the methodology, and the way how land cover changes are monitored are presented. A second objective is the comparison of the LGN6 database with the European CLC2006 database for the Netherlands. Harmonization of both databases is hampered by differences in the level of spatial detail, temporal update frequency and semantics.

18.2 LGN6 and Its History

The Dutch land use database (LGN6) is a grid database of 25*25 m^2 reflecting the land cover for 39 land cover classes including urban areas, forest types, water, crop types and several ecological classes (Hazeu et al. 2010, 2011). Today there are six versions of the LGN database (LGN1..LGN6) with the following reference years 1986, 1992/1994, 1995/1997, 1999/2000, 2003/2004 and 2007/2008. This range of

Table 18.1 Different LGN databases compared on thematic detail, thematic accuracy, images used, integration and monitoring of land cover changes

	LGN1 (1986)	LGN2 (1992)	LGN3 (1997)	LGN4 (2000)	LGN5 (2004)	LGN6 (2008)
Number of classes	17	45	39[a]	39	39	39
Accuracy (%)	67	70–80[b]	85	90[c]	81[c]	85[c]
Number of satellite images	2[d]	>10	17	16[e]	19[e]	19
Number of time steps	1	2–3	3–4	3	2–4	2–4
Integration with other GIS	No	No	Yes	Yes	Yes	Yes
Crop classification base on Top10-vector geometry	No	No	No	Yes	Yes	Yes
Monitoring land cover changes	No	No	No	Yes	Yes	Yes

[a]LGN3: LGN3 consists of 25 classes, LGN3plus consists of 39 classes
[b]LGN2: accuracy with mixed classes
[c]Only crop database
[d]LGN1: exclusive satellite images for completeness
[e]LGN4 and LGN5: without ERS-SAR mosaic

more than 20 years of land cover information creates a unique spatial information source to monitor developments of land cover over time in The Netherlands.

Besides information derived from satellite imagery (mainly Landsat 5/7 and IRS-P6), additional data sources like the Dutch topographical map Top10vector (Kadaster 2003, 2007a, b), "BestandBodemGebruik (BBG)" (CBS 2002), aerial photographs, natural areas database (BKN) (Kramer et al. 2007; Hazeu and Schuiling 2009) and land use information from the Office of Statistics (CBS) were used in the production of LGN6.

During its history the database evolved from an experimental pilot project (LGN1 – 1986) into a widely used and established database (LGN6 – 2007/2008). The first two versions of LGN were produced with a limited nomenclature, mixed classes and a limited accuracy. In LGN3 the nomenclature of 39 classes was consolidated, its thematic accuracy improved and with the introduction of nature classes it strongly became even more useful for environmental analysis. Important improvements in LGN4 were the connection with the Dutch topographical database (Top10vector) which made the classification of the agricultural crops easier and more consistent. The nomenclature (39 classes) and spatial resolution of 25*25 m^2 did not change while monitoring of land cover changes was introduced. LGN4 was used as input database for the production of LGN5. The production methodology for LGN5 and LGN4 are comparable. Emerging user requirements (exchange between databases, interoperability, INSPIRE developments) and increased data availability made it inevitable to adapt the methodology for LGN6 (Hazeu et al. 2011).

A comparison of the different LGN databases shows an increase in thematic detail, in thematic accuracy and in the number of satellite images used for classification (in later versions mainly for the classification of agricultural crops) (see Table 18.1). Figure 18.2 shows the differences in appearance of the different databases. The LGN1 database looks much more patchy then later versions. Also the thematic detail increased from LGN1 onwards.

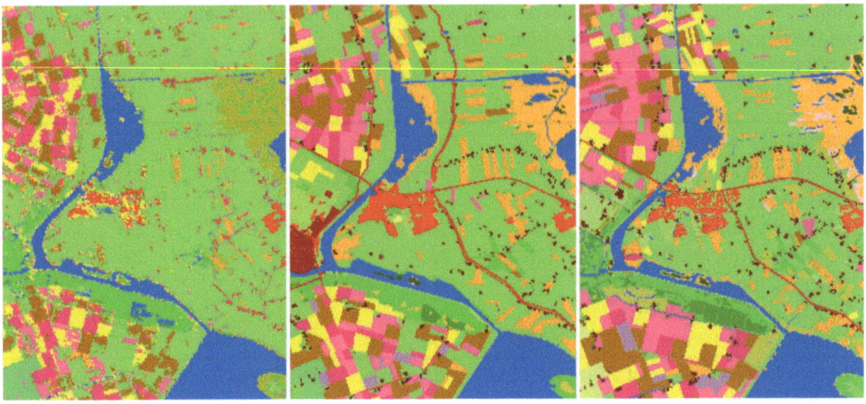

Fig. 18.2 Visual comparison of the same area of land between different LGN versions (from *left* to *right*: LGN1, LGN3plus and LGN6)

From LGN4 onwards a methodology was designed to monitor land cover changes in a consistent and accurate way (De Wit 2003). The LGN nomenclature was aggregated into eight monitoring classes, namely *Urban, Orchards, Greenhouses, Agricultural land, Water, Infrastructure, Forest and Nature* (see Table 18.2). In case of LGN4 and LGN5, the previous LGN3respectively LGN4 version was taken as a basis for the production of the next version to avoid inaccuracies and pseudo-changes due to shifts in geometry. Real changes between the aggregated classes were visually interpreted and manually digitised by comparing multi-temporal images from different acquisition time windows. Improvements or changes that fixed previous misinterpretations were handled as technical changes separately from real changes (De Wit 2003; Hazeu 2006). The methodology to monitor land cover changes gradually changed with the production of LGN6 which will be explained in the methodology section of this chapter.

18.3 Data Sources for LGN6

The following databases were used for the production of LGN6 (see also (Hazeu et al. 2011)):

- Remote sensing: satellite imagery (Landsat5 TM: April 16th, May 2nd, August 6th and September 23rd 2007; May 11th, July 30th and August 31st 2008, IRS-P6: June 11th 2007; May 7th, June 24th and August 30th 2008) and aerial photographs of the year 2006 (spatial resolution of 0.5 m and were acquired during the period April-June 2006).
- Dutch topographical map (Top10vector, version 2006): topographical map sheets at scale 1:10,000 with reference years from 2002 to 2006. Since 1998,

Table 18.2 LGN monitoring classes with corresponding LGN classes (Hazeu et al. 2010)

Agriculture	Grassland, maize, potato, sugar beet, wheat, other crops and flower bulbs, tree nurseries, buildings in agricultural areas
Greenhouses	Greenhouses
Orchards	Orchards, fruit nurseries
Forest	Deciduous forest, coniferous forest
Water	Fresh water, salt water
Infrastructure	Main roads and railways
Urban	Primary and secondary built-up, forest in primary and secondary built-up, grassland in primary and secondary built-up, bare soil in built-up
Nature	Salt marshes, coastal sands, dune areas with low or high vegetation, heath land in coastal areas, drifting sands/river sandbanks, heath land, grassy or very grassy heath land, raised bogs, forest in raised bogs, reeds, forest in swamp areas, other swamp areas, natural grasslands

the entire Netherlands is covered by 1,350 map sheets covering an area of 5 by 6.25 km. The map sheets are updated every 2–4 years which will be increased in the near future. The Top10vector version2006 contains map sheets from 2002 till 2006 and is the last available version (Kadaster 2003, 2007a, b). From then onwards, improvements made the topographical database available as object oriented database (Top10NL). The database is uniform, consistent and covers the complete Netherlands. The database consists of different topographical elements in GML format (e.g. terrain, water, roads, etc.). This database will be used in LGN7 and other future updates of LGN (see Fig. 18.1). Field observations in combination with aerial photography are the basis for the updates.

- Urban areas of 2003: database produced by Ministry of Spatial Planning and Environment (VROM) and based on land use database 2003 (BBG2003) with delineations of urban areas (VROM 2007). Only one version available.
- Land use database of 2003: land use database produced by the Office of Statistics (CBS) with Top10vector database (version 2003) as geometrical basis. The 2003 version (BBG2003) recognises 38 different land uses with special focus on urban (CBS 2002). Figure 18.1 shows a spatial representation of the database. Recently, the 2008 version was published. Other versions are the ones indicating the land use with reference years 1989, 1993, 1996, 2000, 2003, 2006 and 2008 showing an increase in update frequency. The social/economic use of land is the main criteria to classify land into different categories. The updates of the Land Use database are based on the most recent topographical database (Top10vector/Top10NL) in combination with aerial photography and statistical information.
- Natural areas database of 2007: a grid database having Top10vector as basis. The database focuses on the nature areas of the Netherlands and especially coastal and inland dune areas and natural grassland areas (Kramer et al. 2007; Hazeu et al. 2009; Hazeu and Schuiling 2009).
- Previous LGN version (LGN5) (Hazeu 2005, 2006).

18.4 Methodology

Increased data availability and changing user requirements (exchange between databases, interoperability, INSPIRE developments) made it necessary to adapt the methodology to produce LGN6. During the development of the LGN6 database the following pre-conditions were taken into account:

- Dutch topographical database as geometrical basis (instead of former LGN5 database) to harmonize between topographical data and LGN6
- integration with other national databases to avoid redundant data production
- backward compatibility with previous LGN versions

The new approach consisted of two main workflow lines:

1. object-oriented classification: an assignment of LGN classes to Top10vector objects followed by
2. pixel-based classification: a rasterisation process (Hazeu et al. 2010, 2011)

Figure 18.3 respectively Fig. 18.4 are flow diagrams representing these different steps in the production process.

The object based classification started with the aggregation of Top10vector classes into a limited number of LGN classes. Secondly, urban areas from BBG2003/BG2003 were integrated with the Top10vector database (version2006) to define the urban extent. The integration was based on decision rules taking into account Top10vector, the BBG2003/BG2003 and surface area information. Thirdly, Top10vector objects were aggregated into eight LGN monitoring classes. Land cover changes were attributed by comparing 2003/2004 and 2007/2008 imagery with the Top10_upd database as result. The marked land cover changes (Top10_chg) finally appear as a binary mask in the LGN6chg database (Fig. 18.4). The updating methodology is comparable with the update of previous LGN databases and described in detail in De Wit (2003).

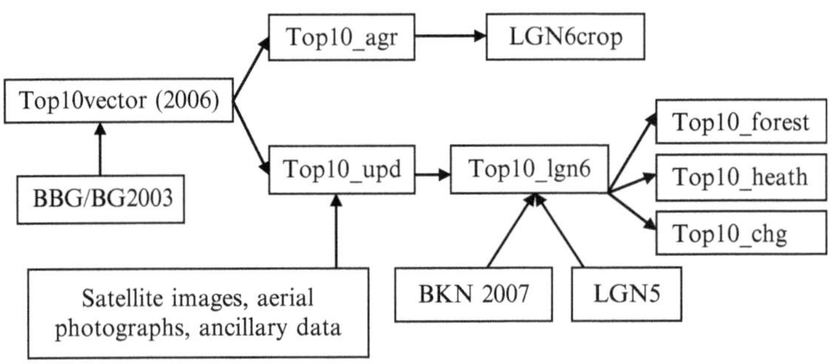

Fig. 18.3 Data flow diagram for the object-oriented LGN6 production phase. The topographical database (Top10vector) is the geometrical basis for updating (Top10_upd) and crop classification (Top10_agr). It is integrated with urban information (BBG/BG2003), land cover changes based on comparison of remote sensing images from different time steps and information on nature areas (BKN2007 and LGN5)

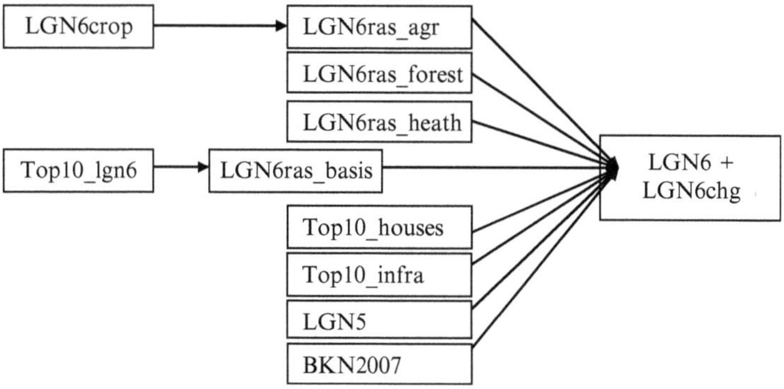

Fig. 18.4 The pixel-based classification as second step in the LGN6 production process. The LGN6 grid cells (LGN6ras_basis) were enriched with additional information from GIS databases (Top10_houses and infrastructure, BKN2007 and LGN5) and remote sensing classifications (forest, heath and crops) resulting in the final database (LGN6 and LGN6chg)

Table 18.3 Databases and remote sensing imagery used in the final thematic assignment of the LGN6 classes (Hazeu et al. 2011)

Top10vector[a]	Greenhouses, orchards, tree and fruit nurseries, main roads and railways
Top10vector buildings	Primary and secondary built-up areas; buildings in agricultural areas
Land use and urban areas of 2003	Urban classes
Natural areas of 2007	Sand and dune classes, swamp classes, natural grasslands
LGN5	Fresh and salt water, salt marshes, raised bogs classes, swamp classes
Satellite imagery[b]	Grassland, maize, potato, sugar beet, wheat, other crops and flower bulbs; forest classes; heath land classes
Aerial photographs[b]	Dune classes

[a]Top10vector geometry and thematic content plays a role for each LGN6 class at an earlier stage
[b]Both remote sensing sources were used to detect land cover changes

The following methodological step was to attach further attributive information to the Top10vector objects on basis of additional databases. LGN5 and Natural Areas database (BKN2007) in combination with Top10vector made it possible to attribute salt marshes, heath land in the coastal areas and raised bog to the Top10vector objects resulting in Top10_lgn6 database (Table 18.3). Finally, forest and heath land areas were selected. In another production step, these areas were classified on basis of satellite imagery into more thematic detailed classes (forest type, cover of grass in heath land areas).

In a separate subsequent step, the agricultural areas were classified. The classification started with the selection of the agricultural parcels of the Top10vector database (pastures and agricultural land) (Top10_agr). Agricultural parcels with more than one agricultural crop were manually subdivided on basis of satellite imagery. A multi-temporal classification of satellite imagery based on NDVI was carried out. The differences in phenology for seven agricultural crops were used to classify them (grassland, maize, potato, sugar beet, cereals, other crops and flower bulbs). A non-supervised classification with manual correction was applied. Classification results (area statistics) were compared with CBS agricultural statistics. Finally, each crop parcel was attributed to the main crop type (LGN6crop database) (De Wit and Clevers 2004; Hazeu 2006).

The second main workflow line (pixel-based classification) started with the conversion of Top10_lgn6 objects into 25*25 m grid cells (lgn6ras_basis) (Fig. 18.4). The location of the grid cells corresponds with the previous LGN versions. Crop information from the LGN6 crop database (LGN6crop) was attributed to the agricultural areas of LGN6 grid database (LGN6ras_basis). Also forest and heath land unsupervised classification results were attributed. The extent and type of different grassland, dune and swamp areas were defined by combining information from topographical data (Top10vector), natural areas (BKN2007) and LGN5 database information. Next, the buildings (Top10vector buildings) were buffered with 10 m, rasterised and attributed to the 25*25 m^2 grid cells. For a selection of main roads and railways rasterisation and attribution was also applied after buffering them with 5–15 m depending on their type. All information was added to the LGN6ras_basis database with the LGN6 database as final result (Fig. 18.4). The LGN6crop and the LGN6chg were validated (see for details (Hazeu et al. 2010, 2011)).

18.5 Results and Discussion

18.5.1 LGN6 and Land Cover Changes

In LGN6 (2007/2008) agricultural pastures are the main land cover in the Netherlands. Together with other agricultural crops they occupy 63.3 % of the Dutch land territory. Urban areas, forests and nature areas occupy 14.2 %, 10.6 % respectively 7.2 % of the land surface (Hazeu et al. 2010, 2011).

Land cover changes are monitored for eight aggregated monitoring classes (see Table 18.2) between LGN5 and LGN6. A total of 259.1 km^2 which is 0.62 % of the land surface has changed between 2003 and 2008. As reported in (Hazeu et al. 2011) almost 50 % of all land cover changes is from agricultural land into urban area. The surface area of real land cover changes for the monitored classes is far below the difference between LGN5 and LGN6 (see Table 18.4). In other words, the real land cover changes are outnumbered by methodological changes (Hazeu et al. 2011). Two significant reasons for this discrepancy are the difference in production methodology between LGN5 and LGN6 together with the fact of not using the former LGN5 as basis for the production of LGN6. In Hazeu et al. (2011) a solution is presented to facilitate modelling and assessment studies as well as statistical comparison of databases produced with different methodologies.

Table 18.4 Areas (km^2) for 8 monitoring classes for LGN5 and LGN6 and the land cover changes (Adapted from Hazeu et al. 2011)

Monitoring classes	LGN5	LGN6	LGN6-LGN5	Real changes	Decrease LGN5	Increase LGN6
Agricultural land	22233.4	21686.6	−546.8	−184.1	193.5	9.4
Greenhouses	152.5	144.0	−8.5	7.1	8.8	15.8
Orchards	291.7	255.3	−36.4	−4.7	5.1	0.4
Forest	3156.8	3544.1	387.3	−8.3	14.0	5.8
Water	7771.4	7984.5	213.0	14.3	2.3	16.6
Urban	5082.9	4768.5	−314.4	136.0	25.6	161.6
Infrastructure	1017.6	734.9	−282.7	4.7	6.2	10.9
Nature	1820.8	2409.3	588.5	35.1	3.6	38.7
Total	41527.1	41527.1	0.0	0.0	259.1	259.1

18.5.2 LGN6 and Corine Land Cover

The CORINE Land Cover (CLC) database is extensively described in Chap. 5 and also appears in various peer-reviewed articles (Büttner et al. 2004; Feranec et al. 2007a, b). CORINE Land Cover is a European land cover vector database characterised by 44 land cover classes (3 level hierarchical nomenclature), minimum mapping units of 25 ha with a width of >100 m and addresses the scale 1:100,000. Three version (1990, 2000, 2006) are available with layers indicating land cover changes of >5 ha. In the near future a new version (CLC2012) will become available. In the Netherlands only 30 out of the 44 CLC class are present (Table 18.5).

The Dutch CLC activities are independent from the production of the national LGN database. One of the reasons is that the long term vision for updating of CLC is not yet stabilised to synchronise it with the national land cover mapping. Also the history of both databases, spatial detail and semantics are different, which makes harmonisation difficult as discussed in e.g. Jansen et al. (2008).

The comparison of the LGN legend with the CLC nomenclature shows that the LGN database has less detail in the urban areas, more detail in agricultural and nature areas. In general, however, LGN has more thematic (and spatial) detail (39 classes) compared with the 30 CLC classes present in the Netherlands. Table 18.6 gives an overview of the relations between LGN and CLC based on their semantics. Mapping of LGN to CLC shows that there are 1:1, 1:m and m:1 relations at CLC level 3. Some examples:

- 1:1 relation: salt marshes (LGN code 30) is 1:1 compatible with *salt marshes* (CLC code 4.2.1),
- 1:m relation: grass in built-up areas (LGN code 23) can be mapped by *construction sites, green urban areas* and *sport and leisure areas* (CLC codes 1.3.3, 1.4.1 respectively 1.4.2) and
- m:1 relation: maize (LGN code 2), sugar beet (LGN code 4) and cereals (LGN code 5) can be merged into the *arable land* (CLC code 2.1.1) or several LGN heathland classes are compatible with *moors and heath lands* (CLC code 3.2.2).

Table 18.5 CLC land cover classes in the Netherlands (Hazeu 2003)

Level 1	Level 2		Code	Level 3 CORINE land cover class
Artificial surfaces	1.1	Urban fabric	1.1.1	Continuous urban fabric
			1.1.2	*Discontinuous urban fabric*
	1.2	Industrial, commercial and	*1.2.1*	*Industrial and commercial units*
		Transport units	*1.2.2*	*Road and rail networks and associated land*
			1.2.3	*Port areas*
			1.2.4	*Airports*
	1.3	Mine, dump and	*1.3.1*	*Mineral extraction sites*
		Construction sites	*1.3.2*	*Dump sites*
			1.3.3	*construction sites*
	1.4	Artificial non-agricultural	*1.4.1*	*Green urban areas*
		Vegetated areas	*1.4.2*	*Sport and leisure facilities*
Agricultural areas	2.1	Arable land	*2.1.1*	*Non-irrigated arable land*
			2.1.2	Permanently irrigated land
			2.1.3	Rice fields
	2.2	Permanent crops	2.2.1	Vineyards
			2.2.2	*Fruit trees and berry plantation*
			2.2.3	Olive groves
	2.3	Pastures	*2.3.1*	*Pastures*
	2.4	Heterogeneous agricultural areas	2.4.1	Annual cops associated with permanent crops
		Agricultural areas	*2.4.2*	*Complex cultivation patterns*
			2.4.3	*Land principally occupied by agriculture with significant natural vegetation*
			2.4.4	Agro-forestry areas
Forests and semi-natural	3.1	Forest	*3.1.1*	*Broad-leaved forest*
Areas			*3.1.2*	*Coniferous forest*
			3.1.3	*Mixed forest*
	3.2	Shrub and/or herbaceous	*3.2.1*	*Natural grasslands*
		Vegetation associations	*3.2.2*	*Moors and heath lands*
			3.2.3	Sclerophyllous vegetation
			3.2.4	*Transitional woodland-scrub*
	3.3	Open spaces with little or no	*3.3.1*	*Beaches, sand, dunes*
		Vegetation	3.3.2	Bare rocks
			3.3.3	Sparsely vegetated areas
			3.3.4	Burnt areas
			3.3.5	Glaciers and perpetual snow

(continued)

Table 18.5 (continued)

Level 1	Level 2		Code	Level 3 CORINE land cover class
Wetlands	4.1	Inland wetlands	*4.1.1*	*Inland marshes*
			4.1.2	*Peat bogs*
	4.2	Coastal wetlands	*4.2.1*	*Salt marshes*
			4.2.2	Salines
			4.2.3	*Intertidal flats*
Water bodies	5.1	Inland waters	*5.1.1*	*Water courses*
			5.1.2	*Water bodies*
	5.2	Marine waters	5.2.1	Coastal lagoons
			5.2.2	*Estuaries*
			5.2.3	*Sea and ocean*

CLC classes in italics are present in the Netherlands

Table 18.6 LGN6 legend and corresponding CORINE land cover classes

Code	Class	CLC class[a]	Code	Class	CLC class[a]
1	Pasture	231	26	Built-up areas outside urban areas	112
2	Maize	211	28	Grass in semi built-up areas[b]	124, 131, 132, 133, 141, 142
3	Potatoes	211	30	Salt marshes	421
4	Sugar beet	211	31	Coastal sands	331
5	Cereals	211	32	Dune areas with low vegetation[c]	321
6	Other agricultural crops	211	33	Dune areas with high vegetation[c]	321
8	Greenhouses	211	34	Heathland in coastal areas	322
9	Orchards	222	35	Drifting sands/river sandbanks	331
10	Flower bulbs	211	36	Heathland	322
11	Deciduous forest	311	37	Grassy heathland	322
12	Coniferous forest	312	38	Very grassy heathland	322
16	Fresh water	511	39	Raised bogs	412
17	Salt water	523, 522	40	Forest in raised bogs	311
18	Urban built-up areas	112, 121, 123, 124	41	Other swamp vegetation	411
19	Semi urban built-up areas	112, 142, 121, 123, 124	42	Reeds	411
20	Forest in built-up areas[c]	141	43	Forest in swamp areas	311
22	Forest in semi built-up areas[c]	141, 142	45	Natural grasslands	321
23	Grass in built-up areas	133, 141, 142	61	Tree nurseries[b]	211
24	Bare soil in built-up areas	133	62	Fruit cultivation[b]	222
25	Main roads & railways	122			

[a]CLC code is explained in Table 18.5 and Chap. 21, LGN-CLC matching based on semantics
[b]New classes in LGN6
[c]Thematic definition of class strongly differs between LGN5-LGN6

Mixed classes like *complex cultivation patterns* (2.4.2), *land principally occupied by agriculture with significant natural vegetation* (2.4.3) and *mixed forest* (3.1.3) are not used in the national LGN database. Also the classes *transitional woodland-shrub* (3.2.4) and *intertidal flats* (4.2.3) do not have an equivalent in the LGN database.

Both CLC and LGN databases have a different format and minimum mapping unit. LGN is a grid database with regular grids of 25*25 m^2; CLC is a vector database with a minimum mapping unit of 25 ha. As a consequence the application scale of CLC and LGN is different. LGN addresses the scale 1:50,000 while CLC addresses the scale 1:100.000. A consequence of the different spatial scales is that a geometrical overlay of LGN and CLC shows a wide variety of LGN classes present in particular CLC polygons.

The temporal synchronisation between national LGN and European CLC releases is largely by accident. The CLC1990 and 2000 database are partly based on satellite imagery with the same acquisition date as LGN1 and LGN4, i.e. the years 1986 and 2000, respectively. CLC2006 has no temporal national equivalent.

Having a single topographical database as a common source of LC/LU information would avoid double work and synchronise the national and European land cover activities. Good examples are United Kingdom with LCM and Germany with DLM-DE (Smith et al. 2007; Arnold 2009). The CLC database would be a spatial and thematic less detailed version of the national land cover database. However, the comparison of these newly defined databases with old versions of LGN or CLC will be hampered by differences in statistics and spatial allocation of land cover. This can be (partly) avoided by mapping separately the real changes and creating a new database with the new geometry in which you introduce for the real changes the land cover of the previous (t-1) database. Another possibility is to create a new database with the old geometry and introduce for the real changed areas the land cover from the newly (t) created database (Hazeu et al. 2011).

18.6 Conclusions

The Netherlands has a long history in land cover/land use mapping. Several country wide databases exist with different history, formats, update frequencies, semantics and spatial detail. The most detailed one in the fields of environment and agriculture is the LGN database. In already more than 20 years a unique series of land cover snapshots has been built up making it possible to have a spatial and statistical overview of developments in the landscape. Meanwhile a lot of experience has been acquired with the mapping of land cover and its changes.

The production of LGN is independent from the Dutch CORINE Land Cover (CLC) activities. One of the reasons is that a long term vision for updating of CLC is not yet stabilised. Synchronisation with the national land cover mapping is therefore not yet possible. Also the history, the spatial and thematic details of both databases are different, which makes harmonisation difficult.

Thematic classes are often not 1:1 exchangeable between databases, which makes the generation of CLC out of LGN (bottom-up approach) difficult without additional information. Ideally a new start for both databases making use of the same topographical baseline would avoid double work. However, backward compatibility needs to be guaranteed.

The temporal synchronisation of national and European land cover activities needs a long term vision. The reference years of the national and European land cover databases have to coincide as much as possible. Ideally, all time snapshots of the less frequent updated database coincide with snapshots of the more frequently updated land cover database.

The integration of national land cover mapping activities (like LGN) with the European CLC activities is a challenge for the future. Spatial, temporal and thematic (semantics) details need to be fine-tuned between MS and EU activities keeping in mind that the time series will be maintained, i.e. the comparison between versions (with different reference years) still will be possible. Several activities (e.g. FP7-HELM project http://www.fp7helm.eu/ and EAGLE working group http://sia.eionet.europa.eu/EAGLE) are running to achieve this goal. The EAGLE concept is a first step for a successful, more efficient and effective land monitoring programme.

Acknowledgements The LGN6 database is produced with help of several colleagues. Especially, I want to acknowledge Rini Schuiling, Judith Oldengarm and Gert van Dorland for their support. Also, CLC2006 was produced in collaboration with colleagues from the Alterra institute. Also special thanks to Chantal Melser for her input regarding the national land use database (BBG).

References

Arnold S (2009) Digital landscape model DLM-DE – deriving land cover information by integration of topographic reference data with remote sensing data. In: Proceedings of the ISPRS workshop on high-resolution earth imaging for geospatial information, Hannover, Germany

Büttner G, Feranec J et al (2004) The CORINE land cover 2000 project. EARSeL eProceedings 3(3):331–346

CBS (2002) Productomschrijving Bestand Bodemgebruik. Centraal Bureau voor de Statistiek (CBS), Den Haag, The Netherlands

Chen Z, Wang J (2010) Land use and land cover change detection using satellite remote sensing techniques in the mountainous Three Gorges Area, China. Int J Remote Sens 31(6):1519–1542

De Wit AJW (2003) Land use mapping and monitoring in the Netherlands using remote sensing data. In: International geoscience and remote sensing symposium (IGARSS), Toulouse

De Wit AJW, Clevers JGPW (2004) Efficiency and accuracy of per-field classification for operational crop mapping. Int J Remote Sens 25(20):4091–4112

Feranec J, Hazeu G et al (2007a) Corine land cover change detection in Europe (case studies of the Netherlands and Slovakia). Land Use Policy 24(1):234–247

Feranec J, Hazeu G et al (2007b) Cartographic aspects of land cover change detection (over- and underestimation in the I&CORINE land cover 2000 project). Cartogr J 44(1):44–54

Hazeu GW (2003) CLC2000 land cover database of the Netherlands: monitoring land cover changes between 1986 and 2000, Alterra-rapport 775. Alterra-WUR, Wageningen, 108 pp

Hazeu GW (2005) Landelijk Grondgebruiksbestand Nederland (LGN5): vervaardiging, nauwkeurigheid en gebruik, Alterra-report 1213. Alterra, Wageningen, 92 pp

Hazeu GW (2006) Land use mapping and monitoring in the Netherlands (LGN5). In: Proceedings of the second workshop of the EARSeL SIG on remote sensing of land use & land cover: application & development, EARSeL, Bonn (Germany), 28–30 Sept 2006

Hazeu GW, Schuiling C (2009) New Dutch approach for land use mapping (LGN6). In: Proceedings of the third EARSeL workshop on land use and land cover, EARSeL, Bonn, Germany, 25–27 Nov 2009

Hazeu GW, Oldengarm J, Clement J, Kramer H, Sanders ME, Schmidt AM, Woltjer I (2009) Verfijning van de Basiskaart Natuur. Segmentatie van luchtfoto's en het gebruik van het Actueel Hoogtebestand Nederland in duingebieden, WOt-rapport 102. Wettelijke Onderzoekstaken Natuur & Milieu, Wageningen

Hazeu GW, Schuiling C et al (2010) Landelijk Grondgebruiksbestand Nederland versie 6 (LGN6): vervaardiging, nauwkeurigheid en gebruik, Alterra-report 2012. Alterra, Wageningen, 132 pp

Hazeu GW, Bregt AK et al (2011) A Dutch multi-date land use database: identification of real and methodological changes. Int J Appl Earth Obs Geoinf 13(4):682–689

Jansen L, Groom G et al (2008) Land-cover harmonisation and semantic similarity: some methodological issues. J Land Use Sci 3(2–3):131–160

Kadaster (2003) Kwaliteit Top10vector. Kadaster, Apeldoorn

Kadaster (2007a) Catalogus Basisregistratie Topografie. Kadaster, Apeldoorn

Kadaster (2007b) Productspecificaties Basisregistratie Topografie. Kadaster, Apeldoorn

Kaufmann RK, Seto KC (2001) Change detection, accuracy, and bias in a sequential analysis of Landsat imagery in the Pearl River Delta, China: econometric techniques. Agric Ecosyst Environ 85(1–3):95–105

Kramer H, Hazeu GW et al (2007) Basiskaart Natuur 2004: vervaardiging van een landsdekkend basisbestand terrestrische natuur in Nederland, WOT Werkdocument 40. Wettelijke Onderzoekstaken Natuur & Milieu, Wageningen, 94 pp

Lambin EF, Baulies X et al (1999) Land-use and land-cover change: implementation strategy. IGBP Secretariat, Stockholm

Smith G, Beare M et al (2007) UK land cover map production through the generalisation of OS MasterMap®. Cartogr J 44(3):276–283

VROM (2007) Begrenzing Bebouwd Gebied 2003. Ministry of Housing, Spatial Planning and Environment (VROM), Den Haag

Chapter 19
The Use of the Land-Cover Classification System in Eastern European Countries: Experiences, Lessons Learnt and the Way Forward

Louisa J.M. Jansen, Alexandru Badea, Pavel Milenov, Cristian Moise, Vassil Vassilev, Ljudmila Milenova, and Wim Devos

19.1 Introduction into Categorisation and the Land-Cover Classification System

The understanding of the interactions between land cover, defined as "*the observed (bio)physical cover on the Earth's surface*" (Di Gregorio and Jansen 2000), and land use, defined as "*the type of human activity taking place at or near the surface*" (Cihlar and Jansen 2001), in their spatial and temporal appearances is fundamental to comprehension of land-use and land-cover change. Land cover can represent an expression (indicator) of human activities and, as such, changes with changes in land use and management. Hence, land cover may form a reference base for applications including forest and rangeland monitoring, production of statistics

L.J.M. Jansen (✉)
Senior Advisor Agriculture, Environment and Land Administration, Arnhem, The Netherlands
e-mail: louisa.jansen@tin.it

A. Badea
Romanian Space Agency, 21-25 Mendeleev st., sector 1, 010362 Bucharest, Romania

Romanian Centre for Remote Sensing Applications in Agriculture (CRUTA), 35-37, Sos. Oltenitei, sector 4, 041293 Bucharest, Romania
e-mail: alexandru.badea@rosa.ro

P. Milenov • W. Devos
MARS Unit, Institute for Environment and Sustainability, European Commission Joint Research Centre, Via E. Fermi 2749, 21027 Ispra, VA, Italy
e-mail: pavel.milenov@jrc.ec.europa.eu; wim.devos@jrc.ec.europa.eu

C. Moise
Romanian Space Agency, 21-25 Mendeleev st., sector 1, 010362 Bucharest, Romania
e-mail: cristian.moise@rosa.ro

V. Vassilev • L. Milenova
Remote Sensing Application Center (ReSAC), 61 Tzar Assen Str., fl. 2, 1463 Sofia, Bulgaria
e-mail: resac@techno-link.com

I. Manakos and M. Braun (eds.), *Land Use and Land Cover Mapping in Europe: Practices & Trends*, Remote Sensing and Digital Image Processing 18,
DOI 10.1007/978-94-007-7969-3_19,

for planning and investment, biodiversity, climate change and desertification control (Di Gregorio and Jansen 1998).

In the 1990s, different groups worked on the development of universally applicable land-cover and land-use categorisations that would contribute to standardisation of the criteria used for description and consequently categorisation, in addition to harmonisation between existing datasets plus harmonisation of change (Jansen 2010). FAO and UNEP emphasized a parameterised approach, i.e. using a set of explicit independent criteria resulting in a flexible data model that can be used as a uniform basis for description, in addition to the use of (part of) these parameters for change detection and monitoring. Thus not class labels are vital but the applied explicit set of parameters (Jansen 2010).

Without categorisation, real landscape phenomena would remain merely a bewildering multiplicity and the precise and unambiguous communication of ideas and concepts concerning these phenomena would be impossible (Shapiro 1959). Categorisation, or classification, is defined as "*the ordering or arrangement of objects into groups or sets on the basis of relationships. These relationships can be based upon observable or inferred properties*" (Sokal 1974). Another, and even earlier, definition by Shapiro (1959) reads "*the sorting of a set of phenomena composed of generally-alike units into classes or kinds, each class or kind consisting of members having definable characteristics in common*" is also interesting but does not underline the importance of relationships. It is important to note that categorisation is an abstraction in the sense that it depicts a representation of the reality (Di Gregorio and Jansen 2000).

The set of diagnostic criteria for the parameterised categorisation approach followed in the Land-Cover Classification System (LCCS), developed in the period 1996–2000 at FAO together with UNEP and financed by the Italian Cooperation (Di Gregorio and Jansen 2000), is based upon examination of criteria commonly used in existing categorisations that identify and describe land cover in an impartial, measurable and quantitative manner (Jansen and Di Gregorio 2002). However, the definition of categorisation provided in FAO (2005) "*classification is an abstract representation of the situation in the field using well-defined diagnostic criteria: the classifiers*" confuses categorisation with an abstract representation of a categorisation example given in Kuechler and Zonneveld (1988) and completely overlooks the fact that categorisation is the basic cognitive *process* of arranging objects into classes or categories, as well as the *act* of distributing objects into classes or categories of the same type (Jansen 2010, p. 24–25).

The parameterised LCCS approach to categorisation aims at a logical and functional hierarchical arrangement of the parameters, thereby accommodating different levels of information, starting with broad-level classes that allow further systematic subdivision into more detailed subclasses. At each level the defined classes are mutually exclusive. Criteria used at one level of the categorisation are not to be repeated at other levels. The increase of detail in the description of a class is linked to the increase in the number of parameters used. In other words, the more parameters are added, the more detailed the class. The class boundary is then defined either by the different number of parameters, or by the presence of one or

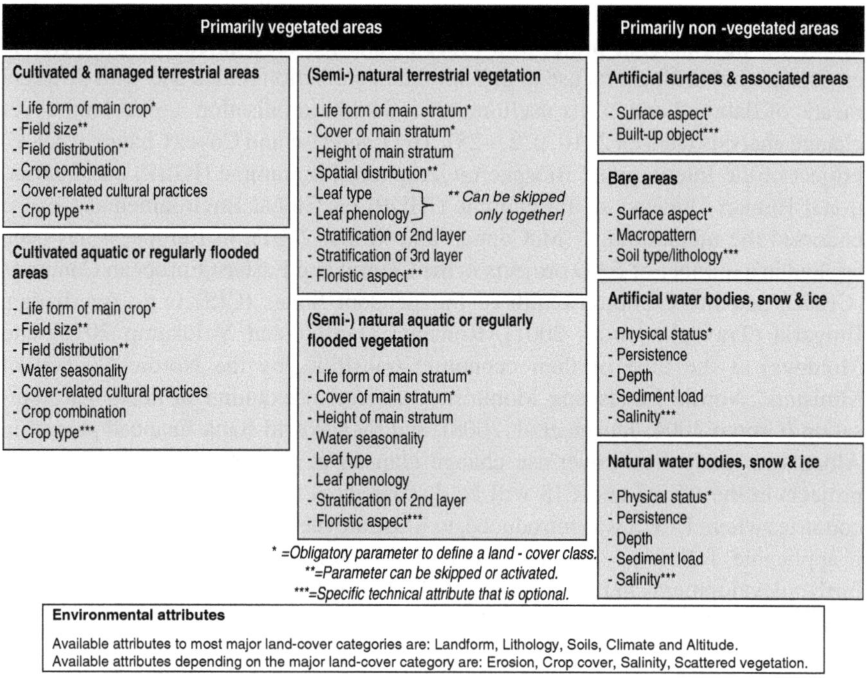

Fig. 19.1 The major land-cover categories of LCCS (version 2.0) grouped under the primarily vegetated and primarily non-vegetated area distinction (Jansen 2010, p. 26)

more different types of parameters. Emphasis is not given to the derived class name, the traditional method, but to the set of parameters used to define this land-cover class (Jansen and Di Gregorio 2002; Jansen 2010). The use of parameters and their hierarchical arrangement is a function of geographic accuracy. The arrangement of parameters will assure at the highest levels of categorisation, i.e. the most aggregated levels, a high degree of geographic accuracy. Since land cover deals with a heterogeneous set of classes, the parameters are tailored to each of the eight major land-cover categories (Fig. 19.1).

LCCS is an *a priori* categorisation system, implying that all combinations of parameters are accommodated in the system independent of scale, method and tools used to identify objects (e.g., human eye, statistics, aerial photographs or satellite remote sensing). Having all pre-defined classes included in the system is the intrinsic rigidity of this type of categorisation. However, it was considered at the time of development the most effective way to produce standardisation of categorisation results between user-communities. In order to assist the latter a dedicated software application was developed (Jansen and Di Gregorio 2002). The use of this software application for harmonisation of existing datasets in a number of Nordic countries has been critically examined by Jansen et al. (2008).

The LCCS categorisation methodology has been tested, modified and validated in several international projects in order to analyse its applicability in different environmental settings, its use at different data collection scales and with different means of data collection, its usefulness for data harmonisation and in land-cover change analysis (Jansen 2010, p. 27–28). The Land Use and Cover Change (LUCC) project of the International Biosphere-Geosphere Programme (IGBP) and International Human Dimensions Programme (IHDP) on Global Environmental Change endorsed the methodology (McConnell and Moran 2001). In Europe, it has been applied in a number of FAO projects in the Central and Eastern European Countries (CEEC) and the Commonwealth of Independent States (CIS) (e.g., Azerbaijan, Bulgaria (Travaglia et al. 2001), Romania (Jansen and Veldkamp 2012) and Moldova) at the time of their economic transition, by the Nordic Council of Ministers' Nordic Landscape Monitoring project to examine in detail harmonisation (Groom 2004; Jansen et al. 2008), and by a World Bank financed project in Albania to study land-cover/use change (Jansen et al. 2006). In this paper the projects in the CEEC and CIS will be discussed, as they were the first European countries where LCCS was introduced, to illustrate the experiences with LCCS and, if applicable, follow-up activities (Sect. 19.2), the lessons learnt (Sect. 19.3), and further developments in Europe (Sect. 19.4). This is followed by suggestions for the way forward at the methodological level (Sect. 19.5).

19.2 The Use of LCCS in the Different CEE and CIS Countries

FAO initiated Technical Cooperation Projects (TCP) in Azerbaijan, Bulgaria, Romania and Moldova shortly after the transition to a market-oriented economy started. Transition from a centrally planned to a market-oriented economy involved 'privatisation' of agricultural lands meaning the shifting of ownership of land from collectives and state to private persons. This land reform is either through restitution, distribution or compensation (Table 19.1). Land is at the centre of sustainable development; this recognition leads to promoting secure tenure rights and equitable access to land (FAO 2012).

Governments in the CEEC and CIS implemented a comprehensive package of social and economic reform policies. Spatial developments were rapid in the period of transition and such developments are related to the reform choices the governments made in these countries (Swinnen 1999; Kuemmerle et al. 2008). These land reforms did not only deal with the transfer of property rights and ownership, they dealt with the structures of the agrarian economy. In rural areas the relation to land has profound implications for agricultural productivity, environmental sustainability and the social and economic status of the rural households. Matching land use and land tenure with the aim to reach a better socio-economic structure therefore becomes crucial (Larsson 2002).

Table 19.1 The privatisation process (Deininger 2003)

Region and country	Potential private ownership	Privatisation strategy	Allocation strategy	Transferability
CEE				
Albania	All land	Distribution (phys.)	Plots	Buy and sell, leasing
Bulgaria	All land	Restitution	Plots	Buy and sell, leasing
Romania	All land	Restitution and distribution (phys.)	Plots	Buy and sell, leasing
CIS				
Azerbaijan	All land	Distribution	Plots (from shares)	Buy and sell, leasing
Moldova	All land	Distribution	Plots (from shares)	Buy and sell, leasing

As FAO had developed the LCCS, it was from the organisational viewpoint the logical approach to apply in these projects in countries that were at the time not European Union (EU) Member States[1] or accession countries, instead of the commonly used EU CORINE Land Cover (CLC). The latter serves as a tool for fulfilling pan-European monitoring needs. The experts worked in more than one FAO project thereby enhancing the cross-fertilisation in the methodologies used and experiences gained.[2]

Land-cover data are seen as a *baseline dataset* for the understanding of how our environment is changing. Thus, knowledge of the actual occurrence of land-cover types ("where" and "what") is essential in order to be able to monitor environmental changes ("why" and "how"). Furthermore, understanding land-*use* change processes of the past ("what was") may contribute to model their effects on land-*cover* patterns in future ("what may be"). Land cover and land use changed dramatically in the transition period, thus the demand for current datasets by the governments was huge to enable them to develop new policy and planning and to make informed decisions.

In the subsequent sections an overview is provided per country of the projects and further activities concerning land-cover/land-use mapping from an operational viewpoint. Especially the efforts to create the technical and human capacities in the countries are highlighted. The successful use of advanced technologies such as remote sensing and GIS depend on these capacities. The science and comparative advantages of the parameterised approach of LCCS are described in more detail

[1] On 1 January 2007 Bulgaria and Romania joined the EU.

[2] The authors were involved in LCCS and the projects described in various ways: Louisa J.M. Jansen is author of LCCS and she was involved in the projects in Albania and Romania; Alexandru Badea worked in the projects in Azerbaijan, Romania and Moldova; Pavel Milenov worked in the projects in Bulgaria and Romania and used LCCS at JRC; Cristian Moise was involved in the projects in Romania; Vassil Vassilev and Ljudmila Milenova were involved in the projects in Bulgaria; Wim Devos uses LCCS at JRC.

Table 19.2 Comparison of two types of data with certain land-cover classes in the Lenkaran District in Azerbaijan

Class	Data obtained from maps (1995) (in ha)	TCP/AZE data (2000) (in ha)
Tea garden	3,3753	8,953
Rice	1,787	927
Orchard	2,826	4,428
Herbaceous crops	956	2,155

elsewhere (Di Gregorio and Jansen 2000; Jansen and Di Gregorio 2002; Jansen et al. 2008). The analysis techniques used in the beginning in each country comprised visual onscreen interpretation of satellite imagery. However, especially in Bulgaria and Romania the approach evolved over time in automated procedures and object-oriented approaches.

19.2.1 Azerbaijan

Azerbaijan was the first country of the CIS where a new land-cover/land-use inventory was executed by the FAO Technical Cooperation Project "*Strengthening Capacity in Inventory of Land Cover/Land Use by Remote Sensing*" (TCP/AZE/8921) starting in July 1999. This new inventory was necessary to see how to balance new owners' rights with the necessity and capacity of the government to regulate land use and development for the best interest of society (Van der Molen and Jansen 2010; Williamson et al. 2010). This new inventory led to a unique georeferenced parameterised dataset using LCCS version 1.0 comprising 45 classes in five categories (e.g., Water, Vegetation, Agriculture, Built-up areas and Bare surfaces) to be used for agricultural planning, forestry and other sectors. In preparing the land-cover dataset at 1:50,000 scale based upon interpretation of LANDSAT satellite data acquired in 1998 and 1999, the project acquired a wealth of information on aspects such as the salt affected areas, destruction of villages and on-going mineral exploitation in the occupied territories. This specific information extracted from the digital dataset, or newly generated with the acquisition of additional satellite images, assisted in provision of a first layer of information in de-mining activities.

The usefulness of the generated digital land-cover dataset was shown by the potential applications, thereby answering requests of different organisations (Badea and Herisanu 2002). For example, following the invitation of the local government of the Lenkaran District, and taking into account the uncontrolled urbanisation (urban sprawl), the land distribution process and the lack of modern measuring tools, the project team realised by image interpretation an inventory of the classes 'Tea plantations', 'Orchards', 'Rice fields' and 'Herbaceous crops' that could be compared with the results obtained using classical methods by a local team of field inspectors. Though the year of data collection differs, significant differences in the perennial crops areas were found (Table 19.2). Thus, the project provided local government with up-to-date information on which to take informed decisions.

After 17 months the established project created a well-equipped remote sensing laboratory at the Azerbaijan National Aerospace Agency (ANASA) that allowed the staff to operate with ease the hardware and software, producing the countrywide land-cover dataset from satellite image interpretation. The local team was then able to offer support to important governmental initiatives in the domain of agriculture, environment and inter-governmental development projects such as the Azerbaijan-Russia transboundary transport corridor, an important part of the Silk Way global infrastructure.

19.2.2 Bulgaria

The FAO Technical Cooperation Project "*Strengthening Capacity in Agricultural Development Through Remote Sensing and GIS*" (TCP/BUL/8922) was executed jointly by FAO and Bulgarian experts from the Remote Sensing Application Centre (ReSAC) at the Bulgarian Aerospace Agency (BASA) from June 1999 onwards. Bulgaria was the first European country where land-cover mapping using LCCS version 1.0 was undertaken (Travaglia et al. 2001; see http://www.fao.org/DOCREP/004/Y0785E/y0785e00.htm).

Land is an important productive resource in Bulgaria. The land reform changed the structure of the agrarian economy. At the time of the implementation of the land reforms neither the size, nor form, nor location of land parcels were issues. As a result, the farm structure deteriorated in most cases. Agricultural, or rural, development was not the main aim of land reform.

This project aimed at the development and testing of remote sensing and GIS methodologies for inventory of land cover/land use in selected districts representative of the agricultural structure and production in the country, to set-up the strategy for the provision of accurate parameterised baseline information on land cover/use and associated land characteristics. This approach enabled the relevant authorities to prepare and implement sound agricultural development plans, including criteria for the land redistribution. A second objective was the strengthening of the remote sensing and GIS capacities at ReSAC, at the time located in BASA. This resulted in a well-structured ReSAC able to provide accurate and timely information to the relevant government agencies.

The project produced initially 14 digital land-cover maps at 1:50,000 scale for selected areas in the country covering an area of 5,600 km^2 (an example is shown in Fig. 19.2). These maps were prepared using LANDSAT satellite data, acquired in 1998 and 1999, as main data source and thus representing the land cover at that time. This resulted in the three selected areas in a class set comprising 49 classes. Different from the approach followed in Azerbaijan were two issues:

1. In Bulgaria each land-cover polygon also contained information on soil type and erosion features from other sources. The resulting comprehensive digital dataset was unique at that time in Bulgaria.

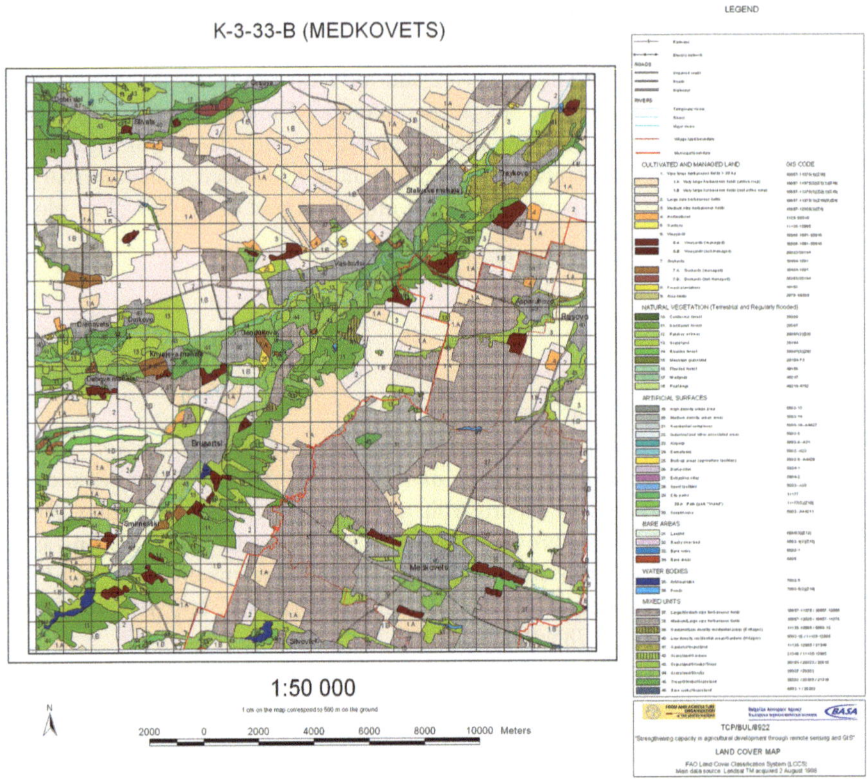

Fig. 19.2 Example of land-cover map at 1:50,000 scale in Bulgaria

2. The project also applied very high resolution IKONOS satellite images, acquired in August 2000, for the inventory of vineyards in the area of Sandanski in the southern part of the Strouma Valley.

The database provided useful information for agriculture, forestry, urban development planning, environmental protection and for many other applications. It was also used for computerised decision-support systems using application-specific models, for instance to calculate erosion hazard risks or changes in water runoff (using data from additional sources) associated with land-cover changes (Travaglia et al. 2001). The project outcome underlined the flexibility and care of the parameterised LCCS methodology, as opposed to land-cover categorisations using a predefined class set such as CLC, commonly used in the Europe Union (EU).

In the years following the FAO project, ReSAC expanded the land-cover mapping activities over the territory of the country (Nedkov et al. 2000; Milenov et al. 2004). More than half of the country was covered with digital land-cover data through various regional projects, such as:

- Biomass assessment in forested areas for the purpose of green energy;
- Impact assessment of the pan-European transport corridors in Bulgaria based on SPOT 5 (5 m resolution) satellite imagery;

- Flood risk analysis;
- Development of the strategy for the Land Parcel Identification System (LPIS) in the context of the European Union's Common Agriculture Policy (CAP) (see also Sect. 19.4);
- Urban planning using land-cover data at 1:5,000 scale using very high resolution satellite imagery; and
- Cross border cooperation (CBC) of ReSAC with Romania in the framework of the FAO and CBC projects.

These initiatives resulted in the decision to develop a national land-cover database, using LCCS as a reference to support the decision-making process at national and regional levels. A nation-wide land-cover dataset, compliant with 1:50,000 cartographic scale, was developed on the basis of LANDSAT 5 satellite images from 2009 with spatial resolution 30 m, using an object-oriented approach. Each land-cover polygon from the dataset represents an 'elementary unit' of the terrain having common (i.e. homogeneous) land cover, elevation and slope aspect characteristics. The information for the landform was taken from the Shuttle Radar Topography Mission (version 4; see http://www2.jpl.nasa.gov/srtm/), available at the Joint Research Centre (JRC) of the European Commission (EC). The reference land-cover dataset was entirely prepared by experts from the Agency for Sustainable Development and Euro-integration – Ecoregions (ASDE) and ReSAC supported by FAO experts. The resulting database was delivered to the Executive Agency Electronic Communication Networks and Information Systems (ECNIS) of the Ministry of Transport, Information Technology and Communications. It is currently available through a national geoportal. This reference land-cover dataset will be used also in accordance with the implementation of the Directive 2/2007/EC of the European Parliament and of the Council of 14 March 2007 establishing an Infrastructure for Spatial Information in the European Community (INSPIRE), the Directive 60/2007/EC of the European Parliament and of the Council of 23 October 2007 on the assessment and management of flood risks and other Directives and EU programmes, including the international project GLOBCOVER, in which the EU, USA, Russia, Canada and other countries participate. In future this land-cover database will be upgraded and enhanced in agreement with the requirements for the 1:25,000 cartographic scale to efficiently contribute to building-up the operational capacity to implement the INSPIRE Directive and the Global Monitoring for Environment and Security (GMES) programme.

In addition to the countrywide BULCOVER dataset (Fig. 19.3), land-cover mapping was executed for regional cities in Bulgaria on the basis of satellite images with very high resolution. These maps are at scale 1:5,000 or 1:10,000 using images with spatial resolution of 0.5–1 m (e.g., IKONOS, Quickbird, Eros and Worldview 1). The class set for these datasets is LCCS-based but adapted to the specific scale and user requirements (Fig. 19.4). Digital maps and databases are available from the beneficiary Ministry of Regional Development and Public Works at the Department Technical Rules and Regulations. These maps were developed in coordination with the Executive Agency ECNIS that created an information system for the harmonisation of spatial data and risk management (Milenova et al. 2003).

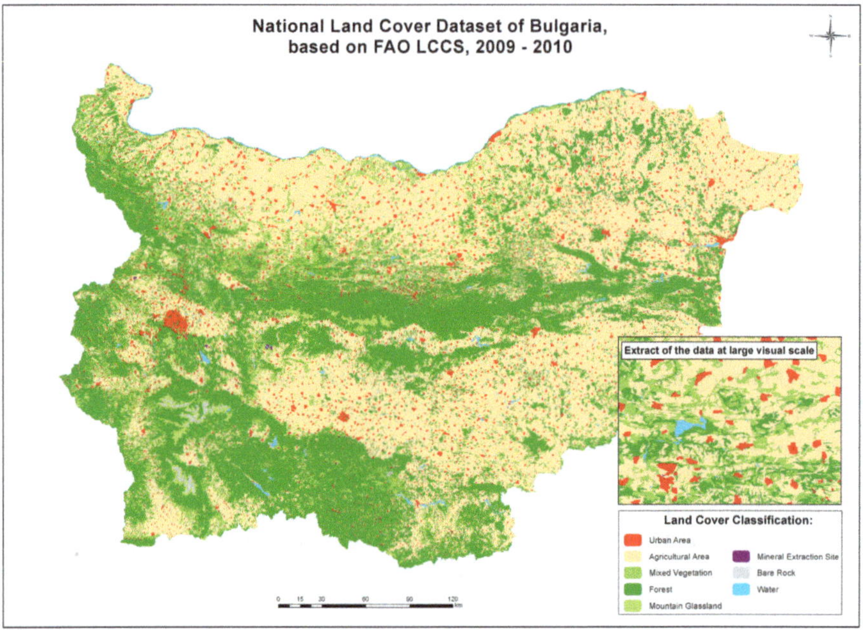

Fig. 19.3 BULCOVER national land-cover dataset: 2009–2010

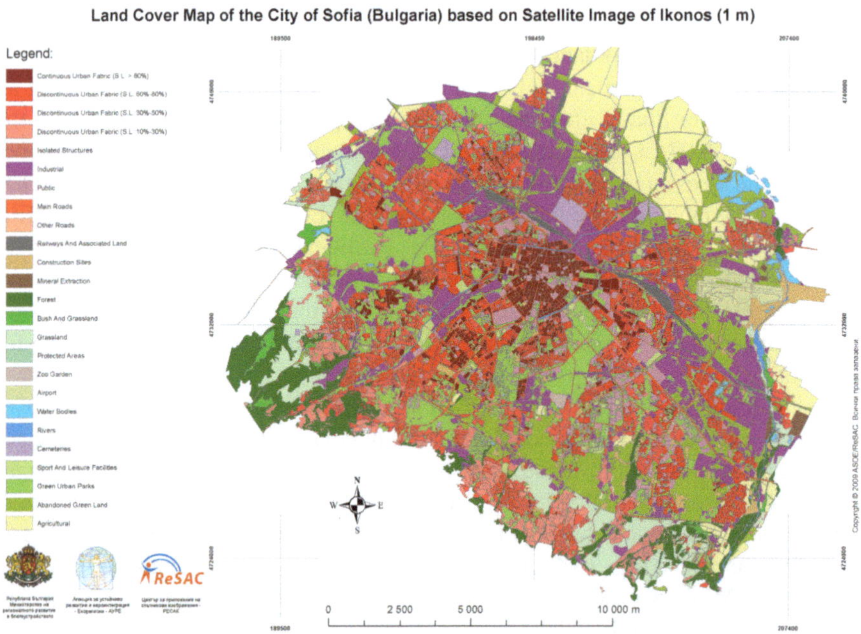

Fig. 19.4 Detailed land-cover map at scale 1:10,000 for 27 district centres based on the BULCOVER LCCS methodology

In 2012, further research on the application of the LCCS methodology was carried out by MSc students from the Geology and Geography Faculty of the Sofia University St Kliment Ohridski and ReSAC. This research concerned the application of the object-oriented image analysis for land cover and analysis of land-cover changes (period 2005–2010).

19.2.3 Romania

With the land reform in 1991, a new era started in Romania concerning land ownership and land use. By imposing a maximum for land restitution and distributing the remaining collective land to farm workers, Romania attempted to combine the strong demand for full property rights from those who formally owned the land (i.e. historical justice) with equity considerations (Swinnen 1999).

Following the successful implementation of the FAO projects in Azerbaijan and Bulgaria, an ambitious undertaking was started in 2002 for the entire territory of Romania in the project "*Land-Cover/Land-Use Inventory by Remote Sensing for the Agricultural Reform*" (FAO/TCP/ROM2801). The project aimed at (1) providing the Ministry of Agriculture with an objective and accurate countrywide database on land cover/land use including larger scale information for areas of particular agricultural interest; (2) strengthening the capacity of the Romanian Centre for Remote Sensing Applications in Agriculture (CRUTA) and other staff from the Ministry of Agriculture to apply internationally recognized methodologies on land-cover/use mapping by satellite remote sensing and GIS technologies; and (3) transferring the know-how and practical applications of high resolution remote sensing, GIS-based methodologies and LCCS to the staff assigned to the project through formal and on-the-job training (see http://www.fao.org/sd/2002/TCPROM2801/fao_tcp_home.htm).

The specific outputs of the activities undertaken during the 18 months lifespan of the project facilitated:

1. The production of a comprehensive digital land-cover dataset at 1:50,000 scale of the whole country and detailed land-cover/use datasets at 1:25,000 scale for few areas of particular interest, using LCCS version 1.0, through satellite image interpretation, integrated with field observations and/or additional information. These outputs became an essential input to the land redistribution and the improved agricultural planning process as these datasets provided detailed information on the various sizes of agricultural fields that occur in different parts of the landscape (Jansen and Veldkamp 2012). To each land-cover polygon additional information on soil types, erosion features (like in Bulgaria) and municipal and/or district boundaries were added.
2. Capacity building in remote sensing and GIS.
3. The transfer of modern technology on satellite image interpretation of land cover/use, based on the considerable FAO experience on this subject, and constituted a good basis for the sustainable capacity and the development of remote sensing applications in the country.

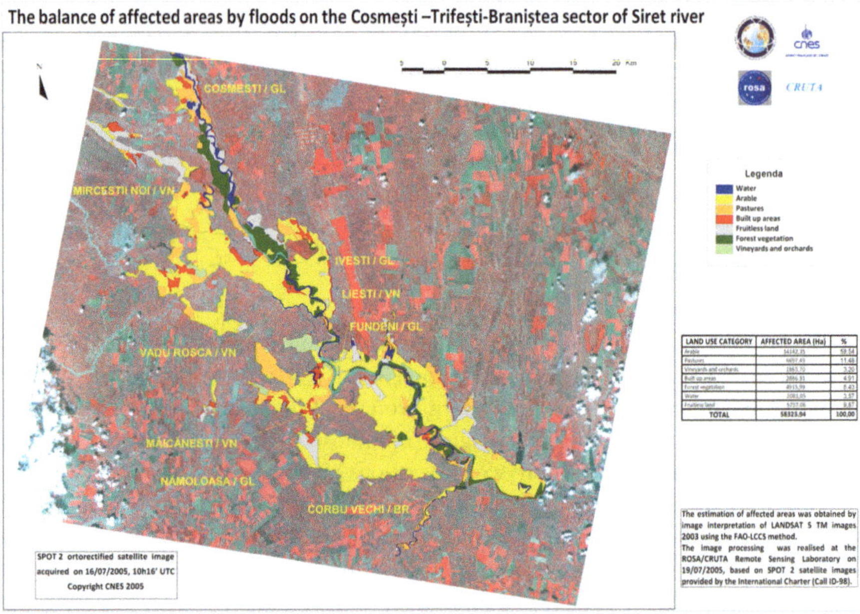

Fig. 19.5 The use of LCCS dataset of 2005 in Romania for generating a thematic map on flood management

Currently, the two LCCS databases (2000 and 2003) created with FAO and with National Research & Development support are used by numerous public institutions and economical agencies for thematic studies. The use of these datasets underlines the accomplishment of the FAO project (Badea et al. 2007, 2008; Docan et al. 2007). The joint team ROSA/CRUTA used the vector database LCCS 2003 for generating a thematic map during the activation of the International Charter on Major Disaster for the floods affecting the lower basin of Siret river in 2005 (Fig. 19.5). The value of the method used for the first time by Romanian experts has been positively appreciated by the final users. Thus, in 2008 and 2010, the maps for flood management included reports about the land-cover/land-use categories affected by the disasters.

Furthermore, since CLC is the commonly used methodology in EU Member States, in Romania two separate groups of experts allowed the creation of datasets based upon CLC and LCCS. Rather than being a duplication of efforts, it allowed the study of the agricultural and forest dynamics in a manner complementary to CLC. The conclusions of the dual CLC/LCCS 2000 analysis imposed the continuation in parallel of the LCCS activities. To ensure the actual use of the LCCS results and their continuity, it was essential to adapt the LCCS 2000 and 2003 databases to be compatible with the INSPIRE Directive requirement that came into force in 2006 for each EU Member State to document the categorisation system used.

Table 19.3 Comparison of the three LCCS datasets in Romania

	Datasets		
	LCCS 2000	LCCS 2003	LCCS 2007
Number of polygons	13,730	15,268	37,973
Minimum (km^2)	>0.01	>0.01	>0.01
Maximum (in km^2)	3,520.7	3,173.9	5,369.6
Sum (in km^2)	41,906.9	41,906.9	41,906.9
Mean (in km^2)	3	2.7	1.1
Number of classes	52	49	56

As logical continuation, the project "*Hybrid Method for Thematic Update of the Land-Use Inventory by Remote Sensing/GIS Technology, Support for the Implementation of the European Agriculture and Environment Programs*" (see http://lccs07.rosa.ro) is continuing the activities developed during the previous years around the group of the Romanian Space Agency (ROSA). This project takes into account the evolution of data acquired by the new earth observation satellites and the advanced methods of the semi-automatic and automatic processing of these data that have become operational. The project allowed both the harmonisation of the LCCS 2000 and 2003 datasets and the preparation of the interpretation of the SPOT 5 satellite image coverage acquired in 2007 (Mamulea and Dana 2008). The main target of this project is to develop a second, scientific level of determination of land-cover/use parameters (Olteanu et al. 2009). This project offered the possibility to obtain the third series of LCCS databases using recent advanced technologies adapted to the Romanian conditions. With this step methods and rules will be set up for processing and validation of data. The first results are expected by late 2013. Having developed such methods, a new coverage can be made every 3–4 years based on the available human resources and at low costs. Furthermore, the third LCCS dataset will enable the testing of mathematical models for the study of landscape dynamics. An evaluation of the limit to which automation can be reached without jeopardising the quality of relevant information is foreseen.

To ensure continuation of the built institutional setting and capacities, two project proposals were submitted to the Seventh Framework Programme in the spirit of the INSPIRE Directive and GMES, the "*Towards an operational GMES Land Monitoring Core Service*" (GEOLAND2) and "*Services and Applications for Emergency Response*" (SAFER). Until present, an important part of the mountainous areas was interpreted. A comparison of the three LCCS datasets for 2000, 2003 and 2007 datasets is shown in Table 19.3. Due to the use of the SPOT 5 multispectral imagery with 10 m resolution the number of polygons increased and new classes were identified.

19.2.4 Moldova

The massive land distribution implemented in Moldova was not accompanied by any initiative to update the information on the land-cover/use changes using modern technologies. For an agricultural country, such as Moldova, knowledge of the extent and location of the cultivated areas in the different regions has not only a basic value for development planning, but also prime economic and social importance. To revitalise the agricultural sector, the Ministry of Agriculture needed timely and accurate information on the distribution of land-cover/use types in the country. Data prepared before the land distribution process were of limited use, as they depicted a situation no longer existent or they were prepared for specific crop suitability studies, while their production was no longer considered necessary.

The assistance provided through the FAO Technical Cooperation Project "*Building Capacity in Inventory of Land Cover/Land Use by Remote Sensing*" (TCP/MOL/2903), started on March 2004, increased the satellite data processing capacity of the State Agency for Land Relations and Cadastre (SALRC), enabling provision of inputs to the Ministry of Agriculture concerning provision of relevant data on land use and the immediate transfer of the generated databases for agricultural planning appropriate to the existing conditions. The specific outputs of the activities undertaken during the 15 months project period were:

1. The production of land-cover data for the whole country at 1:50,000 scale (based on LANDSAT TM image interpretation) and detailed land-cover/use data at 1:25,000 scale for specific areas of particular interest (based on SPOT multispectral satellite data) using LCCS version 1.0 for land-cover categorisation. These outputs constituted an essential input to the land redistribution and the improved agricultural planning process.
2. Production of a comprehensive database for the whole country, by adding to each land-cover polygon from other sources additional information on soil types, erosion features (like in Bulgaria and Romania) and municipal/district boundaries (like in Romania).
3. The transfer of modern technology on satellite image interpretation for land cover/use, specifically based on the considerable FAO experience on this subject. This constituted a sound basis for the sustainable development of remote sensing applications in the country.

The on-the-job training increased capacities in remote sensing and GIS enabling the SALRC staff to provide, through the linkages established by this project, accurate data and services to the Ministry of Agriculture and other institutions. This comprised initially the actual land-cover/use situation, while later it comprised data on water resources, forestry, land degradation and other subjects relevant to the sustainable development of agriculture and renewable natural resources.

This FAO project benefited from the experiences gained in the projects in Azerbaijan, Bulgaria and Romania. In Moldova the reference control points for the satellite imagery were homogenised with those in Romania. In practice, the LANDSAT TM images for 2000 became, for both projects, the reference in the orthorectification process. Due to the common natural conditions of Moldova and the Eastern part of Romania (e.g., geomorphology and vegetation) the class set used in Eastern Romania was adopted in Moldova. The experience of the local staff and the usefulness of the land-cover/use inventories convinced the Government to implement, under the assistance of the European Environment Agency, the project "*Evaluation of the Impact of Land-Use Changes in Moldova*" based on the CLC methodology. Started in 2011, this project is implemented by SALRC in cooperation with the Technical University of Moldova and the Institute of Geography and Ecology using WebGIS technologies (see www.geoportal.md).

19.2.5 Albania

In the turmoil of a rapidly changing economy the Albanian government needed accurate and timely information for management of their natural resources and formulation of land-use policies. The change to a market-oriented economy had also an impact on the natural resources and their management, not only due to privatisation, but also because of the strong land fragmentation as a result of the land distribution and increased urbanisation. The increasing pastoral economy and husbandry caused landscape degradation and natural resources depletion in many regions of the country. Uncontrolled timber harvesting, overgrazing and over-exploitation of wood (in a country with a permanent energy shortage) and other forest products have changed environmental assets. The depletion of forest resources, particularly in accessible areas, had become alarming. Scarce possibilities of control and a lenient policy caused severe, sometimes even irremediable, damages to the natural resources of Albania (Jansen et al. 2006).

The World Bank financed "*Albanian National Forest Inventory*" (World Bank loan/credit 2846 ALB) project, executed in the period 2002–2004, provided a quantitative analysis of spatially explicit land-cover/use change dynamics at national (Fig. 19.6) and district levels in the period 1991–2001 using LCCS version 1.0 for codification of classes, satellite remote sensing (digital LANDSAT 7 Enhanced Thematic Mapper imagery for 2001 and LANDSAT 5 Thematic Mapper for 1991), field survey for data collection and validation of the interpretation, and elements of the object-oriented geo-database approach to handle changes as an evolution of land-cover/use objects (i.e. polygons) over time to facilitate change dynamics analysis. The 1991 polygons are described by what has changed in their state, i.e. the spatial extent of the polygon formed by a set of land-cover/use boundaries and/or the polygon label (land-cover/use class) vis-à-vis their state in the validated 2001 dataset. This allows quick identification of 'hotspots' of change (Jansen et al. 2006). The approach to change analysis was fundamentally different from approaches that delineate the state of land cover at certain times and analyses change dynamics by making overlay procedures in a GIS.

The 1991 and 2001 land-cover/use interpretations apply 35 classes described by LCCS version 1.0. This ensures harmonisation of the data with existing datasets at international level, while at the same time standardising the method used for description of land-cover/use features (Jansen 2006; Jansen et al. 2008). Contrary to the FAO projects previously discussed, the prime focus in this project was on natural resources and not on agriculture. The 2001 digital database provided the

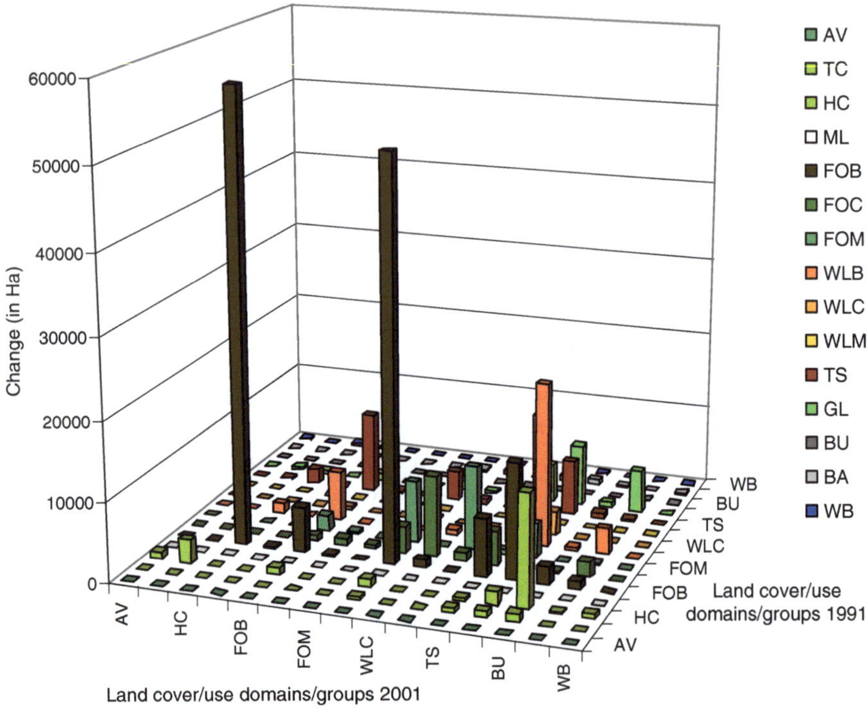

Fig. 19.6 Land-cover/use change by LCCS domain or land cover group (Codes used: *AV* aquatic vegetation, *TC* tree and shrub crops, *HC* herbaceous crops, *ML* managed lands, *FOB* broadleaved forests, *FOC* coniferous forests, *FOM* mixed forests, *WLB* broadleaved woodlands, *WLC* coniferous woodlands, *WLM* mixed woodlands, *TS* thickets and shrublands, *GL* grasslands, *BU* built-up areas, *BA* bare areas, *WB* water bodies) (Jansen et al. 2006)

baseline data in forestry and pasture planning and policy. It was used for the forest inventory in 2003, a grazing impact study (Papanastasis 2003) and in the transfer of responsibility of management of forest areas to communes (Jansen et al. 2006).

19.3 Lessons Learnt

19.3.1 At the Institutional Level

The sequence of FAO Technical Cooperation Projects provided a unique opportunity with a learning curve. The first application of LCCS meant also the first possibility in the region to become acquainted with the new methodology. In each subsequent project the implementation went smoother and new ideas were tested and evaluated. This was strengthened by the fact that experts worked in more than one project. In this manner a cross-fertilisation was generated from which the projects that started later have obviously benefited to a greater extent than those at the start. In 1999 there was, furthermore, the possibility of direct feedback to the LCCS development team so that changes in the methodology could still be implemented.

The FAO projects established a baseline dataset in each country that was used in various, mainly agricultural, applications. With time these applications covered different sectors and the applications became more detailed in nature. Until this day the LCCS methodology is in actual use in order to assist decision makers. The latter have to evaluate what impact a decision may have on future generations. To support decision-making relevant and up-to-date data and information are essential with:

- The capacity to elaborate spatial data in order to create information (e.g., time-series analysis, creation of relationships between various thematic layers, scenario development, impact analysis); and
- The use of powerful models that simulate various scenarios in order to be able to evaluate the effects, risks and advantages of a choice.

The strength in the set-up of the projects lay also in the fact that the focus was not solely on collection of the land-cover/use data, but these were immediately used in an application. In this manner in each country immediate experience was gained in the use of LCCS data in certain applications and the possibility to check if these data comprised the essential classes. Thus, the capacity building was not limited to the use of the LCCS methodology for remote sensing applications, but embraced the analytical (GIS) capacities in the various institutions involved. With this approach the sustainability of the projects has been successfully enhanced.

The World Bank project in Albania had a similar approach in that the land-cover/use data collection was in direct support of the forest inventory and grazing impact study. The World Bank prescribed the use of LCCS in the project. However, the remote sensing and GIS capacities in the country could not be substantially enhanced due to the lack of a dedicated institution or department within the Ministry that comprised a critical mass of experts. Nonetheless, the data collected and the various applications have clearly shown the usefulness of a comprehensive and consolidated approach. The frequency of forest inventories is around 10 years, therefore the near future will disclose if the methodological approach will be repeated.

19.3.2 At the Methodological Level

In the period 1999–2005, the use of LCCS in the FAO and World Bank projects was from the organisational viewpoint a logical decision. LCCS is the methodology developed by FAO and UNEP and therefore an apparent choice in United Nations projects. The World Bank stipulated in the Terms of Reference for the project in Albania the use of LCCS. However, the closer some of the countries came to pre-accession and accession status (e.g., Bulgaria and Romania) the more obvious it would have been to shift the attention, as far as land cover is concerned, to CLC as that is used not only in the EU Member States but by 39 countries in Europe for environmental monitoring purposes. However, in both countries LCCS continued to be used alongside CLC. Since land-cover/use data collection is costly and requires dedicated resources (e.g., specialist in interpretation of satellite images, knowledge of hard and software, analytical skills to operate the GIS, etc.) it is interesting to investigate what the reasons were of using two different categorisations in parallel other than the use of CLC for fulfilling European requirements and LCCS for national purposes.

CLC was originally developed for a smaller group of Member States than the 39 European countries that are using CLC nowadays. Since CLC comprises a pre-defined class set, this meant that over time this set had to be adapted to the environments of the new(er) countries. CLC is not based on a parameterised approach in which new combinations of parameters could be used for definition of new classes. Knowledge and technology have advanced and policy objectives will also have changed over time (Jansen et al. 2008). This means that with time insights are gained and new possibilities are within reach that one wants to use. Therefore, the methodologies applied -independent whether it is general-purpose categorisations like CLC or LCCS or any other- have to be dynamic over time to continue monitoring activities at national and pan-European levels.

The same is true for the use of satellite imagery: in the 1990s use was mainly made of different generations of LANDSAT satellite images, nowadays SENTINEL data from the Copernicus Programme offers a logic continuation from the LANDSAT type satellites but with improved spatial and temporal resolutions, and very high resolution satellite imagery has become available that allows more detail in the applications. Therefore, one observes a shift from countrywide data collections to studies related to specific areas of interest (e.g., in the examples provided for Bulgaria and Romania). At certain periods it is useful to have a countrywide overview and this was certainly the case in the transition period when land-cover/use dynamics were high. In that respect the projects were timely in providing the datasets at a time when a lot of decisions had to be taken in each country. Based on these overviews so-called 'hotspots' of change could be identified where more detailed studies were useful. The projects described show also that the dynamics in agriculture are higher than in forestry and accordingly request a different frequency for up-to-date and reliable data.

Whereas CLC allows adding more detailed levels of classes to the pre-defined class set, LCCS has the advantage that the same concept can be used to generate more detailed classes and these classes have an intrinsic hierarchical order. The latter facilitates regrouping of the classes, this is useful when a study requires various levels of analysis for instance when policies are made at national level and executed at lower levels (e.g., the change analysis at national and district levels in Albania). Especially for monitoring and evaluation purposes the parameterised approach furnishes the parameters that can be measured over time. Thus, it facilitates comparison by parameters over time. The CLC methodology has the advantage to cover all Member States and this is of great value in studies at European level or cross-border studies. However, for specific purposes it may still be required to use different dedicated class sets. For example, Lund (1999) made an inventory of forest definitions all over the world and came across an enormous variety. People look at the world with different views and all of these views are valuable in better understanding our environment.

With the experiences gained in the application of LCCS methodology in land-cover/use data collections, the Eastern European countries can analyse:

- The key factors that may affect the quality of the obtained data (i.e. which classes are determined with what type of precision? What are the problems in obtaining a minimum required level of accuracy? How subjective is satellite image interpretation and does it change over time?); and
- The view and manner of creating the land-cover/use class set, trying to reflect the real world over time.

It would be of particular interest to examine these questions both for CLC and LCCS and compare results.

19.4 Further Developments

19.4.1 Data Standardisation and Harmonisation

In practise, results from different surveys will need to be harmonised over time and space (e.g., in relation to cross-border issues), and reference to existing information is often required to verify new results (e.g., regarding urban sprawl and landscape changes). Data harmonisation, being defined as "*the intercomparison of data collected or organised using different classifications dealing with the same subject matter*" (McConnell and Moran 2001), thus becomes a prerequisite for many data analyses. However, development of the general-purpose LCCS has led to the common belief that once such a categorisation system becomes widely adopted for new surveys the problem of data harmonisation would be overcome because new data sets would be collected using a single standard system allowing direct comparison of new class sets, whilst existing class sets could be 'translated' into the adopted system making possible direct class comparison with new class sets (Jansen et al. 2008). However, this stance that is geared towards data standardisation, defined as "*the use of a single standard basis for classification of a specific subject*" (McConnell and Moran 2001), assumes falsely that the continuous advances in either knowledge, technological developments and/or changing policy objectives will not have any impact on a categorisation framework or its application. With each data collection effort lessons are learnt that leave their imprint on successive efforts (e.g. CLC 1990 versus 2000 (Büttner et al. 2004)). Data standardisation may thus be an unrealistic expectation and only partly feasible with the need for data harmonisation always present (Jansen et al. 2008).

The main documents available that describe the LCCS 1.0 and 2.0 (Di Gregorio and Jansen 2000 updated in FAO 2005) lack a formal definition of the categorisation rules. This represents a problem because, as the software source is not open, there is no possibility to (easily) understand the behaviour of the software application. The underlying logic can only be derived experimentally by using the software intensively and by defining classes step-by-step with the software to know if they are correct (Jansen et al. 2008). This also means that it is not possible to propose modifications as the formal definition of classes is missing and thus it is impossible to adequately describe LCCS (Di Costanzo and Ongaro 2004).

The fundamental structure of LCCS became the International Organization for Standardization ISO 19144–1 'Classification Systems – Part 1: Classification system structure' standard in 2009. This standard establishes the structure of a geographic information categorisation system, together with the mechanism for defining and registering the parameters. It defines the technical structure of a register of parameters in accordance with ISO 19135. It is important to differentiate between this ISO standard and countrywide land-cover/use data collections executed with LCCS versions 1.0 and 2.0. Furthermore, LCCS was used as the basis for the ISO 19144–2 'Classification Systems – Land Cover Meta Language' standard established in 2012. LCML comes with a Unified Modelling Language (UML) diagram. The object-oriented structure of UML is a basis for implementation in languages like Java or C++. In order to use the LCML as a reference for the implementation of land-cover categorisation software application, it is necessary to transform the UML concept model into a computer-oriented format. The Extensible Mark-up Language (XML) schema language from the World Wide Web Consortium has been selected for its immediate compliance with object-oriented structures, for its worldwide diffusion and for its soundness in terms of documentation resources, development and support (see www.glcn.org). With the UML there is finally a formal description of LCCS that provides insights into the categorisation rules and that can form the basis for further developments by scientists and the user community.

As described by Jansen et al. (2008), it may be necessary to 'translate' a class set into a third system, a so-called *reference system* that functions like a bridge between two class sets: each class in the original class sets will find its more or less corresponding class in the reference system. The use of a reference system may be a sensible choice when many class sets are involved as the number of pair-wise class combinations becomes excessive with comparison of n class sets requiring $n(n-1)/2$ comparisons to be made. As Wyatt and Gerard (2001) point out, the use of a reference system requires a single 'translation' from each original class set into the reference system and obviates the need for pair-wise class comparisons between every class set of interest. LCCS would like to function as such a reference system but, as has been pointed out by Jansen et al. (2008) using a number of class sets from the Nordic countries, the system falls short in many ways.

Nonetheless, FAO presents efforts that are limited to a crosswalk 'translation' effort between categorisation systems and/or class sets that ignore the complexity of harmonisation. For example, overviews in which FAO shows that country land-cover *maps* are 'translated' into LCCS (e.g., as shown by Herold et al. 2006) offer the wrong impression of data harmonisation as such efforts have been limited to correspondence of original classes (i.e. legends) with LCCS rather than having examined the full meaning of the data (Jansen et al. 2008).

19.4.2 Monitoring Agriculture with Remote Sensing (MARS)

The 'Land-Cover Meta Language' (LCML) offers ample possibilities to harmonise different land-cover categorisations and class sets. It is also proposed as one possible transformation mechanism in the draft INSPIRE data specification. This will assist to address the need from the EU agricultural policy community for a unambiguous description of the land, exclusively based on its physical nature and independent of land-use considerations, linguistic and/or traditional connotations. This development is a unique opportunity in Europe to apply LCCS and LCML.

The GeoCAP action of the Monitoring Agriculture with Remote Sensing (MARS) Unit of the European Commission Joint Research Centre (JRC), integrated LCCS as a central component of the common methodology for the annual quality assessment of the Land Parcel Identification System (LPIS) that is implemented by all EU Member States. The importance of the LPIS comes from the requirement that it must channel all area-based subsidies under the Common Agricultural Policy (CAP; Council Regulation EC 2009R73) totalling around 41 billion Euro in 2011. For this specific purpose, the LPIS quality can roughly be defined as the ability of the system to fulfil two explicit LPIS functions:

- The unambiguous identification of all declared agricultural parcels by farmer and inspectors; and
- The quantification of all eligible area for farmer declarations and cross-checks during the administrative controls by the Paying Agency in the respective countries.

Failure of LPIS in the unambiguous identification (i.e. the geographical location) induces risks for double declarations of land and for ineffective inspections, while inadequate quantification of eligible area renders the cross-checks ineffective for preventing and identifying over-declarations by farmers. Both failures involve financial risks for the EU Funds. The high financial impact calls for a transparent and pan-European reporting system on the ability of the LPIS to perform these functions with adequate quality.

The definitions and requirements for the geographic data, related to the Integrated Administration and Control System (IACS) and LPIS, are laid down in the CAP Regulations. A key parameter is the definition of the eligible hectares required to activate payment. Council Regulation 73 from 2009 states: "*'Eligible hectare' shall mean any agricultural area of the holding taken up by arable land and permanent pasture or permanent crops,*" or "*parcels taken up by arable land,*

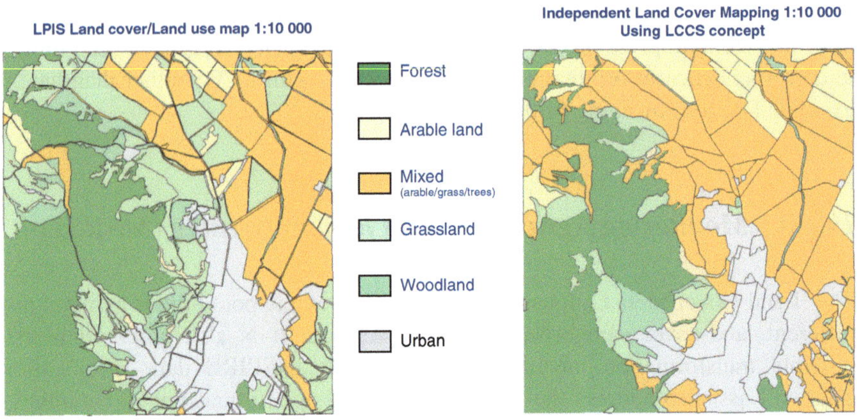

Fig. 19.7 Comparison between the LC/LU information stored in the LPIS (example from a new EU Member State) and a large-scale land cover map

permanent grassland, permanent crops, kitchen gardens or planted with short rotation coppice shall be eligible". However, as legal, financial, land-use or production related criteria can also be considered for granting payment, the key characterisation of land did not always receive due attention. The EU Member States were using different conceptual frameworks in order to define and map eligibility, but all states applied land-cover and land-use related approaches. This is also due to the variety of landscapes, climate, agricultural practices and land management approaches across Europe. Currently there is consensus among the technical actors in the LPIS domain that in the current EU regulation the term 'eligible hectares' is clearly land-cover oriented, while other qualifying conditions for a particular subsidy application are referring to the farmer's socio-economic activity and are thus land-use oriented.

Previous studies at JRC on a number of implementations, showed that the land-related information, stored at reference parcel level in the LPIS, corresponds indeed to the concept of land cover (Fig. 19.7). For spatial mapping to measure eligibility, the land-cover concept has strong advantages over the land-use one, as:

- It provides unambiguous and detailed characterisation of the Earth's surface, exclusively based on the physiognomic-structural (biotic or abiotic) aspect of the land;
- It is the easiest identifiable indicator of human interventions on the land and the resulting changes; and
- It is the main feature constraining the use of land.

The fact that LPIS retains a maximum eligible hectare value derived by delineation requires considering the above advantages and it supports the land-cover conceptual choice. A land-cover categorisation would definitely be the ideal instrument to identify the potential eligibility, expressed by this maximum eligible area.

As laid down in Article 17 of Council Regulation 73/2009, the spatial information stored in the LPIS should be compliant with the cartographic standards applied at a scale of 1:10,000 or larger. Such large-scale land-cover mapping requires a particular approach, taking into account that:

- This deals with much higher information contents than the regional land-cover mapping, traditionally performed at 1:50,000 to 1:100,000 scales.
- There are different landscapes involved, as the EU Member States cover the major part of the European continent.
- The 'eligibility' concept, although applied by the individual EU Member States, should ensure an equal treatment of all European farmers.

Studies showed that a single land-cover class set cannot be realistically applied to the 27 EU Member States at the large scale required for LPIS.

For these reasons, a common quality assessment requires a universal identification and categorisation of the land-cover types (Milenov and Devos 2009, 2012). It was necessary to adopt a systematic framework, or a standard categorisation system, to characterise all agricultural land using a set of well-defined diagnostic criteria. The definition of the land-cover classes and the respective criteria had to be independent of data capture technology and (to some extent) scale. JRC decided to use the LCCS to 'map' this land potentially eligible for payment, as it provided a common language, a common environment and common 'business rules'. LCCS provided the conceptual framework for this comprehensive description of any type of agricultural land cover using a minimum number of parameters. It is flexible in accommodating the variety of country-specific land-cover types and in enabling comparisons between land-cover types. The LCCS methodology also comprises the concept of the variable minimum mapable area and it could handle, in a standardised manner, intrinsically mixed land-cover units, these are important issues in the context of LPIS.

In the framework of the LPIS Quality Assessment, the semantic 'bridge' between land cover and eligibility was implemented through the concept of the so-called 'eligibility profile' (Devos 2011). It is a conversion table allowing mapped land-cover features to be expressed as eligibility for agricultural payments. It quantifies in a single methodology through joining the common class set of land-cover features with the national rules for support schemes applied to the measured areas (Fig. 19.8). It converts thus the results of the land-cover mapping into 'eligible hectares or eligible features found', for each land-cover class defined and coded *a priori* through LCCS in each Member State.

The first year of the LPIS Quality Assessment implementation demonstrated that the introduction of the consistent approach for quantification of the eligible land through the use of standardised and concise land-cover semantics provided extremely valuable feedback to the EC and EU Member States with respect to the understanding of the agricultural landscapes in Europe and contributed to the harmonisation of the land-cover concepts for this community at EU level.

Fig. 19.8 Examples of categorisation of LPIS land cover using LCCS

19.5 The Way Forward to Categorisation

"To classify is human" as Bowker and Star (1999) stated. Few categorisations take formal shape or any formal algorithm, even fewer categorisations are standardised. Yet, we all use (in)formal categorisations on a daily basis, intentionally or inadvertently. The knowledge about which categorisation will be useful under certain conditions and at a given moment is embodied in our responsibilities and routines in a certain context. At the level of policy, categorisation of areas, uses and covers plays an equally important role. The categorisation of an area as either nature reserve or industrial will have a clear impact on future economic decisions. Thus, the relation between categorisation and decision-making may be invisible but is evidently powerful. Nowadays in the information era, scientists work on the design, description and choice of categorisation systems embodying choices that create people's identities but few people realise how much impact a categorisation may have. In the context of land cover, in Europe the CLC contributed in creating a European identity; the LCCS contributes in creating a UN identity (Jansen 2010).

Categorisations embody a worldview and each category and class in it values this specific viewpoint. This is in itself not critical as long as it is recognized that another viewpoint may be silenced, which may be the case if a single categorisation becomes the preferred standard. From the analysis of semantic information and used definitions one can deduce something about this view and the intent of the data producers, but much more transparency is needed. Also more insights in the design of categorisation systems is needed and research examining their impact. The effort of attaching objects to categories and the ways in which those categories are ordered into systems is often disregarded. In the land-cover domain, for instance, several class definitions in CLC (CEC 1999; Bossard et al. 2000) or in LCCS (FAO 2005) are described by taking a bird's eye view, or *map* view, rather than a *geographic entity* view probably because these systems are used in remote sensing (Jansen 2010).

Categorisation facilitates the communication of knowledge concerning specific phenomena (e.g., land use and land cover) between individuals. Ideally categorisations are able to travel across the borders of (scientific) communities, of which the individuals are part, and maintain some sort of constant identity.

The tangible results of categorisation are *classes* and *categories* that serve as the vehicles for communication of meaning (Ahlqvist 2008). *Parameters* used in the categorisation are usually not tangible, simply because they are usually inexplicit rather than explicit. The members in each class or category have definable characteristics in common and with the use of categorisation one can discover general truths related to the distinguished classes or categories rather than to their individual members (Shapiro 1959). Categorisation is, at the same time, a simplification because it represents only part of the complexity of reality (like models represent simplifications of the real world). Different perspectives, or so-called 'scapes', to categorisation can be taken that are all equally valid and valuable (Veldkamp 2009) One needs to recognise, therefore, that no categorisation reflects accurately the social or the natural world (Bowker and Star 1999). Categorisations arise out of social communication needs but they serve specific purposes: not only do they reflect the ideas of a certain community or institution, but they can also be the end-result of negotiating and reconciling individual, group and institutional differences (Ahlqvist 2008). This is particularly true for CLC representing the EU community and LCCS representing the UN community.

Definitions expressed in natural language associated by sub-type/super-type relationships, i.e. hierarchical relationships, are called *terminological ontologies* (Sowa 2000). Almost all land-use and land-cover categorisations to date are terminological ontologies (e.g., CLC and LCCS). Ontology is an explicit specification of a conceptualisation to represent shared knowledge (Gruber 1993; Ahlqvist 2008). Semantic information can be determined from the definitions of the ontology and the representation of categories can be enriched with semantic properties (e.g., purpose, time, location, etc.) and relations (e.g., "is-a", "is-a-part-of", "associated-with", etc.) in order to reveal similarities and heterogeneities (Kavouras et al. 2005). Recognition of semantic heterogeneity is the basis for creating sound data linkages between multiple datasets that are needed for land-change analysis, monitoring and modelling for land-use planning, land development, policy (e.g., EU CAP) and informed decision-making.

As Cihlar and Jansen (2001), Comber et al. (2005, 2007) and Ahlqvist (2008) point out: manifold ways to conceptualise and communicate knowledge exist according to the disciplines of (groups of) experts, professions, etc., so that there are necessarily many-to-many relationships between classes and thus inherent ambiguity in any categorisation. Categorisations contribute to communication of knowledge and in making joint progress in that knowledge by facilitating communication. However, they can only make such contributions by being *dynamic* in nature. By keeping the voices of parameters and their constituents present, as is the case in parameterised categorisations, the maximum flexibility of the system is retained. This includes the key ability to be able to change with changing knowledge, technological developments and changing policy objectives (Jansen 2010). In this respect LCCS has an advantage over CLC.

Collection of *data* leads to the creation of categories. Contrary to old hierarchical class and data sets (or databases), where relations had to be decided once for all the time of original creation, many class and data sets today incorporate object-oriented views whereby different parameters can be selected and combined on the fly for different purposes (Bowker and Star 1999). Object-oriented interpretation exists already in remote sensing (as used in Romania and Bulgaria) and also in the approach to databases (as used in Albania). But what about an object-oriented approach to categorisation? Parametric Object-Oriented Data Models (POODM) should take the place of old-fashioned categorisation systems, like CLC and LCCS, because they allow an unprecedented flexibility and capability in the design and use of very complex information systems and environmental change requires such an information system. Such a POODM should use the UML as standard, something both CLC and LCCS miss (though LCML has) and thereby neglect fulfilling the ISO 19100 standard. These parameterised multi-level class and data sets put more emphasis on the parameters to be used than on the structure in which these are organised. This approach is dynamic, easily adaptable under changing circumstances (Jansen 2010).

Based on the fact that categorisations are dynamic in nature one could argue that the definitions of Shapiro (1959), Sokal (1974), and FAO (2005) or even the definition applied in the documents of ISO/TC 211 should be abandoned in favour of a modified version of the definition of Bowker and Star (1999): *categorisation is a spatial, temporal, or spatio-temporal, and organisational hierarchy based segmentation of the world.* This definition emphasizes that the dimensions of time and space are imperative in determining a categorisation, as well as the organisational level. In the case of a non-hierarchical system one could speak of zero organisational hierarchy, analogue to zero tillage when no tillage occurs.

The most commonly used categorisation systems are hierarchically structured (e.g., plant taxonomy). To many ecologists it has been long apparent that ecological systems are structured as such (Egler 1942; Schultz 1967). Early on it was also acknowledged that "*it is not to be assumed that some one classification will one day be found, and all others will then be abandoned. Each classification serves a certain purpose, and will continue to exist by its own right*" (Egler 1942). Thus, there is not one categorisation that best characterises land cover or land use. In addition, it seems not fruitful to go in search of the one hierarchy because there is no single, *a priori* parameter for developing such a hierarchy. Instead, a number of different hierarchies may be used to address different problems. With standardisation one runs the risk of adopting a categorisation with a determined hierarchy that fits a predetermined purpose. Adopting such a categorisation for another purpose involves working with a system with a bias that might force our thinking into the framework that was designed for, and is probably more appropriate for, another problem area. Currently, there is no *a priori* designation of hierarchy imposed by the social and biophysical sciences in such a way that no other manner of looking at either land use or land cover is feasible or useful. The hierarchy theory also includes that principles developed at one hierarchical level cannot be transposed to higher and lower levels. Clear distinction of type and category within the hierarchy will not lead to more scientific progress. It is the inherent and awkward ambiguities of land cover and land use that should be included in the more innovative approach of using fuzzy set theory as the mathematical theory underlying categorisation.

Thus, to be truly innovative, this means a move away from existing systems like CLC and from LCCS to LCML (Jansen 2010). Together with the methodological way forward is also the potential future use by the land monitoring community of the detailed LPIS datasets (currently these data are unavailable to them), collected each year, to generate land-cover data sets covering the country or areas of specific interest.

Acknowledgements The authors gratefully acknowledge Carlo Travaglia, former Technical Officer at the FAO Environment and Natural Resources Service, who was the driving force for initiating and managing the projects in Azerbaijan, Bulgaria, Moldova and Romania.

References

Ahlqvist O (2008) In search for classification that supports the dynamics of science: the FAO Land-Cover Classification System and proposed modifications. Environ Plan B 35:169–186

Badea A, Herisanu Gh (2002) The FAO concept for land-use and land-cover classification (Sistemul de clasificare Land Use, Land Cover in conceptia FAO). Analele Universitatii Spiru Haret, Seria Geografie No.4, Editura Fundatiei Romania de Maine

Badea A, Dana IF, Moise C, Badea R, Mamulea A (2007) A method for multi-source data exploitation, support for crisis situations management. Case study: floods in Romania 2005–2006. In: Katalinic B (ed) Annals of DAAAM 2007/Proceedings of the 18th international DAAAM symposium. Published by DAAAM International, Vienna, Austria, 23 pp

Badea A, Dana IF, Moise C, Mamulea AA, Olteanu VG, Poenaru VD (2008) Land cover GIS databases, support for the implementation of the European agriculture and environment programs. In: Katalinic B (ed) Annals of DAAAM 2008/Proceedings of the 19th international DAAAM symposium. Published by DAAAM International, Vienna, Austria, 24 pp

Bossard M, Feranec J, Otahel J (2000) CORINE land cover technical guide – addendum 2000. Technical report 40. European Environment Agency, Copenhagen, 105 pp

Bowker GC, Star SL (1999) Sorting things out: classification and its consequences. MIT Press, Cambridge, MA

Büttner G, Feranec J, Jaffrain G, Mari L, Maucha G, Soukup T (2004) The European CORINE land cover 2000 project. XX Congress of the International Society for Photogrammetry and Remote Sensing, Istanbul, Turkey, 12–23 July 2004, 12 pp

CEC (1999) CORINE land cover. Brussels

Cihlar J, Jansen LJM (2001) From land cover to land use: a methodology for efficient land-use mapping over large areas. Prof Geogr 53(2):275–289

Comber A, Fisher P, Wadsworth R (2005) What is land cover? Environ Plan B Plan Des 23:199–209

Comber AJ, Fisher PF, Wadsworth RA (2007) Land cover: to standardise or not to standardise? Comment on 'Evolving standards in land-cover characterisation' by Herold et al. J Land Use Sci 2(4):283–287

Deininger K (2003) Land policies for growth and poverty reduction. World Bank Policy Research Report. June 2003. World Bank/Oxford University Press, 239 pp

Devos W (2011) LPIS quality inspection: EU requirements and methodology, European Commission Joint Research Centre. Available at ftp://mars.jrc.ec.europa.eu/LPIS/Documents/13422_October2011.pdf

Di Costanzo M, Ongaro L (2004) The Land-Cover Classification System (LCCS) as a formal language: a proposal. J Agric Environ Int Dev 98(3–4):117–164

Di Gregorio A, Jansen LJM (1998) A new concept for a land-cover classification system. The Land 2(1):55–65
Di Gregorio A, Jansen LJM (2000) Land-Cover Classification System (LCCS): classification concepts and user manual. FAO/UNEP/Cooperazione Italiana, Rome, 177 pp. (Incl. CD-ROM with application software)
Docan DC, Dana IF, Badea A, Moise C, Badea R (2007) Adopted methods of spatial information analysis developed and implemented in the framework of IACS-LPIS set-up in Romania. In: Katalinic B (ed) Annals of DAAAM 2007/Proceedings of the 18th international DAAAM symposium, Published by DAAAM International, Vienna, Austria
Egler FE (1942) Vegetation as an object of study. Philos Sci 9:245–260
FAO (2005) Land-Cover Classification System – classification concepts and user manual. Software version 2. Environment and Natural Resources Series 8, Rome, 190 pp
FAO (2012) Voluntary guidelines on the responsible governance of tenure of land, fisheries and forests. Final draft 9 March 2012, Rome, Italy, 34 pp
Groom G (ed) (2004) Developments in strategic landscape monitoring for the Nordic countries. ANP2004:705. Nordic Council of Ministers, Copenhagen, 167 pp
Gruber TR (1993) Towards principles for the design of ontologies used for knowledge sharing. Int J Hum Comput Stud 43:907–928
Herold M, Latham JS, Di Gregorio A, Schmullius CC (2006) Evolving standards in land-cover characterisation. J Land Use Sci 1(2–4):157–168
Jansen LJM (2006) Harmonisation of land-use class sets to facilitate compatibility and comparability of data across space and time. J Land Use Sci 1(2–4):127–156
Jansen LJM (2010) Analysis of land change with parameterised multi-level class sets. Exploring the semantic dimension. Ph.D. thesis, Wageningen University, Wageningen, The Netherlands, 229 pp
Jansen LJM, Di Gregorio A (2002) Parametric land-cover and land-use classifications as tools for environmental change detection. Agr Ecosyst Environ 91(1–3):89–100
Jansen LJM, Veldkamp A (2012) Evaluation of the variation in semantic contents of class sets on modelling dynamics of land-use changes. Int J Geogr Inf Sci 26(4):717–746
Jansen LJM, Carrai G, Morandini L, Cerutti PO, Spisni A (2006) Analysis of the spatio-temporal and semantic aspects of land-cover/use change dynamics 1991–2001 in Albania at national and district levels. Environ Monit Assess 119:107–136
Jansen LJM, Groom GB, Carrai G (2008) Land-cover harmonisation and semantic similarity: some methodological issues. J Land Use Sci 3(2–3):131–160
Kavouras M, Kokla M, Tomai E (2005) Comparing categories among geographic ontologies. Comput Geosci 31:145–154
Kuechler AW, Zonneveld IS (eds) (1988) Vegetation mapping, vol 10, Handbook of vegetation science. Kluwer, Dordrecht
Kuemmerle T, Hostert P, Radeloff VC, Van der Linden S, Perzanowski K, Kruhlov I (2008) Cross-border comparison of post-socialist farmland abandonment in the Carpathians. Ecosystems 11:614–628
Larsson G (2002) Matching land use and land tenure arrangements for developing programmes. Landnutz Landentwickl 43:71–77
Lund HG (1999) A 'forest' by any other name…. Environ Sci Policy 2:125–133
Mamulea AA, Dana IF (2008) Combined method for thematic update of the land-use inventory by remote sensing/GIS technology. Support for the Implementation of the European Agriculture and Environment Programs, Rev. CAD *Journal of Geodesy and Cadastre* 8, Alba Iulia, Romania
McConnell WJ, Moran EF (eds) (2001) Meeting in the middle: the challenge of meso-level integration. In: An international workshop on the harmonisation of land-use and land-cover classification, Ispra, Italy. LUCC Report Series No. 5. Anthropological Center for Training and Research on Global Environmental Change – Indiana University and LUCC International Project Office, Louvain-la-Neuve, 17–20 Oct 2000

Milenov P, Devos W (2009) Standardisation of the land-cover classes in the LPIS using FAO Land Cover Classification System (LCCS). In: 33rd international symposium on remote sensing of environment, Stresa, Italy, 4–8 May 2009

Milenov P, Devos W (2012) ANNEX III, Executable Test Suite (ETS), The concept of land cover and 'eligible hectares'. Version 5.2. EUR 25185 EN. Available at http://mars.jrc.ec.europa.eu/mars/content/download/2467/12676/file/LBNA25185ENN.pdf

Milenov P, Vassilev V, Radkov R, Samundzhi V, Nakov D (2004) RS/GIS for environmental and urban management in the Lower Danube region – Rousse pilot study. The Danube and Europe: Integrated Space Applications in the Danube, Mamaia, Romania

Milenova L, Vassilev V, Spiridonov V, Kolev S, Konstantinov V (2003) Concept of integrated system for information and management of forest fires in Bulgaria. In: International space technologies exhibition and conference, Ankara, Turkey, 6–8 May 2003

Nedkov R, Vassilev V, Pironkova Z, Milenov P (2000) Integration of satellite images and GIS technologies for the purposes of land use. J Probl Geogr 1–4:359–363

Olteanu VG, Badea A, Dana IF, Moise C, Jivanescu IE, Poenaru VD (2009) Change detection land cover GIS databases. LCCS and CLC. In: Annals of DAAAM 2009/Proceedings of the 20th international DAAAM symposium, vol 20. Published by DAAAM International Vienna, Austria

Papanastasis V (2003) Special study on grazing impact on wooded lands, including fuel wood consumption assessment. Technical report. World Bank Albanian National Forest Inventory project. Agrotec S.p.A., Rome, 68 pp

Schultz AM (1967) The ecosystem as a conceptual tool in the management of natural resources. In: Cieriacy-Wantrup SV (ed) Natural resources: quality and quantity. University of California Press, Berkeley, pp 139–161

Shapiro ID (1959) Urban land-use classification. Land Econ 35(2):149–155

Sokal R (1974) Classification: purposes, principles, progress, prospects. Science 185 (4157):1115–1123

Sowa JF (2000) Knowledge representation: logical, philosophical and computational foundations. Brooks Cole Publishing Co., Pacific Grove, 594 pp

Swinnen JFM (1999) The political economy of land reform choices in Central and Eastern Europe. Econ Transit 7(3):637–664

Travaglia C, Milenova L, Nedkov R, Vassilev V, Milenov P, Radkov R, Pironkova Z (2001) Preparation of land-cover database of Bulgaria through remote sensing and GIS. FAO environment and natural resources working paper No. 6. FAO, Rome

Van der Molen P, Jansen LJM (2010) Urban and rural land management in the Netherlands: the role of the State as policy maker and landowner. Keynote speech, ARGE/AdV international land management symposium land management strategies for improving urban–rural inter-relationships – best practices and regional solutions, Hanover, Germany, 10–11 May 2010, 20 pp

Veldkamp A (2009) Investigating land dynamics: future research perspectives. J Land Use Sci 4 (1):5–14

Williamson I, Enemark S, Wallace J, Rajabifard A (2010) Land administration for sustainable development. ESRI Press Academic, Redlands, 487 pp

Wyatt BK, Gerard FF (2001) What's in a name? Approaches to the inter-comparison of land-use and land-cover classifications. In: Groom G, Reed T (eds) Strategic landscape monitoring for the Nordic countries. TemaNord 523. Nordic Council of Ministers, Copenhagen, pp 113–121

Part V
Multi-temporal Monitoring in Support of Decision Making and Implementation at Regional, National and Local Scale

Chapter 20
Differentiation of Crop Types and Grassland by Multi-scale Analysis of Seasonal Satellite Data

Thomas Esch, Annekatrin Metz, Mattia Marconcini, and Manfred Keil

20.1 Introduction

The implementation of productive and sustainable cultivation procedures is a major effort regarding the agricultural production in the European Community. However, political, economic and environmental factors impact the cultivation strategies directly and indirectly, and therewith strongly determine the condition and transformation of the cultivated and natural landscape. To assess the actual status, identify basic trends and mitigate major threats with respect to the agricultural production and its impact on the cultural and natural landscape, a frequent and area-wide monitoring of cropland and grassland is required. Satellite-based earth observation (EO) provides ideal capabilities for the area-wide and spatially detailed provision of up-to-date geo-information on the agricultural land use and the properties of the cultivated landscape. A specific benefit of EO is given by analysing multi-seasonal data acquisitions. Intra-annual time series facilitate the analysis of the phenological behaviour of the main crop and grassland types – key information with respect to the characterisation of the land use intensity and its impacts on the environment.

The presented approach focuses on a seasonal analysis of multi-scale EO time series to classify main crop types and differentiate between cropland and grassland for given areas of interest on the basis of field parcels. The areas of interest are typically existing land use/land cover (LULC) data sets (e.g. national topographic

T. Esch (✉) • M. Marconcini • M. Keil
Department Land Surface, German Aerospace Center (DLR), Earth Observation Center (EOC), German Remote Sensing Data Center (DFD), Oberpfaffenhofen, Germany
e-mail: thomas.esch@dlr.de; mattia.marconcini@dlr.de; manfred.keil@dlr.de

A. Metz
Institute for Geoinformatics and Remote Sensing, University of Osnabrück, Barbarastrasse 22b, 49076 Osnabrück, Germany
e-mail: ametz@uni-osnabrueck.de

I. Manakos and M. Braun (eds.), *Land Use and Land Cover Mapping in Europe: Practices & Trends*, Remote Sensing and Digital Image Processing 18,
DOI 10.1007/978-94-007-7969-3_20,

data, CORINE Land Cover, etc.) that show a limited resolution in the semantic and/or spatial domain. Hence, the presented approach is primarily designed to improve the level of thematic/geometric detail for given LULC data sets.

20.2 Seasonal Characteristics of Grassland and Crops

Most established approaches towards the classification of land cover (LC) types from EO data rely on the analysis of the spectral signature. However, the spectral characteristics of grassland and crops show significant variations throughout the vegetation period and their spectra are quite similar for at least some points in time – e.g., depending on the growth states and cultivation forms (cp. Itzerott and Kaden 2006). Cropland shows highly variable seasonal characteristics, whereas grassland features a more continuous seasonal development (Fig. 20.1). Nevertheless, due to more or less distinct intra-class variabilities particular sub-classes might show similar behaviours – e.g. intensively used grassland and certain crops. The main drivers for the different seasonal behaviours are – apart from the local climate at the given geographical region and the weather conditions during the vegetation period – sowing dates, cultivation cycles and forms, and the harvesting times (cropland) or mowing dates (grassland), respectively.

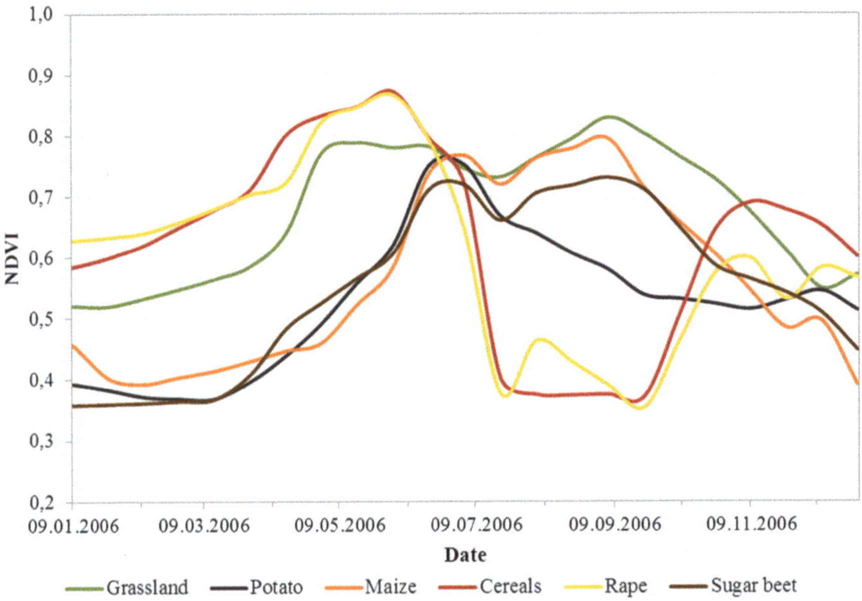

Fig. 20.1 Variation of the Normalized Difference Vegetation Index (*NDVI*) throughout the vegetation period for semi-natural grassland and different crop types derived from data of the MODIS sensor (Modified after Metz 2009)

Considering the ambiguity in the signatures of grasslands and crops during the vegetation period, a robust classification procedure for the identification and separation of grassland and crop types should in particular focus on an analysis of the seasonal development in addition to the interpretation of spectral properties. Studies on regional to global scale have demonstrated that multi-seasonal time series analyses of EO data can serve as an effective instrument for characterising grass- and croplands. Most studies implement optical remote sensing data with a high temporal and a comparably low spatial resolution, e. g. NOAA-AVHRR or MODIS data, to characterise and map crop types based on phenological profiles derived from NDVI or EVI (e.g. Shao et al. 2010; Wardlow and Egbert 2008; Zhang et al. 2003). Several studies successfully implemented high resolution (HR) to very high resolution (VHR) optical imagery, such as RapidEye, SPOT or IKONOS, as they are more applicable to classify crops and grassland in areas with heterogeneous patterns of smaller fields (e.g. Conrad et al. 2010, 2011; Itzerott and Kaden 2006; Singh et al 2011; Turker and Ozdarici 2011). The need of specific images for important points in time during the crop growing cycle and limitations of data acquisition through cloud cover for example, constrict the usage of optical data for operational applications (Blaes et al. 2005). Synthetic Aperture Radar (SAR) systems represent active imaging sensors that can acquire data independently of the weather and environmental conditions – a significant advantage for collecting seasonal time series. Most of the studies employing SAR data for crop and grassland monitoring focus on the analysis of backscattering intensity (as with the NDVI in case of multispectral data, the radar backscattering also correlates with the buildup of biomass) and polarimetric parameters of single scenes, whereas analyses of the seasonal variety and development of these features based on multi-temporal SAR data are still comparably rare (e.g. McNairn et al. 2009; Smith and Buckley 2011; Schuster et al. 2011).

20.3 Multi-scale Analysis of Seasonal Time Series Data – An Example

The main challenge with respect to EO-based seasonal analyses lies in the collection of a cloud-free data base covering key dates and seasons, while at the same time providing enough spatial detail to assure an accurate analysis on field parcel basis. Spatial detail is particularly important when investigating regions with small sized field parcels as they are typical for many rural areas in Central Europe. To address this challenge, we propose one solution combining the interpretation of multi-seasonal high and medium resolution optical data. The basic idea of this approach is to first use one or two HR satellite images (e.g., spring and summer) to properly delineate the different land parcels in the region of interest by means of an image segmentation. Then, medium resolution (MR) satellite scenes are employed in combination with the HR data in order to characterise the seasonal behaviour of the extracted image objects. Here, the use of segments minimizes the effect of mixed pixels that might occur along the borders of field parcels, especially in case of the MR data. MR sensors show a significantly larger swath width compared to HR systems – a property that allows for frequent coverages within comparably short periods of time, therewith increasing the chance to collect a cloud-free seasonal data set. In this study, we introduce a multi-sensor approach combining two HR scenes of IRS-P6 LISS-3 (23.5 m spatial resolution, 120 km swath width) with three MR IRS-P6 AWiFS images (56 m spatial resolution, 700 km swath width). After pre-processing, meaningful land parcels are extracted based on an image segmentation of the LISS-3 data. Next, the three seasonal data takes of AWiFS are used in addition to the LISS-3 images to derive a defined set of seasonality indices for potential cropland or grassland segments. The spatial focus on crop- and grassland is implemented by using a mask derived from vector data of the German Authoritative Topographic-Cartographic Information System (ATKIS). In a next step, thematic point data of the Land Use/Cover Area frame statistical Survey (LUCAS) is applied as sample data to automatically train a tree classifier that finally detects and qualifies grassland as well as different crop types. A schematic view of the entire methodology is provided in Fig. 20.2.

20.3.1 Data, Pre-processing and Derivation of Basic Parameters

The EO data basis includes two LISS-3 scenes and three seasonal AWiFS images (ISRO 2012) for an agricultural region in Mecklenburg Western Pomerania (MV) and Brandenburg (BB), Germany. The area under investigation covers 23,012 km^2 and is dominated by the LC categories agriculture, woodland, settlement and water bodies. The LISS-3 data were acquired on 05.05.06 and 17.07.06 and the AWiFS images on 13.06.06, 17.07.06 and 12.09.06. In addition to the EO images, we use LUCAS point data (Martino and Fritz 2008) collected in 2006 for a training of the classification procedure and later validation of the results. For the study area, the LUCAS data base provided a total of 1,796 points covering a total of 17 classes. For our research we were only interested in the LU/LC categories related to agriculture and grassland. The resulting selection of 1,529 points for 11 classes is listed in Table 20.1, along with a grouping of the single classes into five more general agricultural categories representing main crop types and grassland, which are addresses by the later classification. Finally, ATKIS vector data (AdV 2012) is used to create a mask of agricultural and grassland areas. For that purpose, the polygons of the ATKIS classes for agriculture and grassland are dissolved and merged and then exported as an agriculture mask. The later classification procedure is only applied to those areas assigned as agricultural or grassland areas by the agriculture mask.

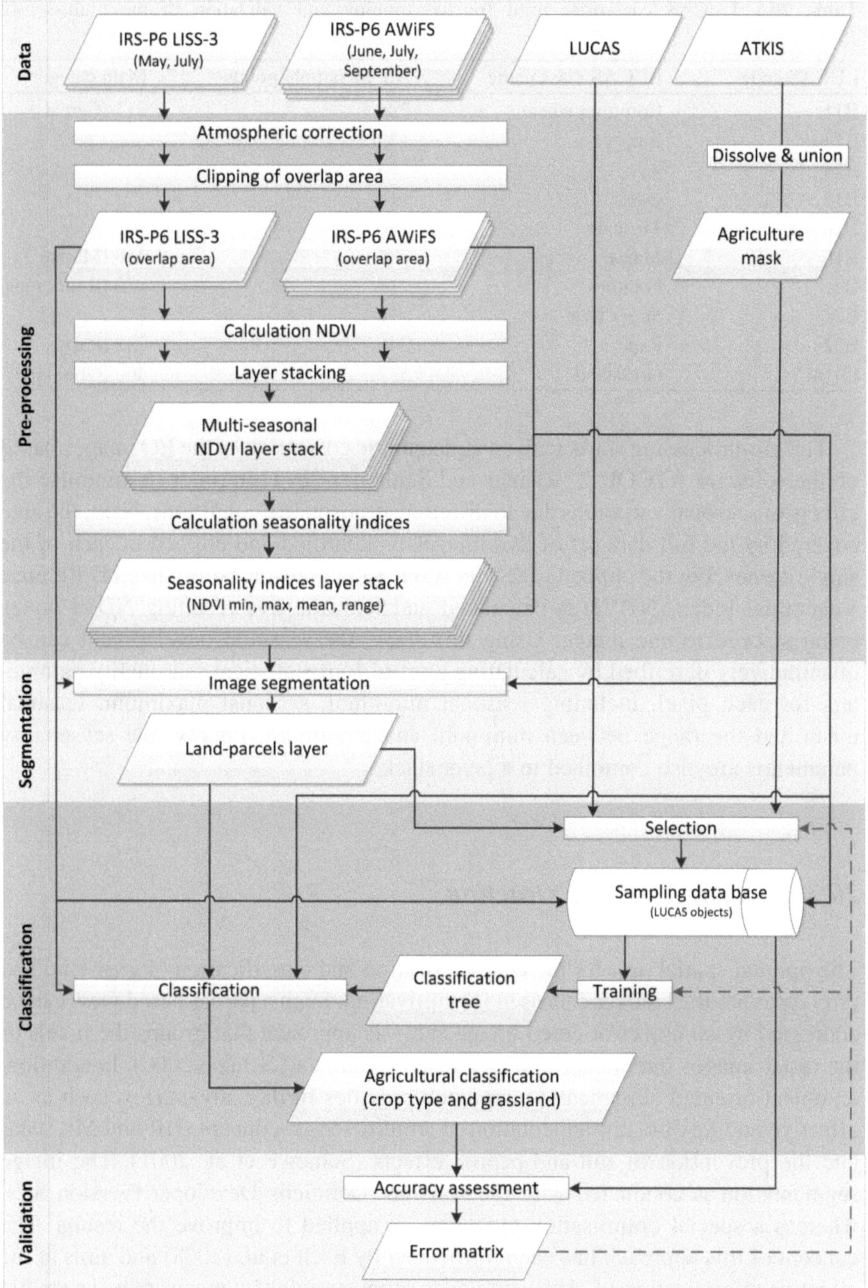

Fig. 20.2 Schematic view on the multi-scale approach towards the analysis of seasonal EO data

Table 20.1 LUCAS categories used for the training and validation of the multi-sensor classification

LUCAS code	LUCAS class name	No. of sample points	Main category
B11	Common wheat	296	211 (Cereals)
B13	Barley	173	
B14	Rye	98	
B15	Oats	14	
B18	Triticale	50	
B16	Maize	84	216 (Maize)
B21	Potatoes	14	220 (Root crops)
B22	Sugar Beat	17	
B32	Rape	246	232 (Rape)
E01/02	Grassland	537	400 (Grassland)

The pre-processing starts with an atmospheric correction of the EO images based on the software ATCOR-2 (Richter and Schläpfer 2011) in order to minimise the effects of seasonal variations due to different atmospheric conditions. Next, the area covered by the full data set of five images is identified and clipped in each of the single scenes. For the clipped LISS and AWiFS data sets the Normalized Difference Vegetation Index (NDVI) is calculated and with the five resulting NDVI layers being stacked to one image. Using this stack, the seasonal development can be quantitatively described by calculating a set of four statistical seasonality parameters for each pixel, including seasonal minimum, seasonal maximum, seasonal mean and the range between minimum and maximum. Finally, the seasonality parameters are also combined to a layer stack.

20.3.2 Land-Parcel Extraction

The optimal spatial unit for the characterisation and classification of grassland and crop types are the land parcels used for cultivation. Such a parcel-based view can be addressed by an object-oriented image analysis approach that groups the pixels of the raster images into meaningful segments (Baatz and Schäpe 2000). In addition, an object-oriented classification approach provides further advantages, such as an effective and flexible implementation of a multi-sensor concept (HR and MR data) and the prevention of salt-and-pepper effects (Schiewe et al. 2001). The image segmentation is conducted with the software Definiens Developer (version 8.7), whereas a special optimisation procedure is applied to improve the results. The concept of this approach has been introduced by Esch et al. (2008) and aims at the effective minimization of over- and under-segmentations by means of a systematic variation of the segmentation parameters. Inputs for the segmentation are the two LISS-3 images and the agriculture mask. In a first step, a segment level is created that exactly reflects the vector geometry of the agriculture mask. Then, the polygons of the mask are subdivided with the segmentation optimisation process that is

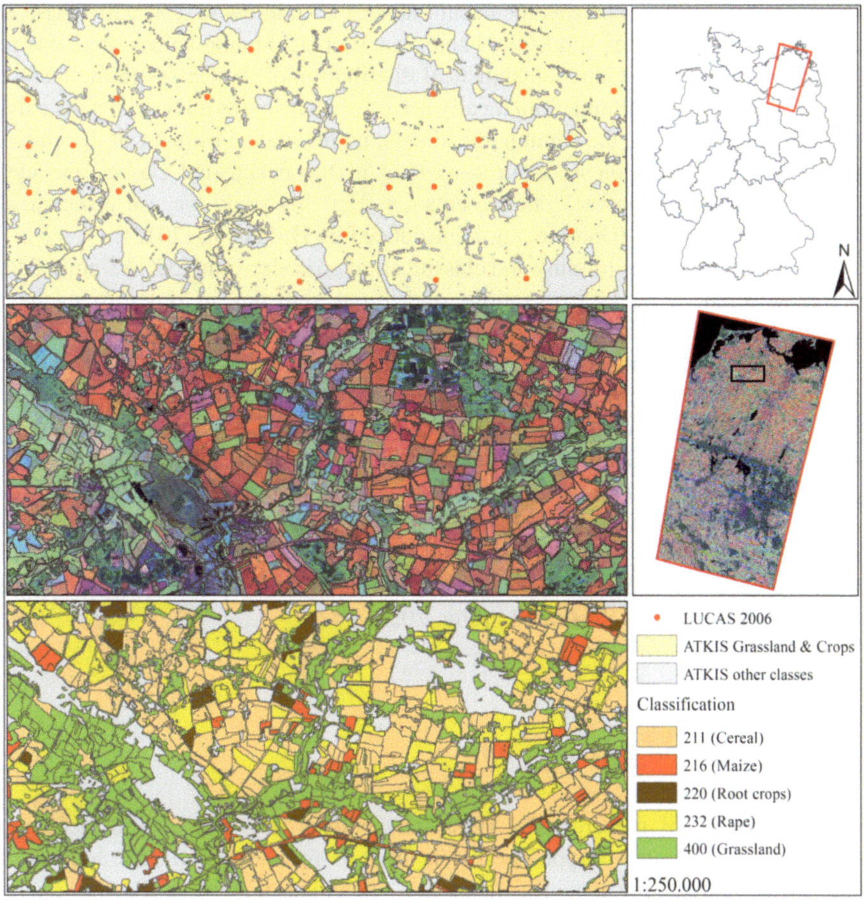

Fig. 20.3 Input reference information for the masking of agricultural areas (Geobasis data © German Federal Agency for Cartography and Geodesy – www.bkg.bund.de) and the automated training of the classification algorithm (*top*); multi-seasonal EO data set with objects resulting from an image segmentation of the corresponding imagery (*center*); result of the classification of main crops and grassland (*bottom*)

conducted solely using the two LISS-3 scenes with five iterations and a scale parameter increasing from 20 to 100 in an interval of 20. The result of the image segmentation is shown in Fig. 20.3.

20.3.3 Identification of Grassland and Main Crops

The classification of grassland and crops starts with the semi-automated generation of a sampling data base for the training of a decision tree classifier – in this case

C5.0 (Quinlan 1992). The specific capabilities of C5.0 allow to deal with classical challenges in time series and LC/LU analyses, such as missing values (e.g., due to cloud cover) or a non-Gaussian statistical distribution of the classes in the feature space. Since the algorithm automatically selects the attributes relevant for the class assignment along with the suitable thresholds, no manual adjustment of the classification parameterisation is required as long as the designated classes are adequately represented by the training data set. For this study we use a technical framework for the C5.0 classification introduced by Huth et al. (2012).

In preparation of the training and classification procedure, the LUCAS point data is pre-selected using GIS-operations and zonal statistics in order to exclude unconfident sample points, e.g., positioned at the border between two land parcels. The pre-selection results in a total set of 1,529 points, from which 50 % are used for training and 50 % for validation. The training process itself is based on all image objects that intersect with a LUCAS point, whereby the C5.0 algorithm automatically creates a classification decision tree that relates properties of each training object to its assigned LUCAS LC/LU class. The set of input features provided for each object includes the spectral bands of the five input scenes as well as the NDVI and seasonality layer stacks. The created classification tree is then applied to all image objects in order to classify grassland and four main crop types. The corresponding result is illustrated in Fig. 20.3.

20.4 Results and Discussion

The assessment of the crop and grassland analysis shows an overall accuracy (OA) of 86.0 % and a Kappa coefficient of 0.79 for the five classes addressed. The error statistic for the classification – including producers (PA) and user accuracies (UA) – is provided in Table 20.2. The outcome indicates that cereals (class 211), rape (class 232) and grassland (class 400) are assigned with PAs and UAs of more than 85 %, whereas maize (class 216) and root crops (class 220) show significantly lower accuracies between 63 and 78 %. The decreased values for these two classes can be attributed to the fact that root crops and maize as well as maize and grassland show a similar seasonal and spectral behaviour resulting in

Table 20.2 Accuracy for classification of main crops and grassland

Class	211	216	220	232	400	Samples	UA
211 (Cereals)	*269*	7	2	10	27	315	85.4
216 (Maize)	1	*33*	1	0	7	42	78.6
220 (Root crops)	0	4	*10*	0	1	15	66.7
232 (Rape)	8	0	1	*110*	4	123	89.4
400 (Grassland)	20	8	1	5	*243*	268	87.3
Samples	298	52	15	125	273	763	
PA	90.3	63.5	66.7	88.0	85.7		**OA: 86.0**

corresponding misclassifications. It has also to be considered that the number of sample points for maize and root crops is quite low so that the error statistics react more sensitive towards slight variations in the number of classification errors. Furthermore, the classifier provides information about the usage and significance of each feature. For this analysis, the NDVI and near-infrared (NIR) of the LISS-3 May as well as the AWiFS September scene had the highest importance, followed by the July AWiFS scene. The June AWiFS scene as well as the seasonality parameters did not provide much additional information to the classification and therefore had a minor influence on the analysis. However, the selection and significance (i.e. temporal) of each feature might vary depending on the geographical region (varying phenology depending on local climate) and the selected crop and/or grassland type.

In addition to the described analysis, we applied a version that aimed at the assignment of all 11 original agricultural LUCAS classes occurring in the study region. This classification shows an OA of 70.2 %. We also compared the results with the outcome of classifications based on one single scene (05.05.2006) and a dual-time data base (05.05.2006, 12.09.2006). Here, OAs of 69.3 % (single-date) and 76.3 % (dual-date) are achieved regarding the five-class-classification.

Considering the seasonal development of major crops and grassland (see Fig. 20.1), it appears that the harvesting period is best fitted to differentiate between cereals and grassland. However, in this phase the NDVI still shows similarities between grassland and maize as well as grassland and root crops. Therefore, the separation of arable crops and grassland requires additional data acquisitions during the spring season. In comparison to arable land, pasture land shows a more constant and homogenous behavior with respect to the development of green vegetation over the entire vegetation period. Looking at the seasonal behavior, the NDVI of grassland shows a distinct increase in spring (March, April) before it levels in early summer, followed by a slight drop during the dry season and a constant decrease from late summer to winter. Meadows affected by hay mowing feature a higher dynamic during the summer period. For arable crops, the vegetation development starts at different time slots. After a significant increase during the growing phase, the NDVI levels or slightly decreases in the maturation before it finally drops significantly during the harvesting period (i.e., cereals and rape).

20.5 Conclusions

The results of the study indicate that seasonal time series analyses afford the accurate identification and characterisation of main crops and grassland. Key issues in this context are the existence of adequate reference information (number of samples and their thematic reliability) and the availability of a constant seasonal and spatial coverage with multi-seasonal satellite data. A limiting factor for the required multi-seasonal EO data analysis is the frequently occurring cloud coverage in time series of optical sensors that clearly constraints the data availability – especially regarding HR imagery and data requirements for specific dates of the year (as explained in the chapter "Multi-scale Analysis of Seasonal Time Series Data"). This factor becomes particularly critical with respect to large-area assessments. The study demonstrated that the combination of HR and MR imagery represents a promising approach to ensure – or at least increase the chances – to achieve a significant coverage with seasonal data while at the same time assuring a high spatial detail of the analysis. An alternative approach to improve the data availability might be the use of radar data since the corresponding systems – e.g., TerraSAR-X, TanDEM-X or Radarsat-2 – can collect data independently of the weather and environmental conditions.

The presented object-based concept possesses effective options for the realisation of a multi-sensor/data, multi-scale and multi-season approach that in turn provides the flexibility which is required for the realisation of multi-seasonal analyses on regional or national scale. At the same time this approach represents a promising methodology for the frequent monitoring of trends and transformations of LC/LU – e.g., extensification or intensification in terms of a conversion from cropland to pastures or vice versa. It also facilitates the (semi-)automated update or thematic extension of existing geo-data layers with respect to general changes in LC/LU or relative changes in the intensity of use or productivity.

Acknowledgements The authors would like to thank the German Federal Agency for Cartography and Geodesy (BKG) for providing GeoBasis-DE data (ATKIS) for this study and the GAF AG and EUROMAP GmbH for the provision of IRS-P6 AWiFS data in the context of the IRS-P6 Scientific Data Pool.

References

AdV (2012) Arbeitsgemeinschaft der Vermessungs-verwaltungen der Länder der Bundesrepublik Deutschland: Erläuterungen zum ATKIS-Objektartenkatalog der Arbeitsgemeinschaft der Vermessungsverwaltungen der Länder der Bundesrepublik Deutschland. http://www.atkis.de. Accessed 28 Apr 2012

Baatz M, Schäpe A (2000) Multiresolution segmentation—an optimization approach for high quality multi-scale image segmentation, Angew. Geographische Informationsverarbeitung XII, J. Strobl et al (eds). Verlag Herbert Wichmann Verlag, Karlsruhe, pp 12–23

Blaes X, Vanhalle L, Defourny P (2005) Efficiency of crop identification based on optical and SAR image time series. Remote Sens Environ 96:352–365

Conrad C, Fritsch S, Zeidler J, Rücker G, Dech S (2010) Per-field irrigated crop classification in arid Central Asia using SPOT and ASTER data. Remote Sens 2(4):1035–1056

Conrad C, Machwitz M, Schorcht G, Löw F, Fritsch S, Dech S (2011) Potentials of RapidEye time series for improved classification of crop rotations in heterogeneous agricultural landscapes: experiences from irrigation systems in Central Asia. In: Proceedings of SPIE, 8174. SPIE remote sensing 2011, 19–22 Sep. 2011, Prague, Czech Republic. ISBN 9780819488015

Esch T, Thiel M, Bock M, Roth A, Dech S (2008) Improvement of image segmentation accuracy based on multi-scale optimization procedure. IEEE Geosci Remote Sens Lett 5(3):463–467

Huth J, Kuenzer C, Wehrmann T, Gebhardt S, Dech S (2012) Land cover and land use classification with TWOPAC: towards automated processing for pixel- and object-based image classification. Remote Sens 4(12):2530–2553

ISRO (2012) http://www.isro.org/satellites/irs-p6resourcesat-1.aspx

Itzerott S, Kaden K (2006) Ein neuer Algorithmus zur Klassifizierung landwirtschaftlicher Fruchtarten auf Basis spektraler Normkurven – an algorithm to classify agricultural crops on the basis of spectral standard curves. – Photogrammetrie, Fernerkundung, Geoinformation 6/2006, pp 509–518, Stuttgart

Martino L, Fritz M (2008) New inside into land cover and land use in Europe – Land use/Cover Area frame statistical Survey: methodology and tools. Eurostat – Statistics in focus 33, European Communities, Luxembourg

McNairn H, Shang J, Champagne C, Jiao X (2009) TerraSAR-X and RADARSAT-2 for crop classification and acreage estimation. In: Proceedings of IEEE International Geoscience and Remote Sensing Symposium (IGARSS) 2009, Cape Town, pp II-898-II-901

Metz A (2009) Knowledge based update of DLM-DE with remote sensing and geodata for deduction of a high resolution land use/land cover mapping using CORINE Land Cover. Diploma thesis, Dresden University of Technology

Quinlan JR (1992) C4.5 programs for machine learning. Morgan Kaufmann, San Mateo

Richter R, Schläpfer D (2011) Atmospheric/topographic correction for satellite imagery (2011), DLR report DLR-IB 565-02/11, Wessling, Germany, pp 202

Schiewe J, Tufte L, Ehlers M (2001) Potential and problems of multi-scale segmentation methods in remote sensing, Geo-Informations-Systeme 6:(34–39)

Schuster C, Ali I, Lohmann P, Frick A, Förster M, Kleinschmidt B (2011) Towards detecting swath events in TerraSAR-X time series to establish NATURA 2000 grassland habitat swath management as monitoring parameter. Remote Sens 3(7):1308–1322

Shao Y, Lunetta RS, Ediriwickrema J, Iiames J (2010) Monitoring agricultural cropping patterns across the Laurentian Great Lakes Basin using MODIS-NDVI data. Int J Appl Earth Obs Geoinf 12(2):81–88

Singh NJ, Kudrat M, Jain K, Pandeya K (2011) Cropping pattern of Uttar Pradesh using IRS-P6 (AWiFS) data. Int J Remote Sens 32(16):4511–4526

Smith AM, Buckley JR (2011) Investigating RADARSAT-2 as a tool for monitoring grassland in western Canada. Can J Remote Sens 37(1):93–102

Turker M, Ozdarici A (2011) Field-based crop classification using SPOT4, SPOT5, IKONOS and QuickBird imagery for agricultural areas: a comparison study. Int J Remote Sens 32(24):9735–9768

Wardlow BD, Egbert SL (2008) Large–area crop mapping using time–series MODIS 250 m NDVI data: an assessment for the U.S. Central Great Plains. Remote Sens Environ 112(3):1096–1116

Zhang X, Friedl M, Schaaf C, Strahler A, Hodges J, Gao F, Reed B, Huete A (2003) Monitoring vegetation phenology using MODIS. Remote Sens Environ 84(3):471–475

Chapter 21
Enhancing Remotely Sensed Low Resolution Vegetation Data for Assessing Mediterranean Areas Prone to Land Degradation

Christof J. Weissteiner, Kristin Böttcher, and Stefan Sommer

21.1 Introduction

Large parts of the Mediterranean region are sensitive eco-regions, which are susceptible to land degradation. It is widely accepted that changes of the vegetation density and structure over time bear important information of land degradation dynamics either caused by natural or man-made processes (e.g. Hanafi and Jauffret 2008). For monitoring the development of vegetation cover on a regional to global scale, archives of satellite data with coarse geometric but high temporal resolution are the preferable choice. These data allow a more continuous spatial and temporal monitoring of the area of interest due to their high revisit rate and long term record of observation. The currently longest back-dating times series with a capability to generate a simple vegetation index and including surface temperature are provided by the NOAA AVHRR sensors.

These long-term remote sensing time series constitute an important basis of land degradation and desertification monitoring and assessment. Land degradation processes can, amongst other techniques, be identified and quantified by change detection analyses in relation to soil/vegetation cover and indirectly land use/land

C.J. Weissteiner (✉)
Via Milite Ignoto 132, 21027 Ispra, Italy

Joint Research Centre (JRC), Institute for Environment and Sustainability,
via E. Fermi 2749, 21027 Ispra, Italy
e-mail: mail@weissteiner.eu

K. Böttcher
Department of Geoinformatics, Finnish Environment Institute,
Mechelininkatu 34 A, 00251 Helsinki, Finland
e-mail: Kristin.Bottcher@ymparisto.fi

S. Sommer
Joint Research Centre (JRC), Institute for Environment and Sustainability,
via E. Fermi 2749, 21027 Ispra, Italy
e-mail: Stefan.sommer@jrc.ec.europa.eu

I. Manakos and M. Braun (eds.), *Land Use and Land Cover Mapping in Europe: Practices & Trends*, Remote Sensing and Digital Image Processing 18,
DOI 10.1007/978-94-007-7969-3_21,

functioning, or by more complex techniques such as desertification syndrome modelling (Hill et al. 2008; Weissteiner et al. 2011a).

The NDVI is one of the most widely used indicators for monitoring the vegetation cover. However, it is not only determined as a pure function of vegetation cover but is also influenced by the background signal by i.e. soil and rock. Soils for instance can show NDVI values up to 0.3 (Price 1993). Especially land degradation vulnerable areas with low vegetation cover, typically the dry land regions in the Mediterranean, are affected by this limitation of detecting sparse vegetation cover. Additionally, the NDVI shows sensitivity to several parameters such as the atmosphere (Gutman 1991), the illumination and the observation geometry (Kaufmann et al. 2000), which is, however, supposed to be partly eliminated by a temporal maximum value compositing of the data (Holben 1986). Furthermore, the NDVI values are platform/sensor dependent due to different spectral system specifications as well as to orbit specific variation of observation geometry which complicates a direct comparison among different sensors.

Due to these problems it is preferable to find a measure for vegetation abundance which would, at least partly, overcome the mentioned problems. Since improved vegetation indices like the JRC-FAPAR (Gobron et al. 2007) are not yet available for the time before 1998, their use for long term assessments (e.g. trend analysis), as needed for land degradation studies, is currently limited. Therefore a long term time series of AVHRR data (MEDOKADS/Mediterranean Extended Daily One Km AVHRR Data Set) (Koslowsky and Bolle 2003; Koslowsky et al. 2005) with a time span of 17 years (for the presented study 1989–2005) was selected as base data set for the derivation of an enhanced vegetation abundance measure.

In this context of data enhancement Linear Unmixing has been recognized a promising approach (Sommer 1999), as hereby not only vegetation cover estimates are provided but also abundances of other applied endmembers and delivers hence data in the sub-pixel domain, which partly compensates for the coarse spatial resolution. The here presented and applied unmixing technique is based on the inverse relationship between NDVI and land surface temperature. Generally, surface temperature (Ts) is observed to be inversely proportional to the amount of vegetation canopy cover and thus to the NDVI. This is due to a variety of factors including latent heat transfer through evapotranspiration, the lower heat capacity and thermal inertia of vegetation compared to soil (Choudhury and Asrar 1989; Goward and Hope 1989).

The applied methodology to derive Green Vegetation Fraction (GVF) is expected to offer higher reliability and robustness than a simple vegetation index. It has been shown that using both NDVI and Ts, allows a characterization of land cover in a more comprehensive and climatically resistant manner than by multitemporal NDVI data alone (Nemani et al. 1993; Ehrlich and Lambin 1996).

In this work methodology and results for the derivation of an improved vegetation measure (compared to NDVI) will be presented. The derived measure is designed to be suitable for the Mediterranean area and to provide a highly reliable time series that can be employed in subsequent analyses. Indeed, two case studies which take advantage of the data set when assessing vegetation based parameters are reported. Compared to FAPAR (available from 1998 onwards) this NOAA AVHRR based time series dates back to 1989 and offers sensor compatibility, i.e. allows to date time spans of 25 years and more to be observed. By generation of useful side-products this vegetation measure achieves further added value.

21.2 Data

The MEDOKADS data set was provided by the Freie Universität Berlin. It consists of radiometrically and geometrically corrected data of all AVHRR channels, including derived data such as NDVI, surface temperature and ancillary data such as viewing angles or illumination geometry data. Despite its relatively coarse spatial resolution of 0.01° (approx. 1 km^2) it does still offer one of the most valuable data sets for long term vegetation analysis.

The AVHRR data set covers the wider Mediterranean region including the North African coast and the Near East region. The base data consisted of 10-day-composites, computed according to Holben and matching the 1st, 11th and 21st of each month. The data is provided in geographic coordinates, WGS-1984.

The time series consists of AVHRR data of different NOAA satellites (NOAA 11, NOAA 14, NOAA 16). Correction algorithms were applied to account for the changes of the short wavelengths channels 1 and 2 (AVHRR instrument changed from NOAA 14 to NOAA 16), the sensor degradation, and the changing illumination and observation geometry due to orbital drift. The BRDF effect was corrected by corrections for the sun zenith angle (cosine correction) and normalization to nadir view conditions. The land surface temperature (Ts) is derived by a split window approach, which uses the difference in brightness temperature between AVHRR channel 4 and 5 to account for the atmospheric effects on Ts. A more detailed description of all pre-processing steps is given in Weissteiner et al. (2008a).

21.3 Strategy

The status and dynamic of vegetation is usually assessed by vegetation indices (e.g. NDVI) when using NOAA-AVHRR data. Limitations for the detection of sparse vegetation with vegetation indices have been widely discussed in the literature. Several alternative indices, like GEMI or SAVI (Huete 1988; Pinty and Verstraete 1992) have been developed to compensate for atmospheric and illumination conditions and soil background reflectance.

Spectral mixture analysis (SMA) has been applied by different authors (Smith et al. 1990; Hill et al. 1995) to detect sparse vegetation cover from Landsat TM images in areas with spectrally diverse substrates. The difficulty encountered in the application of SMA to NOAA AVHRR data is related to the limited number of

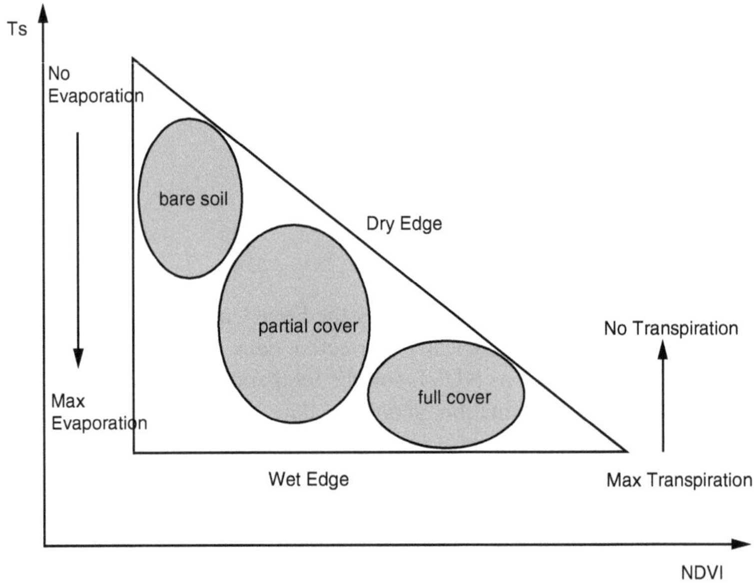

Fig. 21.1 Simplified Ts/NDVI (Lambin and Ehrlich 1996; Sandholt et al. 2002)

reflectance channels recorded, which hence, restricts the separability of different surface materials.

Lambin and Ehrlich (1995) used the ratio between Ts and NDVI for continental scale land-cover classification and demonstrated a clear increase of biome discrimination when integrating both thermal information and vegetation index. A more sophisticated technique to combine Ts and NDVI is described hereafter. In contrast to the classical SMA (Adams et al. 1986), the here presented approach uses a modified unmixing technique with NOAA AVHRR data based on the inverse relationship between vegetation cover and the surface temperature under dry conditions (Sommer 1999).

Due to factors like the latent heat transfer through evapotranspiration, the lower heat capacity and the thermal inertia of vegetation compared to soil, the surface temperature is inversely proportional to the amount of vegetation canopy, which itself is proportional to the NDVI. On small spatial scales the variations of different vegetation species and soil classes can show a high variability regarding surface temperature, while on coarse geometric resolution (e.g. AVHRR) the variation seems to be primarily caused by the vegetation fraction, vegetation physiology and physiognomy being of secondary importance (Nemani et al. 1993). Thus, linear approximations to explain NDVI as well as surface temperature of mixed AVHRR pixel (vegetation and non-vegetation) in relation to vegetation have been given (Fig. 21.1). Fractional cover should predominantly control the position of an AVHRR land surface pixel in the feature space of NDVI and Ts. Water, surface moisture and local meteorological conditions may influence this position further.

The concept of the classical linear spectral mixture analysis is based on the assumption, that the measured surface reflectance of a pixel is equivalent to the sum of the single reflectance from a limited number of pure materials – the so-called endmembers (EM) – depending on their pixel fraction. Mathematically, this assumption can be expressed as:

$$\rho_j = \sum\nolimits_{i=1}^{m} F_i \rho e_{i,j} + \varepsilon_j \tag{21.1}$$

Where ρ_j denotes the reflectance of the mixed spectrum in band j, F_i the fraction of an endmember i of the pixel, $\rho e_{i,j}$ the reflectance of an endmember spectrum i in channel j and ε_j is the residual error in band j. At the same time the proportions of the endmember have to fulfill the sum-to-unity constraint, which can be expressed as:

$$\sum\nolimits_{i=1}^{m} F_i = 1 \tag{21.2}$$

If the reflectance values of the endmembers are known, their fractions can be estimated for each pixel by solving a linear system of equations.

The scheme for the derivation of Green Vegetation Fraction (GVF), (term equally used with normalized vegetation abundance) from the NOAA/AVHRR time series is given in Fig. 21.2. Technically, the approach derives three endmembers for each decade (10 day interval), the non-vegetated EM, the vegetated EM and the cold EM. These EMs represent the edges of the space spanned up by NDVI and surface temperature (Fig. 21.1) and therefore build a distinct and robust model to estimate the vegetation cover. The cold EM works similar to the shade EM in the common SMA, accounting for effects that lower the surface temperature of the surface, including local gradients related to altitude and exposition, temperature variations due to soil moisture, variable evaporation and transpiration respectively and remaining cloud artefacts.

Price (1993) expressed the NDVI of a mixed NOAA/AVHRR pixel as a function of the vegetation cover fraction as

$$NDVI_{pixel} = F\,NDVI_{veg} + (1 - F)\,NDVI_{non-veg} \tag{21.3}$$

where $NDVI_{pixel}$ is the NDVI of a given pixel, F is the vegetation fraction, $NDVI_{veg}$ is the NDVI value for full vegetation and $NDVI_{non\text{-}veg}$ is the NDVI value for a non vegetated surface.

Likewise, the relationship between vegetation cover fraction and surface temperature was stated by Caselles and Sobrino (1989) as

$$Ts_{pixel} = F\,Ts_{veg} + (1 - F)\,Ts_{non-veg} \tag{21.4}$$

where Ts_{pixel} is the surface temperature of a pixel, Ts_{veg} the surface temperature for full vegetation and $Ts_{non\text{-}veg}$ denotes the surface temperature of a non vegetated surface. According to these relationships, the implemented spectral unmixing

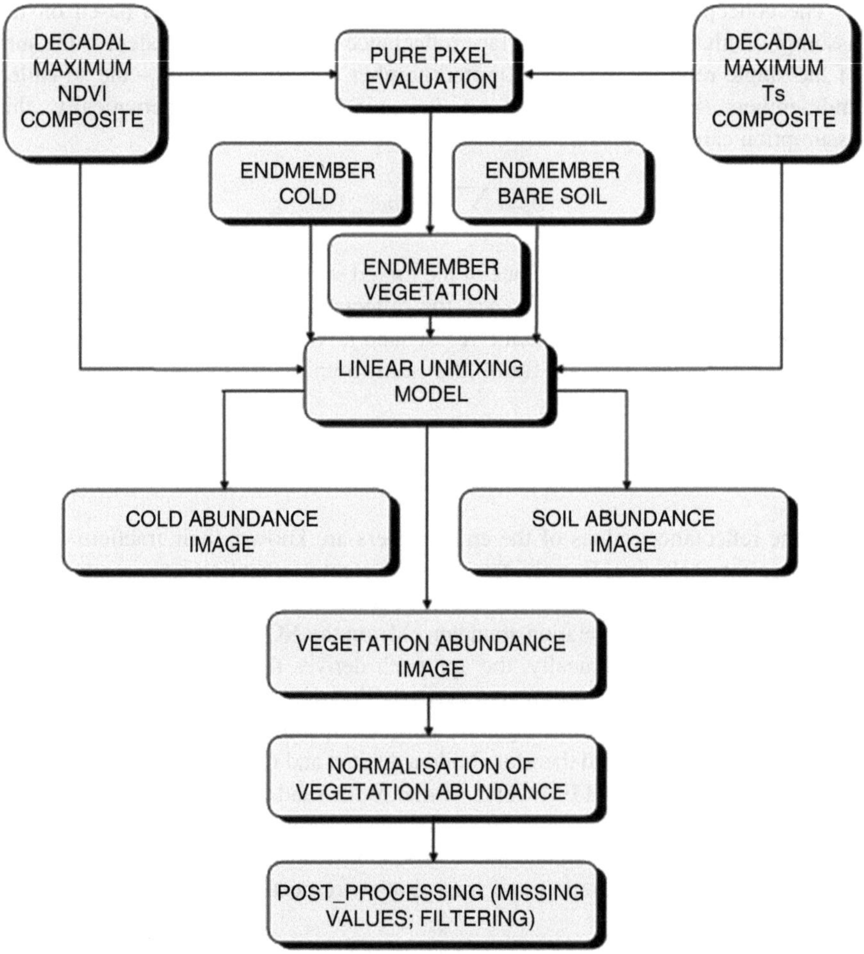

Fig. 21.2 Linear unmixing scheme applied to AVHRR time series for the assessment of vegetation coverage (After Sommer 1999)

approach acts on the assumption that vegetation cover should predominantly control the position of an AVHRR land surface pixel within the feature space formed by NDVI and surface temperature (European Commission 1998), which is illustrated by Fig. 21.3.

21.3.1 Endmember Selection from MEDOKADS Data

For the determination of the three endmembers, the Mediterranean area was subdivided into an eastern and western window (Fig. 21.4) to account for the east-west gradient of the vegetation – surface temperature relationship. To avoid

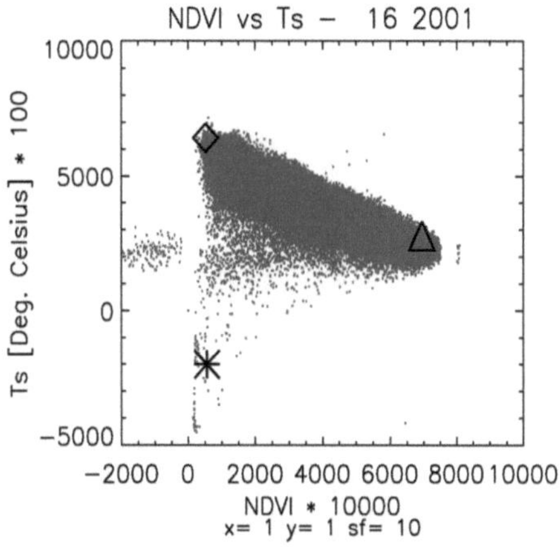

Fig. 21.3 Unmixing triangle for a scatterplot consisting of the NDVI (x-axis) and Ts (y-axis) for the 16th decade 2001 (western Mediterranean window). EMs are depicted as symbols: non-vegetated EM (◊), vegetated EM (Δ) and cold EM (*)

sharp transitions between the separately processed tiles for each pixel a linear interpolation of the derived endmembers was implemented. All criteria for the extraction of endmembers are documented in Table 21.1.

The NDVI-Ts relationship is not constant in time and space due to variable climatic conditions and the changing observation geometry. Therefore the "mixture triangle" was determined for each time step. An automated approach based on synthetic endmembers, following the method described by Stellmes et al. (2005), was applied. Endmembers were either defined upon theoretical considerations or extracted from the data using statistical methods.

In this context the NDVI value of the fully vegetated endmember was set to 0.7, which is the maximum NDVI value of vegetation for not atmospherically corrected NOAA AVHRR data according to literature (Czajkowski et al. 2004). To account for the dependency of the NDVI on the sun zenith angle (Θs) (Singh 1988), the maximum possible NDVI value is corrected for each decade. The relationship between sun zenith angle and maximum NDVI has been determined empirically. The presented approach allowed the extraction of this relationship separately for the periods from 1989–2000 and 2001–2004. The separation into two periods should take into account the varying response of the AVHRR/2 and AVHRR/3 instruments in the red and near infrared band. Since the NDVI of bare soil is almost not affected by Θ_s, its NDVI was extracted at the lower edge of the decadal NDVI distribution (1 % percentile) of the whole Mediterranean dataset. To avoid the influence of remaining cloud pixels on the NDVI of the non-vegetated EM a temperature threshold was introduced (0 °C). The NDVI of the cold EM was extracted separately for the eastern and western window at the lower edge of the decadal NDVI distribution (1 % percentile).

The temperatures of the fully vegetated and the non-vegetated EM were defined through the NDVI-Ts relationship. The relationship, which is described by a linear regression equation, was derived by the automatic approach described by Nemani et al. (1993).

Modifying the initial method described by Stellmes et al. (2005), the temperature of the cold EM was set to a fixed value (20 °C). Unlike other EMs, the cold EM can

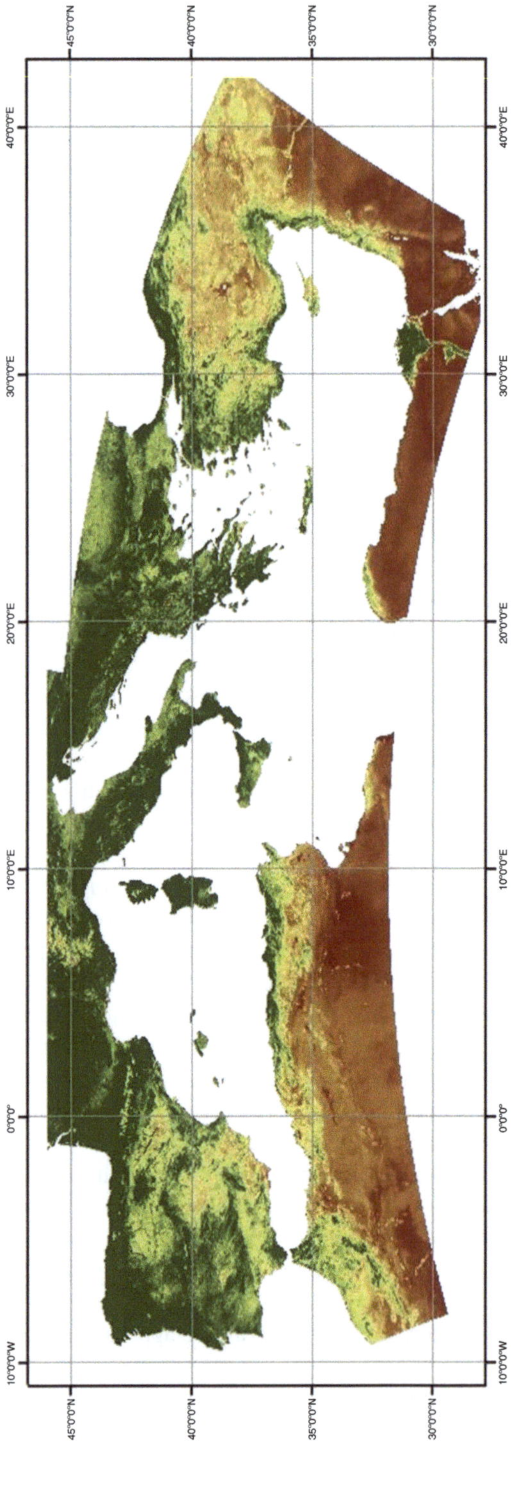

Fig. 21.4 Area covered for the study and subset windows used for the derivation of endmembers. The image depicts the derived average GVF for all years of the time series

Table 21.1 Selection criteria for endmembers in the NDVI-T_S feature space

Endmember	NDVI	Ts
Non-vegetated	Based on percentile at the lower edge of NDVI values (1 %), considering positive NDVI values only. To avoid effects of remaining cloud pixels to the NDVI of the extracted non-vegetated EM, a threshold of 0 °C (lowest allowed value) was applied.	Ts following Nemani (1993). Following this approach, the upperbound pixels of the Ts-NDVI scatter are extracted for each decade and fitted by a linear function. Gain and offset of this function are used to derive the temperature of the given NDVI.
	One window for the whole area	Number of window(s): n = 2
Fully vegetated	Max. NDVI is set to 0.7 and corrected then for the relationship sun zenith angle (corrected for orbital drift) versus NDVI (Holben and Fraser 1984; Singh 1988) – empirical approach, defined separately for the two time periods 1989–2000 and 2001–2004	Ts following Nemani (1993). Following this approach, the upperbound pixels of the Ts-NDVI scatter are extracted for each decade and fitted by a linear function. Gain and offset of this function are used to derive the temperature of the given NDVI.
	Number of window(s): n = 2,	Number of window(s): n = 2,
Cold	Based on percentile at the lower edge of NDVI values (1 %)	−20 °C
	Number of window(s): n = 2,	Number of window(s): n = 2

be considered as an artificial EM and is therefore less dynamic. The chosen temperature value represents occurring extreme values in the population of both windows (Eastern and Western Mediterranean). Linear interpolation was applied for EM outliers in the temporal domain.

21.3.2 *Unmixing*

The computation of proportional abundance can be principally explained and solved with a simple system of linear equations as follows:

$$A \cdot X = B \tag{21.5}$$

where A = m (channels) * n (endmember) matrix of spectral endmembers

X = n * 1 unknown vector of abundances
B = m * 1 observed data vector (mixed pixel NDVI and Ts)

The unknown vector of abundances is determined by inverting the endmember matrix A:

$$X = A^{-1} \cdot B \tag{21.6}$$

A unique solution is possible if the number of spectral endmembers corresponds to the number of spectral bands. Furthermore in the case of an underdetermined problem (this study) the number of unknown endmembers exceeds the number of bands by one, a solution can be found by assuming the set of endmembers is exhaustive (i. e., the sum of the computed endmember fractions is equal to one).

The unmixing procedure resulted in the three abundances images for each decade: the vegetation, non-vegetated (soil) and "cold" abundance.

21.3.3 Normalization of the Vegetation Abundance

The vegetation abundance was normalized according to the assumption that the "cold" component does not change the ratio between the other derived abundances. The abundance of the cold endmember was apportioned to the remaining endmembers by taking into account their fractional abundance through the following factor:

$$F = 1/(1 - F_{cold}) \tag{21.7}$$

This is applied up to a cold abundance of 30 %. If the cold abundance exceeds this value, this GVF modeling approach is rejected as invalid and these pixels are flagged as NODATA. The value 30 % was chosen, as it represents a value, when the cold EM starts getting predominant and this implies that the assumption of an inverse NDVI-Ts-relationship starts getting questionable. It was found for the Iberian Peninsula that the number of pixels above this threshold did not exceed 15 % during the period March–October, averaged over all years 1989–2005 (see Fig. 21.5).

The terms normalized vegetation abundance and green vegetation fraction (GVF) are used equally in this document.

21.3.4 Effect of the Cold Abundance on GVF

Does the cold abundance contribute to an improvement of the GVF and if so, to which extent? As reported earlier (see introduction), several effects may lead to a wrong NDVI. Amongst them, atmospheric effects, which are not eliminated by the maximum value compositing algorithm, play a major role. Generally, clouds and poor atmospheric conditions depress NDVI values. In the case of GVF, this effect should be attenuated by the normalization, when the cold abundance (if being between 0 and 30 %) is re-distributed to the remaining fractions.

The effect was tested for the subset of the Iberian Peninsula for the first decade of May of all years 1989–2005. Out of this subset, only pixels of supposed 'stable' land use types were used. Supposed 'stable' pixels were extracted from unfiltered NDVI data for each decade and pixel, considering the deviation to the long term

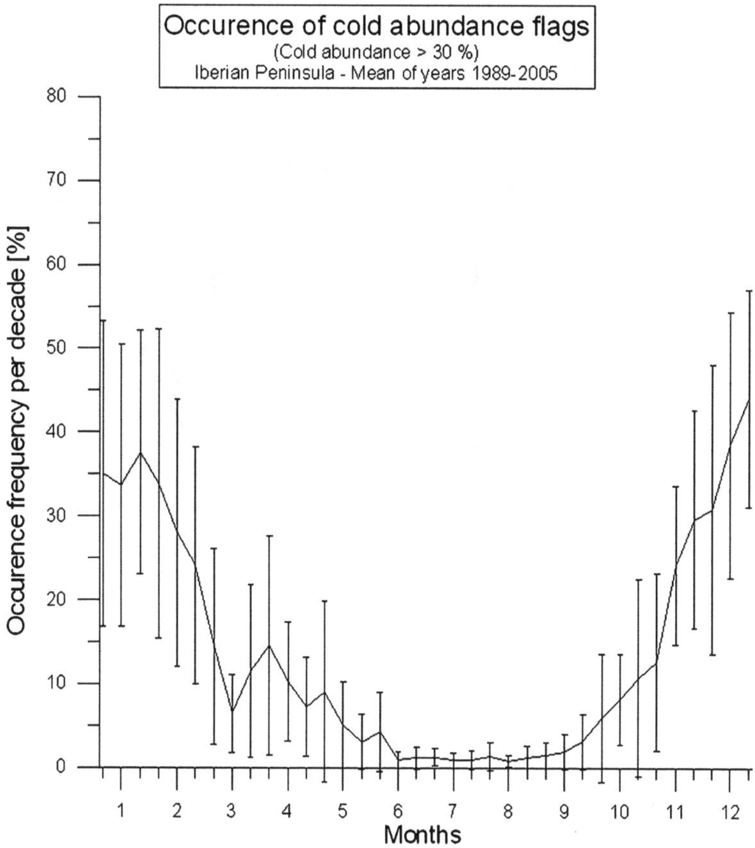

Fig. 21.5 Occurrence of cold abundance greater than 30 % (Iberian Peninsula, average and standard deviation 1989–2005)

median. These represented mainly forests and sclerophyllous vegetation, while urban areas and agricultural land use types were excluded. As Fig. 21.6 shows, both GVF and NDVI show negative deviations from their long term mean in the case of a high cold abundance of 20–30 % (except for the year 2005). However, the negative deviations (depressions) are clearly larger for the NDVI, the difference between GVF and NDVI ranging between −0.1 and −13.0 percentage points, in average over all years 1989–2005 amounting to −5.4 percentage points. It should be noted that the degree of compensation of poor atmospheric conditions via cold abundance is depending on the relation of vegetation and soil abundance: A strong compensation is theoretically possible in the case of abundances closed to 100 %, where the cold abundance has a strong absolute compensation effect with normalisation (see Eq. 21.7). In the case of equal values for soil and vegetation abundance the cold abundance is distributed equally and has hence a minor impact on the absolute value of the single fraction.

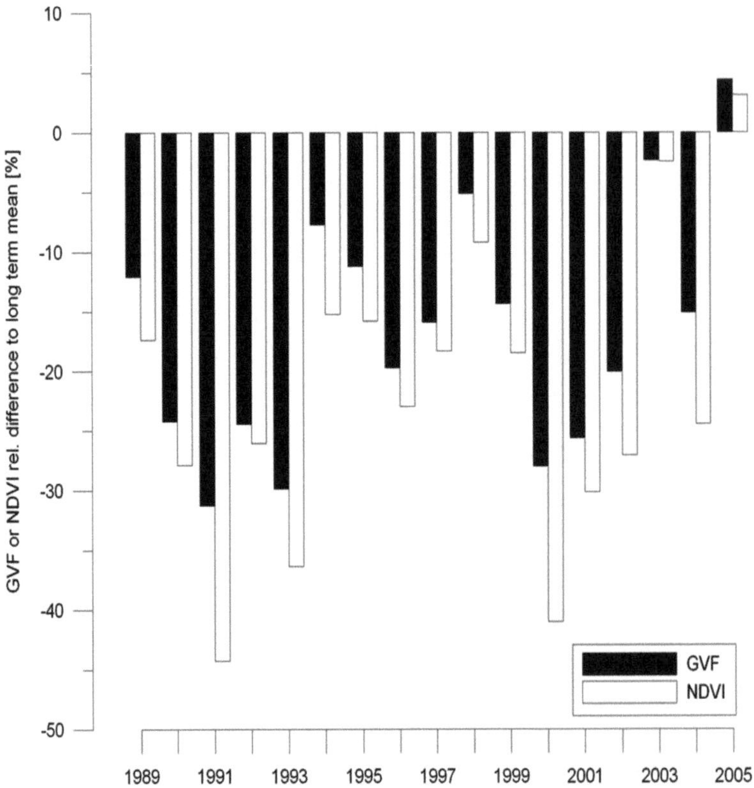

Fig. 21.6 Deviation of unfiltered GVF and NDVI to their long term mean for pixels of a (unfiltered) cold abundance between 20 and 30 %. Only 'stable' pixels (no land use change supposed) were used. Data refers the first decade of May 1989–2005 for the Iberian Peninsula

Keeping in mind, that the pixels used in Fig. 21.6 are supposed to be stable (no land use change, no significant seasonal change), it indicates, that the GVF is much closer to stability than the NDVI, or, in other words, the GVF is less prone to depression by bad atmospheric conditions than the NDVI is.

21.3.5 Post-processing of the GVF Time Series

Post-processing included a procedure to substitute missing data and outliers (both replaced by seasonal means). Data smoothing was done with a *Savitzky-Golay*-filter (Chen et al. 2004). The window size was set to six decades in forward and backward direction.

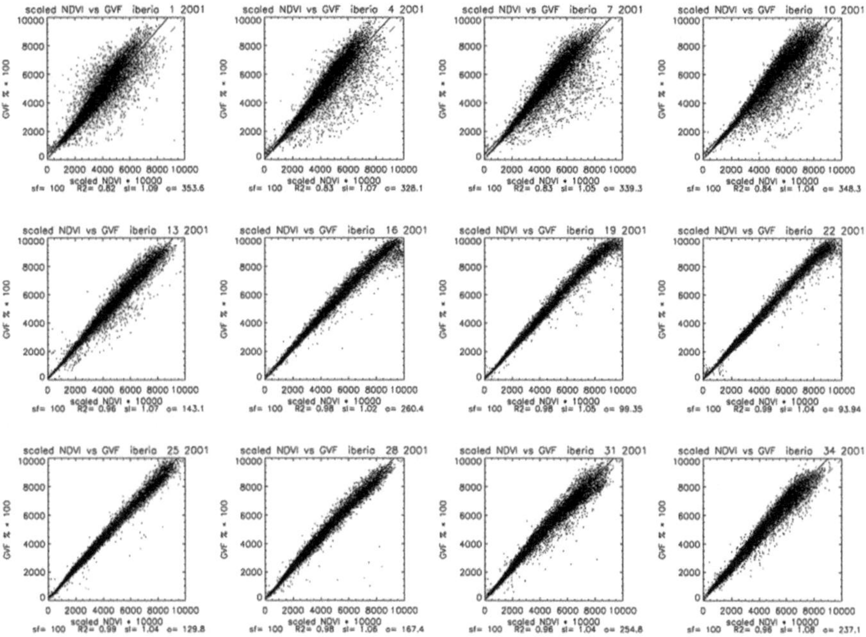

Fig. 21.7 Comparison MEDOKADS scaled NDVI (x-axis) vs. MEDOKADS GVF (y-axis) for 12 decades of the year 2001 (whole Iberian Peninsula). The *continuous line* indicates the correlation between both datasets, while the *dashed line* stands for the 1:1 relationship. Coefficient of determination (R2), slope (sl) and offset (o) of the regression line are indicated

21.3.6 Validation of Results

GVF was tested over different test sites. The aim was to test and compare GVF on sites with different amounts of vegetation cover and vegetation dynamics within the growing cycle. Additionally, GVF was compared with independently derived high and low resolution vegetation cover (fCover).

The agreement between an NDVI, scaled to the same range as GVF, is generally ranging around the 1:1 line (see Fig. 21.7). The agreement is particularly good during the vegetation period. Stronger differences are observed for the winter months, that is when the cold abundance is re-distributed between soil and vegetation abundance. Looking at specific test sites, the GVF was found slightly higher (in particular around the growing cycle peak) than the scaled NDVI, which can be attributed again to the mentioned normalisation.

The presence of the cold endmember has an additional positive effect, when the vegetation abundance is re-scaled to GVF. This step attenuates the typical effect of bad atmospheric conditions, usually depressing NDVI. For the case of GVF this typical depression has been found to be considerably less.

The comparison with independently derived vegetation cover (fCover) estimates from the VALERI[1] network, extracted from SPOT HRV imagery, revealed a linear relationship between GVF and fCover. However, GVF values result in average 15 % higher than fCover.

A comparison with the SPOT VEGETATION derived CYCLOPES fCover product (Baret et al. 2007) revealed again a linear relationship with (again) higher values for GVF.

Apparently, GVF is not identical to the fractional vegetation cover. Which relation exists between the two measures? Relations between a scaled NDVI and fractional vegetation cover are in fact mentioned in the scientific literature. In turn, the close relation between GVF and the scaled NDVI (almost 1:1) allowed an approximate calculation of fractional vegetation cover, as outlined below.

Choudhury et al. (1994) and Gillies and Carlson (1995) independently obtained an identical square root relation between a scaled NDVI and fractional vegetation cover, which was later confirmed by findings of Carlson and Ripley (1997) (see Eq. 21.8). Fractional vegetation cover is defined as green vegetation cover per unit horizontal surface area.

$$Fractional\ vegetation\ cover \approx (NDVI_{scaled})^2 \tag{21.8}$$

The obtained GVF was highly correlated to the scaled NDVI, the relation being almost 1:1, with slightly higher values for GVF, especially in the higher ranges. Assuming a direct 1:1 linear relation between the scaled NDVI and GVF, the square of the 'corrected' GVF should be a real measure of fractional vegetation cover, comparable to the CYCLOPES fCover.

$$Fractional\ vegetation\ cover \approx (GVF)^2 \tag{21.9}$$

For more details on the methodology and the comparison between independent data and GVF it is referred to Weissteiner et al. (2008b).

21.4 Discussion

21.4.1 Advantages of This Unmixing Approach

NOAA AVHRR is the longest available remote sensing time series. It is known that this satellite system is affected by some undesirable properties, e.g. high oscillations due to a bad signal to noise ratio or weak geometrical accuracy. Unlike a simple NDVI, which uses 2 channels only, the Unmixing is based on 4 NOAA

[1] http://www.avignon.inra.fr/valeri/

AVHRR channels and built on a physically based relationship between ratios or combinations of these channels. Certainly, Unmixing is depending on the base data as well as a vegetation index, but mitigates the risk of erroneous data propagation due to its multi-dimensional approach.

Another advantage of the Unmixing has to be regarded the fact, that in the here applied case the technique delivers three basic outcomes, a vegetation abundance, a bare soil abundance and a cold abundance. All of them can be utilized per se and present sub-pixel information. For obvious reasons the main focus of this work was the vegetation abundance, which we normalised (see Eq. 21.7) and hence corrected for the existence of the cold abundance (fourth outcome). The normalised vegetation fraction (GVF) represents a refined product for vegetation estimates, corrected for effects that lower the surface temperature of the surface, e.g. effects related to altitude and exposition, temperature variations due to soil moisture, variable evapo(transpi)ration and remaining atmospheric effects and cloud artifacts. Moreover, the position in the Unmixing triangle allows conclusions about occurring evaporation/transpiration potential.

While direct NDVI transferability of different sensors is critical, the GVF represents a better standardized and hence comparable product, which could be derived from any sensor delivering the necessary input variables NDVI and Ts. This comparability was shown for Landsat TM derived GVF and NOAA AVHRR derived GVF for the Ayora region in Spain by Stellmes et al. (2005). The relationship between both data sets was linear, almost 1:1.

The comparison with independently derived fractional vegetation cover shows clearly, that GVF contains considerable volumetric vegetation information, similar as NDVI does. Although the name Green Vegetation Fraction might be misleading in this sense, it was kept for historical reasons. However, the term 'normalized vegetation abundance' would reflect better the real properties of the parameter. GVF data is higher in comparison to CYCLOPES fCover although these validation data might be too low in general, as stated by the authors themselves (MEDIAS France 2006). GVF may be considered an improved vegetation index which is similar to NDVI, as for the volumetric vegetation information. Most important improvements of GVF in comparison to NDVI are the atmospheric disturbance mitigation and the widening of range (scaling). Also the derivation of useful by-products from the unmixing process and the enhanced cross-comparability of the data are valuable.

The approximation of a real fractional vegetation cover by squaring GVF leads to close results to the CYCLOPES fCover and confirms this approximation technique reported in literature.

21.4.2 Assumptions and Limitations

As reported above, the approach is based on an inverse relationship between NDVI and surface temperature Ts. This inverse relationship is best pronounced during the growing cycle, when some vegetation is on ground and acts as cooling agent. On the contrary, the relationship is not inverted for regions with extreme cold conditions, when vegetation acts as a warming rather than a cooling agent. Hence, this approach is limited to areas of moderate warm or hot temperatures and does not apply, e.g. for high mountainous regions.

Other regions of limited suitability for this approach are wetlands or areas of high moisture content. The presence of surface water biases the relationship NDVI-Ts as the cooling effect does mainly derive from the surface water. These areas, however, are recorded by the system as areas with a high 'cold' abundance and are eventually excluded.

Also, over evergreen forests, without a moisture availability constraint, the Ts-NDVI relation is modified compared to water limited environments. This is often the case in the tropics. At the scale of a decade the slope of the Ts-NDVI relation is positive (Lambin and Ehrlich 1996). However, this does not apply for the Mediterranean area with its classical dryland areas.

A crucial issue for unmixing is the identification of the endmembers (EM). Since scaling of the data is determined by the EMs, these should be chosen with care. In order to guarantee the comparability of the derived data, windows for the extraction of EMs were kept large. Still, the different climatic conditions for Western and Eastern Europe are attempted to be treated differently by the two chosen windows.

The quality of the derived GVF data set depends considerably on the base MEDOKADS data. Although enormous effort was put into correction of the MEDOKADS base data, there are still deficiencies, as reported earlier e.g. the decline of the year 2000 mainly due to the late overpass of NOAA14 in its late operating period. Even though some problems could be mitigated by empirical approaches, some deficiencies linked to atmospheric effects and illumination effects (BRDF) could not be fully corrected. In comparison to the NDVI, the GVF was generally showing a closer temporal profile to the independent data set FAPAR.

An additional factor to consider is the switch from AVHRR/2 to AVHRR/3 instrument, which occurred after the year 2000. The effect of AVHRR/3 is generally a higher NDVI, due to different band widths in the red and infrared wavelength range. The switch can not be easily corrected and was hence not corrected at all for MEDOKADS. The effect is expected to be mitigated by (decadal) Unmixing. A proper demonstration of positive Unmixing effects on the 'instrumental switch' is difficult since the 'orbital drift' effect is overlaid.

21.5 Case Studies

The derived GVF data has been made available to the public and can be downloaded (http://desert.jrc.ec.europa.eu/action/php/index.php?action=view&id=154). The data has been used in a number of studies, ranging from erosion studies on catchment scale to phenology analysis and many more. Two applications, where the data had been of particular importance and usefulness, are mentioned hereafter. These

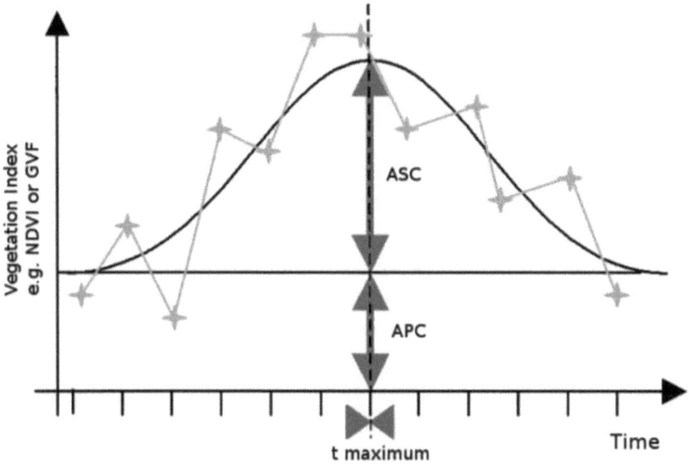

Fig. 21.8 Illustration of observed and modelled permanent and seasonal component by an exemplary typical seasonal run of green vegetation

applications are referring to case studies in the Mediterranean, in one case deriving olive farming intensities and in the second case mapping of rural land abandonment. Both studies are relying on trends of extracted GVF time series parameters, describing the temporal dynamics and phenology related parameters for the years between 1990 and 2004. The parameters are expressing the dynamic (seasonality) as well as stable portions of GVF throughout the year.

Figure 21.8 depicts schematically the dimensions of the extracted vegetation dynamics parameters. The seasonal component is referred to as absolute seasonal component (ASC), and expresses the dynamically changing proportion of the vegetation cycle while the permanent component (absolute permanent component/APC) is formed by the permanent background of the vegetation presence. Both ASC and APC are expressed in the same units as GVF (or any other vegetation index). Besides that, both proportions can be expressed in relative units (%), normalizing them by the overall GVF (e.g. normalized seasonal component/NSC is calculated as ASC/overall GVF). Moreover, the model allows an extraction of the timing of vegetation peak. The extraction of the mentioned parameters can be extracted from large remote sensing time series by the SINFIT model (Weissteiner et al. 2008a).

21.5.1 Olive Farming Intensities

Olive groves are mapped by the European Corine land cover (CLC) survey, which has been carried out in 1990, 2000 and 2006 so far. However, since this survey is based on land-cover, there is no distinction made between varying farming intensities. When in 2007 the High-Nature-Value farmland (HNV) had been to be re-mapped, experts were rising the problem, that HNV in the case of olive groves would need an additional input based on farming intensity, to achieve an accurate assignment (or not) to the HNV areas. Up to that point the assignment to HNV has primarily be done by land-cover data. Due to the above mentioned advantages the GVF data set had been chosen to derive vegetation dynamics parameters, which, together with CLC data to restrict the analysis to mapped olive areas could be successfully used to map olive farming systems in different classes, which, in turn were attributed to the three prevalent olive farming systems according to Beaufoy (2001). The strategy applied is taking advantage from three main points. First, the presence or abundance of the olive grove understory is primarily used to determine the farming intensity, second, the olive tree crown-cover is relatively stable for the majority of cultivations, and third, the olive tree itself is evergreen and hence not significantly disturbing the observations of changing groundcover. If crown-cover and olive tree dynamics can be considered quasi stable, the vegetation dynamics can be attributed to the ground vegetation, which, in turn are supposed to determine the farming type and intensity. Although the spatial resolution of NOAA AVHRR data is around 1 km, the use of ancillary data (CLC), which has been used to restrict the analysis to distinct olive grove areas, and the particular methodology design has allowed a sound assessment of olive groves, which in other terms would require data of a much higher spatial resolution. The analysis was carried out focusing on two distinct target years, which were selected in accordance with the available CLC data (1990 and 2000). To level out undesirable effects, such as exceptional or extreme impacts on vegetation (e.g. droughts), GVF data have been averaged over target and surrounding years. Applying the methodology to two distinct target years allowed a change detection analysis, resulting in further findings such as intensification and extensification strenghts and locations. A subset of the olive farming type classification is depicted in Fig. 21.9. More details can be found at Weissteiner et al. (2011b).

21.5.2 Rural Land Abandonment

The second case study has been using GVF trends over the whole time span of the time series. In this case, trend analysis of vegetation dynamics parameters were used as input for a study of rural land abandonment in the Mediterranean. However, the GVF based trends, are only one part within a set of indicators which were combined to several intermediate levels of meaningful results using appropriate aggregation methods, and were finally combined to the rural land abandonment (RLA) indicator. Similar as for the olive farming intensities, not land-cover but land functionality (relation between land cover, land use and the provision of goods and services by the land system) were used to monitor land change, as proposed by Verburg et al. (2009). These authors could show for a land abandonment study that this process may not have any consequences in land-cover, since poorly represented in this type of data. They state that functionality often requires local and contextual factors

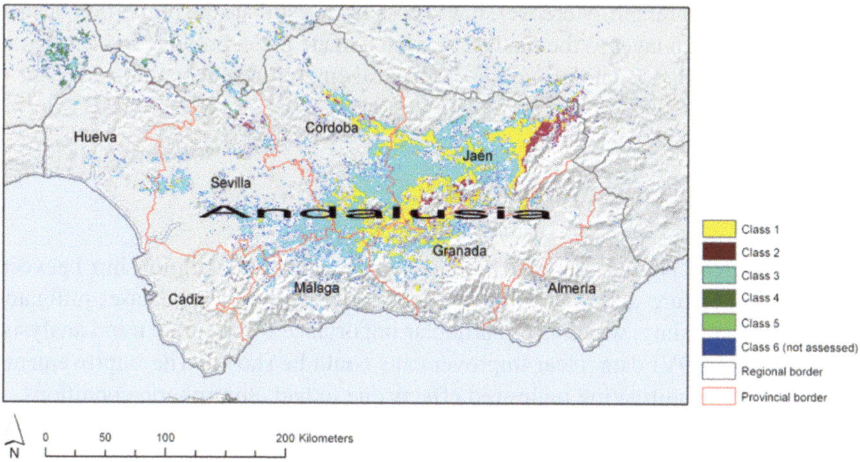

Fig. 21.9 Modeled olive farming classes for Andalusia in Southern Spain (data subset)

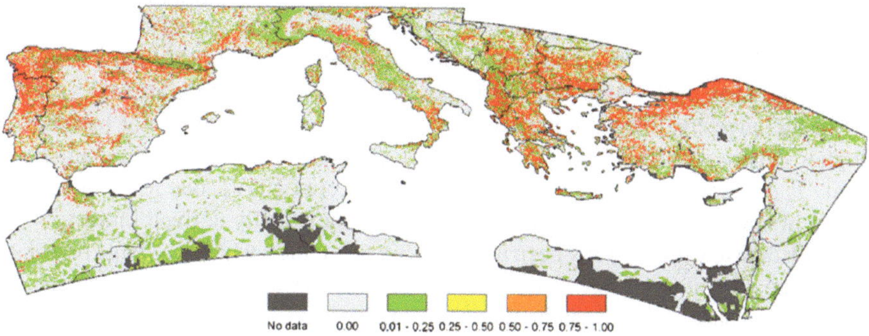

Fig. 21.10 Rural Land Abandonment indicator (*RLA*) for the Mediterranean

synchronously, which can be provided by a set of different data sets of varying nature (i.e. bio-physical and socio-economic data). In the particular case study of RLA, GVF data provided trends of observed vegetation trends (vegetation symptom likeliness indicator – VELI), which were in a later step combined with the land abandonment disposition indicator (LADI), composed of physical environmental conditions (PECI) and socio-economic conditions (SECI), to create the final rural land abandonment index (Fig. 21.10). VELI was created itself by aggregation of three different observed vegetation trends, the seasonal and the permanent proportion of the vegetation cycle, and the total productivity of the annual vegetation cycle, representing a proxy for Net Primary Productivity (NPP). In a similar way PECI and SECI are aggregated group layers (e.g. created out of soil data, economic data, etc.). For more specific details refer to Weissteiner et al. (2011a). By applying modern soft fusion techniques for data aggregation (Bordogna et al. 2012) the final RLA not only delivered a highly detailed map of occurrence of this syndrome in the Mediterranean, but also gave an indication of the syndrome strength. Moreover, it was possible to estimate the proportion of each contributing group layer to the final RLA, which made it e.g. possible to identify the main causes of RLA (socio-economic or physical environmental conditions).

21.6 Conclusions and Outlook

Employing four NOAA AVHRR channels and a physical relationship between surface temperature and NDVI for the generation of this GVF dataset mitigates base data uncertainty, which is of particular importance for accurate trend analysis. Compared to NDVI data, clear improvements could be shown. The improvements are in particular mitigating undesired effects due to bad atmospheric conditions.

The methodology not only delivers a reliable data set of a vegetation measure (GVF) but also abundances of all other employed endmembers (non-vegetated EM, cold EM), each of them utilizable as standalone products.

The GVF or any other EM-abundance represent sub-pixel information, exploiting the limited NOAA AVHRR potentialities to a maximum.

GVF and fractional vegetation cover (fCover) is not identical, since GVF contains also a volumetric component of vegetation. A relationship between GVF and fCover has been found and confirms similar findings in literature.

In this investigation, in comparison to a scaled NDVI, it could not be found a higher sensitivity of GVF for scarcely vegetated areas. However, the higher sensitivity is certainly given in comparison to a non-scaled NDVI, due to the extended range of GVF, exploiting the full margin within statistically derived limits.

GVF data has successfully been used in two case studies, where a reliable base data set for the derivation of status and trends of vegetation was of crucial importance.

The derivation of GVF can be run in an almost operational way. Due to its design GVF is less sensor dependent than NDVI. This enables its use as monitoring measure for long term observations and its application for suitable upcoming sensors.

Acknowledgements Part of this work was performed under the IP DESURVEY project (IP Contract FP6 GCR Programme Contract No 003950) funded by the EC DG RTD 6th Framework Programme. The MEDOKADS data set was prepared and made available by Dirk Koslowsky and colleagues from the Freie Universität Berlin.

References

Adams JB, Smith MO, Johnson PE (1986) Spectral mixture modelling: a new analysis of rock and soil types at the Viking Lander 1 Site. J Geophys Res 91:8098–8112

Baret F, Hagolle O, Geiger B, Bicheron P, Miras B, Huc M, Berthelot B, Niño F, Weiss M, Samain O, Roujean JL, Leroy M (2007) LAI, fAPAR and fCover CYCLOPES global products derived from VEGETATION. Part 1: principles of the algorithm. Remote Sens Environ 110:275–286

Beaufoy G (2001) The environmental impact of olive oil production in the European Union: practical options for improving the environmental impact. 1 February 2007. http://ec.europa.eu/environment/agriculture/pdf/oliveoil.pdf

Bordogna G, Boschetti M, Brivio PA, Carrara P, Stroppiana D, Weissteiner CJ (2012) Handling heterogeneous bipolar information for modelling environmental syndromes of global change. Environ Model Softw 36:131–147
Carlson TN, Ripley DA (1997) On the relation between NDVI, fractional vegetation cover, and leaf area index. Remote Sens Environ 62:241–252
Caselles V, Sobrino JA (1989) Determination of frosts in orange groves from NOAA-9 AVHRR data. Remote Sens Environ 29:23–26
Chen J, Jönsson P, Tamura M, Gu Z, Matsushita B, Eklundh L (2004) A simple method for reconstructing a high-quality NDVI time-series data set based on the Savitzky-Golay filter. Remote Sens Environ 91:332–344
Choudhury BJ, Asrar G (1989) Estimating evaporation and carbon assimilation using infrared temperature data: vistas in modelling. In: Gassem A (ed) Theory and applications of optical remote sensing. Wiley, New York, pp 628–690
Choudhury BJ, Ahmed NU, Idso SB, Reginato RJ, Daughtry CST (1994) Relations between evaporation coefficients and vegetation indices studied by model simulations. Remote Sens Environ 50:1–17
Czajkowski KP, Goward SN, Mulhern T, Goetz SJ, Walz A, Shirey D, Stadler S, Prince SD, Dubayah RO (2004) Estimating environmental variables using thermal remote sensing. In: Quattrocchi DA, Luvall JC (eds) Thermal remote sensing in land surface processes. CRC Press, Boca Raton, pp 11–32
EC (European Commission) (1998) An integrated approach to assess and monitor desertification processes in the Mediterranean basin (DeMon-II). Final report. Directorate General Joint Research Centre, Ispra, Contract no.: 11589-95-12 A2 FP ISP B
Ehrlich D, Lambin EF (1996) Broad scale land cover classification and interannual climatic change. Int J Remote Sens 17:845–862
Gillies RR, Carlson TN (1995) Thermal remote sensing of surface soil water content with partial vegetation cover for incorporation into climate models. J Appl Meteorol 34:745–756
Gobron N, Pinty B, Mélin F, Taberner M, Verstraete MM, Robustelli M, Widlowski J-L (2007) Evaluation of the MERIS/ENVISAT FAPAR product. Adv Space Res 39:105–115
Goward SN, Hope AS (1989) Evapotranspiration from combined reflected solar and emitted terrestrial radiation: preliminary FIFE results from AVHRR data. Adv Space Res 9:239–249
Gutman GG (1991) Vegetation indices from AVHRR: an update and future prospects. Remote Sens Environ 35:121–136
Hanafi A, Jauffret S (2008) Are long-term vegetation dynamics useful in monitoring and assessing desertification processes in the arid steppe, southern Tunisia. J Arid Environ 72:557–572
Hill J, Mégier J, Mehl W (1995) Land degradation, soil erosion and desertification monitoring in Mediterranean ecosystems. Remote Sens Rev 12:107–130
Hill J, Stellmes M, Udelhoven T, Röder A, Sommer S (2008) Mediterranean desertification and land degradation. Mapping related land use change syndromes based on satellite observations. Global Planet Change 64:146–157
Holben BN, Fraser RS (1984) Red and near-infrared sensor response to off-nadir viewing. Int J Remote Sens 5:145–160
Holben BN (1986) Characterization of maximum value composites from temporal AVHRR data. Int J Remote Sens 7:1417–1434
Huete AR (1988) A soil adjusted vegetation index (SAVI). Remote Sens Environ 25:295–309
Kaufmann RK, Zhou L, Knyazikhin Y, Shabanov NV, Myneni RB, Tucker CJ (2000) Effect of orbital drift and sensor changes on the time series of AVHRR Vegetation Index Data. IEEE Trans Geosci Remote Sens 38:2584–2597
Koslowsky D, Bolle HJ (2003) 12 years mediterranean satellite data set and analysis. In: Bolle HJ (ed) Mediterranean climate. Springer, Berlin/Heidelberg/New York, pp 165–177
Koslowsky D, Billing H, Friedrich K (2005) MEDOKADS: a long-term data set for detection and monitoring of desertification risks in the Mediterranean. In: Roeder A, Hill J (eds) Remote sensing and geoinformation in the assessment and monitoring of land degradation and desertification, Trier, Germany, pp 191–198

Lambin EF, Ehrlich D (1995) Combining vegetation indices and surface temperature for land-cover mapping at broad spatial scales. Int J Remote Sens 16:573–579

Lambin EF, Ehrlich D (1996) The surface temperature-vegetation index space for land cover and land-cover change analysis. Int J Remote Sens 17:463–487

MEDIAS France (2006) CYCLOPES (Carbon cYcle and Change in Land Obeservational Products from an Ensemble of Satellites) EGV2-2001-00035 (FP5)- Products read me

Nemani P, Pierce L, Running SW, Goward S (1993) Developing satellite derived estimates of surface moisture status. J Appl Meteorol 32:548–557

Pinty B, Verstraete MM (1992) GEMI: a non-linear index to monitor global vegetation from satellites. Plant Ecol 101:15–20

Price JC (1993) Estimating leaf area index from satellite data. IEEE Trans Geosci Remote Sens 31:727–734

Sandholt I, Rasmussen K, Andersen J (2002) A simple interpretation of the surface temperature/vegetation index space for assessment of surface moisture status. Remote Sens Environ 79:213–224

Singh SM (1988) Simulation of solar zenith angle effect on global vegetation index (GVI) data. Int J Remote Sens 9:237–248

Smith MO, Ustin SL, Adams JB, Gillespie AR (1990) Vegetation in deserts: I. A regional measure of abundance from multispectral images. Remote Sens Environ 31:1–26

Sommer S (1999) Regional desertification indicators. Final report MEDALUS III. ENV4-CT95-0121

Stellmes M, Sommer S, Hill J (2005) Regional vegetation monitoring based on the NOAA AVHRR time series in the Mediterranean Area. In: Erasmi S, Cyffka B, Kappas M (eds) GGRS 2004 – 1st Göttingen GIS & Remote Sensing Days – environmental studies. Göttinger Geographische Abhandlungen, vol 113, pp 124–130

Verburg PH, van de Steeg J, Veldkamp A, Willemen L (2009) From land cover change to land function dynamics: a major challenge to improve land characterization. J Environ Manage 90:1327–1335

Weissteiner CJ, Sommer S, Strobl P (2008a) Time series analysis of NOAA AVHRR derived vegetation cover as a means to extract proportions of permanent and seasonal components at pixel level. EARSEL eProc 7:11–20

Weissteiner CJ, Böttcher K, Mehl W, Sommer S, Stellmes S (2008b) Mediterranean-wide green vegetation abundance for land degradation assessment derived from AVHRR NDVI and surface temperature 1989 to 2005. EUR report, European Commission, IES, Luxembourg, 978-92-79-09777-5, 43 p

Weissteiner CJ, Boschetti M, Böttcher K, Carrara P, Bordogna G, Brivio PA (2011a) Spatial explicit assessment of rural land abandonment in the Mediterranean area. Global Planet Change 79:20–36

Weissteiner CJ, Strobl P, Sommer S (2011b) Assessment of status and trends of olive farming intensity in EU-Mediterranean countries using remote sensing time series and land cover data. Ecol Indic 11:601–610

Chapter 22
Beyond NDVI: Extraction of Biophysical Variables From Remote Sensing Imagery

J.G.P.W. Clevers

22.1 Introduction

By sight humans can use visible differences in the reflection of the sunlight to recognize vegetation colour. Part of the reflected radiation observed with remote sensing (RS) techniques coincides with the part visible to the human eye. When looking at the electromagnetic (EM) spectrum used for optical RS, visible light (VIS) constitutes the first range of wavelengths. In the VIS, ranging from 0.4 to 0.7 μm, various pigments, such as chlorophyll (green), xanthophyll (yellow), and carotene (orange), influence the reflection. This reflectance is a characteristic of an object and it is often plotted against wavelength. It is called the spectral signature. In most plant species two types of chlorophyll (*a* and *b*) determine the reflection, mainly by absorption of blue and red light and to a lesser degree of green light (cf. Fig. 22.1). The energy in these spectral bands is used for the displacement of electrons and initiates the synthesis of carbohydrates from atmospheric CO_2 and absorbed groundwater. Green-yellow chlorophyll *a* is present in all photosynthesizing plants. Higher plants and green algae contain blue-green chlorophyll *b*, although in small quantities. Both chlorophylls absorb the visible light to a large extent, and have two absorption peaks: one in the blue (approx. 0.45 μm) and one in the red (approx. 0.65 μm) region of the EM spectrum. As a result of this and also of the hypersensitivity of the eye to green, vegetation reveals itself to the eye in various shades of green. Subsequently, the peak of the reflectance in the VIS occurs at approx. 0.54 μm. Spectral measurements in the VIS thus may provide information on pigment concentrations of vegetation, although the signal coming from vegetation is relatively low due to the large absorption. This strong absorption also causes that in the VIS the reflectance of only the top canopy layer determines the total reflectance of a vegetation canopy. Soils do not show this strong absorption

J.G.P.W. Clevers (✉)
Centre for Geo-Information, Wageningen University,
P.O. Box 47, 6700 AA, Wageningen, Netherlands
e-mail: jan.clevers@wur.nl

I. Manakos and M. Braun (eds.), *Land Use and Land Cover Mapping in Europe: Practices & Trends*, Remote Sensing and Digital Image Processing 18,
DOI 10.1007/978-94-007-7969-3_22, © Springer Science+Business Media Dordrecht 2014

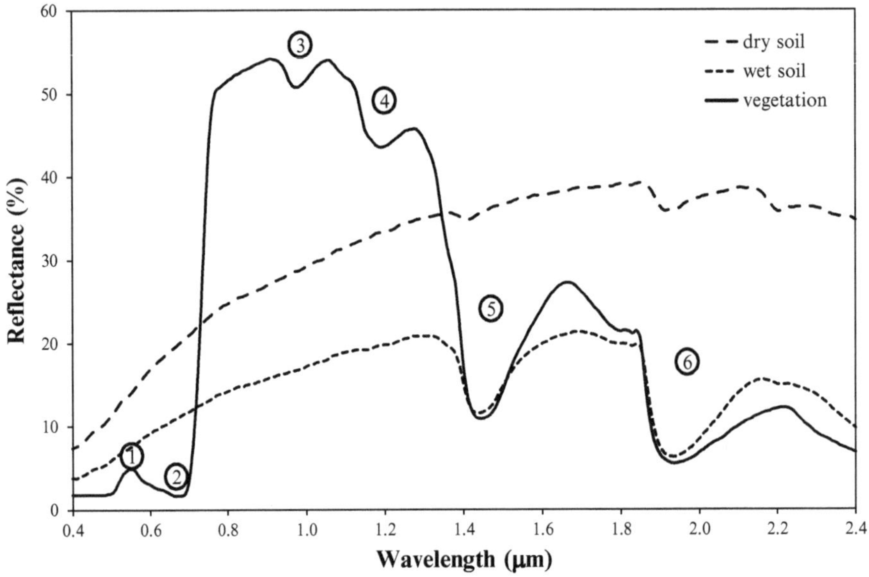

Fig. 22.1 Typical spectral reflectance curves for dry soil, wet soil and vegetation. *1*: 0.54 μm; *2*: 0.65 μm; *3*: 0.97 μm; *4*: 1.20 μm; *5*: 1.40 μm; *6*: 1.90 μm

due to pigments and therefore they mostly show a strong contrast with green vegetation in the VIS. As a result, spectral measurements in the VIS not only provide information on pigments, but also on the percentage of soil covered by vegetation (fCover) and properties related to this quantity, such as the fraction of absorbed photosynthetically active radiation (fAPAR).

The second range of wavelengths from 0.7 to 1.3 μm (near-infrared radiation, NIR) is mainly determined by the absence of absorption by pigments (see Fig. 22.1). This means that the radiation passes through the leaf (the leaf is transparent) or that it is reflected. Approximately 50 % of the NIR radiation is reflected by the leaf. However, this percentage varies widely for different plant species. It has been established for this range of wavelengths that a leaf becomes very transparent if the air channels between the cells of the leaf are filled with fluid. This gave reason to a theory that reflection takes place in the leaf at the transition of air and cell walls (Knipling 1970). Since a green leaf hardly absorbs any NIR radiation, leaves or canopy layers under the top layer contribute significantly to the total measured reflectance. This multiple reflectance denotes the NIR reflectance to be particularly suitable for estimating the so-called leaf area index (LAI) ("counting the number of leaf-layers").

In the third region of wavelengths ranging from 1.3 to 2.5 μm (called middle-infrared (MIR) or shortwave infrared (SWIR)), a great deal of radiation is absorbed by water in the cells (see Fig. 22.1). The figure shows that the major absorption peaks fall at 1.90 and 1.40 μm. It should be pointed out that weak absorption bands of water also occur at 1.20 and 0.97 μm.

Clevers (1999) has shown that the VIS can be considered as one region providing information for estimating plant properties. Since spectral bands (either narrow-band or broad-band) mutually are highly correlated, there is little added value to be expected by combining spectral bands in indices based on just the VIS. The NIR is another principal region for estimating vegetation properties. Again, spectral bands in the NIR are highly correlated and little added value is to be expected by combining spectral bands in just the NIR. A third region is the SWIR, which mainly provides information on water content in vegetation studies, and the same conclusion can be drawn as for the VIS and NIR. Clevers (1999) showed that there is a fourth region that may provide significant information on vegetation properties, in addition to the mentioned three regions. This is the so-called red-edge region exhibiting a steep rise in plant reflectance between 0.67 and 0.78 μm.

In addition to the above-mentioned main spectral regions for deriving vegetation properties, some specific regions related to specific absorption features may be of interest. This is, e.g., used by applying spectroscopy in soil mineralogy. In dry vegetation samples information on nitrogen can be obtained in the SWIR region. E.g., at 1.51 μm an absorption feature occurs due to the first overtone of the N-H band vibration and at 2.18 μm an absorption feature occurs due to the second overtone (Curran 1989). In living plant material these features are obscured by the absorption effects of water. Only the minor water absorption features at 0.97 and 1.20 μm are features that provide specific information on vegetation that is detectable from a remote sensing point of view (see Sect. 22.4.4).

To describe the relationship between spectral measurements and biophysical and chemical variables of vegetation both statistical and physical approaches have been used. As an example of statistical methods, numerous indices have been developed for estimating leaf and canopy properties (Myneni et al. 1995a). Radiative transfer (RT) models are highly suitable for studying the relationship between biophysical variables and reflectance or vegetation indices (VIs) and to study the effect of sources of variability (Combal et al. 2003). Subsequently, RT models may be used to determine 'universal' VIs that are site and species independent by calibrating VIs on large simulated datasets. A good index would be an index only sensitive to the variable of interest and not to other variables (cf. Verrelst et al. 2008). Section 22.2 provides some remarks on using radiative transfer models, whereas Sect. 22.3 gives an overview of the field of vegetation index development.

22.2 Radiative Transfer Models

A number of physically-based models, which account for the interactions of incident radiation with vegetation canopies, have been developed. Radiative transfer (RT) models have been used both in forward and inverse mode. In forward mode, RT model simulation allows validation and intercomparison of different RT model implementations (Myneni et al. 1995b) and sensitivity studies of canopy variables relative to diverse observation specifications, for an improved understanding of the RS signals and an optimized instrument design for future Earth observation systems (Bacour et al. 2002). For the retrieval of vegetation properties, RT models need to be inverted with RS data as input. For a successful inversion, one has to choose an appropriate and well-validated RT model that matches the spatial scale and correctly represents the RT of the observed target (Pinty and Verstraete 1992).

Models are traditionally being developed at the leaf and at the canopy level. Leaf RT models physically simulate reflectance and transmittance of plant leaves, which can be used by canopy level RT models to compute phase functions for multiple scattering. Canopy RT models can be classified either based on their dimensionality or based on the RT solution. In terms of dimensionality, there are two types of models: (i) one-dimensional (1D) models that require homogeneity in the horizontal dimension, such as turbid medium models, and (ii) three-dimensional (3D) models that handle heterogeneity and discontinuity in both horizontal and vertical dimensions (like needed for modeling forests), such as geometric-optical or hybrid models. The latter combine features from turbid medium and geometric models.

The combined PROSPECT leaf optical properties model and SAIL canopy bidirectional reflectance model, also referred to as PROSAIL, is the most popular RT model to study plant canopy spectral and directional reflectance in the solar domain (Jacquemoud et al. 2009). The PROSPECT model is a radiative transfer model for individual leaves. It simulates leaf spectral reflectance and leaf spectral transmittance as a function of leaf chlorophyll content (C_{ab}), leaf water content (C_w) and a leaf structure parameter (N). PROSPECT is also including leaf dry matter (C_{dm}) as a simplification for the leaf biochemistry (protein, cellulose, lignin). The one-layer SAIL radiative transfer model simulates canopy reflectance as a function of canopy parameters (leaf reflectance and transmittance, LAI and leaf inclination angle distribution), soil reflectance, ratio of diffuse/direct irradiation and solar/view geometry (solar zenith angle, zenith view angle and sun-view azimuth angle). It also takes the hot spot effect into account by considering the relative leaf size. This hot spot is a peak in the directional reflectance commonly observed in vegetation canopies when the sun and observer are at the same position, meaning that no shadows are observed. The output of the PROSPECT model can be used directly as input into the SAIL model. As a result, these models can be used as a combined PROSPECT-SAIL model.

Deriving biophysical properties from PROSAIL is feasible due to the relatively small number of input variables required for PROSAIL (Jacquemoud et al. 2000). In principle, inversion is performed by minimizing the difference between simulated and measured reflectance based on some sort of cost function and possible constraints for the model input variables (either set to an a priori value or allowed to vary within a plausible range). For the inversion process, a wide range of minimization techniques have been used: classical iterative optimization, simulated annealing, genetic algorithms, look-up tables, Monte-Carlo Markov chains and generalized likelihood uncertainty estimation. However, classical iterative optimization techniques, look-up tables and neural networks have been the most widely used (Liang 2004). Although the number of input variables is limited, we mostly still are dealing with an underdetermined problem since the number of unknowns to be estimated is larger than the number of independent spectral observations (even in case of hyperspectral sensors). Use of constraints then becomes necessary. Another problem is the ill-posedness of the inversion, meaning that different combinations of input variables yield the same spectral model output. Regularization techniques are then required for obtaining stable solutions. An overview of the state-of-the art is provided by Baret and Buis (2008). For deriving biophysical variables first the top-of-canopy (TOC) is ascertained. Inversion of TOC observations has proven to be successful in the last decades, enabling the production of high level data products (Garrigues et al. 2008). Recently, Laurent et al. (2011) demonstrated the direct use of measured top-of atmosphere (TOA) radiance data to estimate forest biophysical and biochemical variables, by using a coupled canopy–atmosphere radiative transfer model. Advantages of this approach are that no atmospheric correction is needed and that atmospheric, adjacency, topography, and surface directional effects can be directly and more accurately included in the forward modelling.

22.3 Vegetation Indices

Many investigations have been conducted to assess vegetation characteristics, such as biomass and LAI, by means of combinations of reflectances in various spectral bands. Such a combination of reflectance values, the vegetation index (VI), also serves to correct for undesirable influences of varying soil reflectance or atmospheric conditions on the results. These kinds of disturbances are particularly undesirable in spatial and multitemporal analyses. Most commonly used VIs are based on red and NIR spectral bands, because the large difference between red and NIR reflectance of dense green vegetation is a unique feature. Generally, indices are divided into ratio and orthogonal indices. Whereas ratio-based indices are calculated independently of soil reflectance properties, orthogonal indices refer to a base line specific for the soil background. More recently, indices have emerged that can be considered hybrid versions of the classic ratio and orthogonal indices.

The first investigations into vegetation indices concerned the NIR/red ratio by Rouse et al. (1974, 1973). Rouse and his colleagues found this ratio to be suitable – when applied to satellite data – for the estimation of crop characteristics owing to a partial correction for the solar position and atmospheric influence. They also used the normalized vegetation index for the same purpose. Often, this type of vegetation index is called the normalized difference vegetation index (NDVI):

$$NDVI = \frac{(NIR - red)}{(NIR + red)} \tag{22.1}$$

In order to find an index independent of the influence of the soil, Richardson and Wiegand (1977) introduced the so-called perpendicular vegetation index (PVI):

$$PVI = \sqrt{(red_v - red_s)^2 + (NIR_v - NIR_s)^2} \tag{22.2}$$

where subscripts v and s refer to the vegetation and the underlying bare soil, respectively. The increase in the amount of vegetation agreed with the offset of the reflectances perpendicular (orthogonal) to a so-called soil line in a NIR-red feature space plot.

A similar approach for suppressing variations of the soil influence has been developed by Kauth and Thomas (1976). They applied a linear transformation to the four-dimensional data space of Landsat MSS measurements of agricultural regions with various soil types, called the tasselled cap transformation. This transformation of the four Landsat-1 MSS bands resulted in a so-called brightness index dominated by soil differences and a so-called greenness index dominated by green vegetation:

$$\text{Soil brightness} = 0.43 \times \text{MSS4} + 0.63 \times \text{MSS5} + 0.59 \times \text{MSS6} + 0.26 \times \text{MSS7} \tag{22.3}$$

$$\text{Greenness} = -0.29 \times \text{MSS4} - 0.56 \times \text{MSS5} + 0.60 \times \text{MSS6} + 0.49 \times \text{MSS7} \tag{22.4}$$

MSS4: 0.5–0.6 μm, MSS5: 0.6–0.7 μm, MSS6: 0.7–0.8 μm, MSS7: 0.8–1.1 μm.

Later the tasselled cap transformation was also applied to the spectral bands of the Landsat Thematic Mapper (Crist and Cicone 1984). The soil brightness can be considered as a multidimensional soil line and the greenness is orthogonal to this soil line, in essence similar to the PVI concept.

In order to obtain a more precise correction for soil background, Huete (1988) developed the soil adjusted vegetation index (SAVI). This index was further improved by Baret et al. (1989) yielding the transformed soil adjusted vegetation index (TSAVI). Different researchers made further versions of the SAVI, resulting e.g. in an adjusted TSAVI (ATSAVI), second version of SAVI (SAVI2) and a second modified SAVI (MSAVI2) (Broge and Leblanc 2000).

A semi-empirical approach for estimating LAI of a green canopy, introduced by Clevers (1988, 1989), resulted in the so-called weighted difference vegetation index (WDVI). In this model it is assumed that in the multitemporal analysis the soil type is given and the soil moisture content is the only varying property of the soil. For estimating LAI a weighted difference between the measured NIR and red reflectances was ascertained, assuming that the ratio of NIR and red reflectances of bare soil is constant, independent of soil moisture content (which assumption is valid for many soil types). Subsequently, this WDVI was used for estimating LAI according to the inverse of an exponential function. Basically, the WDVI is a 2-dimensional greenness index, and as such also strongly related to the PVI. WDVI is calculated as:

$$WDVI = NIR - C \times red \tag{22.5}$$

$$C = NIR_s / red_s \tag{22.6}$$

where subscript s again refers to soil reflectances.

Up till now, a large set of vegetation indices have been developed, mainly for estimating vegetation cover, LAI, biomass, pigment content, water content and related (indirect) quantities. Various studies have compared many different indices for estimating one of these vegetation variables (Broge and Mortensen 2002; Gong et al. 2003; Haboudane et al. 2004; Schlerf et al. 2005; Thenkabail et al. 2002; Zarco-Tejada et al. 2005). The performance of the various indices always is different, depending on the specific data sets used for the study, resulting each time in different indices as being the best one. This makes it difficult to compare the various studies. One should always consider the theoretical background of an index, its validity range and its purpose, and then use one index as much as possible rendering results that are mutually comparable spatially and temporally. In the next section focus will be on the main biophysical variables that can be estimated with RS techniques.

22.4 Biophysical Variables

22.4.1 Chlorophyll and Nitrogen

In vegetation studies nitrogen and chlorophyll have always played an important role. A sufficient supply of nitrogen is crucial for the biochemistry of plants since nitrogen is an important component in proteins, nucleic acids (e.g., DNA, RNA) and chlorophyll (*a* and *b*). Photosynthesis is the source of energy and of carbon in all organic compounds in plants. This photosynthesis takes place in reaction centers that contain chlorophylls. Plants having a shortage of nitrogen will have a lower chlorophyll concentration resulting into a non-optimal photosynthesis. This results into not only a reduced plant growth but also a reduced carbon fixation. We see that nitrogen and chlorophyll concentrations often are highly correlated in plants (Jongschaap and Booij 2004; Yoder and Pettigrew-Crosby 1995). Actual estimates are relevant for many application fields ranging from local scale such as precision farming up to global scales dealing with the global carbon cycle.

Since the VIS should be considered as one band of information, few vegetation indices have been developed based on bands in the VIS solely. For estimating chlorophyll content actually the main one is the photochemical reflectance index (PRI) as developed by Gamon et al. (1990). The PRI is presented as an index for estimating the green shift, centered near 531 nm, caused by reflectance changes associated with the de-epoxidation of violaxanthin to zeaxanthin. As such this provides information on canopy photosynthesis, in particular the light use efficiency (Gamon et al. 1997). Recently, Garbulsky et al. (2011) provided a review of the scientific literature on the relationship between PRI and photosynthetic efficiency or related variables across a range of plant functional types and ecosystems. However, experiences with the PRI are varying.

Haboudane et al. (2002) gave a typical example how radiative transfer (RT) models can be used for development of an index for estimating the chlorophyll content from hyperspectral data. They established and tested the ratio of two optical indices, namely the transformed chlorophyll absorption in reflectance index, TCARI (Daughtry et al. 2000), and the optimized soil-adjusted vegetation index, OSAVI (Rondeaux et al. 1996). This resulted into the TCARI/OSAVI ratio. Similarly, the ratio of the modified chlorophyll absorption in reflectance index, MCARI, and the OSAVI was tested, MCARI/OSAVI (Daughtry et al. 2000).

The red-edge region as mentioned before is a special region that has often been used for estimating chlorophyll and nitrogen content at both leaf and canopy level. Physically it is the content at canopy level that we expect to estimate with RS. Collins (1978) and Horler et al. (1983) were among the first researchers to point out the importance of the red-NIR wavelength transition for vegetation studies. Both the position and the slope of this red-edge change under stress conditions, resulting in a shift of the slope towards shorter wavelengths (Horler et al. 1983). As an index mostly the position of the inflexion point on the red-NIR slope is used. This is called the red-edge position (REP), and it will be influenced by both the LAI and the chlorophyll concentration (Clevers et al. 2001). It was shown before to be a good estimate for chlorophyll content, but being less sensitive at higher contents. This saturation effect is still a problem. There are various ways to calculate this REP (Clevers et al. 2004). Guyot and Baret (1988) applied a simple linear model to the red-infrared slope. This approach is feasible for satellite data like obtained with the Medium Resolution Imaging Spectrometer, MERIS (Clevers et al. 2002).

Another type of index based on the red-edge slope has been developed specifically with the advent of MERIS: the MERIS terrestrial chlorophyll index, MTCI (Dash and Curran 2004). It is proposed as a better index than the REP.

Wu et al. (2008) suggested to replace the traditional red and NIR spectral bands in indices like MCARI, TCARI and OSAVI by spectral bands in the red-edge region, particularly a band at 705 nm instead of the traditional red band at 670 nm, and a band at 750 nm instead of the band at 800 nm in the traditional MCARI and TCARI. They found that this resulted into indices that have better linearity with chlorophyll content and are thus more suitable. This band replacement is also consistent with the results of the sensitivity analysis by Gitelson and Merzlyak (1996).

Gitelson et al. (2003, 2006a, b) presented a simple index based on a NIR band and a red-edge band (e.g., at 710 nm) to estimate chlorophyll concentration: the so-called chlorophyll index ($CI_{red\text{-}edge} = R_{780}/R_{710}-1$). He also presented a variant using a green band instead of the red-edge band (CI_{green}). Major advantage of these latter two indices would be their linear relationship with chlorophyll and the absence of the saturation effect as obtained with the REP indices.

Clevers and Kooistra (2012) tested the potential of the above-mentioned VIs for retrieving canopy chlorophyll and nitrogen content. The formulae of the indices are given in Table 22.1. Main results are summarized in Table 22.2. They showed through PROSAIL model simulations that out of the above-mentioned VIs the $CI_{red\text{-}edge}$ performed best in estimating canopy chlorophyll content showing a linear

Table 22.1 Vegetation indices evaluated in the study of Clevers and Kooistra (2012)

Index	Formulation
REP	$700 + 40\frac{(R_{670}+R_{780})/2-R_{700}}{R_{740}-R_{700}}$
MTCI	$(R_{754} - R_{709})/(R_{709} - R_{681})$
MCARI/OSAVI	$\frac{[(R_{700}-R_{670})-0.2(R_{700}-R_{550})](R_{700}/R_{670})}{(1+0.16)(R_{800}-R_{670})/(R_{800}+R_{670}+0.16)}$
TCARI/OSAVI	$\frac{3[(R_{700}-R_{670})-0.2(R_{700}-R_{550})(R_{700}/R_{670})]}{(1+0.16)(R_{800}-R_{670})/(R_{800}+R_{670}+0.16)}$
MCARI/OSAVI[705,750]	$\frac{[(R_{750}-R_{705})-0.2(R_{750}-R_{550})](R_{750}/R_{705})}{(1+0.16)(R_{750}-R_{705})/(R_{750}+R_{705}+0.16)}$
TCARI/OSAVI[705,750]	$\frac{3[(R_{750}-R_{705})-0.2(R_{750}-R_{550})(R_{750}/R_{705})]}{(1+0.16)(R_{750}-R_{705})/(R_{750}+R_{705}+0.16)}$
$CI_{red\text{-}edge}$	$(R_{780}/R_{710}) - 1$
CI_{green}	$(R_{780}/R_{550}) - 1$

$R_\lambda$refers to the reflectance factor at wavelength λnm

Table 22.2 Overview of R^2 values of the linear relationships between indices and chlorophyll (PROSAIL) and nitrogen (grass and potato) content (Clevers and Kooistra 2012)

Index	PROSAIL	Grass	Potato
REP	0.49	0.79	0.84
MTCI	0.83	0.80	0.89
MCARI/OSAVI	0.25	0.06	0.09
TCARI/OSAVI	0.39	0.58	0.19
MCARI/OSAVI[705,750]	0.93	0.57	0.87
TCARI/OSAVI[705,750]	0.91	0.75	0.71
CI_{green}	0.94	0.77	0.88
$CI_{red\text{-}edge}$	0.94	0.77	0.88

relationship over the full range of potential values. In contrast, highly non-linear relationships of chlorophyll content with most traditional red-edge indices were found. Subsequently, they performed tests with field data for sampling locations within an extensively grazed fen meadow using ASD FieldSpec measurements and within a potato field with a Cropscan radiometer for estimating canopy nitrogen content. Also at these study sites the $CI_{red\text{-}edge}$ was found to be a good and linear estimator of canopy nitrogen content (no chlorophyll was measured) for both the grassland site and the potato field (Clevers and Kooistra 2012). Currently, this approach can, e.g., be applied with MERIS, Hyperion and RapidEye data and with the upcoming Sentinel-2 and -3 systems. An example of the relationship between $CI_{red\text{-}edge}$ and nitrogen content of potatoes is shown in Fig. 22.2. For a more detailed analysis of the data and description of results the reader is referred to Clevers and Kooistra (2012).

22.4.2 Vegetation Cover Fraction (fCover) and fAPAR

As stated in the introduction, radiation in the VIS can be used by plants for photosynthesis. Therefore, this is called photosynthetically active radiation (PAR). The rate of photosynthesis can be calculated from the amount of absorbed PAR (the APAR) and the photosynthesis-light response of individual leaves.

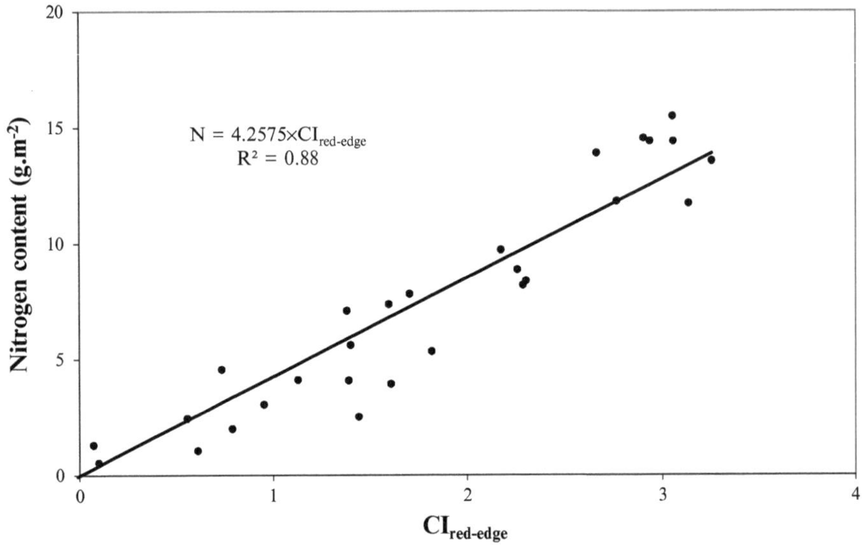

Fig. 22.2 Relationship between $CI_{red\text{-}edge}$ and canopy nitrogen content for a potato study site in The Netherlands (Clevers and Kooistra 2012)

Estimating APAR over some time interval (e.g., daily APAR) requires both incident PAR and the (average) fraction of APAR by vegetation, which is called fAPAR. Daughtry et al. (1992) showed that daily APAR may be reliably computed from measurements of the fAPAR near solar noon and daily incident PAR. The fAPAR is considered one of the main essential climate variables related to the terrestrial ecosystem (WMO/IOC 2010). It is strongly correlated to the vegetation cover fraction (fCover) (Bacour et al. 2006). fCover corresponds to the gap fraction of vegetation either measured from above or from below in the nadir viewing direction. fAPAR, conversely to fCover, depends on the illumination conditions.

The large contrast in reflectance between bare soil and vegetation in the VIS will be most suitable for estimating fCover and fAPAR. Since this contrast not only depends on the amount of vegetation but also on the moisture content of the soil, a single band is not suitable and a vegetation index should be used. Again, due to the strong mutual correlation of bands in the VIS, mostly a combination of VIS and NIR bands is used. Clevers et al. (1994) showed that a linear relationship may be assumed between WDVI or NDVI and fAPAR. External factors such as soil brightness, diffuse/direct irradiation ratio and solar zenith angle only have a minor influence on such a relationship between WDVI and fAPAR. Moreover, leaf parameters such as chlorophyll content, mesophyll structure and hot spot parameter (see Sect. 22.2) also have quite a small influence for green leaves. The leaf angle distribution (LAD) is the main parameter influencing the relationship between WDVI and fAPAR. So, for different LADs different regression functions should be used. Although the relationship between NDVI and fAPAR is slightly less influenced by LAD and solar zenith angle, important disturbing factors are the soil brightness and leaf chlorophyll content. Similar results have been found in other studies (Asrar et al. 1992; Goward and Huemmrich 1992). So, the WDVI is to be preferred over the NDVI.

22.4.3 Leaf Area Index

Leaf area index (LAI) is defined as the total one-sided green leaf area per unit ground area and it is regarded a very important canopy characteristic because photosynthesis takes place in the green plant organs. Moreover, it is also considered an essential climate variable (WMO/IOC 2010). Most of the vegetation indices mentioned in Sect. 22.2 have been used for estimating LAI. Clevers and Verhoef (1993) have performed an extensive sensitivity analysis using a combination of the PROSPECT leaf reflectance model and the SAIL canopy reflectance model towards the relationship between WDVI and LAI. As expected according to the Lambert-Beer law for light extension in a canopy, this is an exponential relationship. The influence of external factors such as soil brightness, diffuse/direct irradiation ratio and solar-view geometry hardly have an effect on the relationship between WDVI and LAI. Moreover, leaf parameters such as chlorophyll content, mesophyll structure and hot spot parameter also have only a small influence for green leaves at near nadir observation (like is the case for many satellite observations). The main variable influencing the relationship between WDVI and LAI is the leaf angle distribution (LAD). So, for different LADs different regression functions should be used. An example of calibration lines for estimating LAI using the WDVI for a range of agricultural crops in the Netherlands is provided by Bouman et al. (1992). Figure 22.3 shows an example of the WDVI – LAI relationship for barley from the original WDVI paper of Clevers (1989). He applied the inverse of a special case of the Mitscherlich function (Mitscherlich 1920) for estimating LAI:

$$LAI = -1/\propto \times Ln(1 - WDVI/WDVI_{\infty}) \qquad (22.7)$$

22.4.4 Canopy Water

Currently, one of the main scientific issues in studying global climate change is to understand the role of terrestrial ecosystems and the changes they may undergo. The water cycle is one of their most important characteristics (ESA 2006). In this respect, the canopy water content (CWC) is of interest, also in view of the water use efficiency of plants. As stated in Sect. 22.1, the SWIR region of the EM spectrum mainly is sensitive for canopy water. Water absorption features as a result of absorption by O-H bonds in liquid canopy water can be found at approximately 0.97, 1.20, 1.45 and 1.95 μm (Curran 1989). The features at 1.45 and 1.95 μm are

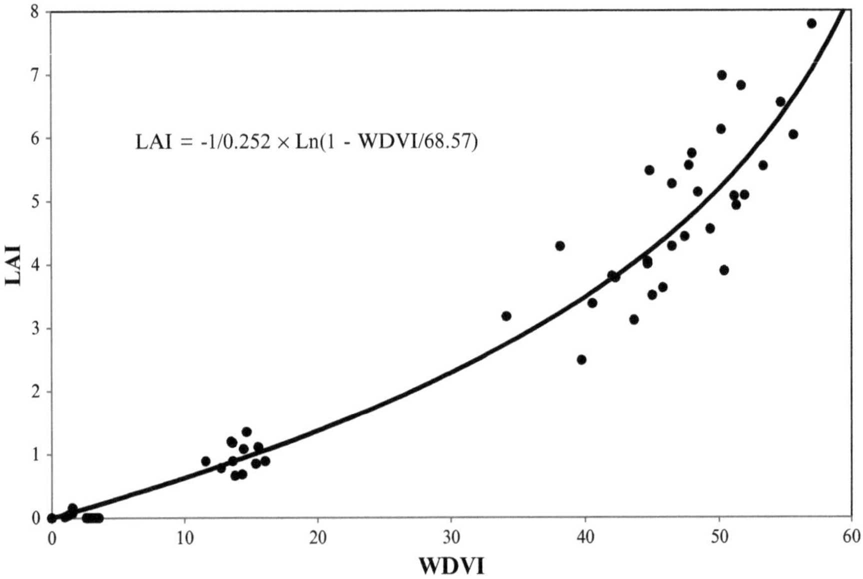

Fig. 22.3 Regression of LAI on WDVI for a field trial with a barley crop at the vegetative stage (Clevers 1989)

most pronounced. However, when using remotely sensed observations, one should also consider water vapour in the atmosphere, which also results in several absorption bands in the infrared part of the spectrum. Main atmospheric absorption features occur around 1.40 and 1.90 μm. As a result, those bands will result in very noisy measurements and should not be used for remote sensing. Spectral bands outside these main features in the shortwave infrared (SWIR) region are suited for the remote sensing of canopy water content (Tucker 1980). Landsat Thematic Mapper band 5 (1.55–1.75 μm) was designed because of this sensitivity to canopy water content. Also Thematic Mapper band 7 (2.08–2.35 μm) is sensitive to canopy water content. Various broad-band vegetation indices are based on these wavelength regions. One of the first ones is the infrared index (II) as defined by Hardisky et al. (1983).

The canopy water absorption features at 0.97 and 1.20 μm are not that pronounced, but still clearly observable (Danson et al. 1992; Sims and Gamon 2003). Therefore, these offer interesting possibilities for deriving information on canopy water content. In these regions one should consider the water vapour band absorptions at 0.94 and 1.14 μm when observing through the atmosphere (Gao and Goetz 1990; Iqbal 1983). One can notice that the centers of the liquid water bands (in the canopy) are shifted to longer wavelengths as compared to the corresponding water vapour band centers. Due to the development of imaging spectrometers, accurate measurements on these minor absorption features in the near-infrared (NIR) have become feasible.

Various spectral techniques, based on the water absorption features at 0.97 and 1.20 μm, have been proposed to estimate CWC. Often these techniques are equivalent to those applied to the chlorophyll absorption feature in the red part of the electromagnetic spectrum. Thus far, approaches based on spectral indices, continuum removal and derivative spectra have been studied in literature. Concerning spectral indices, Peñuelas et al. (1993, 1996) focused on the 0.95–0.97 μm slope and defined the so-called water band index (WI) as the ratio between the reflectance at 0.97 μm and the one at 0.90 μm (as a reference wavelength). Gao (1996) defined the normalized difference water index (NDWI), analogously to the well-known normalized difference vegetation index (NDVI), by using the 1.20 μm feature and 0.86 μm as a reference wavelength. In addition, a continuum removal approach can be applied to the two absorption features at about 0.97 and 1.20 μm. This is a way of normalizing the reflectance spectra (Kokaly and Clark 1999). The maximum band depth, the area under the continuum, and the band depth normalized to the area (Curran et al. 2001) have been used thus far for estimating foliar biochemicals like chlorophyll. Few studies have applied this to the water absorption features at 0.97 and 1.20 μm (Kokaly et al. 2003; Stimson et al. 2005).

Danson et al. (1992) showed that the first derivative of the reflectance spectrum corresponding to the slopes of the absorption features provides better correlations with leaf water content than those obtained from the direct correlation with reflectance. Rollin and Milton (1998) found moderate correlations between the first derivative at the left slope of both absorption features and CWC for a grassland site in the UK. Clevers et al. (2008) applied derivatives in a preliminary study at the field and airborne level. These studies showed that spectral derivatives at the slopes of the 0.97 μm and (to a lesser extent) 1.20 μm absorption feature have good potential as predictors of CWC. Recently, Clevers et al. (2010) showed that the first derivative of the reflectance spectrum at wavelengths corresponding to the left slope of the minor water absorption band at 0.97 μm was highly correlated with CWC. PROSAIL model simulations showed that it was insensitive to differences in leaf and canopy structure, soil background and illumination and observation geometry. However, these wavelengths are located close to a water vapour absorption band at about 0.94 μm (Gao and Goetz 1990). In order to avoid interference with absorption by atmospheric water vapour, the potential of estimating CWC using the first derivative at the right slope of the 0.97 μm absorption feature were studied by Clevers et al. (2010). Their results of PROSAIL simulations showed a linear relationship between the first derivative over the 1015–1050 nm interval and CWC. This result was confirmed, e.g., using an ASD FieldSpec spectroradiometer for a range of grassland plots at a fen meadow. Consistency between simulation results and actual field data confirmed the potential of using simulation results for calibrating the relationship between the first derivative and CWC. An example of this is provided in Fig. 22.4.

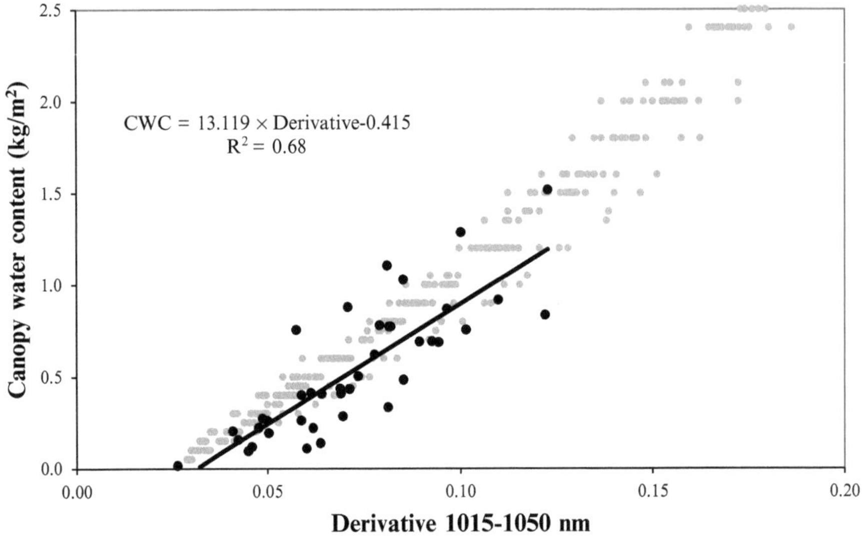

Fig. 22.4 Relationship between first derivative of canopy reflectance over the interval 1015–1050 nm and canopy water content for a fen meadow study site in The Netherlands (Clevers et al. 2010). At the background the simulated relationship for PROSAIL with varying input parameters is shown

22.5 Outlook

Within Europe the application of Earth observation data, particularly as acquired in the reflective solar domain, is reaching a state of maturity. Especially the availability of data will boost applications. For successful applications the user requirements in terms of spatio-temporal continuity, consistency and quality of products have to be fulfilled. Users require univocal numbers. Currently a multitude of satellite data is available already, and this availability will increase enormously in the near future. ESA has for long focused on research instruments, but is now developing five new missions called Sentinels specifically for the operational needs of the Copernicus programme (previously known as GMES). For land applications using the solar reflective domain, in particular two systems are relevant. Sentinel-2 (equipped with the Multi Spectral Instrument MSI) will provide images with high spatial, spectral and temporal resolution and it aims at ensuring continuity of Landsat and SPOT (Système pour l'Observation de la Terre) observations. In addition, it incorporates three new spectral bands in the red-edge region, which are very important for retrieval of chlorophyll (Delegido et al. 2011). It has a swath width of 290 km by applying a total field-of-view (FOV) of about 20°. Sentinel-3 is a medium resolution land and ocean mission, to be seen as a continuation of the Envisat mission. The Ocean and Land Color Instrument (OLCI) has a swath width of 1,270 km (FOV of about 68°, but slightly tilted) and a spatial resolution of 300 m. It will provide data continuity of MERIS on Envisat. Both Sentinel-2 and Sentinel-3 missions are based on a constellation of two satellites each in order to fulfill revisit and coverage requirements, providing robust datasets for Copernicus services.

As discussed in this chapter, RS can provide a number of key biophysical and biochemical products of vegetation, such as the fraction of absorbed photosynthetically active radiation (fAPAR), leaf area index (LAI), chlorophyll content and water content. The first two have been identified as essential climate variables (ECVs) by the UNFCCC and are key variables that are both used in surface process models and retrieved from remote sensing observations in the reflective solar domain. Various algorithms are used to derive these products, but they mostly relate to nadir or directionally normalized observations (Baret et al. 2007). For operational RS applications multisensor data usage will be required to increase the number of observations within a given time period, particularly relevant in regions with frequent cloud cover (Verger et al. 2011). This will result in an increase of high-quality data in time-series for monitoring activities. However, instability of retrieval algorithms to directional effects will degrade the accuracy of derived products.

A major research item for the coming years is assessing the anisotropic reflectance behavior of vegetation and soils, as described by the bidirectional reflectance distribution function (BRDF). With the advent of remote sensing systems with off-nadir viewing capabilities like SPOT and commercial high-spatial resolution systems (like IKONOS, QuickBird, GeoEye and WorldView), and sensors with wide FOV (like the OLCI on Sentinel-3) information on the BRDF characteristics of surface features is becoming very important for the retrieval of surface parameters. Moreover, directional information may also be significant for sensors with a limited FOV (like the MSI on Sentinel-2) for accurate retrieval of surface parameters. As a result, information on the BRDF of targets is relevant for normalizing images taken under different illumination and/or viewing conditions (Baret et al. 2007), but on the other hand it has been shown that multi-angular observations provide additional information that can be used to improve the accuracy of retrieved products (Coburn et al. 2010; Laurent et al. 2011; Verger et al. 2011). The BRDF of surface targets contains information on structure and composition that cannot be inferred from spectral properties alone (Barnsley et al. 1994). As such, it provides an additional dimension to remote sensing observations.

References

Asrar G, Myneni RB, Choudhury BJ (1992) Spatial heterogeneity in vegetation canopies and remote sensing of absorbed photosynthetically active radiation: a modeling study. Remote Sens Environ 41:85–103

Bacour C, Jacquemoud S, Tourbier Y, Dechambre M, Frangi JP (2002) Design and analysis of numerical experiments to compare four canopy reflectance models. Remote Sens Environ 79:72–83

Bacour C, Baret F, Béal D, Weiss M, Pavageau K (2006) Neural network estimation of LAI, fAPAR, fCover and LAI×Cab, from top of canopy MERIS reflectance data: principles and validation. Remote Sens Environ 105:313–325

Baret F, Buis S (2008) Estimating canopy characteristics from remote sensing observations: review of methods and associated problems. In: Liang S (ed) Advances in land remote sensing: system, modeling, inversion and application, pp 173–201

Baret F, Guyot G, Major DJ (1989) TSAVI: a vegetation index which minimizes soil brightness effects on LAI and APAR estimation. In: Digest – international geoscience and remote sensing symposium (IGARSS), Vancouver, 10–14 July 1989, pp 1355–1358

Baret F, Houlès V, Guérif M (2007) Quantification of plant stress using remote sensing observations and crop models: the case of nitrogen management. J Exp Bot 58:869–880

Barnsley MJ, Strahler AH, Morris KP, Muller JP (1994) Sampling the surface bidirectional reflectance distribution function (BRDF): 1. evaluation of current and future satellite sensors. Remote Sens Rev 8:271–311

Bouman BAM, Van Kasteren HWJ, Uenk D (1992) Standard relations to estimate ground cover and LAI of agricultural crops from reflectance measurements. ISPRS J Photogramm Remote Sens 4:249–262

Broge NH, Leblanc E (2000) Comparing prediction power and stability of broadband and hyperspectral vegetation indices for estimation of green leaf area index and canopy chlorophyll density. Remote Sens Environ 76:156–172

Broge NH, Mortensen JV (2002) Deriving green crop area index and canopy chlorophyll density of winter wheat from spectral reflectance data. Remote Sens Environ 81:45–57

Clevers JGPW (1988) The derivation of a simplified reflectance model for the estimation of leaf area index. Remote Sens Environ 25:53–69

Clevers JGPW (1989) Application of a weighted infrared-red vegetation index for estimating leaf area index by correcting for soil moisture. Remote Sens Environ 29:25–37

Clevers JGPW (1999) The use of imaging spectrometry for agricultural applications. ISPRS J Photogramm Remote Sens 54:299–304

Clevers JGPW, Kooistra L (2012) Using hyperspectral remote sensing data for retrieving total canopy chlorophyll and nitrogen content. IEEE J Sel Top Appl Earth Obs Remote Sens 5:574–583

Clevers JGPW, Van Leeuwen HJC, Verhoef W (1994) Estimating the fraction APAR by means of vegetation indices: a sensitivity analysis with a combined PROSPECT-SAIL model. Remote Sens Rev 9:203–220

Clevers JGPW, Verhoef W (1993) LAI estimation by means of the WDVI: a sensitivity analysis with a combined PROSPECT-SAIL model. Remote Sens Rev 7:43–64

Clevers JGPW, de Jong SM, Epema GF, van der Meer F, Bakker WH, Skidmore AK, Addink EA (2001) MERIS and the red-edge position. Int J Appl Earth Obs Geoinf 3:313–320

Clevers JGPW, De Jong SM, Epema GF, Van der Meer FD, Bakker WH, Skidmore AK, Scholte KH (2002) Derivation of the red edge index using the MERIS standard band setting. Int J Remote Sens 23:3169–3184

Clevers JGPW, Kooistra L, Salas EAL (2004) Study of heavy metal contamination in river floodplains using the red-edge position in spectroscopic data. Int J Remote Sens 25:3883–3895

Clevers JGPW, Kooistra L, Schaepman ME (2008) Using spectral information from the NIR water absorption features for the retrieval of canopy water content. Int J Appl Earth Obs Geoinf 10:388–397

Clevers JGPW, Kooistra L, Schaepman ME (2010) Estimating canopy water content using hyperspectral remote sensing data. Int J Appl Earth Obs Geoinf 12:119–125

Coburn CA, Van Gaalen E, Peddle DR, Flanagan LB (2010) Anisotropic reflectance effects on spectral indices for estimating ecophysiological parameters using a portable goniometer system. Can J Remote Sens 36:S355–S364

Collins W (1978) Remote sensing of crop type and maturity. Photogramm Eng Remote Sens 44:42–55

Combal B, Baret F, Weiss M, Trubuil A, Mace D, Pragnere A, Myneni R, Knyazikhin Y, Wang L (2003) Retrieval of canopy biophysical variables from bidirectional reflectance – using prior information to solve the ill-posed inverse problem. Remote Sens Environ 84:1–15

Crist EP, Cicone RC (1984) Application of the Tasseled Cap concept to simulated Thematic Mapper data. Photogramm Eng Remote Sens 50:343–352

Curran PJ (1989) Remote sensing of foliar chemistry. Remote Sens Environ 30:271–278

Curran PJ, Dungan JL, Peterson DL (2001) Estimating the foliar biochemical concentration of leaves with reflectance spectrometry testing the Kokaly and Clark methodologies. Remote Sens Environ 76:349–359

Danson FM, Steven MD, Malthus TJ, Clark JA (1992) High-spectral resolution data for determining leaf water content. Int J Remote Sens 13:461–470

Dash J, Curran PJ (2004) The MERIS terrestrial chlorophyll index. Int J Remote Sens 25:5403–5413

Daughtry CST, Gallo KP, Goward SN, Prince SD, Kustas WP (1992) Spectral estimates of absorbed radiation and phytomass production in corn and soybean canopies. Remote Sens Environ 39:141–152

Daughtry CST, Walthall CL, Kim MS, Brown de Colstoun E, McMurtrey JE III (2000) Estimating corn leaf chlorophyll concentration from leaf and canopy reflectance. Remote Sens Environ 74:229–239

Delegido J, Verrelst J, Alonso L, Moreno J (2011) Evaluation of sentinel-2 red-edge bands for empirical estimation of green LAI and chlorophyll content. Sensors 11:7063–7081

ESA (2006) The changing Earth. In: Battrick B (ed) ESA Publication. ESA, Noordwijk, p 83

Gamon JA, Field CB, Bilger W, Björkman O, Fredeen AL, Peñuelas J (1990) Remote sensing of the xanthophyll cycle and chlorophyll fluorescence in sunflower leaves and canopies. Oecologia 85:1–7

Gamon JA, Serrano L, Surfus JS (1997) The photochemical reflectance index: an optical indicator of photosynthetic radiation use efficiency across species, functional types, and nutrient levels. Oecologia 112:492–501

Gao BC (1996) NDWI – A normalized difference water index for remote sensing of vegetation liquid water from space. Remote Sens Environ 58:257–266

Gao BC, Goetz AFH (1990) Column atmospheric water vapor and vegetation liquid water retrievals from airborne imaging spectrometer data. J Geophys Res 95:3549–3564

Garbulsky MF, Peñuelas J, Gamon J, Inoue Y, Filella I (2011) The photochemical reflectance index (PRI) and the remote sensing of leaf, canopy and ecosystem radiation use efficiencies: a review and meta-analysis. Remote Sens Environ 115:281–297

Garrigues S, Lacaze R, Baret F, Morisette JT, Weiss M, Nickeson JE, Fernandes R, Plummer S, Shabanov NV, Myneni RB, Knyazikhin Y, Yang W (2008) Validation and intercomparison of global Leaf Area Index products derived from remote sensing data. J Geophys Res G Biogeosci 113, art no. G02028

Gitelson AA, Merzlyak MN (1996) Signature analysis of leaf reflectance spectra: algorithm development for remote sensing of chlorophyll. J Plant Physiol 148:494–500

Gitelson AA, Gritz Y, Merzlyak MN (2003) Relationships between leaf chlorophyll content and spectral reflectance and algorithms for non-destructive chlorophyll assessment in higher plant leaves. J Plant Physiol 160:271–282

Gitelson AA, Keydan GP, Merzlyak MN (2006a) Three-band model for noninvasive estimation of chlorophyll, carotenoids, and anthocyanin contents in higher plant leaves. Geophys Res Lett 33, art. no. L11402

Gitelson AA, Viña A, Verma SB, Rundquist DC, Arkebauer TJ, Keydan G, Leavitt B, Ciganda V, Burba GG, Suyker AE (2006b) Relationship between gross primary production and chlorophyll content in crops: implications for the synoptic monitoring of vegetation productivity. J Geophys Res D Atmos 111, art. no. D08S11

Gong P, Pu RL, Biging GS, Larrieu MR (2003) Estimation of forest leaf area index using vegetation indices derived from Hyperion hyperspectral data. IEEE Trans Geosci Remote Sens 41:1355–1362

Goward SN, Huemmrich KF (1992) Vegetation canopy PAR absorptance and the normalized difference vegetation index: an assessment using the SAIL model. Remote Sens Environ 39:119–140

Guyot G, Baret F (1988) Utilisation de la haute resolution spectrale pour suivre l'etat des couverts vegetaux. In: Proceedings of the 4th international colloquium 'spectral signatures of objects in remote sensing', Aussois, France: ESA, Paris, pp 279–286

Haboudane D, Miller JR, Tremblay N, Zarco-Tejada PJ, Dextraze L (2002) Integrated narrow-band vegetation indices for prediction of crop chlorophyll content for application to precision agriculture. Remote Sens Environ 81:416–426

Haboudane D, Miller JR, Pattey E, Zarco-Tejada PJ, Strachan IB (2004) Hyperspectral vegetation indices and novel algorithms for predicting green LAI of crop canopies: modeling and validation in the context of precision agriculture. Remote Sens Environ 90:337–352

Hardisky MA, Klemas V, Smart RM (1983) The influence of soil salinity, growth form, and leaf moisture on the spectral radiance of Spartina alterniflora canopies. Photogramm Eng Remote Sens 49:77–83

Horler DNH, Dockray M, Barber J (1983) The red edge of plant leaf reflectance. Int J Remote Sens 4:273–288

Huete AR (1988) A soil-adjusted vegetation index (SAVI). Remote Sens Environ 25:295–309

Iqbal M (1983) An introduction to solar radiation. Academic, Ontario

Jacquemoud S, Bacour C, Poilve H, Frangi JP (2000) Comparison of four radiative transfer models to simulate plant canopies reflectance: direct and inverse mode. Remote Sens Environ 74:471–481

Jacquemoud S, Verhoef W, Baret F, Bacour C, Zarco-Tejada PJ, Asner GP, François C, Ustin SL (2009) PROSPECT + SAIL models: a review of use for vegetation characterization. Remote Sens Environ 113:S56–S66

Jongschaap REE, Booij R (2004) Spectral measurements at different spatial scales in potato: relating leaf, plant and canopy nitrogen status. Int J Appl Earth Obs Geoinf 5:205–218

Kauth RJ, Thomas GS (1976) The tasseled cap – a graphic description of the spectral-temporal development of agricultural crops as seen by Landsat. In: Proceedings of the symposium on machine processing of remotely sensed data, 4B, Purdue University, West Lafayette, pp 41–51

Knipling EB (1970) Physical and physiological basis for the reflectance of visible and near-infrared radiation from vegetation. Remote Sens Environ 1:155–159

Kokaly RF, Clark RN (1999) Spectroscopic determination of leaf biochemistry using band-depth analysis of absorption features and stepwise multiple linear regression. Remote Sens Environ 67:267–287

Kokaly RF, Despain DG, Clark RN, Livo KE (2003) Mapping vegetation in Yellowstone National Park using spectral feature analysis of AVIRIS data. Remote Sens Environ 84:437–456

Laurent VCE, Verhoef W, Clevers JGPW, Schaepman ME (2011) Inversion of a coupled canopy-atmosphere model using multi-angular top-of-atmosphere radiance data: a forest case study. Remote Sens Environ 115:2603–2612

Liang S (2004) Quantitative remote sensing of land surfaces. Wiley, Hoboken

Mitscherlich A (1920) Das Liebigsche Gesetz vom Minimum und das Wirkungsgesetz der Wachstumsfaktoren. Naturwissenschaften 8:85–88

Myneni RB, Hall FG, Sellers PJ, Marshak AL (1995a) Interpretation of spectral vegetation indexes. IEEE Trans Geosci Remote Sens 33:481–486

Myneni RB, Maggion S, Iaquinta J, Privette JL, Gobron N, Pinty B, Kimes DS, Verstraete MM, Williams DL (1995b) Optical remote sensing of vegetation: modeling, caveats, and algorithms. Remote Sens Environ 51:169–188

Peñuelas J, Filella I, Biel C, Serrano L, Save R (1993) The reflectance at the 950–970 nm region as an indicator of plant water status. Int J Remote Sens 14:1887–1905

Peñuelas J, Filella I, Serrano L, Save R (1996) Cell wall elasticity and water index (R970 nm/R900 nm) in wheat under different nitrogen availabilities. Int J Remote Sens 17:373–382

Pinty B, Verstraete MM (1992) On the design and validation of surface bidirectional reflectance and albedo models. Remote Sens Environ 41:155–167

Richardson AJ, Wiegand CL (1977) Distinguishing vegetation from soil background information. Photogramm Eng Remote Sens 43:1541–1552

Rollin EM, Milton EJ (1998) Processing of high spectral resolution reflectance data for the retrieval of canopy water content information. Remote Sens Environ 65:86–92

Rondeaux G, Steven M, Baret F (1996) Optimization of soil-adjusted vegetation indices. Remote Sens Environ 55:95–107

Rouse JW, Haas RH, Schell JA, Deering DW (1973) Monitoring vegetation systems in the Great Plains with ERTS. In: Earth resources technology satellite-1 symposium, Goddard Space Flight Center, Washington, DC, pp 309–317

Rouse JW, Haas RH, Deering DW, Schell JA, Harlan JC (1974) Monitoring the vernal advancement and retrogradation (green wave effect) of natural vegetation. In: NASA/GSFC type III final report, Greenbelt, MD, p 371

Schlerf M, Atzberger C, Hill J (2005) Remote sensing of forest biophysical variables using HyMap imaging spectrometer data. Remote Sens Environ 95:177–194

Sims DA, Gamon JA (2003) Estimation of vegetation water content and photosynthetic tissue area from spectral reflectance: a comparison of indices based on liquid water and chlorophyll absorption features. Remote Sens Environ 84:526–537

Stimson HC, Breshears DD, Ustin SL, Kefauver SC (2005) Spectral sensing of foliar water conditions in two co-occurring conifer species: Pinus edulis and Juniperus monosperma. Remote Sens Environ 96:108–118

Thenkabail PS, Smith RB, De Pauw E (2002) Evaluation of narrowband and broadband vegetation indices for determining optimal hyperspectral wavebands for agricultural crop characterization. Photogramm Eng Remote Sens 68:607–621

Tucker CJ (1980) Remote sensing of leaf water content in the near infrared. Remote Sens Environ 10:23–32

Verger A, Baret F, Weiss M (2011) A multisensor fusion approach to improve LAI time series. Remote Sens Environ 115:2460–2470

Verrelst J, Schaepman ME, Koetz B, Kneubühler M (2008) Angular sensitivity analysis of vegetation indices derived from CHRIS/PROBA data. Remote Sens Environ 112:2341–2353

WMO/IOC (2010) Implementation plan for the global observing system for climate in support of the UNFCCC (2010 Update). Report GCOS-138/GOOS-184/GTOS-76/WMO-TD/No. 1523, p 180

Wu C, Niu Z, Tang Q, Huang W (2008) Estimating chlorophyll content from hyperspectral vegetation indices: modeling and validation. Agr Forest Meteorol 148:1230–1241

Yoder BJ, Pettigrew-Crosby RE (1995) Predicting nitrogen and chlorophyll content and concentrations from reflectance spectra (400–2500 nm) at leaf and canopy scales. Remote Sens Environ 53:199–211

Zarco-Tejada PJ, Berjon A, Lopez-Lozano R, Miller JR, Martin P, Cachorro V, Gonzalez MR, de Frutos A (2005) Assessing vineyard condition with hyperspectral indices: leaf and canopy reflectance simulation in a row-structured discontinuous canopy. Remote Sens Environ 99:271–287

Chapter 23
Land Transformation Processes in NE China: Tracking Trade-Offs in Ecosystem Services Across Several Decades with Landsat-TM/ETM+ time Series

Joachim Hill, Marion Stellmes, and Changyao Wang

23.1 Introduction

The concept of ecosystems services and functions has been central to the establishment of the Millennium Ecosystem Assessment reports (Millennium Ecosystem Assessment (MEA) 2005). Ecosystem functions refer to the habitat, biological or system properties or processes of ecosystems. Ecosystem goods and services represent the benefits human populations derive, directly or indirectly, from ecosystem functions. The ecosystem services consist of flows of materials, energy, and information from natural capital stocks which combine with manufactured and human capital services to produce human welfare (Costanza et al. 1997). There is increasing consensus about the importance of incorporating the valuation of ecosystem services into resource management decisions, but quantifying the levels and values of these services has proven difficult (Nelson et al. 2009).

As human impact on ecosystems became increasingly pronounced, the focus in earth-system science is more and more on the dynamics of social–ecological systems (Ellis and Ramankutty 2008; Lambin and Meyfroidt 2010). Land use practices have not only affected global and regional climate due to the emission of relevant greenhouse gases but also by altering energy fluxes and water balance (Baldocchi et al. 2004; Bormann et al. 2007; Foley et al. 2005). Changes in factors that indirectly affect ecosystems, such as population, technology, and land use practices frequently lead to changes in factors directly affecting ecosystems. This implies that land use and land use change processes are directly linked to ecosystem

J. Hill (✉) • M. Stellmes
Environmental Remote Sensing & Geoinformatics, Faculty of Regional and Environmental Sciences, Trier University, 54286 Trier, Germany
e-mail: hillj@uni-trier.de

C. Wang
State Key Laboratory of Remote Sensing, Chinese Academy of Science,
Institute of Remote Sensing and Digital Earth, P. O. Box 9718, 20 Datun Road,
Beijing 100101, People's Republic China

I. Manakos and M. Braun (eds.), *Land Use and Land Cover Mapping in Europe: Practices & Trends*, Remote Sensing and Digital Image Processing 18,
DOI 10.1007/978-94-007-7969-3_23,

services and their modification. Land use change and associated alterations of habitat structure as well as release of substances like fertilizers, pesticides, air pollutants and waste water impact on various ecosystems goods and services, amongst them biodiversity, substance flows, water and air quality, soil properties and disease vectors may force ecosystem services to change and thereby affect human well-being (DeFries et al. 2004; Foley et al. 2005; Lambin and Geist 2006; Vitousek et al. 1997).

Foley et al. (2005) have argued that land use often includes conflicting aspects of development. Many land use practices are absolutely essential (they provide critical natural resources and ecosystem services, such as food, fiber, shelter, and freshwater), but at the same time may degrade the ecosystems and services, upon which we depend. Typically, progress toward one objective such as increasing food production develops at the cost of other objectives such as conserving biological diversity or improving water quality. Particularly countries with urgent development needs may see themselves confronted with substantial challenges in designing appropriate and sustainable land use concepts. Management decisions generally involve trade-offs among ecosystem services, which are to be balanced with respect to societal objectives, i.e. to reduce negative environmental impacts of land use while maintaining economic and social benefits (Kareiva et al. 2007; Nelson et al. 2009). In this respect, a quantitative and scientifically based assessment of trade-offs is an essential prerequisite for sound decision-making (DeFries et al. 2004). Satellite remote sensing represents one of the most efficient data sources and analytical approaches for assessing the current spatial extent and condition of ecosystems and, owing to the available long observation records, for identifying trends in ecosystem conditions and services (Millennium Ecosystem Assessment 2005). Advances in remote sensing technologies over the past few decades now enable repeated and calibrated observations of the Earth's surface. The potential to apply these data for assessing trends in ecosystem condition is only beginning to be realized.

China's cultivated land has been undergoing dramatic change along with its rapidly growing economy and population. The impact of land use transformation on food production at the national scale, however, is poorly understood due to the lack of detailed spatially explicit agricultural productivity information on cropland change and crop productivity (Gao et al. 2006). In many parts of the country, such as Inner Mongolia in NE China, land degradation processes are associated with these land use transformations and have become an important issue (Chen and Tang 2005). The availability of long records of satellite observations has opened new perspectives on assessing consequences of these land transformation processes on ecosystem services and functions.

23.2 Background and Objectives

Conversion of natural or semi-natural ecosystems to grow crops, raise animals, obtain timber, and build cities is one of the foundations of human civilization. As a result of this transformation a range of other ecosystem functions, such as the provisioning of freshwater, regulation of climate and biogeochemical cycles, maintenance of soil fertility and others will be altered. Balancing the inherent trade-offs between satisfying immediate human needs and maintaining other ecosystem functions requires quantitative knowledge about ecosystem responses to land use (DeFries et al. 2004). Ecological knowledge and process understanding is one crucial prerequisite for assessing such trade-offs. However, equally important are precise estimates of changing land use proportions and land use intensity over time, together with selected key attributes being considered suitable surrogates for specific ecosystem services and functions. In many cases, land use decisions and associated trade-offs manifest themselves on regional scale levels which necessarily require the availability of spatially explicit information as provided by the existing range of satellite observation systems.

Land cover changes can be distinguished in two major groups: (i) conversions and (ii) modifications (Coppin et al. 2004; Lambin and Geist 2006). Land use transformations usually involve the replacement of one land use/land cover class with another one (e.g. shrub lands with arable land), whereas modifications are usually related to gradual changes within one thematic class (e.g. shrub encroachment within natural ecosystems). Besides land use change in terms of conversion also modifications of land use might profoundly alter the performance and health of ecosystems and therefore, have to be considered when assessing ecosystem services over time. Therefore, the assessment of both conversions and modifications is important to provide a comprehensive picture of land use/cover changes. In particular, the assessment of modifications is a crucial element in dryland areas because changes related to land degradation are often associated with modifications of the existing land use system (Lambin and Geist 2006; Lambin and Geist 2001); they include for instance vegetation cover loss due to overgrazing.

As mentioned before remote sensing images can provide valuable surrogates linked to land use changes and cover both, conversions and modifications which can be related to the condition of ecosystem services. While incomplete spatial coverage, infrequent temporal coverage, and large data volumes have precluded global analysis of land cover change, data acquired by medium resolution (10–100 m) optical sensors onboard satellites (e.g. Landsat, SPOT, ASTER) have been the primary sources for identifying land cover change in particular locations. The spatial resolution of the imagery has made the Landsat archives an invaluable information source for science, management, and policy development. Landsat imagery of some form has been collected since 1972, resulting in the longest continuously acquired collection of space-based terrestrial observation. Further, the opening of the entire U.S. Geological Survey (USGS) Landsat archive in 2008 and 2009, which made all of the USGS Landsat imagery freely available through a web-portal (glovis.usgs.gov) (Woodcock et al. 2008), has resulted in an increased capacity to undertake ambitious analyses of terrestrial dynamics across large areas, and in using dense time series of imagery (Vogelmann et al. 2012; Wulder et al. 2012).

The preparation and analysis of Landsat time series covering several decades imposes specific requirements with regard to data calibration and standardization, selection of appropriate reference periods, extraction of meaningful indicator variables (such as biophysical vegetation attributes, soil characteristics, surface water) and the identification of changing land use patterns. If these conditions are met, Landsat observations can provide at least some quantifiable and measurable indicators for assessing and understanding temporal trends and changes of important ecosystem functions (e.g. Cohen and Goward 2004).

The scientific basis for incorporating natural capital into resource- and land use decisions on a large scale is still in its infancy (Daily et al. 2009). However, Nelson et al. (2009), for example, have proposed a spatially-explicit modeling approach to predict changes in ecosystem services based on available spatial data layers. Usually, the most important spatial data input to ESS assessment models are maps of land use and land use change. However, the focus on conversions of land cover (change of land use and cover type) often disregards modifications within existing land use systems (e.g. intensification of arable land) and their consequences for ecosystem performance and the sustainable provision of ecosystem services.

Instead of modeling future scenarios, this study aims at considering these issues and explores how satellite-derived indicators of dynamic land surface properties may be directly transferred into surrogates of spatial-temporal changes of relevant ecosystem services (i.e. biomass production, dune fixation and groundwater recharge) by taking the example of a study site in Inner Mongolia, China. It aims at demonstrating the impacts of governmental management policies on land use change and its impact on the long-term availability of important ecosystem services. The study focuses on following objectives:

- Identify and evaluate suitable remote sensing based surrogates that are able to assess modifications within the dominant land use systems of the study area
- Evaluate the status of ecosystems and their capacity to provide ecosystem services based on the identified modifications of land use.

The use of associated models for translating these findings into quantitative biophysical or monetary terms is neglected here and remains reserved to future extensions of this work.

23.3 Study Case: Horqin Sandy Lands

Since the early 1980s, the unprecedented combination of economic and population growth in China has led to substantial land transformations across the nation. Horqin Sandy Lands with the core study site of Naiman County (Fig. 23.1), located in the agro-pastoral zone between the Inner Mongolian Plateau and the Northeast Plains in China (42°41′ – 45°45′ N, 118°35′ – 123°30′ E), make in this respect no exception.

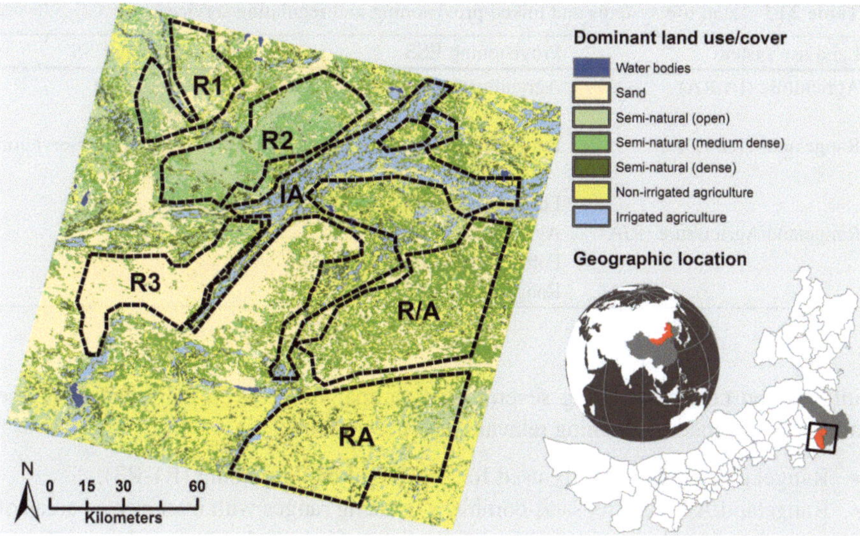

Fig. 23.1 Location of Naiman County (part of Jirem Prefecture) in Inner Mongolia (China); the coverage of Landsat frame used in this work is inserted in the map of administrative units (in *red*). The MODIS-derived land use map with major land use strata (*black*-rimmed areas) is shown in the *upper* part of the figure

23.3.1 Geographical Characteristics

Covering more than 42,000 km^2 Horqin Sandy Lands forms one of the major sand areas in Northern China and includes an important part of Inner Mongolian grasslands. The vegetation consists predominantly of shrubs and perennial grasses, trees are found in places where water conditions are favorable. Climatically, the region is part of continental drylands with hot and short summers and very cold winters. Mean annual temperature minima and maxima are −8.8 °C in January and 30.4 °C in July, respectively; mean annual precipitation is 375 mm with nearly 80 % concentrating in the months from June to September. An important characteristic is rainfall irregularity: annual precipitation varied from 205 to 679 mm year^{-1} in the observation period from 2000 to 2008 (NOAA's National Climate Data Center, NCDC, http://www.ncdc.noaa.gov/). These physiographic characteristics, including easily erodible loess soils and mobile sand dunes, render the area prone to pronounced land degradation processes (Chen and Tang 2005).

Figure 23.1 illustrates the current distribution of land use classes within the study site. Dominating land use and cover classes are intensively used irrigated as well as rainfed arable land, semi-natural areas mainly serving as range lands, as well as extensive sand dune systems. Based on satellite-derived land use maps and the

Table 23.1 Land use systems and linked provisioning and regulating services

Land use system	Provisioning ESS	Regulating/supporting ESS
Agriculture (IA,RA)	Agricultural production	Climate & air regulation
	Forest production	Carbon sequestration
Rangelands (R1-R3)	Rangeland production	Habitat & biodiversity preservation
		Groundwater recharge
	Forest production	Dune fixation
Rangeland/Agriculture (R/A)	Agricultural production	
	Forest production	
	Rangeland production	

information collected during several field campaigns between 2008 and 2011 the area was stratified by defining relevant land use systems:

- Rangeland, predominantly used for sheep and cattle grazing (R1-R3),
- Rangeland/Agriculture, sand-dominated grazing ranges with interspersed areas of agricultural land use with varying proportions of irrigated and rainfed crops (R/A),
- Agriculture, either dominated by irrigated (IA) or rainfed crop systems (RA).

23.3.2 Land Use History and Ecosystem Services

Against this background, this study focuses on the evaluation of eight ecosystem services of major relevance in the study site. Following the proposed scheme by the MEA study (2005), these services are classified into provisioning (agricultural production, forest production, and rangeland production), supporting (groundwater recharge, dune fixation) and regulating services (habitat and biodiversity preservation, carbon sequestration and climate, and air regulation). Each of the existing land use systems involves a characteristic assembly of the services (Table 23.1), which may be directly and indirectly linked to anthropogenic activities and changing land use preferences.

The approximate reconstruction of relevant ecosystem services performance under traditional land use at the beginning of the nineteenth century, and at the time period where this study departs (Fig. 23.2b) is primarily based on information provided by Jiang (1999, 2002, 2006), Tong et al. (2004), Wu and de Pauw (2010) and Zhu and Zou (1988). We address a relative evaluation of ecosystem performance by assigning scores between zero and one, one being the optimum provisioning of a service, zero no provisioning, respectively. However, whereas the provisioning services as well as the supporting services may often be directly linked to suitable proxies, the regulating services are usually a more complex mixture and interaction of several factors, especially when regarding the whole range of ecosystems present in a study area. This affects the valuation of an actual status of ecosystem services as well as the estimation of change rates due to land use conversions and modifications.

Foley et al. (2005) have illustrated the provisioning of multiple ecosystem services under different land use regimes through 'spider' diagrams, in which the condition of important ecosystem services is indicated along several axes. Picking up on this concept, we attribute starting scores (approximated based on the above cited literature) to the main provisioning services "Agricultural Production" (AP = 0.4), "Forest Production" (FP = 0.4), "Rangeland Production" (RP = 0.95) and "Groundwater Recharge" (GR = 1.00). At this point we wish to emphasize that in this study the ecosystem service "Rangeland Production" is not related to the amount of livestock but to the capacity of the ecosystem to provide net primary productivity to feed the livestock.

In the absence of better data, regulating services such as "Carbon Sequestration" (CS) are approximated by an area-weighted linear combination of the service scores for rangeland, agriculture, and forest productivity [CS = 0.2 + 0.8 • (0.2•AP + 0.5•FP + 0.3•RP)], assuming that long-term storage in forest ecosystems (FP) is the most important component. The lower proportional weights for agriculture (AP) respond to the fact that carbon incorporated in plants is released back as CO_2 into the atmosphere through the food web within the year (Piao et al. 2009); rangelands (RP) are considered slightly more favorable in this context since higher amounts of biomass are permanently present. Soils are the largest source of uncertainty in the terrestrial carbon balance of China; in the cold drylands of Horqin, a base value of 0.2 is attributed to soil carbon pools and considered roughly constant over time, although it is understood that also the soil carbon storage changes over time. For example, the amount of root input and crop residue has augmented in agricultural areas in more recent times (Piao et al. 2009).

Similarly, an approximate service score is also defined for "Habitat & Biodiversity Preservation" (HBP) where proportional contributions from "Rangeland Production" and "Forest Production", "Dune Fixation" (assuming that fixed dunes favor plant and animal biodiversity) and "Groundwater Recharge" (owing to its impact on aquatic and wetland ecosystems) are considered; again, a constant base value for biodiversity services (0.2) is assumed to exist independently from other contributions, i.e. [HBP = 0.2 + 0.8 • (0.1•AP + 0.4•FP + 0.3•RP + 0.2•GW)]. An adequate definition of "Climate and Air Regulation" (CAR) is difficult; assuming a more or less constant base level of 0.5, the hypothesis is that "Dune Fixation" and "Groundwater Recharge" are the main control parameters, hence [CAR = 0.5 + 0.5 • (0.5•DF + 0.5•GW)]. The implemented approach is simplistic and especially the evaluation of the regulating services is based on subjective reasoning and leaves room for improvement. Nevertheless, the major intention of this study was to demonstrate how an integrated estimation in the performance of all major ecosystems, including feedback mechanisms, might be implemented.

The resulting spider diagram (Fig. 23.2a) approximately describes an ecosystem that once supported a sustainable nomadic production system: extensive areas with sandy soils or sand dunes hosted drought-resistant shrubs and numerous grass species which provided top quality pasture for sheep grazing. Remnant forest patches provide evidence of previous landscapes in the forest-steppe transitional bioclimatic vegetation zone, being highly supportive for carbon sequestration,

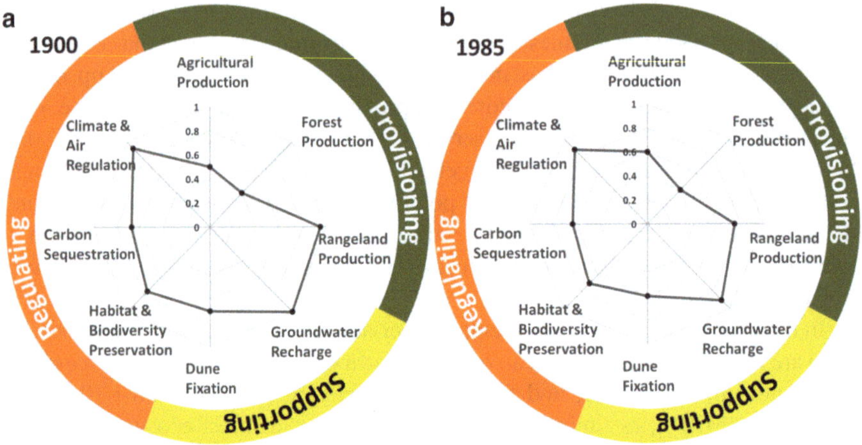

Fig. 23.2 Synthesis of important ecosystem services in Horqin Sandy Lands for the early twentieth century (**a**) and around 1985 (**b**) in the form of spider diagrams proposed by Foley et al. (2005)

regional air quality regulation (suppression of dust movements through dune stabilization) and the preservation of habitats and biodiversity. Most remarkably, the sandy soils with their high infiltration rates supported a rich store of ground water, sometimes only a few meters below ground (e.g. Jiang 2002). The diagram thus characterizes a landscape where rangeland productivity and groundwater recharge capacities are close to maximum performance while agricultural productivity is limited (mainly due to climatic risks of drought); forested areas are limited to isolated patches.

The main aspect of land degradation in this area is what Chinese authors term "sandy desertification" (including sand dune reactivation, shifting sand dunes and sands spreading into grasslands, wind erosion in dry farmland), which develops on sand sheets and fans of quaternary origin. Although the role of drought has not been entirely clarified most authors suggest that human activity and cultural change are key determinants of these processes. Main drivers emerged from the transformation of the traditional Mongolian grazing society: Chinese immigrants brought with them a cultural tradition that is rooted in farming, a practice further promoted by the socialist regime (Jiang 2002). In the 1950s and 1960s local farmers were forced to give up their nomadic way of life and to settle in small villages, hamlets or individual farms. Together with a moderate increase in number of livestock this policy has increased grazing pressure around the newly established settlements and brought the large scale pastoral movements between seasonal pastures to an end (Sneath 1998). Reclaiming rangelands in areas of marginal rainfall and strong winds for agricultural production has caused the destruction and loss of topsoil, while the concentration of cattle in the remaining rangeland areas increased stocking rates above sustainability thresholds (Zhu and Zou 1988). The consequences have negatively impacted several inter-related ecosystem services (Fig. 23.2b).

It is important to understand that the increasing extent of degradation of grassland ecosystems in Inner Mongolia was strongly influenced by political decisions.

Ever since, national policies, such as Deng Xiaoping's reforms during the first period of new policy formulation (1979–1985), but also decisions and regulations issued by regional governments have encouraged both new and intensified land use practices (expansive groundwater-based irrigation) as well as restoration and protection measures (enclosures of pasture land, grazing regulation, tree planting campaigns) (Jiang 2006; Wu and de Pauw 2010).

The new economic policy established in the 1980s, which allowed individuals to profit directly from increased meat or wool production, sharply fostered the pressure on the land resources and resulted in intensified agricultural land use and large scale overgrazing (Butterbach-Bahl et al. 2011). According to satellite-based estimates which are claimed to be a magnitude more precise than official statistics (Liu et al. 2005a, b) more than 70 % of the increased cropland area occurred in the two provinces of Heilongjiang and Inner Mongolia. Conversion from grasslands to croplands dominated the land transformation in Inner Mongolia. Although less densely populated than Southern and Central China, population density in Inner Mongolia is considered very high, relative to the inherent low productivity of arid and semi-arid zones. Today, irrigated agriculture in the study region is dominated by corn, areas with rain fed agriculture by winter wheat and vegetables.

23.4 Material and Methods

An extremely important component in identifying cumulative impacts of land transformation processes is the availability of sufficiently long observation records. Land degradation processes, for example, have been conceptualized as a pathological process of multi-annual land-cover dynamics. Following this concept it is almost mandatory to cover time spans on the scale of decades and decouple changes on the long run from the impact of short-term fluctuations driven by seasonal pulses or single events. Since required observation periods typically exceed the life-span of individual satellites one needs to establish calibrated archives which include data from identical or spectrally comparable observation systems. Connecting Landsat data from different archives (e.g. USGS and Remote Sensing Ground Station in Beijing, RSGS) with the scope of characterizing surface properties across several decades with calibrated data products still remains a challenge.

This study uses a dataset of 30 Landsat-5-TM and Landsat-7-ETM+ scenes from USGS and RSGS covering the period from 1987 to 2007 (Table 23.2). The majority of images were obtained from the USGS archive (www.glovis.usgs.gov); some scenes not available there have been acquired from the Remote Sensing Satellite Ground Station in Beijing (http://www.fas.org/spp/guide/china/agency/rsgs.htm). So far, 11 scenes (indicated in bold letters in Table 23.2) from a phenological

Table 23.2 Acquisition dates of the Landsat 5-TM and Landsat 7-ETM+ images used in this study

Date	Sensor	Provider	Date	Sensor	Provider
11/08/1987	**L5-TM**	**RSGS**	25/10/2000	L7-ETM+	USGS
26/06/1988	L5-TM	RSGS	19/04/2001	L7-ETM+	USGS
22/08/1991	L5-TM	RSGS	13/05/2001	L5-TM	RSGS
27/08/1993	**L5-TM**	**RSGS**	21/05/2001	L7-ETM+	USGS
15/09/1994	**L5-TM**	**RSGS**	25/08/2001	L7-ETM+	RSGS
18/09/1995	**L5-TM**	**RSGS**	**02/09/2001**	**L5-TM**	**RSGS**
04/09/1996	**L5-TM**	**RSGS**	12/10/2001	L7-ETM+	USGS
05/09/1999	**L7-ETM+**	**USGS**	11/07/2002	L7-ETM+	USGS
13/09/1999	L5-TM	RSGS	20/08/2002	L5-TM	RSGS
23/10/1999	L7-ETM+	USGS	**13/09/2002**	**L7-ETM+**	**USGS**
31/03/2000	L7-ETM+	USGS	25/04/2003	L7-ETM+	USGS
02/05/2000	L7-ETM+	USGS	**02/09/2004**	**L7-ETM+**	**RSGS**
13/07/2000	L5-TM	RSGS	15/10/2005	L5-TM	RSGS
14/08/2000	L5-TM	RSGS	**16/09/2006**	**L5-TM**	**USGS**
07/09/2000	L7-ETM+	USGS	**03/09/2007**	**L5-TM**	**USGS**

window in the late summer period (late August–September) have been used for studying land change processes.

While high-precision geometric ortho-rectification is routinely implemented in various processing chains, radiometric corrections still requires attention, because distorted image radiometry may significantly impact on the identification of gradual ecosystem changes. Variations in solar illumination conditions, atmospheric scattering, atmospheric absorption and detector performance result in differences in radiance values unrelated to the reflectance of land cover. Traditionally, authors have argued that absolute radiometric quantification is an expensive, time consuming and likely untenable goal for land-cover and land use monitoring (Rogan and Chen 2004). However, successful efforts in monitoring the instrument calibration of the different Landsat imagers (e.g. Chander et al. 2009; Markham and Helder 2012) have substantially increased the options for creating calibrated satellite observations. In combination with efficient implementations of validated radiative transfer models such as SMAC (Rahman and Dedieu 1994), ATCOR (Richter 1996), AtCPro (Hill et al. 1995; Hill and Sturm 1991; Röder et al. 2005), 6S (Kotchenova et al. 2006; Vermote et al. 1997) and LEDAPS (Ju et al. 2012) it has become feasible to produce calibrated reflectance spectra, from which meaningful biophysical indicator variables can be retrieved.

23.4.1 Ortho-Rectification

All images where ortho-rectified using SRTM-derived digital elevation data and an automatized image correlation approach for identifying the required ground control points (GCPs) (Hill and Mehl 2003). Based on up to 500 GCPs per image the average registration error (RMSE) for each scene did not exceed 0.25 pixels (i.e. 7.5 m). The Chinese Gauss-Kruger Projection System (Zone 21) was used in this study.

23.4.2 *Atmospheric Correction*

Relative radiometric normalization using pseudo-invariant features (PIFs) is traditionally used for standardizing image time series (Schott et al. 1988). More recently, improved procedures have been proposed which combine radiative transfer models and image-based retrievals of atmospheric parameters with non-lambertian topographic correction (e.g. Vicente-Serrano et al. 2008). Kaufman et al. (1997) have suggested efficient approaches for deriving image-based estimates of the aerosol optical depth (τ_{aer}) but it is much more difficult to obtain reasonable estimates for the absorption optical depth of atmospheric water vapour (τ_{H2O}). The fundamental problem here is that in dryland ecosystems (typically characterized by high albedo surfaces) good estimates of atmospheric water vapour concentrations are more important than aerosol effects.

The processing concept developed for this study is based on the AtCProradiative transfer code (Hill and Sturm 1991), originally inspired by the formulation of 5S (Tanré et al. 1990), and consecutively upgraded with additional features (e.g. Hill et al. 1995). AtCPro accounts for atmospheric extinction processes as a function of sensor and terrain altitude and provides corrections for atmospheric absorption, scattering, pixel adjacency effects and terrain-dependent illumination effects. A wide concentration range for absorbing gases (H_2O, O_3, CO_2, CH_4, and O_2) is taken into account by sampling pre-calculated tables generated with MODTRAN-4 (Berk et al. 1999).

The atmospheric parameters (both τ_{aer} and τ_{H2O}) for correcting the Landsat time series were retrieved by combining AtCPro with a non-linear optimization algorithm (a modified Powell's direction set method) (Press et al. 1992). The principle of this method is to match a series of spectrally contrasting Landsat DNs with the corresponding surface reflectance by simultaneously adjusting aerosol and water vapor parameters for the radiative transfer until optimum convergence is reached. Reference spectra can be imported either from direct reflectance measurements or, as in this study, by combining the MODIS reflectance product and temporally invariant surfaces (for processing Landsat images collected before 2000).

This approach has been validated by comparing water vapour retrievals for 21 Landsat-TM/ETM+ acquisitions from the period 2000–2010 with simultaneously recorded MODIS-Terra estimates (MOD05_L2) which are derived with a differential absorption technique; the product has been reported to reach accuracies better than 0.2 g/cm^2 (e.g. Vermote et al. 2002; Vermote and Kotchenova 2008) and is therefore considered an excellent validation reference. The comparison between the τ_{H2O}-retrievals for the Landsat images and the corresponding MODIS reference

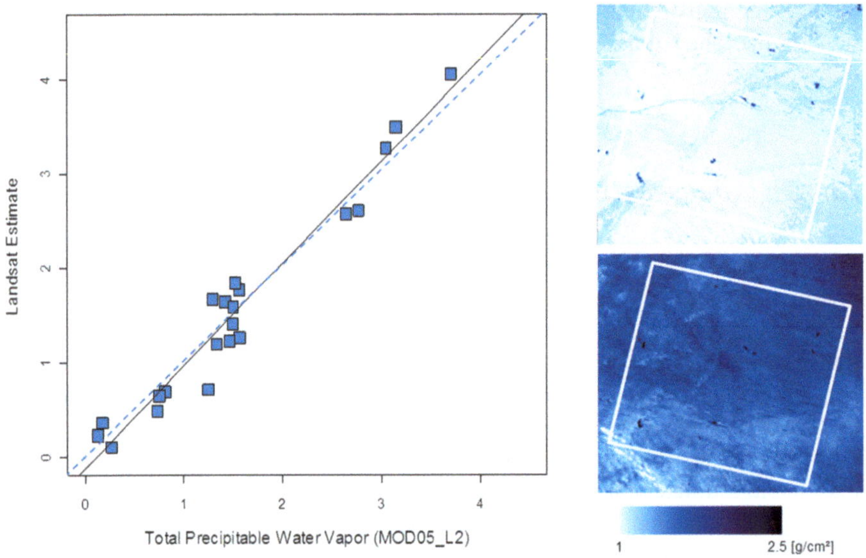

Fig. 23.3 Comparison between the estimated water vapor concentrations (AtCPro/Powell Optimization) and the corresponding MODIS-Terra "Precipitable Water" product (MOD05_L2); the one-to-one line is shown for comparison. The MODIS reference values are derived from the statistical average inside the Landsat frame extension (*right* part of figure)

produced a residual standard error of only 0.2473 g cm^{-2}, a coefficient of determination of 0.9521 and the regression coefficient was significant at the 5 %-level (Fig. 23.3); this proves that the proposed method is suited to reliably retrieve the critical water vapour optical depth parameter for a precise atmospheric correction of extended Landsat-TM/ETM+ time series.

23.4.3 Spectral Indicators for Changing Land Surface Conditions

As stated in the objectives the main goal was to characterize trade-offs between selected ecosystem services (biomass production, dune fixation and groundwater recharge) by assessing gradual changes in land surface conditions. This implies selecting a range of spectral indicators which can be associated with these services. Spectral Mixture Analysis (SMA) has been advocated as an efficient method to computationally decompose spectra into proportions of pure spectral components (end-members) (e.g. Schowengerdt 1997; Smith et al. 1990), which can be conveniently used for analyzing temporal trends (Röder et al. 2008; Vogelmann et al. 2012). SMA builds on the basic assumption that most of the spectral variation in multi-spectral images is caused by mixtures of a limited number of surface materials (i.e. vegetation, soil, shade) with different reflectance properties; they commonly mix at the sub-pixel scale, producing mixed-pixel spectra.

In first approximation, spectral mixing is modeled as a linear combination of pure component ("end-member") spectra, such that

$$\mathrm{R}_i = \sum\nolimits_{j=1}^{n} F_j \cdot RE_{ij} + \varepsilon_i \tag{23.1}$$

R_i is the reflectance of the mixed spectrum in band i, RE_{ij} the reflectance of the end-member spectrum j in band i; F_j denotes the fraction of end-member j, and ε_i the residual error in band i. Linear mixing assumes that the surface components are large and/or opaque enough to allow photons to interact with only one component. Conceptualizing radiative transfer processes to be additive, spectra can be unmixed by inverting the linear mixing equation. The objective is to isolate the spectral contributions of surface materials ("end-member abundance") before these can be edited and recombined to produce thematic maps. End-member abundance maps derived from simple linear mixture models using a common set of end-member spectra provide the convenience and inter-comparability of standard land cover metrics (e.g. NDVI) while retaining benefits of physically based estimates. Several studies (Elmore et al. 2000; Small 2004) suggest that these include good agreement with ground based measurements and output units corresponding to physical properties of land cover that can be used directly as input to land surface process models.

SMA offers the additional advantage that several indicators for land surface properties can be obtained simultaneously. The core hypothesis here was to parameterize a linear mixture model with spectral surrogates of the targeted ecosystem services. Firstly, this implies selecting Green (i.e. photosynthetic) Vegetation (GV) as one of the spectral end-members to account for changes in biomass production.

Secondly, in the Horqin study area one of the most significant surface properties related to degradation processes is the presence of Mobile Sand (MS) (primarily due to the re-mobilization and dislocation of quaternary sands), assuming that intensified grazing with too high stocking rates is a major socio-economic driver behind this process. Ultimately, as it can be assumed that the sinking water table, primarily triggered by ground water extraction, might affect the spatial extension of lakes, ponds, bogs and swamps, Water (W) was selected as the third end-member component. Figure 23.4 shows an example of the three layers (GV, MS, W) produced by SMA for one of the Landsat acquisition dates.

23.4.4 Trend and Change Analysis

The temporal coverage available from the Landsat sensor family usually confines time series analyses to the transient or linear component (e.g. Röder et al. 2008; Vogelmann et al. 2012); in case sufficiently dense time series are available the use

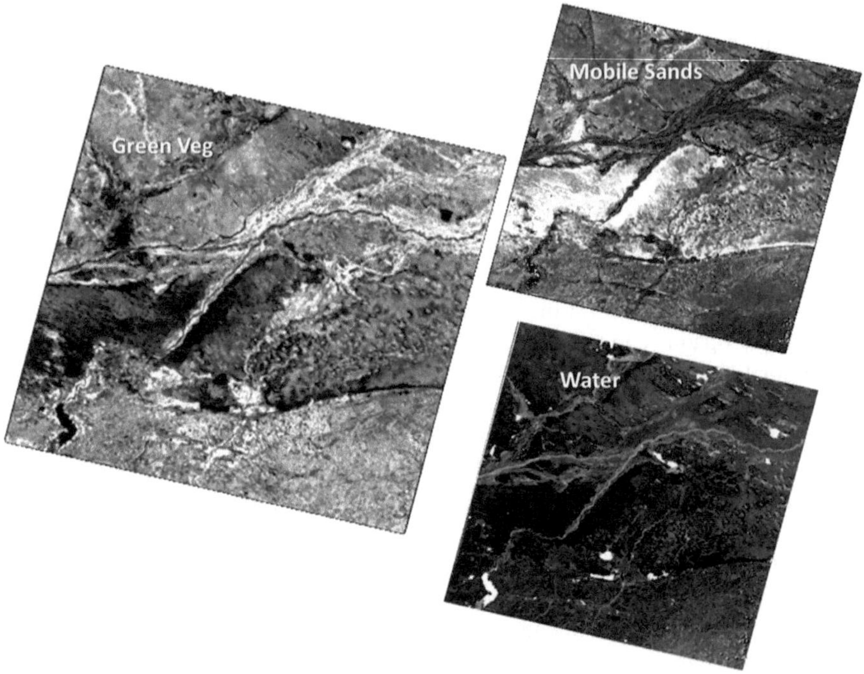

Fig. 23.4 Abundance estimates for the Landsat image from 27 Aug 1993 derived from linear unmixing with three endmembers (GV, MS and W)

of piece-wise linear trends (Verbesselt et al. 2010) might be more efficient. In this study we applied a linear regression function to each pixel, yielding a function of the type

$$y = g \cdot t + o \tag{23.2}$$

where y represents the proportional abundance of an end-member at date t (image acquisition date in days since the first observation), g is the regression gain and o the corresponding intercept. The latter characterizes the level of end-member abundance at the starting date of the observation period, while the gain is an expression of general direction and magnitude of the temporal dynamics during the observation period. As these regression parameters are calculated on a per-pixel basis, the temporal development can be illustrated by corresponding maps.

Additional output parameters are the coefficient of determination (R^2) and the significance for the retrieved regression coefficient (a two-sided t-test with $\alpha = 10\,\%$).

For reasons discussed later the linear trend analysis was applied only to the SMA-derived theme layers of mobile sand and green vegetation. Figure 23.5 illustrates typical results for the sand abundance (cover in per cent) over time: permanently active dune areas (1) expose a stable/invariant behavior on a level

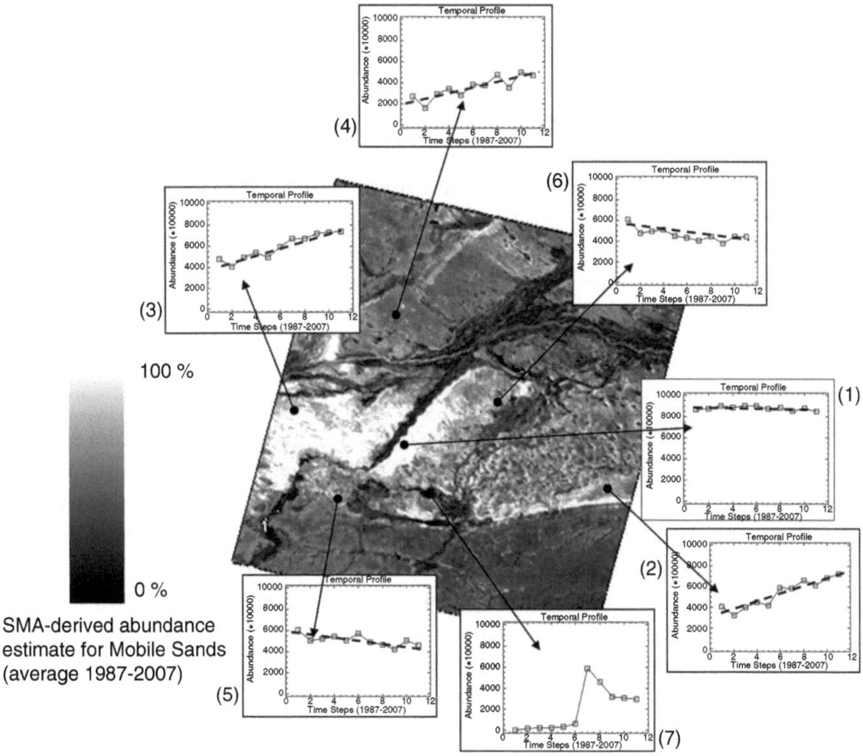

Fig. 23.5 Linear trends of SMA-derived estimates for the abundance of Mobile Sand, based on 11 time steps for selected reference areas (image backdrop is the average SMA-derived abundance of Mobile Sand for the period 1987–2007)

between 85 and 95 %, while areas with varying grades of increasing or decreasing sand proportions produce corresponding trend lines (2–6). Area (7) presents a special case: the discontinuous pattern is caused by the rapid drying of an extended lake area between 2000 and 2001. The mobile sand proportion first increases due to the exposed substrate, but then decreases again while the former lake bed is successively integrated into the agricultural production area (with gradually increasing vegetative cover). Such cases can be excluded from the trend analysis based on statistical tests of significance.

This example suggests that open water areas in the study region are affected by abrupt and sudden alterations rather than changing gradually over longer time periods. In these conditions we considered trend analysis not an efficient assessment strategy, in particular because statistical tests of significance tend to become meaningless. It was therefore decided to map the presence and extension of lakes, ponds, wetlands and swamps based on assessing their entire area at selected observation times (1987, 1995, 2001, 2006 and 2010).

23.4.5 Approximate Valuation of Ecosystem Services

The proposed approach for valuating trade-offs between ecosystem services is derived from blending the long-term change of the spectral surrogate for a specific ecosystem service with the spider diagram concept of Foley et al. (2005). If we consider the length of a specific ESS indicator at time t_0 (= time of last assessment available) being proportional to the theoretical optimum ($P_{max} = 1.0$), the updated indicator length at time t_{act} is determined by adjusting its end position in accordance with the proportional change rate of the associated endmember abundance during the observation period, i.e.

$$P_{t_{act}} = P_{t_0} + P_{t_0} \cdot [(g \cdot t_{act})/o] \tag{23.3}$$

where P_{t_0} is the ESS indicator length at time t_0, $P_{t_{act}}$ the updated ESS indicator length at the end of the observation period (t_{act}) (P_{t_0} and $P_{t_{act}}$ assume values between 0 and 1.0). The spatial context of selected ESS indicators is accounted for by using the median of $P_{t_{act}}$ within representative sampling areas (e.g. land use or land cover strata); in case a land use stratum is composed of several spatial entities (such as in the case of the rangeland stratum in this study) an adjustment of the corresponding ESS indicator is obtained from an area-weighted sum of the respective median values.

Similar to the initial assignment of starting scores for major ecosystem services for the beginning of the observation period (compare Sect. 23.3) remote sensing based surrogates might be directly intertwined with changes of ecosystem services whereas the development of other scan only be roughly estimated with a high uncertainty due to a lack of external data. In the context of this study one would, for example, conclude that the ESS "Crop Production" has improved in those pixel locations inside a stratum "Agriculture" where the productivity indicator (i.e. the SMA-derived abundance for Green Vegetation) has increased in comparison to the initial value of the time series (i.e. all positions where $[(g \cdot t_{act})/o] > 0$). In case a median value of +0.25 is found, the previous ESS indicator (which was at 50 % of a theoretical optimum) would thus be adjusted towards a value of 62.5 % and indicate that this ecosystem service has substantially improved during the observation period. In contrast, carbon sequestration is a function with a high complexity that is related to primary productivity which can be estimated by remote sensing but is also to a large degree dependent on other factors like for instance agricultural practices.

23.5 Results and Discussion

The linear trend analysis of SMA-derived abundance estimates for Mobile Sand and Green Vegetation within the observation period (1987–2007) provides clear evidence for substantial land cover changes in the study region. With the objective to uncover possible trade-offs between ecosystem services associated to specific

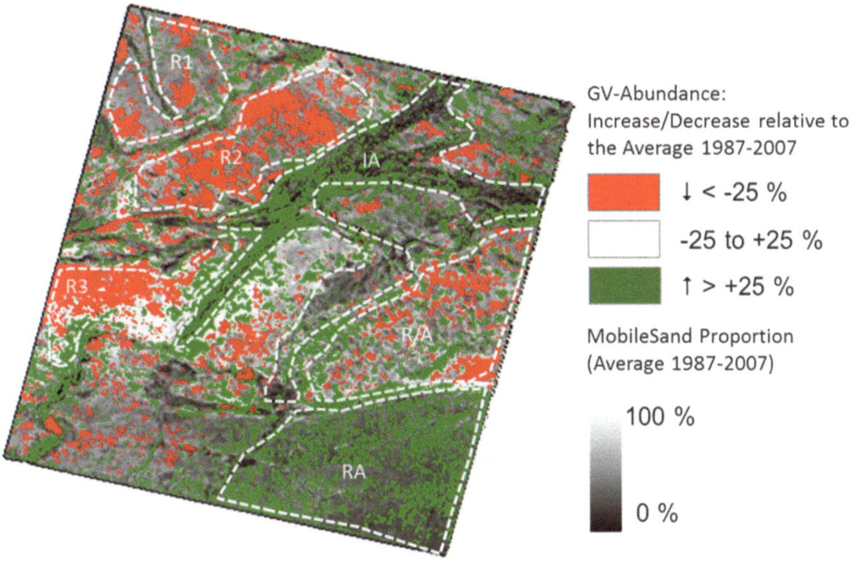

Fig. 23.6 Areas with increasing and decreasing trends of Green Vegetation abundance (1987–2007), overlaid on the land use strata described in Sect. 23.3. Colors are discussed with more detail in the text; b/w image backdrop is the long-term average of the mobile sand abundance

land units, the evaluation was conducted in relation to relevant land use strata (compare Fig. 23.2).

Figure 23.6 presents these strata as an overlay to the trend map of SMA-derived values of the proportional changes in GV abundance between 1987 and 2007: green areas indicate an increase of more than 25 %, red areas a decrease of more than 25 % of the initial GV abundance in 1987; the range between −25 and +25 % has no color assigned and might be considered as uncertainty corridor. It is evident that the agricultural strata have consistently increased or at least maintained greenness during the observation period (1987–2007), while the grazing ranges (R1-R3) show a significant loss of greenness proportions. Rangelands with interspersed agricultural areas (R/A) exhibit a patchwork of areas with positive and negative changes in GV abundance (Fig. 23.6).

A more differentiated analysis is derived from box plots of statistically significant trends which emphasize the statistical distribution of the proportional increase or decrease of end-member abundances within selected strata. Inside the three rangeland sub-areas (R1-R3) more than 50 % of the GV abundance changes are negative, suggesting that the presence of green biomass has thoroughly diminished within the observation period (Fig. 23.7). In comparison, the mixed rangeland and agriculture stratum (R/A) still has a negative median but exhibits a substantially larger statistical variance; this is consistent with the spatial pattern of negative and positive GV trends, either associated with sandy rangeland or interspersed agricultural areas of increasing productivity (Figs. 23.7 and 23.8). Both agricultural strata

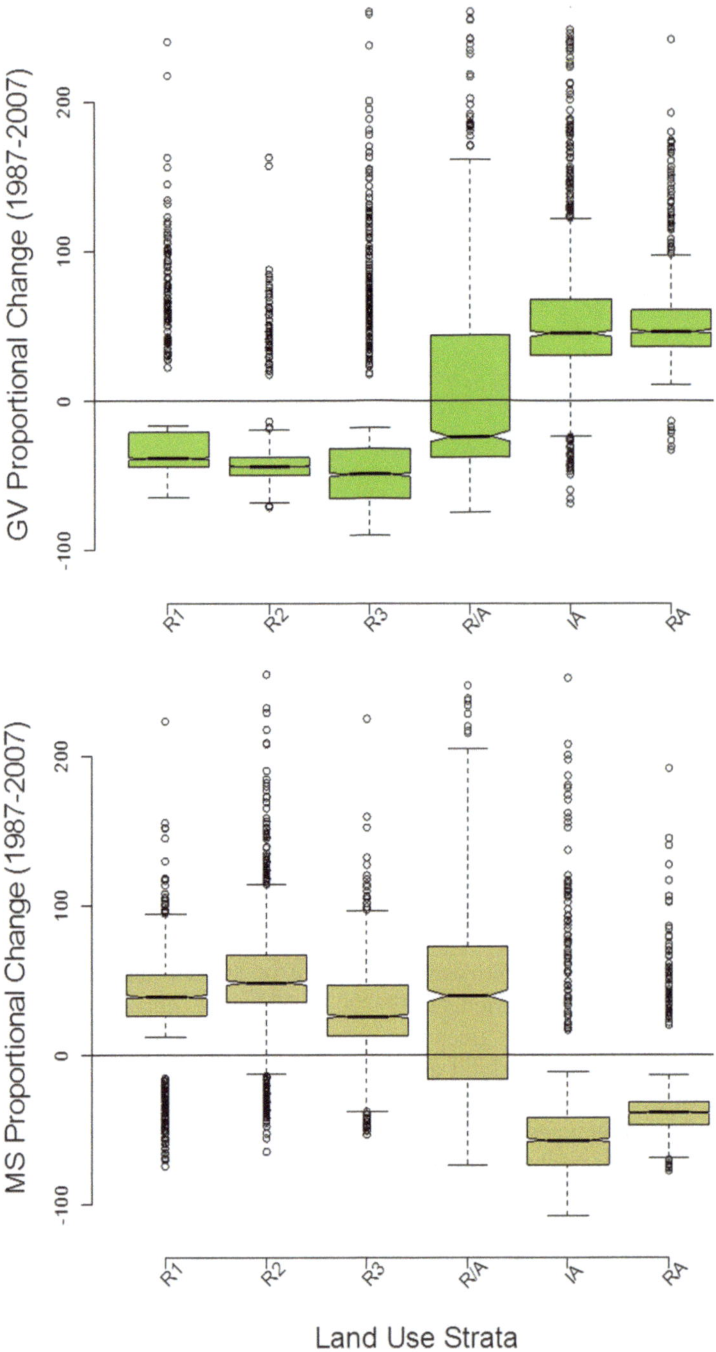

Fig. 23.7 Aerial view of the mixed rangeland/agriculture stratum (R/A) with its characteristic patchwork of sandy rangelands and intensifying agricultural land use (Photo © J. Hill 2008)

Fig. 23.8 Box plots of proportional change (in per cent, relative to of the starting date of the time series) in the SMA-derived abundance of Green Vegetation (GV) and Mobile Sand (MS) over a 20-year observation period (1987–2007)

(IA and RA), however, are characterized by statistical distributions where the majority of proportional change values are positive, indicating that the abundance of green biomass has substantially increased since the mid-1980s (Fig. 23.7). Not only is the interpretation, that these findings suggest substantial improvements of agricultural productivity, in agreement with the constantly growing area under irrigation. It is also supported by statistical data for Naiman County which, for example, demonstrates an increase of fertilizer consumption (from 500 to 75,000 t) and farm machine power (from 50,000 to 425,000 kW) during the observation period (Helldén 2010).

The changing presence of water at the land surface is not as much a continuous process but characterized by abrupt changes as water bodies tend to disappear within few years. Since it builds on the assumption of gradually changing end-member proportions, linear trend analysis of SMA-derived abundance estimates is therefore not considered as suitable for characterizing such phenomena. Instead, the reduction or complete disappearance of open water surfaces within the observation period was characterized by mapping their spatial extension approximately along 5-year observation intervals (1987, 1995, 2001, 2006 and 2010). The water map for each date was produced by applying a consistent threshold to the Water abundance image (i.e. $F_{water} > 0.65 \rightarrow$ open water). It is found that the

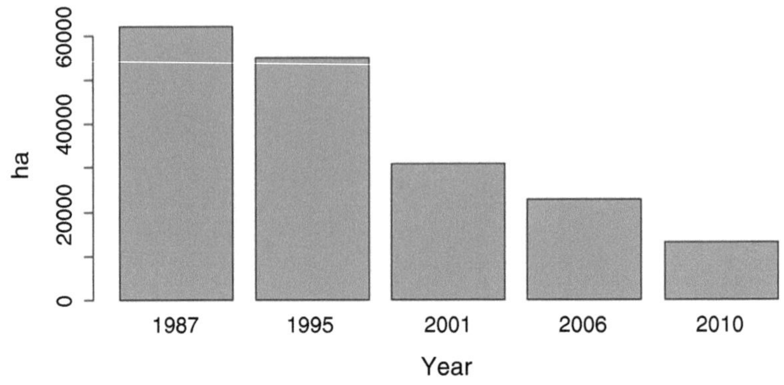

Fig. 23.9 Change in total area (in ha) covered by water (lakes, ponds and rivers) within the Landsat-TM/ETM+ scene area, based on thresholding the SMA-derived abundance estimate for water

SMA-detected acreage of lakes, ponds, bogs and swamps has diminished from approximately 62,000 ha (1987) to 22,800 ha in 2006, i.e. a reduction of more than 60 % has taken place during the observation period (Fig. 23.9).

This drastic decrease is only explained by an accelerated decline of the water table, where the most rapid change occurred between 1995 and 2001. However, including a more recently acquired scene from 2010 indicates that the drying of lakes and ponds has not yet come to an end, as groundwater exploitation and the expansion of irrigation agriculture into former rangelands is continuing. These findings are strongly supported by available statistical data: the number of power wells in Naiman County almost continuously increased from approximately 2,000 (1985) to slightly more than 10,000 in 2007 (an increase of >800 %), and the irrigated surface has grown from roughly 24,000 to 88,000 ha (Helldén 2010).

Evaluating the identified changes of land surface characteristics in the framework of ecosystem services requires a contextual evaluation of the derived indicators. In this sense a preliminary and simplified rating scheme was designed, which aims at linking image-derived indicators to those ecosystem services for which they have a specific expressive capacity (Table 23.3). Ponding water on the land surface, for example, used to be a strong indicator for a shallow water table and, since these ponds and lakes persisted over time, provided evidence for high ground water recharge rates. Over the past 20 years, these areas have shrunk by more than 60 %. Since this is a strong indicator for a declining water table, the ESS-Indicator for "groundwater recharge" was adjusted accordingly (Fig. 23.10). It should be noted that the reduction of this service does not necessarily imply a sensible restriction, because ground water extraction may continue as long as the capacity of pumps is sufficient to deliver water from increasingly deeper levels. However, the reduced service clearly indicates that compensating measures (introduction of more efficient irrigation practices and technologies, changes in crop varieties, water assignment schemes, etc.) need to be prepared with urgency. The indirect effects upon associated ecosystem services result from the adjustment rules introduced and explained earlier.

Table 23.3 Proposed linkage between SMA-derived indicators of changing land surface conditions in the available time window (1987–2007) and the associated ecosystem services (based on Eq. 23.2)

ESS	Descriptive indicator	Proportional change in abundance and stratum		ESS indicator length adjustment
Groundwater recharge	Total area covered by water (ponds, lakes, rivers)	**– 63.20 %**		**0.90 – 0.57** ↓
Dune fixation	SMA-derived abundance of "Mobile Sand" within strata R1-R3	R1: +38.86 % R2: +48.40 % R3: +29.06 %	**39.23 %**	**0.70 – 0.27** ↓
Range productivity	SMA-derived abundance of "Green Vegetation" within strata R1-R3	R1: –38.41 % R2: –43.92 % R3: –49.38 %	**44.52 %**	**0.80 – 0.36** ↓
Crop productivity	SMA-derived abundance of "Green Vegetation" within strata IA and RA	IA: + 44.82 % RA: + 46.12 %	**45.49 %**	**0.65 + 0.30** ↑

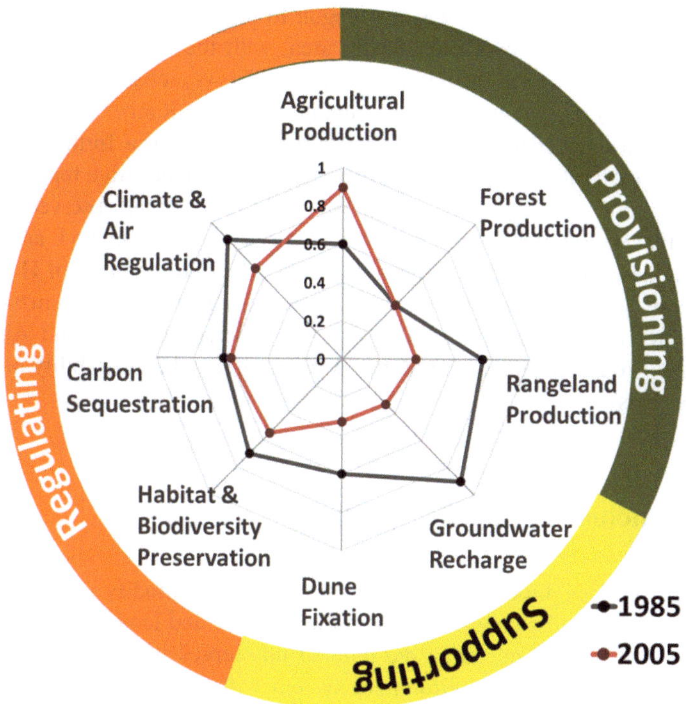

Fig. 23.10 Spider diagram displaying selected ecosystem services and their changes within the observation period (1987–2007)

The indicators for ecosystem services such as “Dune Fixation”, “Range Productivity” and “Crop Productivity” were adjusted with regard to the median values of proportional abundance changes within the relevant land use strata. The specific relevance for certain ecosystem services follows the concept that “Dune Fixation” and “Rangeland Productivity” is intrinsically linked to the rangeland stratum, while “Crop Productivity” is evaluated based upon the agricultural strata solely. Representative values for strata composed of several spatial entities (such as the rangeland stratum which is composed by R1, R2 and R3) were obtained by calculating the area-weighted sum of the corresponding median values.

Following this type of reasoning, the increasing abundance of Mobile Sand in the rangeland strata suggests a certain loss in the capacity of stabilizing sand dunes. The co-occurring decrease in Green Vegetation abundance indicates that this has developed in parallel to a reduction of grassland biomass (with negative implications for a second service, namely “Range Productivity”). Accordingly, the increase in the abundance of Green Vegetation within irrigated and non-irrigated agricultural land over the past 20 years is interpreted as an increase in “Crop Productivity”.

A synoptic representation of these changes in the form of spider diagrams suggests that over the past 20 years the ESS “Agricultural Production” has been optimized at the cost of other services, primarily “Groundwater Recharge” (Fig. 23.10). More recently, agricultural land use has also begun to penetrate into former rangelands, causing a reduction of available grazing area, with the potential consequence of increasing stocking rates in the remaining ranges. This is probably one of the reasons why the ecosystem services “Dune Fixation” and “Range Productivity” are experiencing a notable reduction, the latter being related to a further reduction in “Carbon Sequestration”. Additionally, these changes have negative repercussions on other ecosystem services, such as “Forest Production” (it is observed in several locations that willow trees, traditionally used for harvesting firewood, are dying-off due to the increasing distance to the water table), the “Preservation of Habitats and Biodiversity” (loss of wetlands as well as increasing pressure on grazing ranges leads to exchange and/or loss of species) and regional “Climate and Air Quality Regulation” (reduced capacity for dune fixation triggers the availability of particles for dust storms).

23.6 Conclusions and Perspectives

Ecosystem services, biodiversity conservation, and commodity production values are a function of land characteristics and the land use and land use change pattern. The process of land change analyzed in this study represents a significant example for transforming a human-environment system with limited resource availability into an alternative state. It can be considered representative for many parts of arid and semi-arid China. After 1973 the objective of improving rural livelihoods has been pursued by a combination of incentives aimed at increasing agricultural productivity and enforced regulations directed towards protecting rangeland resources at risk (e.g. Gao et al. 2006; Jiang 2002, 2006; Wu and de Pauw 2010). The main strategy was to render agricultural production less dependent on climatic risks (i.e. drought) by increasing the proportion of irrigated areas. The water table being just a few meters below surface, this prime objective could be addressed with simple technologies and moderate investments. Additionally, private initiative has been encouraged by modified land leasing concepts and by facilitating access to investments, agricultural mechanization and fertilizers. This decision has unleashed a process which traded different ecosystem services against each other, primarily by optimizing the service "Crop Production" at the cost of "Groundwater Recharge Capacities", and dragging behind a number of negative impacts on interrelated ecosystem services. It is shown in this study that long time series of satellite observations can be used to trace both, the impact of the rapidly declining water table but also the improving productivity of the agricultural system over an observation period of 20 years. However, trend analysis of suitable indicators of specific land surface conditions also revealed more subtle indications for increasing environmental risks: the accessibility of ground water resources facilitates the expansion of agricultural activities into formerly rangeland-dominated ecosystems and/or the reactivation of already existing agricultural land with marginal productivity. Not only did this increase the exploitation of groundwater resources, but caused a reduction of the area available for grazing sheep and cattle. In combination with the implemented legal restrictions in accessing certain parts of rangelands this inevitably led to increasing stocking rates in the remaining rangelands. Not unexpectedly, the analysis of the satellite observations covering the past 20 years provides evidence for a decrease in rangeland productivity (negative GV trend), which combines with indicators for an increasing disturbance of formerly consolidated sand and loess soils (positive MS trend).

The study confirms that Landsat time series are a unique source for deriving broad scale information about gradual changes of land surface properties at landscape level, very difficult to obtain through other approaches. Substantial progress is demonstrated with respect to composing geometrically and radiometrically calibrated time series of Landsat observations. However, while available processing chains have reached considerable maturity, there is a need to focus on the development of advanced interpretation concepts able to cope with the full complexity of land transformation processes, involving both conversions and modifications of land use and cover. Adjusting specific ESS indicators in the spider diagrams in proportion to satellite-derived indicators, for example, is one option for generating a synoptic perspective on land transformation processes and associated trade-offs in ecosystem services. It is also shown how the underlying processes are linked to changing capacities of ecosystems to provide goods and services, thereby providing essential insights for designing adapted management concepts.

It is of course acknowledged that the linear adjustments and scaling functions applied in this study are somewhat arbitrary and leave room for improvements and more differentiated scaling concepts (e.g. Barrett et al. 2005). The disappearance of water bodies, for example, may be related to their original water depth thus that their gradual drying out over time could be quantified as descent rates of the water table. Knowing these rates may enable improved predictions of future water availability or account for additional investments imposed by the necessity of increasingly powerful pumps, thereby closing the conceptual gap to associating monetary costs to specific ecosystem services. However, the main objective here was to demonstrate that EO satellites may contribute to assessing changes of ecosystem services and functions beyond supplying simple land use or cover maps.

In spite of the principal feasibility in pairing SMA-derived end-member abundance with selected ecosystem services, it should also not be overlooked that simple three-end-member models are rarely sufficient to adequately characterize complex land surface conditions. In our study case, for example, the inspection of images representing the modeling error suggests that using a single end-member (Mobile Sand) for representing a range of non-photosynthetic background materials is a fairly rough approximation. Efficient multiple end-member models (e.g. Rogge et al. 2006) may offer challenging perspectives to be explored in this context. Also, discontinuities and breaks in the behavior of surface conditions over time have so far been neglected, but might be addressed by using improved trend analysis concepts (e.g. Verbesselt et al. 2010). It is evident that the evaluation of trends in land surface conditions largely benefits from spatial stratification concepts. In this study meaningful strata have been outlined based on few available base maps and visual interpretation of Landsat imagery, additionally fuelled by information collected during several field visits. Given the recent advances in pattern recognition algorithms, it might be expected that automatic mapping strategies together with the availability of large volumes of freely accessible image archives may efficiently assist in such tasks.

Acknowledgements This study would not have been possible without the invaluable support Dr. Du Zitao (Institute for Remote Sensing Applications, Chinese Academy of Sciences, Beijing) provided during several field visits to Horqin Sandy Lands. The discussions with Prof. Ulf Helldén (University of Lund) and Dr. Achim Röder (University of Trier) were essential in sharpening the authors' perception of land transformation processes and for developing conceptual views discussed in this study. The support of Wolfgang Mehl (European Commission, Joint Research Centre, Ispra, Italy) in implementing a semi-automatic processing chain for high-precision geocoding is gratefully acknowledged. Part of this research was financially supported by the European Commission through funding the project "DeSurvey" (Integrated Project contract No. 003950). This support is gratefully acknowledged.

References

Baldocchi DD, Xu L, Kiang N (2004) How plant functional-type, weather, seasonal drought, and soil physical properties alter water and energy fluxes of an oak–grass savanna and an annual grassland. Agr Forest Meteorol 123:13–39

Barrett DJ, Hill MJ, Hutley LB, Beringer J, Xu JH, Cook GD, Carter JO, Williams RJ (2005) Prospects for improving savanna biophysical models by using multiple-constraints model-data assimilation methods. Aust J Bot 53:689–714

Berk A, Anderson GP, Bernstein LS, Acharya PK, Dothe H, Matthew MW, Adler-Golden SM, Chetwynd JH, Richtsmeier SC, Pukall B, Allred CL, Jeong LS, Hoke ML (1999) MODTRAN4 radiative transfer modeling for atmospheric correction. SPIE

Bormann H, Breuer L, Gräff T, Huisman JA (2007) Analysing the effects of soil properties changes associated with land use changes on the simulated water balance: a comparison of three hydrological catchment models for scenario analysis. Ecol Model 209:29–40

Butterbach-Bahl K, Kögel-Knabner I, Han X (2011) Steppe ecosystems and climate and land-use changes – vulnerability, feedbacks and possibilities for adaptation. Plant and Soil 340:1–6

Chander G, Markham B, Helder DL (2009) Summary of current radiometric calibration coefficients for Landsat MSS, TM, ETM+, and EO-1 ALI sensors. Remote Sens Environ 113:893–903

Chen Y, Tang H (2005) Desertification in North China: background, anthropogenic impacts and failures in combating it. Land Degrad Dev 16:367–376

Cohen WB, Goward SN (2004) Landsat's role in ecological applications of remote sensing. Bioscience 54:535–545

Coppin P, Jonckheere I, Nackaerts K, Muys B, Lambin E (2004) Digital change detection methods in ecosystem monitoring: a review. Int J Remote Sens 25:1565–1596

Costanza R, d'Arge R, de Groot R, Farber S, Grasso M, Hannon B, Limburg K, Naeem S, O'Neill RV, Paruelo J, Gaskin RG, Sutton P, van den Belt M (1997) The value of the world's ecosystem services and natural capital. Nature 387:253–260

Daily GC, Polasky S, Goldstein J, Kareiva PM, Mooney HA, Pejchar L, Ricketts TH, Salzman J, Shallenberger R (2009) Ecosystem services in decision making: time to deliver. Front Ecol Environ 7:21–28

DeFries R, Foley JA, Asner GP (2004) Land-use choices: balancing human needs and ecosystem function. Front Ecol Environ 2:249–257

Ellis EC, Ramankutty N (2008) Putting people in the map: anthropogenic biomes of the world. Front Ecol Environ 6:439–447

Elmore AJ, Mustard JF, Manning SJ, Lobell DB (2000) Quantifying vegetation change in semiarid environments: precision and accuracy of spectral mixture analysis and the normalized difference vegetation index. Remote Sens Environ 73:87–102

Foley JA, DeFries R, Asner GP, Barford C, Bonan G, Carpenter SR, Chapin FS, Coe MT, Daily GC, Gibbs HK, Helkowski JH, Holloway T, Howard EA, Kucharik CJ, Monfreda C, Patz JA, Prentice C, Ramankutty N, Snyder PK (2005) Global consequences of land use. Science 309:570–574

Gao J, Liu Y, Chen Y (2006) Land cover changes during agrarian restructuring in Northeast China. Appl Geogr 26:312–322

Helldén U (2010) Application of the Lund University Coupled Desertification Model (LUCDM) in Naiman County, Jirem Prefecture, Inner Mongolia, China. In: DeSurvey-IP. A surveillance system for assessing and monitoring desertification, deliverable 1.8.3.1. Lund, Sweden, pp 1–35

Hill J, Mehl W (2003) Geo- und radiometrische Aufbereitung multi- und hyperspektraler Daten zur Erzeugung langjähriger kalibrierter Zeitreihen. Photogrammetrie Fernerkundung Geoinformation 1:7–14

Hill J, Sturm B (1991) Radiometric correction of multi-temporal Thematic Mapper data for use in agricultural land-cover classification and vegetation monitoring. Int J Remote Sens 12:1471–1491

Hill J, Mehl W, Radeloff V (1995) Improved forest mapping by combining corrections of atmospheric and topographic effects. In: Askne J (ed) Proceedings of the 14th EARSeL symposium on sensors and environmental applications of remote sensing. Göteborg, Sweden: A.A. Balkema, Rotterdam/Brookfield, pp 143–151

Jiang H (1999) The Ordos Plateau of China: an endangered environment. United Nations University Press, Tokyo/New York/Paris

Jiang H (2002) Culture, ethnology, and nature's changing balance. Sandification on Mu Us Sandy Land, Inner Mongolia, China. In: Reynolds JF, Stafford-Smith M (eds) Global desertification. Do humans cause deserts? Dahlem University Press, Berlin, pp 181–196

Jiang H (2006) Decentralization, ecological construction, and the environment in post-reform China: case study from Uxin Banner, Inner Mongolia. World Dev 34:1907–1921

Ju J, Roy DP, Vermote E, Masek J, Kovalskyy V (2012) Continental-scale validation of MODIS-based and LEDAPS Landsat ETM+ atmospheric correction methods. Remote Sens Environ 122:175–184

Kareiva PM, Watts S, McDonald R, Boucher T (2007) Domesticated nature: shaping landscapes and ecosystems for human welfare. Science 316:1866–1869

Kaufman YJ, Wald A, Remer LA, Gao B, Li R, Flynn L (1997) The MODIS 2.1 μm channel – correlation with visible reflectance for use in remote sensing of aerosol. IEEE Trans Geosci Remote Sens 35:1286–1298

Kotchenova SY, Vermote EF, Matarrese R, Klemm FJ (2006) Validation of a vector version of the 6S radiative transfer code for atmospheric correction of satellite data. Part I: path radiance. Appl Optics 45:6762–6774

Lambin EF, Geist HJ (2001) Global land-use and land-cover change: what have we learned so far. Glob Change Newsl 46(6):27–30

Lambin EF, Geist HJ (2006) Land use and land cover change. Local processes and global impacts. Springer, Berlin/Heidelberg/New York

Lambin EF, Meyfroidt P (2010) Land use transitions: socio-ecological feedback versus socio-economic change. Land Use Policy 27:108–118

Liu J, Liu M, Tian H, Zhuang D, Zhang Z, Zhang W, Tang X, Deng X (2005a) Spatial and temporal patterns of China's cropland during 1990–2000: an analysis based on Landsat TM data. Remote Sens Environ 98:442–456

Liu J, Tian H, Liu M, Zhuang D, Melillo JM, Zhang Z (2005b) China's changing landscape during the 1990s: large-scale land transformations estimated with satellite data. Geophys Res Lett 32:L02405

Markham BL, Helder DL (2012) Forty-year calibrated record of earth-reflected radiance from Landsat: a review. Remote Sens Environ 122:30–40

Millennium Ecosystem Assessment (2005) Ecosystems and human well-being: current state and trends. Island Press, Washington, DC/Covelo/London

Nelson E, Mendoza G, Regetz J, Polasky S, Tallis H, Cameron DR, Chan KMA, Daily GC, Goldstein J, Kareiva PM, Lonsdorf E, Naidoo R, Ricketts TH, Shaw R (2009) Modeling multiple ecosystem services, biodiversity conservation, commodity production, and tradeoffs at landscape scale. Front Ecol Environ 7:4–11

Piao S, Fang J, Ciais P, Peylin P, Huang Y, Sitch S, Wang T (2009) The carbon balance of terrestrial ecosystems in China. Nature 458:1009–1013

Press W-H, Teukolsky SA, Vetterling WT, Flannery BP (1992) Numerical recipes in C. The art of scientific computing. Cambridge University Press, Cambridge, MA

Rahman H, Dedieu G (1994) SMAC: a simplified method for the atmospheric correction of satellite measurements in the solar spectrum. Int J Remote Sens 15:123–143

Richter R (1996) A spatially adaptive fast atmospheric correction algorithm. Int J Remote Sens 17:1201–1214

Röder A, Kümmerle T, Hill J (2005) Extension of retrospective datasets using multiple sensors. An approach to radiometric intercalibration of Landsat TM and MSS data. Remote Sens Environ 95:195–210

Röder A, Udelhoven T, Hill J, del Barrio G, Tsiourlis G (2008) Trend analysis of Landsat-TM and -ETM+ imagery to monitor grazing impact in a rangeland ecosystem in Northern Greece. Remote Sens Environ 112:2863–2875

Rogan J, Chen D (2004) Remote sensing technology for mapping and monitoring land-cover and land-use change. Prog Plan 61:301–325

Rogge DM, Rivard B, Zhang J, Feng J (2006) Iterative spectral unmixing for optimizing per-pixel endmember sets. IEEE Trans Geosci Remote Sens 44:3725–3736

Schott JR, Salvaggio C, Volchok WJ (1988) Radiometric scene normalization using pseudoinvariant features. Remote Sens Environ 26:1–16

Schowengerdt RA (1997) Remote sensing. Models and methods for image processing. Academic, San Diego/London

Small C (2004) The Landsat ETM+ spectral mixing space. Remote Sens Environ 93:1–17

Smith MO, Ustin SL, Adams JB, Gillespie AR (1990) Vegetation in deserts: I. A regional measure of abundance from multispectral images. Remote Sens Environ 31:1–26

Sneath D (1998) State policy and pasture degradation in Inner Asia. Science 281:1147–1148

Tanré D, Deroo C, Duhaut P, Herman M, Morcrette JJ, Perbos J, Deschamps PY (1990) Description of a computer code to simulate the signal in the solar spectrum: the 5S code. Int J Remote Sens 11:659–668

Tong C, Wu J, Yong S, Yang J, Yong W (2004) A landscape-scale assessment of steppe degradation in the Xilin River Basin, Inner Mongolia, China. J Arid Environ 59:133–149

Verbesselt J, Hyndman R, Zeileis A, Culvenor D (2010) Phenological change detection while accounting for abrupt and gradual trends in satellite image time series. Remote Sens Environ 114:2970–2980

Vermote EF, Kotchenova S (2008) Atmospheric correction for the monitoring of land surfaces. J Geophys Res Atmos 113(D23):D23S90

Vermote EF, Tanré D, Deuze JL, Herman M, Morcrette JJ (1997) Second simulation of the satellite signal in the solar spectrum, 6S – an overview. IEEE Trans Geosci Remote Sens 35:675–686

Vermote EF, El Saleous N, Justice C (2002) Atmospheric correction of the MODIS data in the visible to middle infrared: first results. Remote Sens Environ 83:97–111

Vicente-Serrano SM, Pérez-Cabello F, Lasanta T (2008) Assessment of radiometric correction techniques in analyzing vegetation variability and change using time series of Landsat images. Remote Sens Environ 112:3916–3934

Vitousek PM, Kooney HA, Lubchenco J, Melillo JM (1997) Human domination of Earth's ecosystems. Science 277:494–499

Vogelmann JE, Xian G, Homer C, Tolk B (2012) Monitoring gradual ecosystem change using Landsat time series analyses: case studies in selected forest and rangeland ecosystems. Remote Sens Environ 122:92–105

Woodcock CE, Allen R, Anderson M, Belward A, Bindschadler R, Cohen W, Gao F, Goward SN, Helder D, Helmer E, Nemani R, Oreopoulos L, Schott J, Thenkabail PS, Vermote EF, Vogelmann J, Wulder MA, Wynne R (2008) Free access to Landsat imagery. Science 320:1011

Wu W, de Pauw E (2010) Policy impacts on land degradation: evidence revealed by remote sensing in Western Ordos, China. In: Zdruli P, Pagliai M, Kapur S, Faz Cano AF (eds) Land degradation and desertification: assessment, mitigation and remediation. Springer, Dordrecht/Heidelberg/London/New York, pp 219–233

Wulder MA, Masek JG, Cohen WB, Loveland TR, Woodcock CE (2012) Opening the archive: how free data has enabled the science and monitoring promise of Landsat. Remote Sens Environ 122:2–10

Zhu, Z, Zou B (1988) Desertification and rehabilitation – a case study in Horqin Sandy Land. Institute of Desert Research, Academia Sinica, Lanzhou. Research report, 1–113

Chapter 24
Carbon Stock Estimation of Tropical Forests on Borneo, Indonesia, for REDD+

Sandra Englhart, Jonas Franke, Vanessa Keuck, and Florian Siegert

24.1 Introduction

Tropical forests store huge amounts of carbon (C), the majority (78 %) in aboveground trunks, branches, and leaves as well as in belowground roots (22 %) (Saatchi et al. 2011b). Forested tropical peatlands accumulate additional carbon in belowground peat deposits which are sustained by intact forests. In Indonesia, approximately 55–58 GtC is stored belowground in peatlands and 18.6 GtC aboveground in forests (Jaenicke et al. 2008; Koh et al. 2009; Baccini et al. 2012). Peatlands are often drained, deforested or burned for industrial agricultural development, such as the establishment of oil palm and pulp wood plantations, which causes massive carbon emissions that are released to the atmosphere (Hooijer et al. 2010).

Emissions from deforestation and forest degradation in Southeast Asia, including tropical peatland burning and oxidation amounted to 23 % of total anthropogenic CO_2 (carbon dioxide) emissions worldwide between 1997 and 2006 (Van der Werf et al. 2009). Hooijer et al. (2012) estimated the carbon loss from converting peat swamp forests into agriculture to be on average 100 tCO_2 per hectare per year annualized over 25 years. Through these processes, Indonesia became one of the largest CO_2 emitters worldwide.

Considering these high emission rates, projects aiming at forest conservation offer good prospects for climate change mitigation in developing countries. One example is REDD+, which aims at reducing emissions from deforestation and forest degradation, conservation of forest carbon stocks, sustainable management

S. Englhart (✉) • J. Franke • V. Keuck
Remote Sensing Solutions GmbH, Isarstr. 3, 82065 Baierbrunn, Germany
e-mail: englhart@rssgmbh.de

F. Siegert
Remote Sensing Solutions GmbH, Isarstr. 3, 82065 Baierbrunn, Germany

Biology Department II, GeoBio Center, Ludwig-Maximilians-University, Großhadener Str. 2, 82152 Planegg-Martinsried, Germany

I. Manakos and M. Braun (eds.), *Land Use and Land Cover Mapping in Europe: Practices & Trends*, Remote Sensing and Digital Image Processing 18,
DOI 10.1007/978-94-007-7969-3_24,

of forest land and enhancement of forest carbon stocks (Campbell 2009). REDD+ was approved at the United Nations Climate Change Conferences in 2009 and 2010, the year 2011 was the starting point for the development of a worldwide forest monitoring system. The system considers current technical capabilities to monitor greenhouse gas emissions and removals from deforestation, reforestation and degradation activities in forest land remaining forest land (GOFC-GOLD 2011). REDD+ intends for conditional payments to countries reducing emissions, and conditional payments from national levels to forest stewards reducing emissions (Campbell 2009). Implementation of REDD+ policies depends on accurate and precise estimates of emissions avoided at national scale. A nationwide monitoring is needed to prevent leakage within a country, where reduced deforestation or forest degradation could occur in one part of the country but increase in another part through displaced activities (DeFries et al. 2007).

In May 2010, a contract between the Indonesian and Norwegian government was signed providing one billion US$ for a cooperation on REDD+. A major component is a moratorium on new agricultural and logging licenses, which aims to support Indonesia's goal of reducing national emissions by 26 % until 2020 and to prepare Indonesia to draw payments from industrial nations via the REDD+ scheme (Sloan et al. 2012).

A crucial element for REDD+ is the estimation of current carbon stocks. The Intergovernmental Panel on Climate Change (IPCC) provides Guidelines for National Greenhouse Gas Inventories which refer to two basic inputs for calculating greenhouse gas inventories: activity data, which specify the extent of deforestation, reforestation and forest degradation/enhancements in unit area, and emission factors, which describe emissions/removals of greenhouse gases per unit area (IPCC 2006). Uncertainties of both activity data and emission factors are an important element of greenhouse gas inventories to identify the contributions to the overall accuracy (Grassi et al. 2008). The guidelines include three different tiers which represent the level of methodological complexity: Tier 1 uses IPCC default values to estimate emissions, Tier 2 requires country specific carbon data, and Tier 3 is based on a detailed national inventory.

The most accurate way of aboveground biomass (AGB) and carbon stock retrieval using Tier 3 are forest inventories which include field based measurements (e.g. diameter at breast height (DBH), tree height, tree species-specific wood density) to estimate biomass on the basis of allometric equations (Brown 1997; Chave et al. 2005). The carbon stock is generally derived from AGB estimates by assuming a carbon content of dry biomass of 50 % (Goetz et al. 2009). Albeit this method provides very precise AGB values, it is time and cost consuming, difficult to implement in remote areas and most importantly lacks information on the spatial variability (Lu 2006). Monitoring AGB by remote sensing is less accurate, but the major advantage is the ability to generate spatially explicit and potentially 'wall-to-wall' carbon stock estimations in large and remote areas. Most approaches are based on an assessment of historic, current and future deforestation rates based on detectable changes in forest areas using satellite or airborne data (Boettcher et al. 2009). Moderate to coarse resolution data (e.g. MODIS) are usually used for global carbon stock estimation (Baccini et al. 2012) and medium resolution data (e.g. ALOS PALSAR) are usually selected for AGB estimations at national or regional scales (Englhart et al. 2012; Ryan et al. 2012). Monitoring systems that allow for credible measurements, reporting and verification (MRV) of forest carbon stocks and their changes in REDD+ project sites are among the most crucial elements for a successful implementation of REDD+.

Different approaches have been developed to assess AGB by remote sensing using forest inventory AGB data. The stratify & multiply approach is a simple way to retrieve a carbon stock map by linking a single AGB value (or a defined range, ideally derived from field inventories) determined for a specific vegetation type to a remote sensing based land cover map (Goetz et al. 2009). A major disadvantage of this discrete approach is lack of information on the spatial variance of AGB within one land cover class.

A more sophisticated technique is the continuous approach, at which radiometric satellite measurements are calibrated to field based AGB measurements to derive a wall-to-wall AGB estimation map also indicating the spatial variation of the carbon stocks (Goetz et al. 2009). Examples using optical and SAR (Synthetic Aperture Radar) imagery for a continuous estimation of AGB are described in the following paragraph.

Multispectral satellite images have been widely used to derive AGB in tropical regions using vegetation indices (Zhang and Run-Guo 2009; Li et al. 2010), spectral signatures (Baccini et al. 2008; Tangki and Chappell 2008; Li et al. 2010; Avitabile et al. 2012), image texture (Lu 2005; Wijaya et al. 2010; Nichol and Sarker 2011; Sarker and Nichol 2011), and spectral mixture analysis (SMA) (Soenen et al. 2010). Major constraints of using optical data for AGB estimation are the saturation of the signal in high biomass ranges, the dependence on daylight, and the obstruction by clouds, which is a crucial point in tropical regions.

Active SAR systems can operate day and night while penetrating through haze, smoke, and clouds. The correlation of backscatter signal and biomass is mainly dependent on wavelength, polarization, and incidence angle. Longer wavelengths have been proven to be more useful for AGB estimation because of an increasing backscatter range with changing biomass (Luckman et al. 1997; Lu 2006) and a higher saturation level in regard to the biomass range (Saatchi et al. 2011a; Englhart et al. 2011).

In order to compare the performance of SAR and multispectral satellite data for AGB estimations using the continuous approach, a case study was established in a tropical peat swamp forest area in Central Kalimantan, Indonesia. The potential of different approaches and data sources are demonstrated in the context of REDD+. The objectives of this case study are therefore (i) an evaluation of either different SAR frequencies and polarizations or SMA MF fractions of a multispectral RapidEye scene to estimate AGB by the continuous approach and (ii) a comparison of continuous and discrete (stratify & multiply) AGB estimation using the example of RapidEye data.

24.2 Methodology

24.2.1 Study Area

The study area is located east of Palangka Raya, the capital of the province Central Kalimantan, on the island of Borneo, Indonesia (Fig. 24.1).

Peat swamp forest is the predominant vegetation type besides riparian forest, forest areas heavily degraded by fire (shrubs and regrowing forest) and seasonally flooded wetlands. The peatlands have been drained and deforested, mainly for agricultural development and plantations (Hooijer et al. 2010) and recurrent fire events destroyed approximately 50 % of the forest cover in the past decade. The most severe impact was caused by the Mega Rice Project (MRP), conceptualized by the Indonesian government in 1995 in order to convert an area of one million hectare for rice cultivation through the construction of about 4,000 km of drainage and irrigation channels in peatlands (Page et al. 2002; Boehm and Siegert 2004).

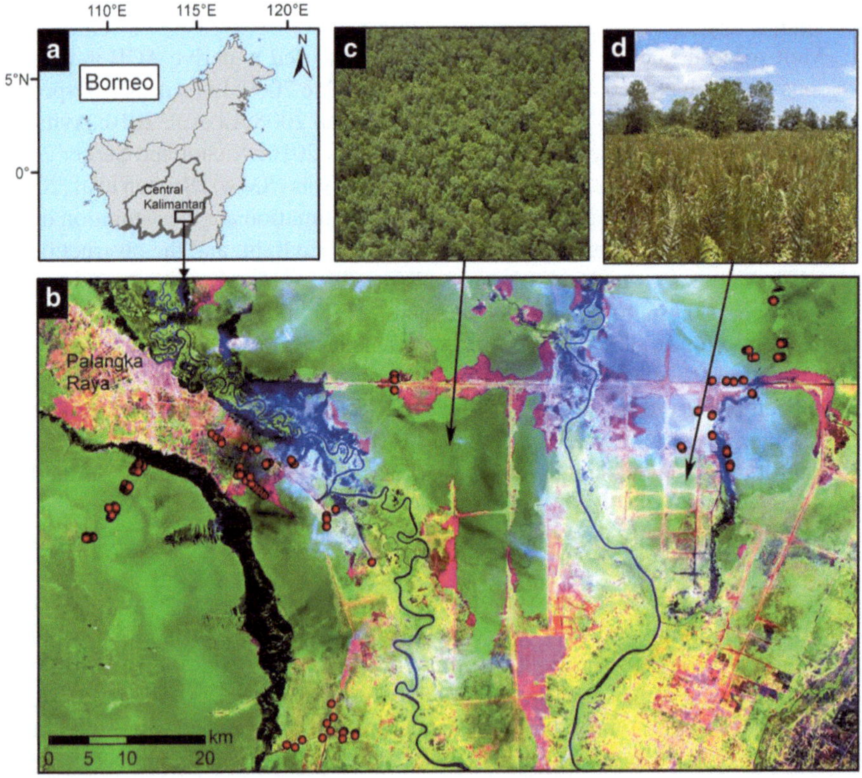

Fig. 24.1 (**a**) Overview map of the study area in Central Kalimantan on Borneo. (**b**) Landsat 5 TM satellite image from 10/02/2010 (bands R:5, G:4, B:3) showing the study area. Field inventory locations are depicted in red. (**c**) shows an aerial photo of an intact forest and (**d**) depicts a degraded regrowing area (© F. Siegert, S. Englhart); typical locations are indicated by *arrows*

24.2.2 Data

24.2.2.1 Field Inventory Data

Field inventory measurements were collected in the years 2008, 2010 and 2011. Forest inventory plots with different plot sizes were established in forested and regrowing areas (Fig. 24.1b depicts the plot locations). The sample plot design is following the guidelines provided by Pearson and Walker (2005).

For regrowing forests, a 20 $\times$ 50 m rectangular plot was chosen and all trees within this area were sampled. For forests, three circular nested plots with radii of 4, 14, and 20 m were sampled. In each nest, trees of a certain DBH were measured depending on degradation intensity: 2–10 cm or 5–20 cm (within the 4 m radius), 10–20 cm or 20–50 cm (within 14 m radium), and greater than 20 or 50 cm (within 20 m radius).

Within regrowing and forested plots, following parameters were collected: DBH, tree height, and species of all trees in order to estimate their wood density. Tree specific wood densities were derived from databases provided by Chudnoff (1984), World Agroforestry Centre (2011), and IPCC (2006). If the tree species could not be identified, an average wood density for Asian tropical trees of 0.57 Mg m^{-3} was applied (Brown 1997).

AGB was calculated using a combination of allometric models from Hughes et al. (1999) for saplings (if DBH < 5 cm and height ≤ 1.3 m) or trees (if DBH < 5 cm and height > 1.3 m) and Chave et al. (2005) for moist tropical forest stands including DBH and tree height (if DBH ≥ 5 cm and height > 1.3 m).

All field inventory plots were monitored for possible changes and were removed if there were any. Altogether 107 plots were sampled, 48 plots in regrowing vegetation ranging from <0.1 to 19.6 t/ha and 59 plots in forested areas ranging from 8.7 to 458.4 t/ha.

24.2.2.2 SAR Data

X-, C- and L-band data in HH and HV polarizations were investigated for their potential for estimating AGB. Four TerraSAR-X (X-band) ScanSAR HH polarized images with a pixel spacing of 8.25 m and an incidence angle of 34.2° (acquired on: 11/07/2011, 22/07/2011, 13/08/2011, and 24/08/2011) and three RADARSAT-2 (C-band) standard mode HH and HV polarized images with a pixel spacing of 8.00 m and an incidence angle of 36.6° (22/07/2011, 15/08/2011, and 08/09/2011) were analyzed. In addition, six ALOS PALSAR (L-band) fine beam HH and HV polarized images with a pixel spacing of 12.5 m and an incidence angle of 38.8° were evaluated which covered three time stamps, as two different paths were necessary to cover the whole study area (30/06/2010, 17/07/2010, 15/08/2010, 01/09/2010, 30/09/2010, 17/10/2010). All images were acquired during the dry season (May to October) to minimize any impact of rainfall and soil moisture. No major land cover changes took place between 2010 and 2011.

Preprocessing of SAR images included standard radiometric calibration for sigma naught (Fritz and Eineder 2009; Luscombe 2009; Shimada et al. 2009), frost speckle filtering with a moving window of 7 × 7 pixel, and co-registration.

Multi-temporal backscatter coefficients were used because averaging backscatter in time reduces speckle without losing spatial resolution. In a previous study, we found that multi-temporal backscatter values were superior to mono-temporal values for estimating AGB (Englhart et al. 2011).

24.2.2.3 Multispectral Imagery

A RapidEye scene was used to analyze the potential for estimating AGB using multispectral data. RapidEye data has a pixel resolution of 5 m and contains five spectral bands. An image recorded on 21/06/2010 was preprocessed including a geometric and atmospheric correction using ATCOR (Richter 1997). Spectral Mixture Analysis (SMA) is a suitable technique to derive forest parameters from remote sensing data (Souza et al. 2005; Asner et al. 2009; GOFC-GOLD 2011). Forests degraded by selective logging or fire are characterized by mixed pixels due to the observed mixed reflectance from different constituents of the earth's surface, i.e. green vegetation (GV), non-photosynthetic vegetation (NPV), soil, shade, etc., within the area of one image element (pixel). The advantage of SMA is that the abundance of sub-pixel components can be estimated as continuous values. A special type of SMA, the Mixture Tuned Matched Filtering (MTMF) was applied. A detailed description of SMA and MTMF is provided by Adams et al. (1986), Williams and Hunt (2002) and Mundt et al. (2007). The result of the MTMF is a grey-scale matched filtering (MF) fraction image representing the estimated relative degree to which each pixel matches the reference spectrum (Williams and Hunt 2002). The reference spectra of GV, soil and NPV were generated by manually creating areas containing these reference spectra and storing the information in a spectral library. The MF fractions derived from the RapidEye image were scaled to values between 0 and 1, where 1 indicated a perfect match of the pixel spectrum to the reference spectrum.

24.2.3 AGB Estimation

24.2.3.1 Continuous Approach for AGB Estimation Using SAR and Multispectral Data

The relationship between each of all available SAR backscatter or multispectral MF fractions and AGB was analyzed using spatially averaged signals over a grid with a cell size of 40 × 40 m. This size was chosen as it is similar to the size of the biggest nested field plot (radius 20 m). Due to the saturation of the satellite signals in high biomass ranges, only field data smaller than 300 t/ha were used for the regression modeling. The AGB field data was randomly split up to be used for training (85 % of all data) and validating (15 %) the AGB models. For SAR images, all 98 AGB reference samples were usable while for RapidEye only 53 AGB samples were available due to clouds and area coverage. In a first step, the relationship of each single SAR and multispectral input signal and AGB was analyzed. Different curve progressions were examined, but the exponential one yielded in all cases the best results. Based on these results, a combined regression model was tested. Therefore, a least-square multivariate linear regression was conducted using exponential values of either SAR backscattering coefficients or multispectral MF fractions of GV, soil and NPV.

24.2.3.2 Comparison Continuous vs. Discrete AGB Estimation Using Optical Imagery

A comparison of continuous and discrete AGB estimations was conducted on the basis of RapidEye. The continuous AGB estimation is based on MF fractions per pixel and the discrete AGB estimation is derived from a land cover classification linking a single AGB value derived from field inventory data to the individual classes of a land cover classification. The RapidEye land cover classification is based on an object based classification approach of the scene from 21/06/2010 (Fig. 24.3). A hierarchical rule set defining the different classes was applied using spectral, geometric, thematic and topologic criteria.

In total, eleven land cover classes were defined whereby only six are relevant for carbon stock comparison (primary and secondary peat swamp forest, riparian forest/agroforestry, bush/shrubs/regrowth, grassland/fern/agriculture, and recently burned/sparse regrowth). For a quantitative accuracy assessment of the land cover classification, 75 sample plots (located in peat swamp forest, bush/shrub/regrowth, and recently burned/sparse vegetation), mapped according to the LCCS standard (land cover classification system) and recorded with differential GPS, were evaluated. Furthermore, a geo-tagged video, recorded during a flight over the study area in 2010, was analyzed for the existing land covers and their spatial pattern and compared to the final land cover map. The achieved overall accuracy was 87.8 %.

24.3 Results

24.3.1 Continuous Approach for AGB Estimation Using SAR and Multispectral Data

The relationships between SAR backscattering coefficients of X-, C- and L-band data and RapidEye based MF fractions and AGB were investigated separately and

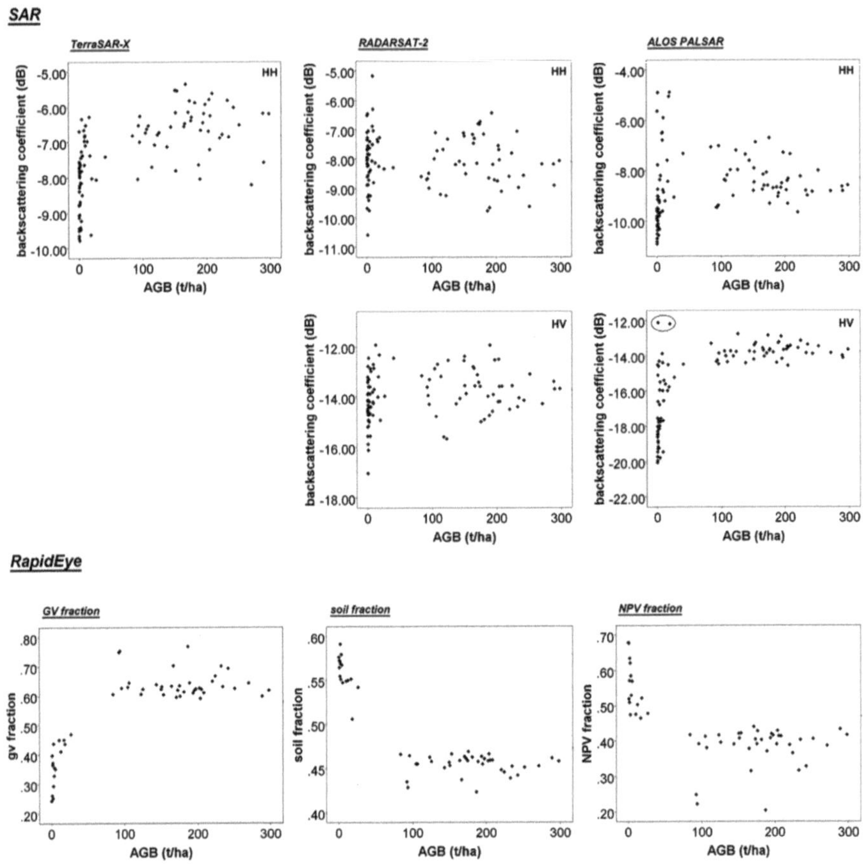

Fig. 24.2 Scatterplots showing AGB derived from field inventory measurements versus SAR backscatter signal of TerraSAR-X, RADARSAT-2 and ALOS PALSAR (*upper panel*: HH polarization, *middle panel*: HV polarization; n = 98) and RapidEye GV, soil, and NPV MF fractions (*lower panel*; n = 53). The *circle* in the ALOS PALSAR HV scatterplot indicates double bounce backscatter on a recently burned area

are depicted in scatterplots in Fig. 24.2. A significant correlation between SAR backscatter and AGB was only found with TerraSAR-X HH and ALOS PALSAR HV polarized data. Two of the field inventory plots were located in a recently burned area and are visible in the ALOS PALSAR HV polarized scatterplot (Fig. 24.2, ALOS PALSAR HV circle). Trunks of dead trees scattered on the ground cause double bounce backscatter and thus high backscatter signals there. RADARSAT-2 HH and HV and ALOS PALSAR HH polarized backscatter showed no correlation to AGB. All investigated RapidEye-derived MF fractions showed a correlation to AGB (Table 24.1). The exponential dependency is also linked to saturation in higher biomass ranges as the curve flattens in higher biomass ranges. The saturation effect is visible in the scatterplots depicted in Fig. 24.2, showing similar backscatter or MF fraction values from approximately 100 t/ha upwards.

Table 24.1 Results of regression modeling (reg.) and subsequent independent validation (val.) of AGB values on the basis of TerraSAR-X HH and ALOS PALSAR HV polarized data (n = 98) or RapidEye-based GV, soil, and NPV fractions (n = 53)

Input data	r^2 (reg.)	r^{22} (val.)	*RMSE* [t/ha] (val.)	*RMSE* [%] (val.)
TerraSAR-X HH	0.51	0.30	81.71	110
ALOS PALSAR HV	0.68	0.69	63.77	86
TerraSAR-X HH & ALOS PALSAR HV	0.68	0.50	60.92	82
RapidEye GV	0.86	0.86	37.11	37
RapidEye soil	0.89	0.82	38.11	38
RapidEye NPV	0.70	0.87	67.87	67
RapidEye GV & soil & NPV	0.92	0.83	44.17	44

On the basis of the exponential dependencies, a multivariate linear regression using exponential values of SAR signals and multispectral MF fractions was also analyzed. Table 24.1 depicts the regression and independent validation results of all investigated relationships. The multivariate regression models using either SAR backscatter coefficients or multispectral MF fractions turned out to be more accurate for AGB estimation than regression models using only a single variable (Table 24.1). A regression model, which combined SAR backscatter and multispectral MF fractions, was also evaluated but the achieved accuracy was not higher than the multivariate MF fractions regression model and was therefore not further analyzed. The independent validation demonstrates that AGB estimates derived from optical MF fractions are more accurate, resulting in higher coefficients of determination (r^2) and lower root mean square errors (*RMSEs*). The multivariate SAR AGB model was more accurate than the single variable TerraSAR-X or ALOS PALSAR model in terms of the *RMSE*. The multivariate RapidEye model achieved a higher r^2 from the regression modeling, but the results of the independent validation showed slightly lower r^2 and slightly higher *RMSEs* than the GV and soil fraction AGB models. The NPV model is the least accurate model of the multispectral MF fractions with a high *RMSE*.

In order to qualitatively test these validated regression models for AGB prediction of larger areas, they were applied to remote sensing data of a 1,893 km^2 test area in Central Kalimentan, Borneo (Fig. 24.3). AGB estimations are shown in aggregated classes from <5 t/ha (dark red) to >300 t/ha (dark green). Estimations exceeding 300 t/ha were not further differentiated due to the saturation effect of the satellite signal in the higher biomass range. TerraSAR-X AGB estimations were more reliable in low biomass ranges and less reliable in high biomass ranges and ALOS PALSAR AGB estimations showed opposite results. Due to double bounce on recent burn scars, the ALOS PALSAR based regression model clearly overestimates AGB in these areas (see Fig. 24.3 arrow). AGB estimations derived from the combined TerraSAR-X and ALOS PALSAR regression model are higher in the low biomass range than derived from either TerraSAR-X or ALOS PALSAR. This

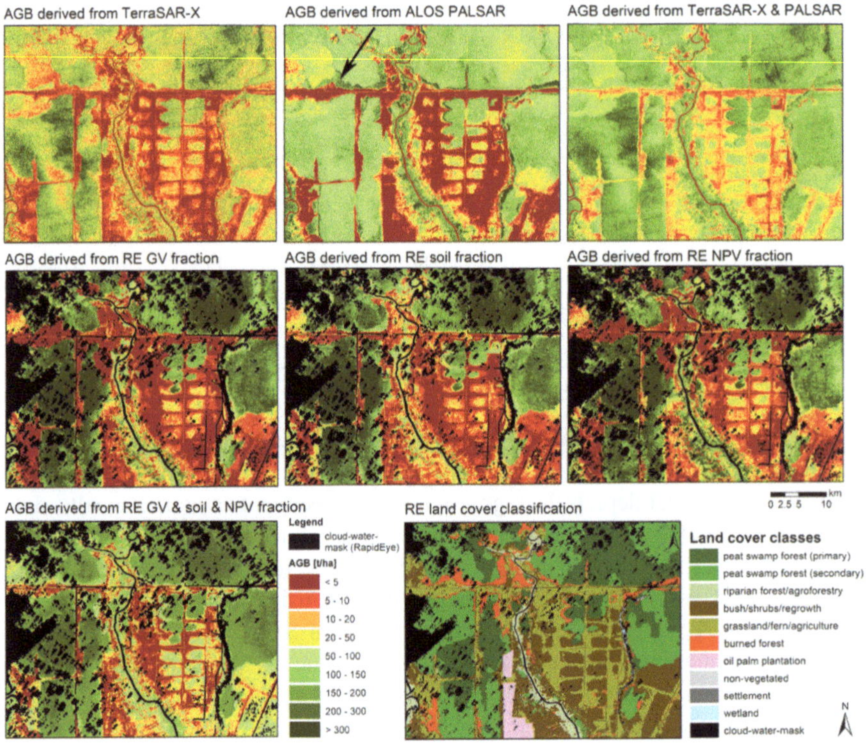

Fig. 24.3 Estimated continuous AGB maps of the different regression models. A detailed RapidEye (RE) land cover classification of 21/06/2010 is shown for a comparison to the discrete AGB estimation method for which this map was used (example given in Fig. 24.4). The *arrow* indicates a fresh burn scar on which AGB is overestimated

model also overestimates AGB on burned areas due to double bounce of the ALOS PALSAR HV polarized signal.

Estimating AGB from RapidEye imagery is clearly spatially limited by clouds which cover 12.8 % of the study area. AGB estimations based on GV and NPV MF fractions are very similar. AGB predicted by soil fractions is overestimated in burned areas, but apart from that, similar to GV and NPV estimated AGB, which was expected due to the fact that there is a linear relation between these three fractions as a result of SMA. The multivariate RapidEye model estimates AGB to be higher in low biomass ranges than the single variable MF fractions models. In burned areas, AGB is again overestimated.

24.3.2 Comparison Continuous vs. Discrete AGB Estimation Using Multispectral Data

AGB estimates of the continuous approach applying the multivariate multispectral model were compared to discrete AGB estimates derived from a land cover classification (Fig. 24.3). Table 24.2 depicts the median and the standard deviation values of the estimated AGB for each land cover class. The number of 40 × 40 m grid cells used for deriving these averaged figures from the estimated AGB values (continuous approach) or used field inventory plots (discrete approach) is given in brackets. Every predicted AGB value within one land cover class was used for the calculation of the continuous AGB mean value. The mean AGB value of the discrete approach was calculated on the basis of all available field inventory plots within each class. Two additional field inventory plots of riparian forest which were located outside of the study area were used because no field inventory plot was located inside the study area. The continuous AGB estimates of primary and secondary peat swamp forest are similar (196.7 and 183.1 t/ha, respectively) with a relatively low standard deviation (25.2 and 26.7 t/ha) whereas the values of the discrete AGB estimate have a wider difference (220.2 and 178.4 t/ha) with a higher standard deviation (73 and 73.3 t/ha), although the number of field inventories is much lower than the number of continuous AGB estimations. The field inventory values indicate high biomass variability within these two classes. AGB of riparian forest/agroforestry is difficult to compare, as only two field inventory plots were available which were both located in riparian forest outside of the study area while agroforestry is not included in the field data. Agroforestry and riparian forest cannot be differentiated in this region using RapidEye for the land cover classification and were therefore combined. These facts may explain the wide difference between continuous and discrete AGB estimates (110.7 and 294.2 t/ha) of riparian forest/agroforestry. The class bush/shrubs/regrowth shows higher AGB values with a higher variability (indicated by a higher standard deviation) in the continuous method (37.7 t/ha, std dev: 33.1 t/ha) compared to the discrete approach (12.6 t/ha, std dev: 8.3 t/ha). It is assumed that the biomass variability of this class is not correctly represented by three field inventory plots. AGB of the class grassland/fern/agriculture was estimated very similar by the continuous and discrete method (3.8 and 3.3 t/ha) but there is a higher standard deviation in the continuous AGB estimations. AGB estimates on recently burned/sparse regrowth areas are extremely overestimated by the continuous approach, which is also indicated in Fig. 24.3.

Figure 24.4 depicts an area of active logging showing the continuous and discrete AGB estimation, the land cover classification used for the discrete AGB estimation and the true color RapidEye image. The narrow logging tracks (2–10 m wide) are clearly visible within the RapidEye scene (purple color) and the resulting biomass loss from logging is indicated in the continuous AGB map. The land cover classification and the derived discrete AGB estimation do not differentiate such different levels of degradation within the same land cover class.

24.4 Discussion

Different SAR frequencies and polarizations, as well as multispectral MF fractions were analyzed for their potential for estimating AGB. The most accurate SAR-based AGB regression model was found to be a multivariate TerraSAR-X

Table 24.2 Comparison of continuous and discrete AGB estimates: median values and standard deviations of AGB for different land cover classes

AGB (t/ha) *standard deviation (t/ha)*	Peat swamp forest (primary)	Peat swamp forest (secondary)	Riparian forest/agroforestry	Bush/shrubs/regrowth	Grassland/fern/agriculture	Recently burned/sparse regrowth
Continuous	**196.7**	**183.1**	**110.7**	**37.7**	**3.8**	**87.7**
	25.2	*26.7*	*27.6*	*33.1*	*30.7*	*35.2*
	(n = 183,435)	(n = 255,159)	(n = 25,371)	(n = 187,209)	(n = 143,331)	(n = 45,238)
Discrete	**220.2**	**178.4**	**294.2**	**12.6**	**3.3**	**0.7**
	73.0	*73.3*	*36.7*	*8.3*	*10.4*	*8.4*
	(n = 15)	(n = 24)	(n = 2[a])	(n = 3)	(n = 5)	(n = 6)

AGB values are either derived from a continuous approach using MF fractions or from a discrete approach by linking average field biomass values to each land cover class. The number of 40 × 40 m grid cells used for averaging AGB estimations (continuous approach) or field inventory data (discrete approach) are given in brackets (n)

[a]Outside of the study area

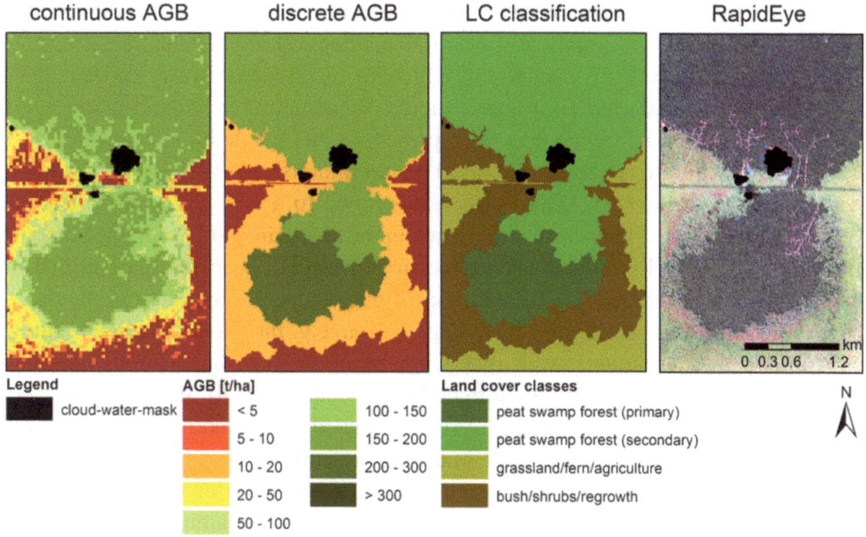

Fig. 24.4 Example of a logging area showing continuous and discrete AGB estimations, RapidEye land cover classification (LC classification) used for the discrete AGB estimation and the true color RapidEye scene from 21/06/2010

HH and ALOS PALSAR HV model. The combined multispectral MF fraction model based on RapidEye derived GV, soil and NPV MF fractions was the most accurate AGB model in terms of *RMSE*. Combining multispectral MF fractions and SAR backscatter was not found to add an additional value for AGB estimation, because a higher accuracy than of the multivariate MF model could not be achieved. Both multivariate SAR and multispectral AGB models overestimate AGB in recently burned areas (3 years and less after the fire event). The combined TerraSAR-X and ALOS PALSAR AGB model is negatively influenced by double bounce from the ALOS PALSAR signal occurring in recently burned areas. The RapidEye MF fraction model is affected by AGB overestimations in case of deriving AGB from soil fractions in burned areas. The soil fraction is relatively low in these areas due to the large amount of NPV (dead trees and trunks, standing and lying) and GV (rapid regrowth of young trees, fern and bushes). Frequent cloud cover hampers the acquisition of multispectral imagery and repetitive AGB maps are therefore very difficult to produce, which can only be overcome by very short repetition cycles. The combined SAR model estimates AGB are not as accurate as the combined multispectral model, but SAR is able to penetrate clouds, smoke and haze and repetitive area coverage can be achieved. Since SAR signals are influenced by water, it is very important to use multi-temporal SAR data acquired during the dry season to minimize any impact of precipitation or soil moisture.

A further crucial point is the AGB field reference data set. The range between 20 and 80 t/ha is underrepresented in the reference data set (Fig. 24.2) as such areas are very rare and difficult to access. On the one hand, regrowing areas contain mostly AGB values lower than 20 t/ha due to recurrent fire events since the risk of a repeated fire increases dramatically after one fire event (Langner and Siegert 2009). On the other hand, selective logged or degraded forest areas mostly contain AGB values higher than 80 t/ha. To improve the AGB estimation models, it is necessary to expand the AGB reference data set representing the whole biomass range.

The most accurate AGB model, based on continuous multivariate MF fractions, was compared to discrete AGB estimates assigned to a land cover classification. Based on the authors' long-term experience in the study area, it is supposed that the real biomass variability of primary peat swamp forest is not correctly described by the continuous approach in high biomass ranges due to the saturation effect. However, AGB loss caused by severe degradation, e.g. intensive selective logging, is visible in the continuous AGB map (Fig. 24.4). AGB of bush/shrubs/regrowth and grassland/fern/agriculture is estimated very accurate. AGB of recently burned/sparse regrowth areas was extremely overestimated, most likely due to low soil fractions caused by rapid regrowth and large quantities of dead woody debris. Considering these results, a combination of continuous and discrete AGB estimates appears to be the most promising approach. AGB in forest areas with high AGB values such as primary peat swamp forest, as well as recently burned/sparse regrowth areas is therefore suggested to be estimated by the discrete approach, whereas AGB in areas with low AGB values such as secondary peat swamp forest, riparian forest/agriculture, bush/shrubs/regrowth and grassland/fern/agriculture should be estimated via a continuous approach.

24.5 Conclusions

Using a combination of continuous and discrete AGB estimations based on remote sensing and detailed field inventory data meets the Tier 3 definition of the IPCC guidelines, which describes the most complex methodological level of AGB estimation (IPCC 2006). The combined approach of continuous and discrete AGB estimation overcomes the problem of saturation in higher biomass ranges. It has still to be proven if AGB of areas that experienced low impact degradation can be differentiated from secondary forest despite the saturation occurring in high biomass ranges.

The challenge of carbon stock monitoring is to avoid gaps, achieve high accuracies and observe defined time intervals thereby meeting the requirements of REDD+. In general, field inventories are the most accurate way to estimate AGB but they are time and cost consuming and the single point based measurements do not describe the spatial variability. In contrast, remote sensing is advantageous to estimate AGB over large areas although field inventory measurements are mandatory. LiDAR (Light detection an ranging) provides the best input data for deriving accurate AGB estimations due to the information that can be obtained on the vertical structure and height of the vegetation (Koch 2010). Associated large data volumes and costs usually limit the spatial extent and a repetitive application. But

LiDAR measurements are suitable to provide accurate, numerous AGB estimations including the spatial variability (providing a powerful basis) for an up-scaling with multispectral or SAR data using the presented approach for large-scale sub-national and national AGB carbon stock monitoring.

Acknowledgements The authors would like to thank the German Space Agency (DLR) and Canadian Space Agency (CSA) for providing TerraSAR-X and RADARSAT-2 data within the SOAR-DLR initiative (ID 5035), as well as the Japan Aerospace Agency (JAXA) for supplying ALOS PALSAR data (PI project No. 211). RapidEye data were provided by the German Aerospace Center (DLR) via the RESA – RapidEye Science Archive with funds from the German Federal Ministry of Economics and Technology (proposal no 267). Special thanks to Suwido Limin and his team from the Centre for International Co-operation in Management of Tropical Peatland (CIMTROP) in Palangka Raya for the logistical support during the field surveys. Furthermore, we acknowledge FAUNA & FLORA International (FFI) for providing aboveground biomass data of riparian forests.

References

Adams JB, Smith MO, Johnson PE (1986) Spectral mixture modeling – a new analysis of rock and soil types at the Viking Lander-1 Site. J Geophys Res Solid Earth Planets 91:8098–8112

Asner GP, Knapp DE, Balaji A, Paez-Acosta G (2009) Automated mapping of tropical deforestation and forest degradation: CLASlite. J Appl Remote Sens 3:1–24

Avitabile V, Baccini A, Friedl MA, Schmullius C (2012) Capabilities and limitations of Landsat and land cover data for aboveground woody biomass estimation of Uganda. Remote Sens Environ 117:366–380

Baccini A, Laporte N, Goetz SJ et al (2008) A first map of tropical Africa's above-ground biomass derived from satellite imagery. Environ Res Lett 3:9–17

Baccini A, Goetz S, Walker W et al (2012) Estimated carbon dioxide emissions from tropical deforestation improved by carbon-density maps. Nat Clim Change 2:182–185

Boehm HDV, Siegert F (2004) The impact of logging and land use change in Central Kalimantan, Indonesia. Int Peat J 12:3–10

Boettcher H, Eisbrenner K, Fritz S et al (2009) An assessment of monitoring requirements and costs of "Reduced Emissions from Deforestation and Degradation". Carbon Balance Manage 4:1–14

Brown S (1997) Estimation biomass and biomass change of tropical forests: a primer. FAO For Pap 134:1–55

Campbell BM (2009) Beyond Copenhagen: REDD plus, agriculture, adaptation strategies and poverty. Glob Environ Change Hum Policy Dimens 19:397–399

Chave J, Andalo C, Brown S et al (2005) Tree allometry and improved estimation of carbon stocks and balance in tropical forests. Oecologia 145:87–99

Chudnoff M (1984) Tropical timbers of the world, vol 607, Agriculture handbook. US Department of Agriculture, Forest Service, Forest Product Laboratory, Madison

DeFries R, Achard F, Brown S et al (2007) Earth observations for estimating greenhouse gas emissions from deforestation in developing countries. Environ Sci Pol 10:385–394

Englhart S, Keuck V, Siegert F (2011) Aboveground biomass retrieval in tropical forests – the potential of combined X- and L-band SAR data use. Remote Sens Environ 115:1260–1271

Englhart S, Keuck V, Siegert F (2012) Modeling aboveground biomass in tropical forests using multi-frequency SAR data – a comparison of methods. IEEE J Sel Top Appl Earth Obs Remote Sens 5:298–306

Fritz T, Eineder M (2009) TerraSAR-X ground segment basic product specification document. Doc: TX-GS-DD-3302. http://sss.terrasar-x.dlr.de/pdfs/TX-GS-DD-3302.pdf
Goetz SJ, Baccini A, Laporte N et al (2009) Mapping and monitoring carbon stocks with satellite observations: a comparison of methods. Carbon Balance Manage 4:2
GOFC-GOLD (2011) A sourcebook of methods and procedures for monitoring and reporting anthropogenic greenhouse gas emissions and removals caused by deforestation, gains and losses of carbon stocks in forests remaining forests, and forestation. GOFC-GOLD Report version COP17-1, GOFC-GOLD Project Office, Natural Resources Canada, Alberta, Canada
Grassi G, Monni S, Federici S et al (2008) Applying the conservativeness principle to REDD to deal with the uncertainties of the estimates. Environ Res Lett 3:1–12
Hooijer A, Page S, Canadell JG et al (2010) Current and future CO_2 emissions from drained peatlands in Southeast Asia. Biogeosciences 7:1505–1514
Hooijer A, Page S, Jauhiainen J et al (2012) Subsidence and carbon loss in drained tropical peatlands. Biogeosciences 9:1053–1071
Hughes RF, Kauffman JB, Jaramillo VJ (1999) Biomass, carbon, and nutrient dynamics of secondary forests in a humid tropical region of Mexico. Ecology 80:1892–1907
IPCC (2006) IPCC guidelines for national greenhouse gas inventories. Prepared by the national greenhouse gas inventories programme. In: Eggleston HS, Buendia L, Miwa K, Ngara T, Tanabe K (eds) IGES, Japan
Jaenicke J, Rieley JOO, Mott C et al (2008) Determination of the amount of carbon stored in Indonesian peatlands. Geoderma 147:151–158
Koch B (2010) Status and future of laser scanning, synthetic aperture radar and hyperspectral remote sensing data for forest biomass assessment. ISPRS J Photogramm Remote Sens 65:581–590
Koh LP, Butler RA, Bradshaw CJ (2009) Conversion of Indonesia's peatlands. Front Ecol Environ 7:238
Langner A, Siegert F (2009) Spatiotemporal fire occurrence in Borneo over a period of 10 years. Glob Chang Biol 15:48–62
Li H, Mausel P, Brondizio E, Deardorff D (2010) A framework for creating and validating a non-linear spectrum-biomass model to estimate the secondary succession biomass in moist tropical forests. ISPRS J Photogramm Remote Sens 65:241–254
Lu D (2005) Aboveground biomass estimation using Landsat TM data in the Brazilian Amazon. Int J Remote Sens 26:2509–2525
Lu D (2006) The potential and challenge of remote sensing based biomass estimation. Int J Remote Sens 27:1297–1328
Luckman A, Baker J, Kuplich TM et al (1997) A study of the relationship between radar backscatter and regenerating tropical forest biomass for spaceborne SAR instruments. Remote Sens Environ 60:1–13
Luscombe A (2009) Image quality and calibration of radarsat-2. IEEE Int Geosci Remote Sens Symp IGARSS 2:II757–II760
Mundt JT, Streutker DR, Glenn NF (2007) Partial unmixing of hyperspectral imagery: theory and methods. In: Proceedings of the American Society of Photogrammetry and Remote Sensing, Tampa
Nichol JE, Sarker M (2011) Improved biomass estimation using the texture parameters of two high-resolution optical sensors. IEEE Trans Geosci Remote Sens 49:930–948
Page SE, Siegert F, Rieley JO et al (2002) The amount of carbon released from peat and forest fires in Indonesia during 1997. Nature 420:61–65
Pearson T, Walker SBS (2005) Sourcebook for land use, land-use change and forestry projects. Winrock International, Little Rock
Richter R (1997) Correction of atmospheric and topographic effects for high spatial resolution satellite imagery. Int J Remote Sens 18:1099–1111
Ryan CM, Hill T, Woollen E et al (2012) Quantifying small-scale deforestation and forest degradation in African woodlands using radar imagery. Glob Chang Biol 18:243–257

Saatchi S, Marlier M, Chazdon RL et al (2011a) Impact of spatial variability of tropical forest structure on radar estimation of aboveground biomass. Remote Sens Environ 115:2836–2849

Saatchi SS, Harris NL, Brown S et al (2011b) Benchmark map of forest carbon stocks in tropical regions across three continents. Proc Natl Acad Sci U S A 108:9899–9904

Sarker LR, Nichol JE (2011) Improved forest biomass estimates using ALOS AVNIR-2 texture indices. Remote Sens Environ 115:968–977

Shimada M, Isoguchi O, Tadono T, Isono K (2009) PALSAR radiometric and geometric calibration. IEEE Trans Geosci Remote Sens 47:3915–3932

Sloan S, Edwards DP, Laurance WF (2012) Does Indonesia's REDD+ moratorium on new concessions spare imminently threatened forests? Conserv Lett 5:222–231

Soenen SA, Peddle DR, Hall RJ et al (2010) Estimating aboveground forest biomass from canopy reflectance model inversion in mountainous terrain. Remote Sens Environ 114:1325–1337

Souza CM, Roberts DA, Cochrane MA (2005) Combining spectral and spatial information to map canopy damage from selective logging and forest fires. Remote Sens Environ 98:329–343

Tangki H, Chappell NA (2008) Biomass variation across selectively logged forest within a 225-km (2) region of Borneo and its prediction by Landsat TM. For Ecol Manage 256:1960–1970

Van der Werf GR, Morton DC, DeFries RS et al (2009) CO_2 emissions from forest loss. Nat Geosci 2:737–738

Wijaya A, Liesenberg V, Gloaguen R (2010) Retrieval of forest attributes in complex successional forests of Central Indonesia: modeling and estimation of bitemporal data. For Ecol Manage 259:2315–2326

Williams AP, Hunt ER (2002) Estimation of leafy spurge cover from hyperspectral imagery using mixture tuned matched filtering. Remote Sens Environ 82:446–456

World Agroforestry Centre (2011) Wood density database. http://www.worldagroforestrycentre.org/Sea/Products/AFDbases/WD/. Accessed 27 June 2013

Zhang Z, Run-Guo Z (2009) Modelling the spatial distribution of aboveground biomass based on vegetation index in a tropical forest in Bawangling, Hainan Island, South China. Chinese J Plant Ecol 33:833–841

Author Index

I. Manakos and M. Braun (eds.), *Land Use and Land Cover Mapping in Europe: Practices & Trends*, Remote Sensing and Digital Image Processing 18,
DOI 10.1007/978-94-007-7969-3, © Springer Science+Business Media Dordrecht 2014

Subject Index

I. Manakos and M. Braun (eds.), *Land Use and Land Cover Mapping in Europe: Practices & Trends*, Remote Sensing and Digital Image Processing 18,
DOI 10.1007/978-94-007-7969-3, © Springer Science+Business Media Dordrecht 2014

MIX
Papier aus verantwortungsvollen Quellen
Paper from responsible sources
FSC® C105338

If you have any concerns about our products,
you can contact us on
ProductSafety@springernature.com

In case Publisher is established outside the EU,
the EU authorized representative is:
Springer Nature Customer Service Center GmbH
Europaplatz 3, 69115 Heidelberg, Germany

Printed by Libri Plureos GmbH
in Hamburg, Germany